Fundamental Design of Steelmaking Refractories

Fundamental Design of Steelmaking Refractories

Debasish Sarkar
National Institute of Technology, Odisha, India

WILEY

This edition first published 2023

Registered Office
John Wiley & Sons, Inc., 111 River Street, Hoboken, NJ 07030, USA

For details of our global editorial offices, customer services, and more information about Wiley products visit us at www.wiley.com.

Wiley also publishes its books in a variety of electronic formats and by print-on-demand. Some content that appears in standard print versions of this book may not be available in other formats.

Library of Congress Cataloging-in-Publication Data
Names: Sarkar, Debasish, 1972- author. | John Wiley & Sons, publisher.
Title: Fundamental design of steelmaking refractories / Debasish Sarkar.
Description: Hoboken, NJ : JW-Wiley, 2023. | Includes bibliographical references and index.
Identifiers: LCCN 2022057612 (print) | LCCN 2022057613 (ebook) | ISBN 9781119790730 (hardback) | ISBN 9781119790846 (pdf) | ISBN 9781119790853 (epub) | ISBN 9781119790860 (ebook)
Subjects: LCSH: Refractory materials. | Steel--Metallurgy.
Classification: LCC TN677.5 .S279 2023 (print) | LCC TN677.5 (ebook) | DDC 669/.82--dc23/eng/20230103
LC record available at https://lccn.loc.gov/2022057612
LC ebook record available at https://lccn.loc.gov/2022057613

Cover Image: © donatas1205/Shutterstock
Cover Design: Wiley

Set in 9.5/12.5pt STIXTwoText by Integra Software Services Pvt. Ltd, Pondicherry, India

Allocating
to my
Jovial and Eternal 'Madhava'

Contents

Preface

The first part of the book accentuates the valuable basics of "*Heat and Mass Transfer*," "*Equilibrium and Nonequilibrium Phases*," "*Packing, Stress, and Defects in Compaction*," "*Degree of Ceramic Bonding*," "*Thermal and Mechanical Behavior*," and "*High-Temperature Refractory Corrosion*," including relevant finite element analysis in the perception of composition design, manufacturing, and failure mechanism of steelmaking refractories. While considering the steelmaking refractories, detailed "*Operation and Refractories for Primary Steelmaking*," "*Operation and Refractories for Secondary Steelmaking*," "*Precast and Purging System*," "*Refractories for Flow Control*," "*Refractories for Continuous Casting*," and "*Premature Refractory Life by Other Parameters*" are essential to acme. These issues have been discussed in the second half of the book to fulfill the academic demand of undergraduate, postgraduate, and research scholars of ceramic engineering, metallurgical engineering, and mechanical engineering outlets who want to nurture in the refractory and steel sectors. The description of such cumulative basic knowledge, collective shop floor data, and relevant failure analysis criteria makes sense and eventually stimulates the awareness of how to grasp and analyze a particular class of refractory for steelmaking.

Refractory production, as fighting fit as their consumption, includes a certain degree of heat and mass transfer. Preliminary from the thermodynamics, heat and mass transfer mechanisms are being described, and eventually, an analogy is drawn in *Chapter 1*. In-situ phase formation during manufacturing and transformation in the presence of impurities are common phenomena in refractory; thus fundamentals of binary and ternary equilibrium phases and non-equilibrium phases are described in *Chapter 2*. Optimum compaction and load are a prerequisite to press organic-bonded refractories. A low load regime results in low green density, whether high load beyond critical stress consequences spring back and expedite lamination that eventually produces defect and early stage failure during the maneuver. Such phenomena are deliberated in *Chapter 3*. Industrial-scale production demands a uniform temperature distribution throughout the kiln to form adequate ceramic bonding or sintering of compact mass otherwise results in premature refractory failure. In this regard, *Chapter 4* describes the initial and final stages of sintering, densification, grain growth, and their shape in the matrix. Even with refractory processing failure, meticulous thermal and mechanical stress cracking, severe wear aggravated by abrasion, and corrosion are unavoidable in refractory practice and applications. For these concerns, *Chapter 5* highlights the thermal and mechanical behavior, and *Chapter 6* underscores the high temperature corrosion mechanism with a relevant model.

In the face of refractory consumption in iron production and transportation, billions of tons of refractories are used for primary steelmaking, secondary steelmaking, and continuous casting around the rondure. Several classic oxides and non-oxide shaped or unshaped refractories are needed because of their operational features, steel and slag chemistry, and tailor-made demand. In this context, how the MgO-C refractory protects by slag, failure due to thermal gradient, slag-refractory corrosion, and eventually their maintenance practice by monolithic is illustrated in *Chapter 7*. Steel ladle transports molten steel from BOF to tundish through secondary steelmaking processes. The different working lining is evident in the account of different steel grades, and this refractory manufacturing to the mode of failure is discussed in *Chapter 8*. Processing and probable failure of monolithic either as a back-up lining of ladle, the roof of EAF, or different precast shapes including tundish dam, impact pad, and well-block is discussed in *Chapter 9*. Continuous gas purging through monolithic porous plug maintain homogenized steel chemistry; thus an exhaustive performance and failure analysis of a different class of porous plug are also deliberated. The flow control mechanism has been discussed in *Chapter 10* before an elaborate analysis of the continuous casting refractories in *Chapter 11*. Occasionally, premature refractory failure depends on several unknown and invisible factors encountered and discussed in *Chapter 12*.

Acknowledgment

I earnestly acknowledge all academicians, researchers, students, and industry personalities who have contributed to this field. Considerably chosen are several classic pieces of work, including industry practice data, and I have prepared the manuscript with all relevant references and probable copyright permissions. Special thanks to my dearest family members, fellow friends, students, and colleagues who have allowed the discussion and space to prepare the manuscript.

About Author

Debasish Sarkar is a Professor at the Department of Ceramic Engineering, National Institute of Technology, Rourkela, Odisha, India. By 2007 Prof. Sarkar had finished his PhD and held a visiting researcher position at the Korea Research Institute of Standards and Science, Daejeon, South Korea. Gaining adequate experience in research, he received the Materials Research Society of India (MRSI) Medal in 2016. He is avid about research, and his publications number more than a hundred, including peer-reviewed international journals, book chapters, books, and national and international patents.

Debasish has a wide horizon of collaboration with refractory and steel sectors around the Globe that eventually aligns the content according to industry demand. Refractory development with life assessment and performance monitoring are his recent time interests. Professor Sarkar is consistently engaged as an advisor for the industry sponsored project on 'Steelmaking Refractories and Flow Control'.

He was designated head of the "Centre for Nanomaterials" at NIT Rourkela and established an excellent team to accomplish high-end research and products. As an experienced researcher in nanostructured ceramics and ceramic processing, he is also a mentor for the 'Miniaturization of Ceramic Components' to support the industry in every way viable.

1

Heat and Mass Transfer

1.1 Introduction

Commercial grade raw material calcination in a rotary kiln is common to make an economical product. The end product feeding regions are divided into several segments to achieve calcined outcomes like dolomite, spinel, mag-chrome, bauxite, etc. Ineffective heat distribution and mass-transfer during rotation result in an immature and uncalcined product that eventually reduces the properties of the refractories. While considering the nonuniform heat distribution in a kiln, there is always a chance to form a premature uncalcined product with a moisture-absorbing tendency; dolomite is an excellent example. Similarly, incomplete spinel conversion from a mixture of MgO and Al_2O_3 in synthetic raw materials eventually enhances the volume expansion in spinel-based refractory, as well as the improper calcination of natural bauxite facilitates forming a porous mass that absorbs more water during the castable casting results in premature failure.

Organic binder essentially develops a polymeric network and binding refractory grains, but a gradual formation of ceramic bonding requires a predefined temperature in the firing. Tunnel or batch kilns are used in the refractory industry to achieve prerequisite temperatures for high-performance working lining, backup lining, and continuous casting refractories. A highly efficient kiln may provide uniform heat distribution all along the sides from top to bottom, sidewall burner zone to opposite side wall burner zone, car to car, and brick stack to stack; otherwise, some bricks are overfired or underfired. Low temperature is responsible for higher porosity, low bulk density, and cold crushing strength. In contrast, overfired may produce unexpected expansion/shrinkage behavior and low eutectic phases in the presence of impurities.

Precast and Refractory lining preheating in the steel industry shop floor is essential before processing molten steel. Process efficiency depends on the refractory material properties like density, thermal conductivity, and specific heat. However, the foremost thing is the employed process temperature and how

Fundamental Design of Steelmaking Refractories, First Edition. Debasish Sarkar.

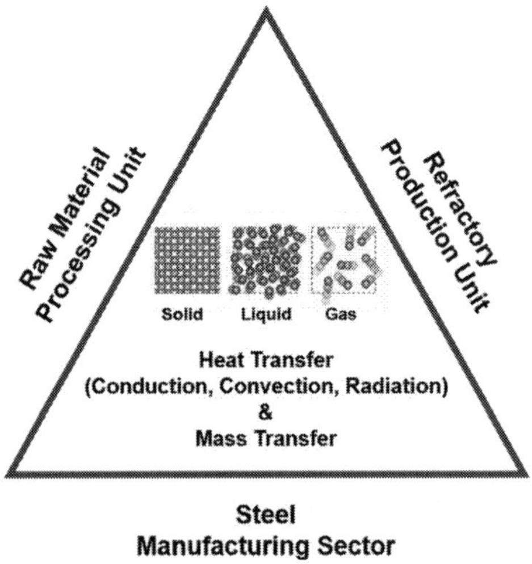

Figure 1.1 A schematic representation of heat and mass-transfer in refractory processing, production, and application sectors. The atomic arrangement defines solid, liquid, and gas states and possible separation under heat.

it is transferred effectively throughout the system. Continuous argon gas purging through a porous plug facilitates homogenous molten steel composition and temperature throughout the vessel. This multiphase heat transfer behavior differs from heat transfer in a single-phase solid or gaseous environment. Moreover, there is enough scope for heat loss from steel vessels (BOF, LF, Tundish) either through the open space or refractory lining. Different refractory quality experience a temperature gradient due to differences in thermal conductivity from working to safety lining in a vessel.

Raw material calcination, shaped refractory firing, refractory preheating, maintaining uniform temperature in the entire mass of molten steel, and temperature loss through working lining to the outer vessel shell are critical issues and worthy of analyzing the perspective of heat and mass-transfer phenomena. A schematic representation of such a relation is given in Figure 1.1. The inset solid to liquid to gas conversion depends on the temperature that usually enhances kinetic energy and diffusion of atoms. Thus, heat and mass-transfer fundamentals are considered a starting discussion to explore the properties and analysis for the designing refractory for the steel sector.

1.2 Energy Conservation

Industry measures the temperature profile in a kiln or furnace, but heat distribution is technically responsible for device efficiency. The temperature is different from the heat. The heat transfer is distinctly different from thermodynamics.

Temperature (T, °C) measures the quantity of energy influenced by the molecules of a material or system and predicts both the degrees of hotness and direction of heat transfer. Heat (Q, Joule) measures thermal energy transfer from a hotter body to a colder one. Thus, heat transfer describes how much heat, transfer rate, and resultant temperature distribution are inside the body. However, thermodynamics is concerned with the equilibrium state of matter, the amount of heat transferred (dQ), the amount of work done (dW), and the final form of the system. In practice, heat transfer is complementary and an extension of thermodynamics.

a) Control Volume (Open system)

The first law of thermodynamics defines the conservation of total energy and can only change if it crosses the boundaries. A region of fixed mass known as the closed system allows the transfer of heat through boundaries and work done on or by the system, resulting in the total energy changes of the system, as illustrated in Equation (1.1) and Figure 1.2a.

$$\Delta E_{\text{system}} = Q - W \tag{1.1}$$

Where total energy change stored in the system is ΔE_{system}, net heat transferred to the system is Q, and new work done by the system is W. In the same chronology, control volume (or open system) region has many possible mass entries. Herein, the "many" terminology refers to the possibility of the energy it carries with it when entering and leaving the mass in the control volume, known as energy advection, which facilitates energy across the boundaries of a control volume. Thus, energy can enter and leave the control volume due to heat transfer from the boundaries, work done on or by the control volume, and energy advection. The total energy combines internal energy and mechanical energy (cumulative potential and kinetic energy). Internal energy can be further subdivided by thermal energy and other forms of internal energy, such as chemical and nuclear energy. Thermal and mechanical forms of energy are a prime concern in heat transfer. The sum of these energies is not conserved because of the possibility of conversion between other forms of energy and thermal energy. However, the stored energy rate must be balanced during inflow and outflow energy in the control volume.

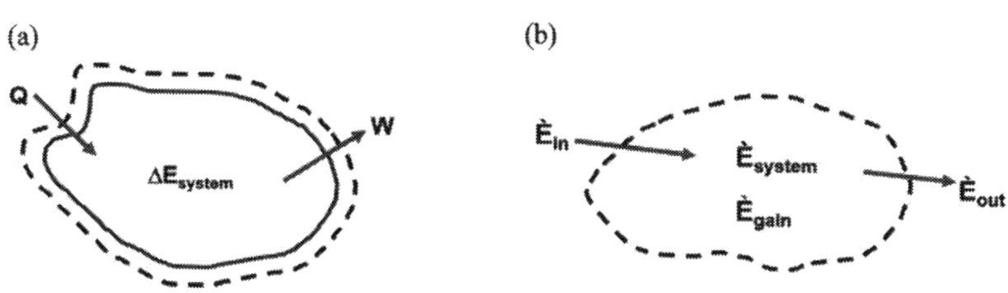

Figure 1.2 (a) Closed system with a time interval (Δt), (b) control volume instantly.

Thus, the change in mechanical and thermal energy stored (ΔE_{stored}) over the time interval Δt and generation of thermal energy (E_g) in the control volume is:

$$\Delta E_{stored} = E_{in} - E_{out} + E_g \tag{1.2}$$

The incidence rate of stored energy ($\grave{E}_{stored}$) increment in the controlled volume is represented in Figure 1.2b and calculated by Equation (1.3):

$$\grave{E}_{stored} = \frac{dE_{stored}}{dt} = \grave{E}_{in} - \grave{E}_{out} + \grave{E}_g \tag{1.3}$$

Total mechanical energy represents the sum of KE (kinetic energy $= \frac{1}{2}mv^2$, where m is the mass and v is the velocity, respectively); PE (potential energy $= mgh$, where g and h are the gravitational acceleration and vertical distance, respectively); and U (internal energy). However, the internal energy (U) consists of:

i) A *sensible* component comprised of rotational, translational, and vibrational motion of the molecules/atoms of the matter;
ii) Component of *latent* indicates the intermolecular forces influencing the changes in phase between vapor, liquid, and solid states;
iii) Accounts of *chemical* components for stored energy in the chemical bonds in atoms; and
iv) The accounts of *nuclear* components for the binding forces in the nucleus.

In a heat transfer study, sensible (U_{sen}) and latent (U_{lat}) are the most important internal energy components and together are known as thermal energy (U_t); thus, $U_t = U_{sen} + U_{lat}$. Sensible and latent energy is responsible for temperature (although it depends on pressure) and phase change, respectively. For example, in the control volume region, if the material changes from solid to liquid represents melting, and liquid to vapor indicates vaporization, boiling, and evaporation, the latent energy increases. Conversely, condensation (vapor to solid) or solidification and freezing (liquid to solid) decreases the latent energy. Thus, stored energy (E_{stored}) is, $E_{stored} = KE + PE + U_{sen} + U_{lat}$. The KE and PE are often small, and if there is no phase change, both mechanical energy and U_{lat} can be neglected, and thus, $E_{stored} = U_{sen}$; if no temperature change, $E_{stored} = U_{lat}$.

The inflow and outflow incidence in the control volume region is initiated from a surface; thus, it is a surface phenomenon. The process exclusively on the control surface is ordinarily proportional to the surface area. Therefore, the concepts of inflow and outflow of energy terms include the transfer of heat associated with solid medium (conduction), liquid presence (convection), vacuum (radiation), and work interactions that occur at the boundaries of the system. If mass enter or leave the control volume boundary, both the inflow and outflow advected mechanical and thermal energy in the control volume. For instance, if entering mass-flow rate $\dot{m}$, then the control volume is the product of $\dot{m}\,(u_t + ½v^2 + gh)$, where u_t represents the thermal energy per unit mass, and $(½v^2 + gh)$ is the mechanical energy per unit mass.

b) Surface Energy Balance

Energy conservation at the surface is an important aspect. As shown in Figure 1.3, the control surfaces are assumed on either side of the bold physical boundary where no mass or volume is enclosed. Herein, Equation (1.3) is no longer valid and only deals with surface phenomena considering the energy storage and generation concept. Under this circumstance, the conservation becomes:

$$\grave{E}_{in} - \grave{E}_{out} = 0 \tag{1.4}$$

Although thermal energy generation may occur in the medium, the control surface energy balance is not affected and holds steady-state and transient conditions. On the unit area basis, three heat transfer terms are shown for the control surface in Figure 1.3.

Under this circumstance, the conduction is in between medium and control surface (q''_{cond}), the surface to a fluid (q''_{conv}) indicates convection, and exchange of the net radiation between the surface to the surroundings (q''_{rad}) can be balanced by Equation (1.5). In brief, the heat flux (q/A, q = heat rate, A = area) for conduction, the $q''_{cond} = k\,(\Delta T/L)$:

$$q''_{cond} - q''_{conv} - q''_{rad} = 0 \tag{1.5}$$

where k = thermal conductivity (W/m.K), ΔT, and L are temperature differences and thickness of the medium, respectively.

In convection, $q''_{conv} = h\,(T_S - T_\infty)$, where h = heat transfer coefficient (W/m^2.K), T_S is surface temperature, and T_∞ is fluid temperature. In radiation, $q''_{conv} = \varepsilon\sigma(T^4_s - T^4_{sur})$, where ε is emissivity ($0 \le \varepsilon \le 1$), σ = Stefan–Boltzmann constant ($\sigma = 5.67 \times 10^{-8}$ W/m^2.K^4), T_{sur} is the surrounding temperature. From

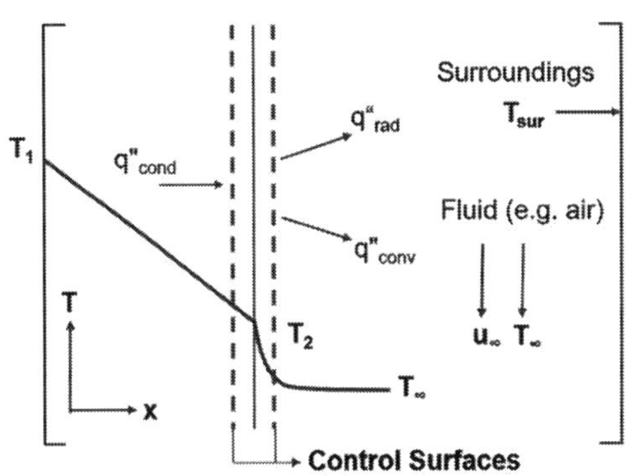

Figure 1.3 The energy balance for energy conservation at the surface of a medium.

a thermodynamic perspective, it is reasonably clear why conduction, convection, and radiation are important aspects to explain heat transfer mechanisms and energy balance. Thus, the origin and details of each mechanism are explained in the next section.

1.3 Conduction

High temperature aggravates molecular energies, and collision facilitates energy transfer to adjacent low energy molecules. This incidence develops thermal gradient and heat flow from higher to lower temperature regimes in the conduction process. Despite a collision, the random motion of molecules is also responsible for energy transfer, but a collision accelerates the energy transfer rate. This class of net energy transfer is nothing but the diffusion of energy. Eventually, the heat transfer by conduction relies on direct contact between atoms or molecules that can also be materialized in gas and liquid, not only in solid. Molecules are more closely packed in liquids than gas, and their interactions are more frequent and substantial. Similarly, conduction attributes to atomic activity in the translational motion of free *e* in a solid.

1.3.1 Basic Concept and Properties

In consideration of Fourier's law, high school science describes simple steady-state conductive heat flow relation by heat flux, thermal gradient, and thermal conductivity of materials. Consider a steel cylinder insulated on its lateral surface, as illustrated in Figure 1.4. The heat flow occurs from one end to another when end faces have different temperatures, say T_1 and T_2, where $T_1 > T_2$ follows the one-direction flow, and x represents direction only.

The temperature difference induces heat transfer. The rate of heat transfer (q_x) depends on the variables ΔT—temperature difference, Δx—cylinder length, A—area of cross-section, and can be represented by Equation (1.6).

$$q_x \infty A \frac{\Delta T}{\Delta x} \quad (1.6)$$

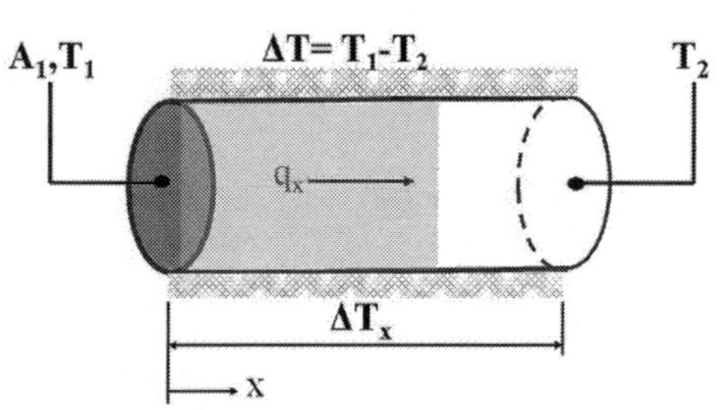

Figure 1.4 Schematic representation of steady-state heat conduction experiment.

This heat transfer rate becomes lower for alumina (Al_2O_3), and lowest for high-density polyethylene (HDPE) compared to equal A, Δx, and ΔT of the steel cylinder. Thus, this proportionality converts into a coefficient, k, known as thermal conductivity (W/m.K) and Equation 1.6 to be:

$$q_x = kA\frac{\Delta T}{\Delta x} \tag{1.7}$$

Consider Figure 1.5a, and solving Equation 1.7, when $\Delta x \rightarrow 0$, the heat rate Equation 1.7 becomes Equation 1.8, in which the negative sign is mandatory because the flow of heat is always in the temperature decreasing direction; thus, temperature gradient dT/dx is −ve, and q''_x is +ve.

$$q_x = -kA\frac{dT}{dx} \tag{1.8}$$

Or the isothermal surfaces are planes normal to the x-direction, and the heat flux:

$$q''_x = \frac{q_x}{A} = -k\frac{dT}{dx} \tag{1.9}$$

Heat flux is a vector quantity, and a general statement of Fourier's law of conduction follows Equation (1.10), where T(x,y,z) is the scalar temperature field:

$$q'' = -k\left(i\frac{\partial T}{\partial x} + j\frac{\partial T}{\partial y} + k\frac{\partial T}{\partial z}\right) \tag{1.10}$$

In more generalized terms this equation can be written as:

$$q''_n = -k\frac{\partial T}{\partial n} \tag{1.11}$$

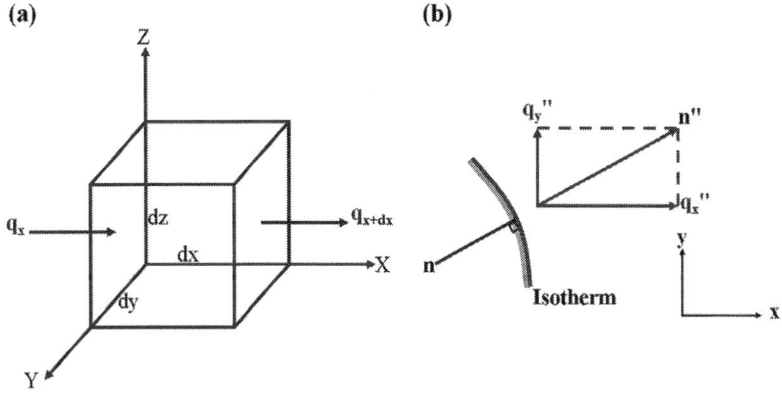

Figure 1.5 (a) Heat flow along x-direction in a cubic system, (b) heat flux vector for a two-dimensional system.

Where, q''_n is the heat flux in n direction, normal to an isotherm for two coordinate systems, as shown in Figure 1.5b. Table 1.1 provides the probable thermodynamic data for the materials associated with refractory analysis. Detailed fundamental concepts of thermal conductivity and the effect of temperature on conductivity variation are discussed in Chapter 5, Section 5.4.

Heat transfer analysis requires many thermophysical properties, including two distinct groups, thermodynamic and transport properties of a matter. The transport properties have the diffusion rate coefficients such as thermal conductivity (k) for the heat transfer and kinematic viscosity (ν), for momentum transfer. In other words, the density (ρ) and specific heat (c_ρ) refer to the thermodynamic properties. Product ρc_ρ (J/m^3.K), known as volumetric heat capacity, is essential for measuring thermal energy storage ability. High-density materials have low

Table 1.1 Thermal properties of different materials.

	K (W/m.K)			
Material	**100°C**	**1000°C**	**ρ(g/cc)**	**c_ρ(J/kg.K)**
SiO_2	2	2.5	2.65	732
Fe_2O_3	8.5	2.38	5.25	104
TiO_2	6.7	3.35	4.25	690
MnO	1.3	1.24	5.43	611
CaO	15.48	7.84	3.32	418
MgO	38	7.1	3.58	890
Cr_2O_3	32	10	5.23	529
B_4C	30	42	2.51	1000
Steel	13.2	30.8	7.86	502
Al_2O_3	30	6.3	3.99	160
ZrO_2	1.9	2.3	6.09	450
SiC	370	30	3.22	344
Graphite	179.9	62.7	2.27	711
Al_4C_3	107.8	98	2.97	–
Si_3N_4	19.66	15.90	3.19	712
H_2O	0.025	0.121	1	4184
O_2	0.032	0.036	0.00143	917
CO_2	0.023	0.072	1.56	709
Ar	0.019	0.058	1.784	519

specific heat, act as a perfect energy storage media, and have value in $\rho c_p > 1$ MJ/m^3·K. In contrast, gases are low-grade energy storage media in the range of $\rho c_p \approx 1$ kJ/m^3·K. With the help of thermal conductivity, density, and specific heat, the thermal diffusivity $\alpha = k/(\rho c_p)$ implies the ability to conduct thermal energy relative to its ability to store thermal energy, an essential phenomenon during sintering. High diffusivity refers to rapid change in their thermal environment to maintain equilibrium. Selecting reliable data for a particular phase is critical to drawing a convincing conclusion during heat transfer analysis.

1.3.2 One-Dimensional Steady-state Conduction

Heat transfer by diffusion is directly proportional to the temperature gradient (except thermal radiation), and it is expected that heat flux transportation will vary in one-dimensional (x), two-dimensional (x,y), and three-dimensional (x,y,z). The majority of the heat transfer occurs in one-dimension: heat transfer from the inner wall to the outside of a kiln when two sides maintain a temperature gradient. Two-dimensional heat transfer is a vital function of the corresponding cross-section without variation of the third dimension. Fins extended surfaces are used for heat transfer between a hot body and its surroundings, an excellent example of two-dimensional flow. However, in many cases, the one coordinate direction heat flow is oversimplified and necessary to account for multidirectional effects.

Where one-dimension steady-state conduction is concerned, two types of heat transfer phenomena may be observed, "without internal heat generation (heat dissipation through a refractory wall)" and "with internal heat generation (during energy conversion from electrical or chemical to thermal energy)." Refractory lining for a tunnel kiln is an excellent example of the heat transfer process with no heat generation possible through a plane wall. However, contact resistance may appear during such transfer because of different phases (e.g., air) within composite layers. In Figure 1.6a, two fluids are separated by a single plane wall at different temperatures. Heat transfer from the hot fluid at $T_{\infty 1}$ to the inner wall surface by convection at T_{s1}, by conduction through the wall, and further outer wall surface at T_{s2} to the cold fluid at $T_{\infty 2}$ through the convention.

For one-dimensional ($\rightarrow x$), steady-state conduction in a plane wall with no heat generation and constant thermal conductivity (k), the temperature varies linearly with x. Thus, the conduction heat transfer rate (from Equation 1.8) is:

$$q_x = -kA\frac{dT}{dx} = \frac{kA}{L}(T_{s1} - T_{s2}) \tag{1.12}$$

Herein, the prescribed surface temperatures at $x = 0$ and $x = L$ are boundary conditions, although this temperature strictly depends on the homogeneity of fluid temperature. Analogous to electrical resistance, the thermal resistance for

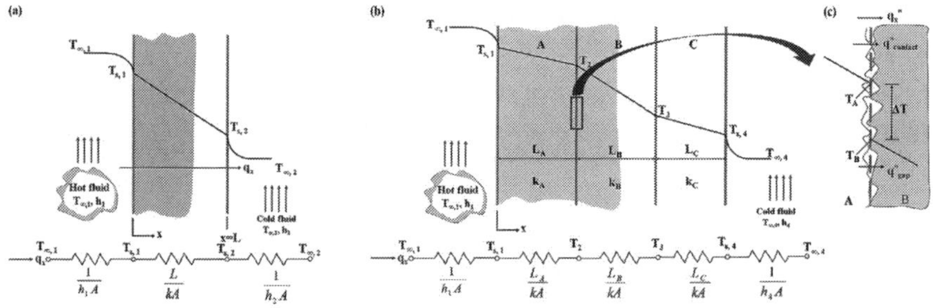

Figure 1.6 Heat transfer through (a) a plane wall, (b) equivalent thermal circuit for a series composite wall (example of kiln lining), (c) temperature drop due to thermal contact resistance.

conduction in a plane wall is associated with thermal conductivity and can be written as:

$$R_{t,cond} = \frac{\left(T_{s1} - T_{s2}\right)}{q_x} = \frac{L}{kA} \tag{1.13}$$

Thermal conductivity (k) of insulating solid depends on heat capacity per unit volume (C), average phonon velocity (v), and mean free path (l). The C is reduced by the system's porosity volume fraction (p), and the resultant k becomes; where C^{o} is the heat capacity at $p = 0$. Thus, k can be represented as Equation (1.14):

$$k = \frac{1}{3} vlC_o(1-p) \tag{1.14}$$

From Equation (1.14), it is obvious why porous ceramics are preferred as insulating surfaces to reduce heat loss from the kiln's outer wall. Thermal resistance is often associated with heat transfer through convection to a surface and conductive resistance. According to Newton's law of cooling, the heat rate q may be represented by Equation (1.15), where h is the heat transfer coefficient in convection:

$$q_x = hA(T_s - T_\infty) \tag{1.15}$$

The thermal resistance for convection is:

$$R_{t,conv} = \frac{\left(T_s - T_\infty\right)}{q} = \frac{1}{hA} \tag{1.16}$$

The heat transfer rate is constant throughout the system; thus q_x becomes:

$$q_x = \frac{\left(T_{\infty 1} - T_{s1}\right)}{\frac{1}{h_1 A}} = \frac{\left(T_{s1} - T_{s2}\right)}{\frac{L}{kA}} = \frac{\left(T_{s2} - T_{\infty 2}\right)}{\frac{1}{h_2 A}} \tag{1.17}$$

In consideration of overall temperature difference $T_{\infty 1} - T_{\infty 2}$, and the total thermal resistance R_{tot}, the heat transfer rate may be written as:

$$q_x = \frac{\left(T_{\infty 1} - T_{\infty 2}\right)}{R_{tot}} \tag{1.18}$$

Where R_{tot}:

$$R_{tot} = \frac{1}{h_1 A} + \frac{L}{kA} + \frac{1}{h_2 A} \tag{1.19}$$

Exchange of radiation between surface and surroundings may also be important when h for convection is very small. The thermal resistance for radiation $R_{t,rad} = (T_s - T_{sur})/q_{rad} = 1/(h_r A)$, where h_r:

$$h_r \equiv \varepsilon\sigma(T_s + T_{sur})(T_s^2 + T_{sur}^2) \tag{1.20}$$

Radiation of surface and resistances due to convection act parallel and can be combined to obtain a single, significant surface resistance if $T_\infty = T_{sur}$. In the exact chronology, the heat transfer rate (q_x) and overall heat transfer coefficient (U) may also be estimated for composite walls placed in a series connection, as shown in Figure 1.6b. The rate of heat transfer is:

$$q_x = \frac{\left(T_{\infty 1} - T_{\infty 4}\right)}{\frac{1}{h_1 A} + \frac{L_A}{k_A A} + \frac{L_B}{k_B A} + \frac{L_C}{k_C A} + \frac{1}{h_4 A}} \equiv UA\Delta T \tag{1.21}$$

Where

$$U = \frac{1}{R_{tot} A} = \frac{1}{\frac{1}{h_1 A} + \frac{L_A}{k_A A} + \frac{L_B}{k_B A} + \frac{L_C}{k_C A} + \frac{1}{h_4 A}} \tag{1.22}$$

The above equation is valid when heat flow presumed that surfaces normal to the x direction are isothermal, but R_{tot} is different when surfaces parallel to the x direction and the assumed adiabatic system.

Another critical factor, gap influenced by surface roughness filled by air, between the composite system interface is temperature drop, as shown in Figure 1.6c. This temperature change is known as thermal contact resistance $R_{t,c}$. The resistance per unit area of the interface is:

$$R''_{t,c} = \frac{T_A - T_B}{q''_x} \tag{1.23}$$

Conductive heat transfer follows through relatively minute contact spots, whereas convection and radiation across the gaps result in two parallel resistances. The contact resistance may also be reduced by increasing the thermal conductivity of entrapped fluid or maximizing the contact spots, nothing but R_a's reduction (a statistical roughness parameter) value.

The heat transfer rate and resistance behavior are different for a cylindrical rotary kiln, a circular cross-section, as shown in Figure 1.7a. Cylindrical and spherical systems experience radial direction heat transfer, and thus the same one-directional approach may be employed to determine the heat transfer behavior.

Employing Fourier's law in circular geometry analogous to the single plane wall, the conductive heat transfer rate through radial mode (q_r) is:

$$q_r = \frac{2\pi Lk\left(T_{s1} - T_{s2}\right)}{\ln\frac{r_2}{r_1}} \tag{1.24}$$

And the thermal resistance:

$$R_{t,cond} = \frac{\ln\frac{r_2}{r_1}}{2\pi Lk} \tag{1.25}$$

Similarly, the q_r and U may also be calculated for the composite wall from Figure 1.7b:

$$q_r = \frac{T_{\infty 1} - T_{\infty 4}}{\frac{1}{2\pi r_1 L h_1} + \frac{\ln\frac{r_2}{r_1}}{2\pi k_A L} + \frac{\ln\frac{r_3}{r_2}}{2\pi k_B L} + \frac{\ln\frac{r_4}{r_3}}{2\pi k_C L} + \frac{1}{2\pi r_4 L h_4}} = UA(T_{\infty 1} - T_{\infty 4}) \tag{1.26}$$

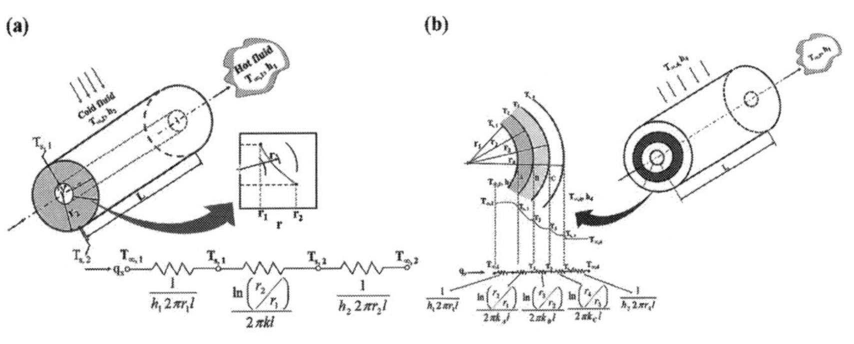

Figure 1.7 Heat transfer through (a) single hollow cylinder with convective surface conditions, (b) a composite cylindrical wall (an example of working, backup linings, and shell of a rotary kiln).

The overall coefficient may be defined as:

$$U_1A_1 = U_2A_2 = U_3A_3 = U_4A_4 = \left(\sum R_{tot}\right)^{-1} \tag{1.27}$$

An alternative integral approach can also estimate the one-dimensional heat transfer rate (q_x) for other shapes like conical systems (e.g., nose-tip of monoblock stopper). In consideration of:

Figure 1.8, any differential element dx, $q_x \rightarrow q_{x+dx}$, even if the area varies with position $A(x)$, and the thermal conductivity varies with temperature, $k(T)$. As the q_x is constant, we can express Fourier's law in the integral form:

$$q_x \int_{x_o}^{x} \frac{dx}{A(x)} = \int_{T_o}^{T} k(T)dT \tag{1.28}$$

In order to obtain the heat transfer rate for the nonuniform dimension, the cross-section area as a function of x (x_o to x) and change the thermal conductivity with temperature (T_o to T) is required. If area A is constant and thermal conductivity is independent of temperature, Equation (1.28) reduces to Equation (1.29):

$$\frac{q_x \Delta x}{A} = -k\Delta T \tag{1.29}$$

An alternative integral method effectively solves diffusion rate equations with a one-dimensional steady-state limiting condition with no heat generation. A simplified integral approach may estimate the heat transfer rate for the sphere (Figure 1.9). The appropriate form of Fourier's law for one-dimensional steady-state heat transfer rate for the sphere is Equation (1.30), where $A = 4\pi r^2$ is the area normal to the direction of heat transfer.

$$q_r = -kA\frac{dT}{dr} = -k(4\pi r^2)\frac{dT}{dr} \tag{1.30}$$

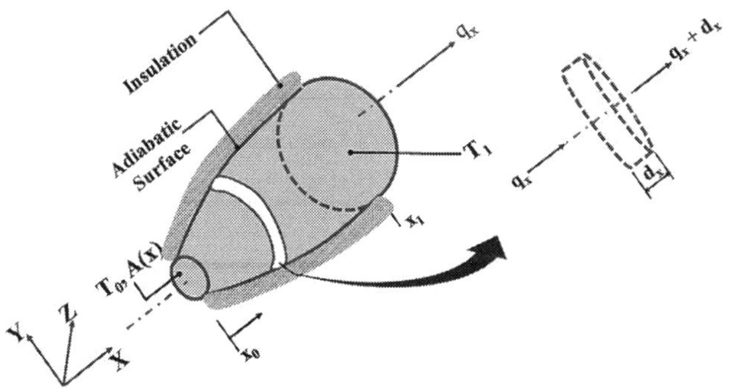

Figure 1.8 Schematic representation of conical nose-tip to determine constant conductive heat transfer rate through the tip.

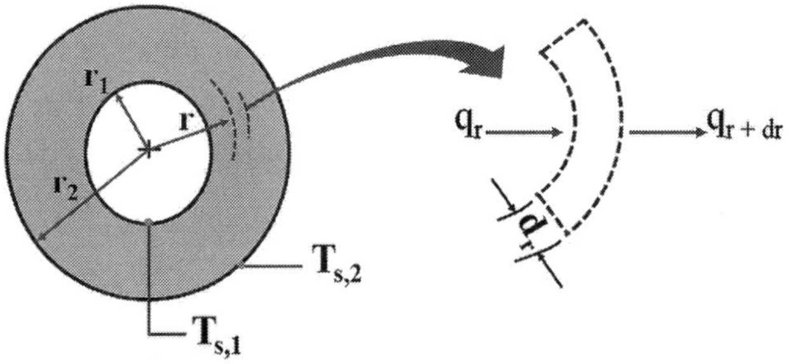

Figure 1.9 Schematic representation of the conduction in a spherical shell.

If q_r is constant and independent of r:

$$\frac{q_r}{4\pi}\int_{r_1}^{r_2}\frac{dr}{r^2} = -\int_{T_{s1}}^{T_{s2}} k(T)dT \tag{1.31}$$

If thermal conductivity, k, is constant:

$$q_r = \frac{4\pi Lk\left(T_{s1} - T_{s2}\right)}{\dfrac{1}{r_1} - \dfrac{1}{r_2}} \tag{1.32}$$

And the thermal resistance $R_{t,\ cond}$ for sphere:

$$R_{t,cond} = \frac{1}{4\pi k}\left(\frac{1}{r_1} - \frac{1}{r_2}\right) \tag{1.33}$$

Other parameters for the spherical composite wall may also be determined as a plane wall and cylinders.

A cumulative list of one-dimensional steady-state heat transfer parameters for three geometries is given in Table 1.2. Herein, the ΔT refers to the difference in temperature $T_{s1} - T_{s2}$ between inner and outer identified surfaces.

1.3.3 Two-Dimensional Steady-state Conduction

One-dimensional steady-state concept allows heat transfer through one coordinate system, but the two-dimensional heat transfer provides better insight and a more accurate heat transfer analysis in advance. Consider a uniform prismatic solid as shown in Figure 1.10, in which the top and bottom are insulated and two sides open with a temperature difference of $T_1 > T_2$, and heat transfers from surface 1 to 2. In consideration of Equations (1.10) and (1.11), the local heat flux in the solid is a resultant vector (q'') of the heat flux of x (q_x'') and y (q_y'') direction, which is anywhere perpendicular to the constant temperature lines.

Table 1.2 One-dimensional steady-state with no heat generation.

	Plane wall	Cylindrical wall	Spherical wall
Heat equation	$\frac{d^2T}{dx^2}=0$	$\frac{1}{r}\frac{d}{dr}\left(r\frac{dT}{dr}\right)=0$	$\frac{1}{r^2}\frac{d}{dr}\left(r^2\frac{dT}{dr}\right)=0$
Temperature distribution	$T_{s,1}-\Delta T\frac{x}{l}$	$T_{s,2}+\Delta T\frac{ln\left(\frac{r}{r_2}\right)}{ln\left(\frac{r_1}{r_2}\right)}$	$T_{s,1}-\Delta T\left[\frac{1-\left(\frac{r_1}{r}\right)}{1-\left(\frac{r_1}{r_2}\right)}\right]$
Heat flux (q'')	$k\frac{\Delta T}{L}$	$\frac{k\Delta T}{rln\left(\frac{r_2}{r_1}\right)}$	$\frac{k\Delta T}{r^2\left(\frac{1}{r_1}\right)-\left(\frac{1}{r_2}\right)}$
Heat rate (q)	$kA\frac{\Delta T}{L}$	$\frac{2\pi Lk\Delta T}{ln\left(\frac{r_2}{r_1}\right)}$	$\frac{4\pi k\Delta T}{\left(\frac{1}{r_1}\right)-\left(\frac{1}{r_2}\right)}$
Thermal resistance ($R_{t,\,cond}$)	$\frac{L}{kA}$	$\frac{ln\left(\frac{r_2}{r_1}\right)}{2\pi Lk}$	$\frac{\left(\frac{1}{r_1}\right)-\left(\frac{1}{r_2}\right)}{4\pi k}$

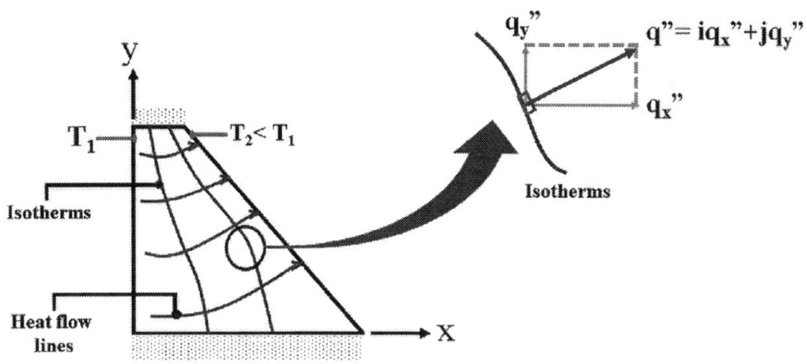

Figure 1.10 Schematic representation for two-dimensional conduction.

The adiabatic surface is usually defined as a boundary where no conduction is across a heat flow line. Thus, the heat equation for an x-y plane is:

$$\frac{\partial^2 T}{\partial x^2}+\frac{\partial^2 T}{\partial y^2}=0 \tag{1.34}$$

Under steady-state conditions, no heat generation and constant thermal conductivity are essential aspects that eventually assist in estimating the "temperature distribution" from the heat equation and "heat flux components" in a two-dimensional system.

The exact mathematical solution of two-dimensional steady-state heat transfer can be determined by analytical, graphical, or numerical methods. In the case of simple geometry, two-dimensional and three-dimensional conduction problems may be easily overcome by using well-documented heat diffusion equations, expressed in terms of shape factor, S. However, numerical methods like finite-difference, finite-element with relevant boundary conditions may be used for complex-shaped two-dimensional and three-dimensional geometries to estimate the heat transfer problems. A dimensionless shape factor method may be the rapid solution for the two-dimensional steady-state heat transfer rate calculation by Equation (1.35); where ΔT_{1-2} is the difference in temperature between two boundaries:

$$q = Sk\Delta T_{1-2} \tag{1.35}$$

And the two-dimensional conduction resistance may be expressed as:

$$R_{t,cond(2D)} = 1/kS \tag{1.36}$$

The shape can be obtained analytically and geometrically for several two-dimensional and three-dimensional systems, and a few classic information related to refractory analysis are summarized in Table 1.3.

It isn't easy to estimate the shape factor in some critical conditions. Solving the finite-difference equations by using a numerical method for a nodal network's discrete points is necessary. Further, it is also mandated to consider that various heat transfer processes are time-dependent. Such transient or unsteady incidences arise when the system boundary conditions are changed. For example, a continuous temperature of surface alteration is a common intermittent hot metal billet phenomenon when it comes out from the caster and is exposed to a cool airstream in which the energy is transmitted from its surface to its surroundings through convection and radiation. The conduction process persists from the hot billet interior to the surface until a steady-state condition is reached. The time-temperature history is key to fabricating new materials with tailor-made properties to control the resultant 'performance' of such a process. Considering different heat transfer solutions and numerical approaches, one can estimate and control the properties [1–3].

1.4 Convection

In the previous section, a limited convection heat transfer was considered to emphasize the boundary condition for conductive problems. Usually, convection heat transfer occurs between surface and moving fluid (e.g., air, water, molten metal, etc.) when both have different temperatures. Thus, convection includes energy transfer due to the random motion of adjacent fluid molecules (conduction or diffusion) and

Table 1.3 Conduction shape factors for practical geometries.

System	Schematic	Restrictions	Shape factor $S = q/k(T_1 - T_2)$
Vertical cylinder in a semi-infinite medium	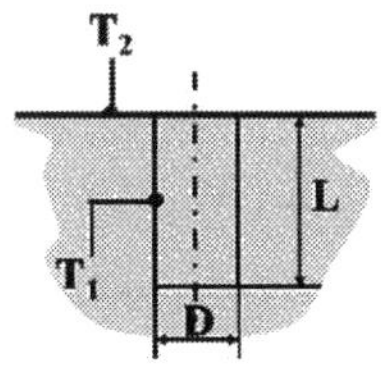	L > D	$\frac{2\pi L}{\ln(4L/D)}$
Circular cylinder of length L centered in a square solid of equal length	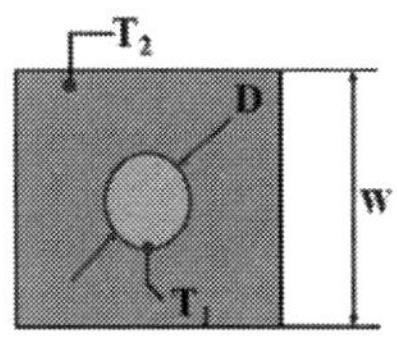	w > D L > w	$\frac{2\pi L}{\ln(1.08w/D)}$
Conduction through the edge of adjoining walls	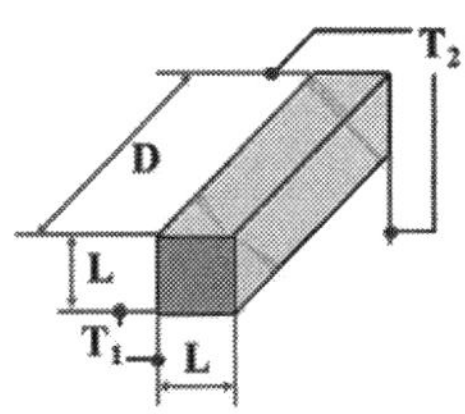	D > 5L	0.54D
Conduction through the corner of three walls with a temperature difference ΔT_{1-2} across the walls	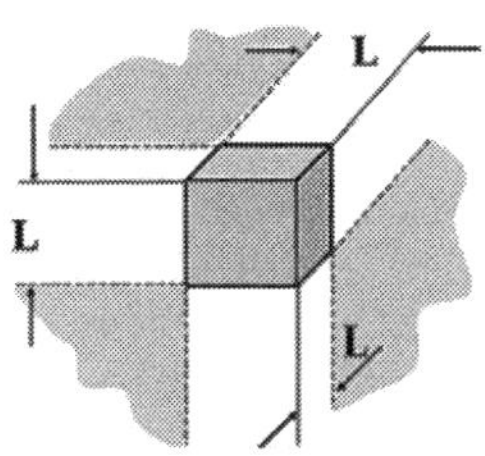	L—length and width of the wall	0.15L
Square channel of length L	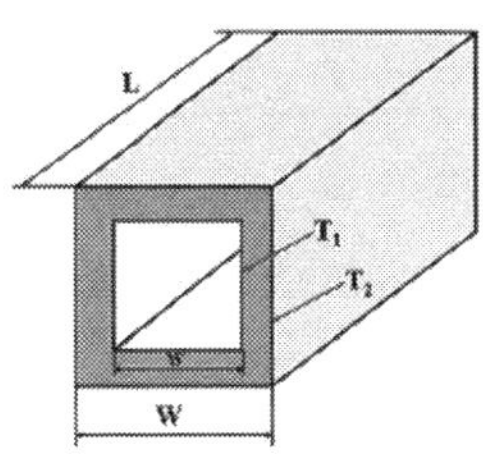	$\frac{W}{W} < 1.4$ $\frac{W}{W} > 1.4$ L > W	$\frac{2\pi L}{0.785\ln(W/w)}$ $\frac{2\pi L}{0.930\ln(W/w) - 0.050}$

bulk fluid (advection) due to force and consequent slippage that eventually forms a gradient. Adjacent to the surface, the molecular velocity and temperature remain nearly constant, but both vary and form as a free stream with boundary layers beyond a certain distance. Consequently, fundamentals including a brief of boundary layers, heat transfer coefficients, laminar and turbulent velocities, internal and external flow, and convection without forced velocity are pointed out.

1.4.1 Boundary Layers

Understanding the boundary layer provides insight into the mass- and heat-transfer between the surface and fluid flowing on it. Convection experiences boundary layers of thermal, concentration, and velocity, and their relationships with heat, mass-transfer, and friction are encountered.

Fluid molecule assumes zero velocity contact to the surface that retards the motion adjoining fluid layer due to shear stress τ parallel to the fluid velocity (u). Such an effect exists up to a certain distance, $y = \delta$, as shown in Figure 1.11. The velocity gradually increases and becomes free stream (∞) with increasing y and follows $u = 0.99u_\infty$ beyond the distance δ. It develops a velocity boundary layer variation with distance from the surface.

This narrow region thickness $\delta = Re_L^{-0.5}$ has sharp velocity gradient. This region is known as hydrodynamic boundary layer region, where Re_L is known as Reynolds number, defined as:

$$\mathrm{Re}_L = \frac{inertia}{viscous\ force} = \frac{u_\infty \mathrm{L}}{\vartheta} = \frac{\rho u_\infty \mathrm{L}}{\mu} \tag{1.37}$$

where, L = length scale, ρ = fluid density, μ = dynamic viscosity of fluid, ϑ = momentum diffusivity or kinematic viscosity. Minute disruptions occur in any flow, which can be intensified to create a turbulent flow. A small Reynolds number (Re_L) experiences high viscous force compared to inertia forces that restrict the laminar to turbulent flow amplification. As Re_L rises at a fixed surface spot, viscous forces have low prominence compared to inertia forces. Therefore, the viscosity of possessions doesn't enter the free stream as far as possible, diminishing the δ value. Fluid flow transport adjacent to surface experiences shear stress, and thus friction, and this friction coefficient (C_f) for external flow depends on:

$$C_f \equiv \frac{\tau_s}{\rho u_\infty^2 / 2} \tag{1.38}$$

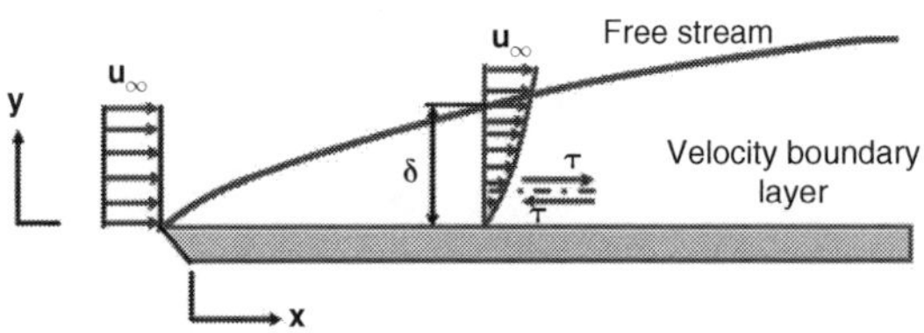

Figure 1.11 Velocity boundary layer on a flat surface.

Assume a Newtonian fluid, the surface shear stress at the surface edge is:

$$\tau_s = \mu\left(\frac{\partial u}{\partial y}\right)_{\gamma=0} \tag{1.39}$$

Usually, the velocity gradient at the surface depends on the distance x from the plate's leading edge in a velocity boundary layer. Eventually, the gradient of velocity and shear stress becomes negligible compared to the adjacent surface in the free stream zone, where fluid viscosity is predominant to encounter the flow analysis.

Similar to the velocity gradient, a sharp thermal gradient experience in the convection process differs from the fluid stream and surface temperature. The temperature profile along the y-axis, $T_y = T_\infty$. However, fluid contact with the plate achieves thermal equilibrium, as shown in Figure 1.12. These fluid molecules exchange heat energy to the adjoining fluid layer, form a thermal boundary with a thickness of δ_t along the y-axis, and follow temperature variation $(T_s - T) = (T_s - T_\infty)\,0.99$ as similar as velocity difference.

The thermal boundary layer δ_t thickness is equal to $(Re_L.\ Pr)^{-0.5}$, where Pr is the Prandtl number, $c_{p,f}$ = specific heat, and k_f = thermal conductivity of the fluid and can be represented by:

$$\mathrm{Pr} = \frac{\textit{momentum diffusivity}}{\textit{thermal diffusivity}} = \frac{\vartheta}{\alpha} = \frac{\mu/\rho}{k_f/\rho c_{p,f}} = \frac{\mu c_{p,f}}{k_f} \tag{1.40}$$

Momentum diffusivity becomes dominant when $Pr \gg 1$, but the heat diffuses fast compared to fluid velocity in a small value. It implies $\delta_t < \delta$ for oil, $\delta_t \approx \delta$ for gas, and $\delta_t > \delta$ for liquid metal flow adjacent to a flat surface. As the Reynolds number increases, it can also be seen that the boundary layers become narrow, the temperature gradient becomes high, and increases the rate of heat transfer. The local heat flux at any distance x from the leading surface edge strongly influenced by the wall temperature gradient $\partial T / \partial y$ and can be written as:

$$q_s'' = -k_f\left(\frac{\partial T}{\partial y}\right)_{y=0} \tag{1.41}$$

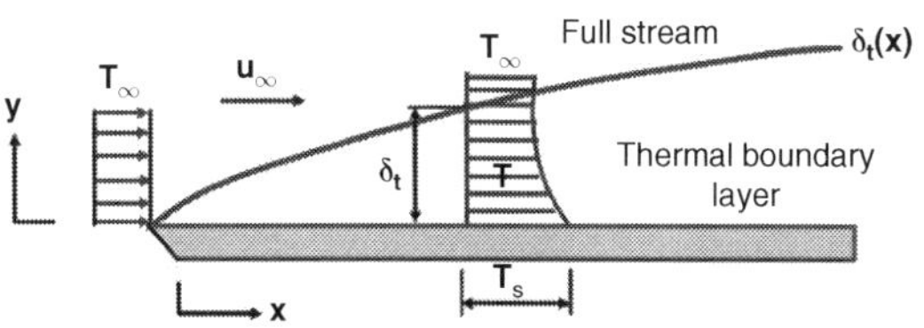

Figure 1.12 Thermal boundary layer on a flat surface.

Employing Newton's law of cooling, the heat flux can be rewritten as similar to Equation (1.15), where $(T_s - T_\infty)$ is constant and independent of x:

$$q''_s = h(T_s - T_\infty) \tag{1.42}$$

From Equations (1.41) and (1.42), we obtain:

$$h = \frac{-k_f\left(\partial T/\partial y\right)_{y=o}}{\left(T_s - T_\infty\right)} \tag{1.43}$$

From Equation (1.43), the magnitude of thermal gradient decreases with increasing x, as well as the reduction of q''_s, and h with increasing of x.

The boundary layer of concentration and mass-transfer of convection exist if the surface species concentration differs from its free stream concentration. In tunnel kiln firing, continuous firing and releasing several volatile and gaseous species results in different near-to-kiln walls and interiors that eventually produce a concentration gradient. Consider a binary mixture of chemical species A (e.g., O_2) and B (e.g., CO_2) that flows over a surface, as shown in Figure 1.13.

The molar concentration (kmol/m^3) of species A at the surface is $C_{A,s,}$ and in the free stream, it is $C_{A,\infty}$. Convection develops a concentration boundary layer from the surface to free stream with a concentration gradient of $(C_{A,s} - C_A) = 0.99\ (C_{A,s} - C_{A,\infty})$, with having a thickness of δ_c, equal to $(Re_L.\ Sc)^{-0.5}$, where Sc is Schmidt number, D_{AB} = mass-diffusivity, where Sc is:

$$Sc = \frac{viscous\ diffusion\ rate}{molecular\left(mass\right) diffusion\ rate} = \frac{\vartheta}{D_{AB}} = \frac{\mu}{\rho D_{AB}} \tag{1.44}$$

The Schmidt number measures mass-transport's momentum and relative effectiveness by diffusion in the concentration and velocity boundary layers. Therefore, mass-transfer through convection in laminar flow determines the thickness of boundary layer concentration and relative velocity, $\delta_t/\delta \approx Sc$. The ratio of Pr and Sc (δ_t/δ_c) known as Lewis number (Le) is relevant to any condition requiring simultaneous transfer of mass and heat through convection. Thus, it is the measure of the relative concentration and thermal boundary layer thickness. Analogous to heat flux $N''_{A,s}$, molar flux (kmol/s.m^2) at the surface can be

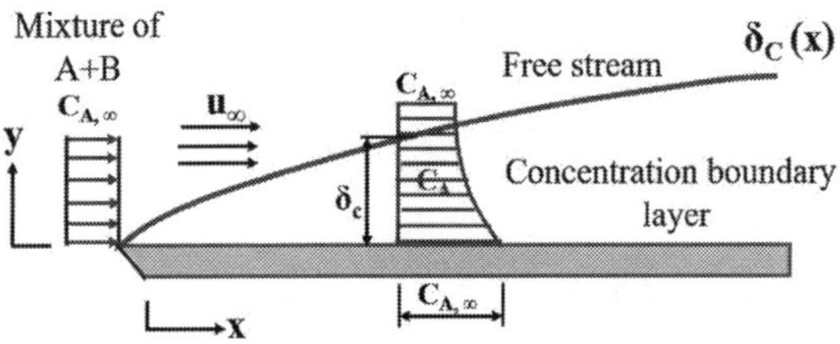

Figure 1.13 Species concentration boundary layer on a flat surface.

represented through Fick's law (Equation 1.45) and Newton's law (Equation 1.46) of cooling, where D_{AB} is the binary diffusion coefficient (mass-diffusivity), and h_m is the convection mass-transfer coefficient:

$$N''_{A,s} = -D_{AB}\left(\frac{\partial C_A}{\partial y}\right)_{y=0} \tag{1.45}$$

$$N''_{A,s} = h_m\left(C_{A,s} - C_{A,\infty}\right) \tag{1.46}$$

From Equations (1.45) and (1.46), the h_m becomes:

$$h_m = \frac{-D_{AB}\dfrac{\partial C_A}{\partial y}}{\left(C_{A,s} - C_{A,\infty}\right)} \tag{1.47}$$

Therefore, the concentration boundary layer conditions strongly affect the surface gradient of concentration and mass-transfer coefficient, hence the species transfer rate in the boundary layer.

Three boundary layers may exist together if temperature and species concentrations are different from the surface to free stream during fluid flow over the surface. The boundary layer velocity is the extent of δ (x) and is characterized by a velocity gradient. Shear stresses, thermal boundary layer δ_t (x) is characterized by temperature gradient and heat transfer, and the concentration gradient and species transfer describe concentration boundary layer $\delta_c(x)$. However, the boundary layers hardly develop at the equivalent rate, and the values of δ, δ_t, δ_c at a given location are not the same. The critical boundary layer parameters are coefficient of friction (C_f), coefficients of heat, and mass-transfer convection h and h_m, respectively.

1.4.2 Laminar and Turbulent Flow

It is essential to determine whether the boundary layer is laminar or turbulent flow before deciding the convection problem. Surface friction and the transfer rates of convection strongly depend on which of these conditions occurred.

Figure 1.14a represents a boundary layer formation for both laminar and turbulent flow on a flat plane. The velocity components in the direction of x and y, characterize the motion of the fluid. While the fluid moves away from the surface through x-direction, three distinct zone forms, laminar (1st), laminar to the turbulent transition zone (2nd), and turbulent zone (3rd). Fluid flow is highly ordered within the laminar boundary layer and maintained uniform streamlines along with fluid particles (molecules) move. The local shear stress of the surface reduces with increasing x (Equation 1.39). This behavior continues up to the transition zone where laminar flow converts to turbulent flow, where fluid coexists the dual character of both laminar and turbulent flow.

Beyond the transition zone, the turbulent boundary is highly irregular, and random results bulk three-dimensional flow without maintaining a two-dimensional streamline like laminar flow. In the chaotic flow condition in the turbulent zone, pressure and velocity variations arise at any point within the turbulent boundary layer. Three distinct regions appear as a function of distance from the surface. The "viscous sublayer" is dominated by diffusion where velocity profile is nearly linear, adjoining "buffer sublayer" in which diffusion and turbulence coexist, followed by dominated transport due to the turbulent mixing within the turbulent zone.

A competitive boundary layer of laminar and turbulent profiles for the x-component of the velocity are represented in Figure 1.14b, representing cumulative flow behavior in Figure 1.14a. Laminar to turbulent flow transition is finally due to triggering mechanisms, including unsteady flow that develops naturally within the fluid or minor disturbances within many typical boundary layers.

The onset of turbulence depends on the amplification of triggering mechanisms that eventually depend on the Reynolds number (Equation 1.37). The Reynolds number is small when the inertia forces are negligible compared to viscous forces. The flow remains laminar when the disturbances are dissipated. The inertia forces are high enough for a large Reynolds number to amplify the triggering mechanisms, and a transition to turbulence occurs. Beyond a critical distance x_c, along x-axis persist laminar flow, and the transition begins beyond it.

The fluid flow nature has a significant outcome on mass- and heat-transfer rates. Analogous to the velocity of the laminar layer, the species, and thermal boundary layers propagate in the streamwise (increasing x) direction, gradients of temperature, and species concentration in the fluid at $y = 0$ decrease in the streamwise direction. From Equations (1.43) and (1.47), the mass and heat transfer coefficients also decrease with increasing x. Mixing of turbulence promotes a large concentration gradients of species and temperature adjacent to the solid surface and an equivalent increase in the mass- and heat-transfer coefficients across the region of transition. The cumulative thickness of velocity boundary layer δ and the heat transfer coefficient of local convection h represents a characteristic behavior, as shown in Figure 1.15.

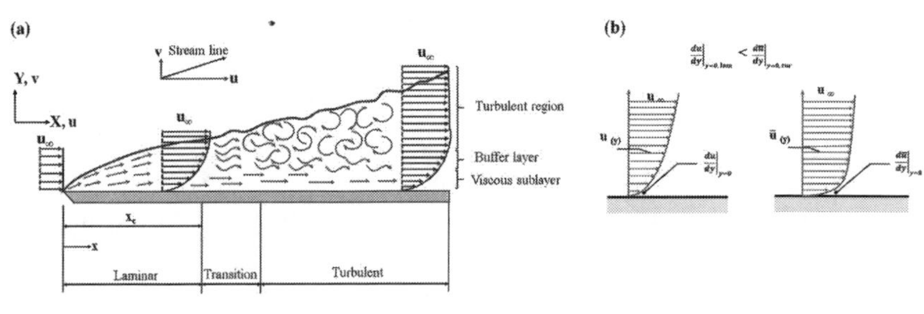

Figure 1.14 (a) Development of velocity boundary layer on a flat plate, (b) comparison of laminar and turbulent velocity boundary layer profiles for the same free stream velocity.

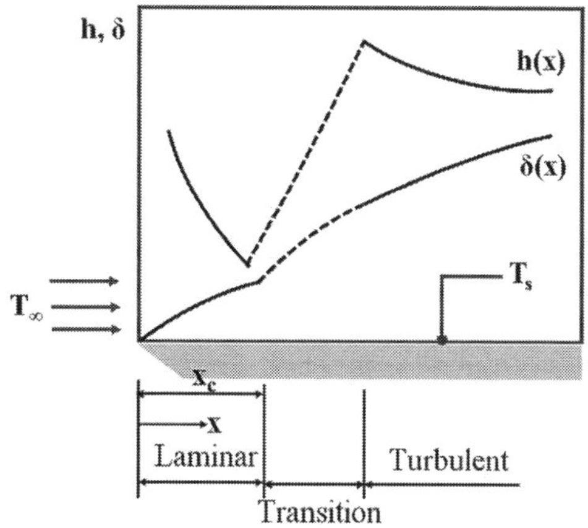

Figure 1.15 Variation of velocity boundary layer thickness δ and the total heat transfer coefficient h for flow over an isothermal flat plate.

The resulting variations in the thickness of the velocity, the species, and the temperature boundary layer appear to be much smaller in the turbulent flow because turbulence causes mixing, which decreases the role of conduction and diffusion in deciding the species and thermal boundary thickness.

1.4.3 Free and Forced Convection

The basic concepts of velocity, thermal, and concentration boundary layer thickness; surface friction coefficient; heat and mass-transfer coefficient; laminar and turbulent flow behavior, and their transition zone in two-dimensional and three-dimensional convection processes; and how these are related to physical parameters such as density, viscosity, thermal conductivity, specific heat, diffusivity, and dimensionless number *Re*, *Pr*, and *Sc* are discussed in the previous section. Bulk fluid motion facilitates the heat transfer and convection mechanism. Convection is further subdivided into two categories: free (or natural), and another is forced convection, based on how the action of fluid originated. In free convection, any fluid motion is initiated by natural means, such as the differences in density and effect of buoyancy, and the rise and fall of warmer and cooler fluid, respectively. Whereas in forced convection, the motion of the fluid is forced to flow over a surface or in a circular section by external agency means a fan or pump. Such a convection process can be observed when air is supplied or extracted by a blower or suction pump into a tunnel kiln to maintain heat and mass-transfer, turbulence in tundish, and subsequent consistent flow control during casting, etc. Forced convection creates a more uniform temperature throughout the entire region. Essential information for the convection in the confined region may provide insight into this understanding and is discussed in the next section.

The convection heat transfer from or to the surface of a material may be calculated using Equation (1.15), and empirical correlation:

$$Nu = cRe^n Pr^m \tag{1.48}$$

Where Nusselt number (Nu) is equal to hL/k, the Reynolds number (Re) is $\rho uL/\mu$, and the Prandtl (Pr) is $c_p\mu/k$. From Equation (1.49), the free convection coefficient can be estimated:

$$Nu = cRa^n Pr^m \tag{1.49}$$

The Rayleigh number (Ra) is $(Ts - T\infty)\beta g\rho L^3/(\alpha\mu)$, where β is the volumetric thermal expansion coefficient, g is the gravitational acceleration, ρ is density, l is length, and α is thermal diffusivity. The constants c, n, and m for different geometry and flow conditions are available in the literature [4, 5].

1.4.4 Flow in Confined Region

Few real-life incidences represent an example of forced convection during the external flow of fluid. However, several incidences, including molten steel casting, experience confined region flow in a tube such as a slide gate, shroud, subentry nozzle, etc. The boundary layer develops on a surface in external flow and continues without external constraints. In contrast, the pipe's fluid flow cannot form the boundary layer; rather, the hydrodynamic layer's velocity affects the confined flow region. The convection heat transfer coefficients and energy balance are presented for internal flow in the circular channel, considering velocity and thermal boundary layer effects.

Figure 1.16a represents the internal fluid flow entrance and fully developed laminar flow regions with a uniform velocity in a circular tube of radius r_o. Viscous force facilitates the formation of a boundary layer with increasing x, which provokes shrinkage of inviscid flow region and subsequent boundary layer merger [6].

Beyond this merger, viscous effects extend over the entire cross-section. In such conditions, the velocity profile becomes unaltered with increasing x, and the flow is fully developed, known as hydrodynamic entry length, x_h. The fully developed velocity profile is parabolic for laminar flow in a circular tube, whereas the profile is flatter in the radial direction during turbulent flow. The Reynolds number determines the fully developed internal flow exists as laminar or turbulent flow. The critical Reynolds number ($Re_{D,c}$) for the internal flow near to 2300, where, $Re_{D,c} \equiv \rho u_m D/\mu \approx 2300$, $u_m = mean\ fluid\ velocity$, $D = tube\ diameter$. For laminar flow ($Re_D < 2300$), the hydrodynamic length can be calculated from $(x_{e,h}/D)_{lam} \approx 0.05Re_D$, whereas $(x/D) > 10$ indicates the fully developed turbulent flow.

The velocity varies with cross-section and a mean velocity u_m dealing the mass-flow rate during internal flow in the tube. The mass-flow rate is the product of fluid density ρ, mean velocity u_m, and cross-sectional area A_c $(= \pi D^2/4)$, and the Reynolds number can be expressed as Equations (1.50) and (1.51), respectively:

$$\dot{m} = \rho u_m A_c \tag{1.50}$$

$$Re_D = \frac{4\dot{m}}{\pi D\mu} \tag{1.51}$$

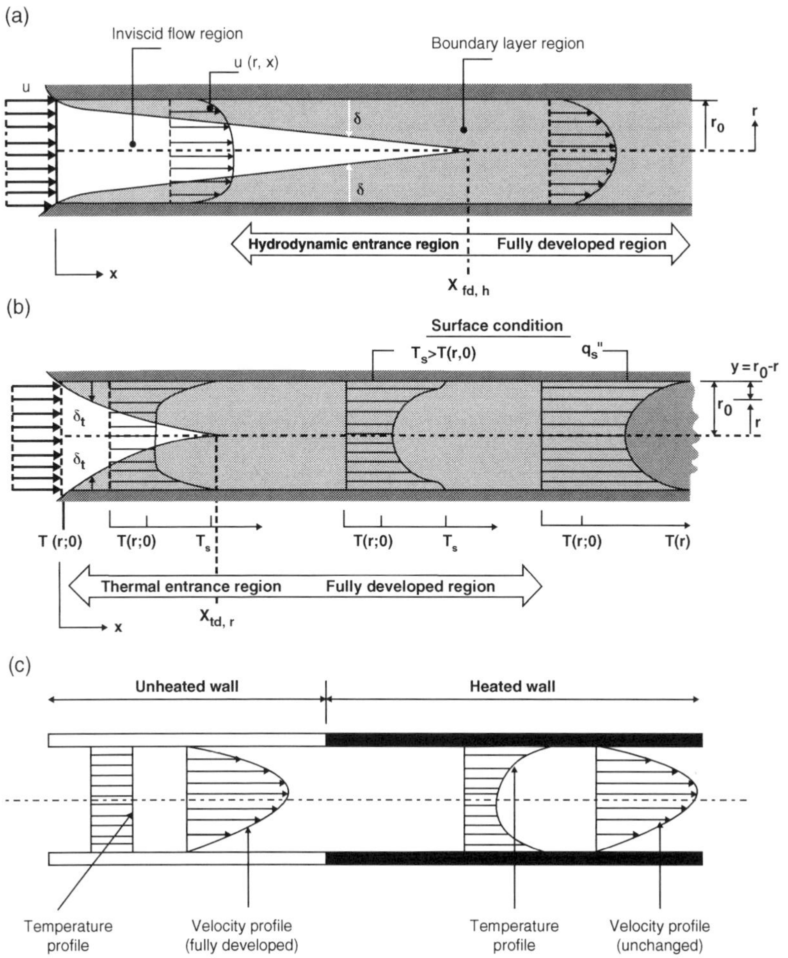

Figure 1.16 (a) Laminar, hydrodynamic boundary layer development in a circular tube, (b) thermal boundary layer development in a heated circular tube, (c) variation of temperature and velocity profile for an unheated and heated wall.

The velocity profile may readily be determined for the laminar flow of an incompressible, constant property fluid fully developed region. The resultant velocity concerning *r* may also be determined from Equation (1.52):

$$u(r) = 2u_m\left[1-\left(\frac{r}{r_o}\right)^2\right] \tag{1.52}$$

Pressure drop is a common incidence due to friction, and Moody (or Darcy) friction factor *f* for fully developed laminar flow becomes:

$$f = 64/\mathrm{Re}_D \tag{1.53}$$

however, the analysis of friction factors for turbulent flow relies on experimental data. In approximate, the f for turbulent flow follows the relation:

$$f = (0.790\, ln\, Re_D - 1.64)^{-2},\ where\ 3000 \le Re_D \le 5\times10^6 \tag{1.54}$$

Despite fluid velocity gradient, a thermal boundary layer is shown in Figure 1.16b when fluid enters at a uniform temperature $T(r,0)$ less than surface temperature and gradually increases with x. The shape of the hydrodynamically fully developed temperature profile $T\,(r,x)$ differs according to whether a uniform surface temperature (UST) or uniform heat flux (UHF) is maintained. The characteristic velocity and temperature profile concerning the heating condition is shown in Figure 1.16c. For laminar flow, the thermal entrance length may be expressed as:

$$\left(\frac{x_{e,t}}{D}\right)_{lam} \approx \begin{cases} 0.033 Re_D \mathrm{Pr}\ for\ UST \\ 0.043 Re_D \mathrm{Pr}\ for\ UHF \end{cases} \tag{1.55}$$

We can generalize, $(x_{e,t}/D) = 0.05\ Re_D\ Pr$. Under the circumstance of $Pr > 1$, the hydrodynamic boundary layer develops more rapidly than the thermal boundary layer x_h (velocity profile) $< x_{e,t}$ (thermal profile), then $\delta > \delta_T$ at any section; while the inverse is true for $Pr < 1$. For extensive Prandtl number fluids, such as oils ($Pr \ge 100$), $x_{e,h}$ is much smaller than $x_{e,t}$, and it is reasonable to assume a fully developed velocity profile throughout the thermal entry region. In contrast, conditions are nearly independent of the Prandtl number for turbulent flow, and to a first approximation, $(x_{e,t}/D) = 10$. The bulk mean temperature (T_m) at any section (x) is given by:

$$T_m = \frac{2}{u_m r_o^2} \int_0^{r_o} u(r)T(r)rdr \tag{1.56}$$

By employing Newton's law of cooling, the heat flux is $q''_s = h(T_s - T_m)$, where h is the local convection heat transfer coefficient. The value of T_m increases with x if heat transfer is from the surface to the fluid ($T_s > T_m$); it decreases with x if the opposite is true ($Ts < Tm$).

In a confined region, as shown in Figure 1.17, the $T_m\,(x)$ varies with position along the x, and the total q_{conv} is related to the difference in temperatures at the tube inlet ($T_{m.i}$) and outlet ($T_{m,o}$), constant flow rate ṁ and specific heat capacity of the fluid. In steady-state flow conditions, the convection heat transfer can be written as Equation (1.57):

$$q_{conv} = \dot{m}c_p\left(T_{m,o} - T_{m,i}\right) \tag{1.57}$$

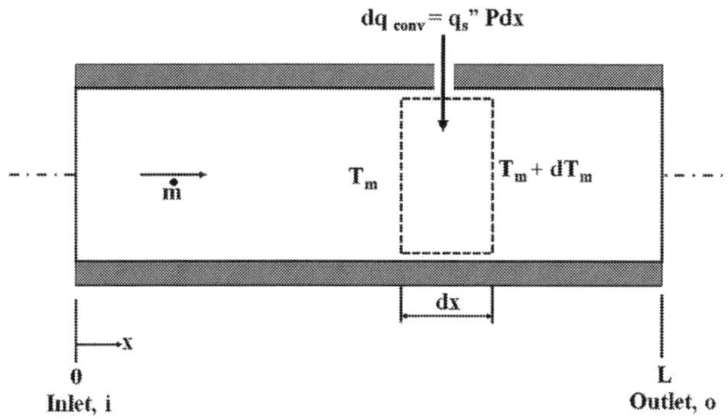

Figure 1.17 Control volume for flow in a confined region.

The rate of thermal energy advection integrated over the cross-section:

$$dq_{conv} = q_s''Pdx = \dot{m}c_\rho\left[\left(T_m + dT_m\right) - T_m\right)\right] \tag{1.58}$$

The heat flux $q_s'' = dq_{conv} / Pdx$, where P = *surface perimeter* = πD for a circular tube. From the combination of Equations (1.42) and (1.58), we may obtain:

$$\frac{dT_m}{dx} = \frac{q_s''P}{\dot{m}c_\rho} = \frac{P}{\dot{m}c_\rho}h\left(T_s - T_m\right) \tag{1.59}$$

This expression is a beneficial result, from which the axial variation of Ts may be determined. If $T_s > T_m$, heat is transferred to the fluid, and T_m increases with *x; if* $T_s < T_m$, the opposite is true. The convection coefficient is constant in the fully developed region, and it varies with *x* in the entrance region. The T_m *(x)* depends on two special cases, constant surface heat flux and constant surface temperature.

For constant surface heat flux, the q_s'' is independent of *x*, and follows $q_{conv} = q''_s(P.L)$ and equating with Equation (1.57), followed by integrating results

$$T_m\left(x\right) = T_{m,i} + \frac{q''_sP}{\dot{m}c_\rho}x \tag{1.60}$$

$$q''_s = constant$$

Equation (1.60) reveals that the mean temperature varies linearly with *x* along the tube and same time variation of ($T_s - T_m$) with *x* as shown in Figure 1.18a. Due to the significant value of *h* near the entrance, the difference is negligible at starting

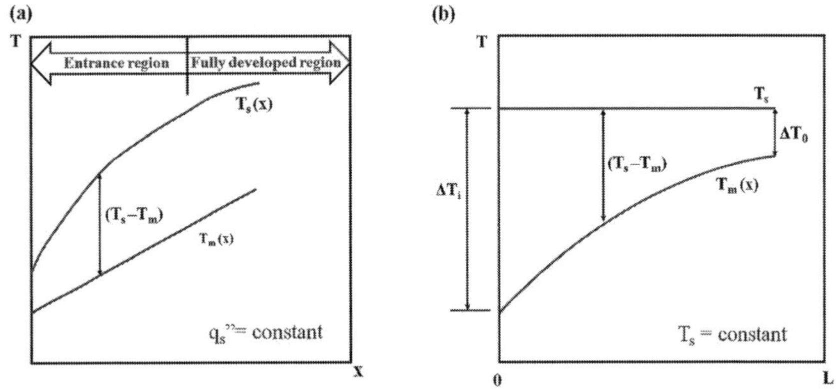

Figure 1.18 Axial temperature variation for heat transfer in a tube (a) constant surface heat flux, (b) constant surface temperature.

but increases with x, the increase owing to the reduction in h that occurs as the boundary layer grows. In reality, the h is independent of x in the fully developed region; thus the $(T_s - T_m)$ is independent of x in this region.

While considering the uniform temperature of the surface, the total rate of heat transfer and axial distribution follows exponential decay and depends on the average coefficient of heat transfer $(\bar{h})$, the resultant relation can be expressed as:

$$\frac{\Delta T_o}{\Delta T_i} = \frac{T_s - T_m(x)}{T_s - T_{m,i}} = \exp\left(-\frac{Px}{\dot{m}c_\rho}\bar{h}\right) \tag{1.61}$$

$$T_s = constant$$

Thus, the difference in temperature $(T_s - T_m)$ decays exponentially with distance along the tube axis, and thus the axial surface and distributions of average temperature are as shown in Figure 1.18b.

Despite the circular cross-section, several engineering applications demand convection transport in noncircular tubes. Many of the circular tube results may be applied by using an effective diameter as the characteristic length to determine the Re_D and Nu_D. In a simple way, this effective diameter of hydraulic is defined as:

$$D_h \equiv \frac{4A_c}{P} \tag{1.62}$$

Where A_c and P are the cross-sectional flow area and the wetted perimeter, respectively. For laminar flow, circular tube correlations are less accurate when the system has sharp corners. For such cases, the details of the Nusselt number corresponding to fully developed conditions may be obtained from detailed analysis in literature [7].

1.5 Radiation

Conduction and convection both process experience temperature gradient; however, the radiation is an important heat transfer mode for much industrial heating, cooling, and drying processes as well as energy conversion involved with fossil fuel combustion and solar radiation.

1.5.1 Basic Concepts

All forms of matter emit radiation. The emission mechanism is related to energy released due to transitions or oscillations of the electrons present in the material. These oscillations are in turn persistent by the emission of internal energy as thermal radiation, and therefore the temperature of the matter. Hence, the emission of thermal radiation with thermally excited conditions within the matter. At elevated temperatures, the emission is a volumetric phenomenon; however, the discussion only focuses on radiation through the surface. Accordingly, radiation emitted from a solid or a liquid originates from molecules within an approximate distance of 1 μm from the exposed surface.

Despite the situation in nanoscale or microscale devices, emission from a solid or a liquid toward adjacent gas or a vacuum can be viewed as a surface phenomenon. Emission does not require the presence of matter, may happen in a vacuum; thus radiation is the result of the propagation of collective particles known as *photons* or quanta or defined as the propagation of electromagnetic waves. For radiation propagating in a particular medium, the wavelength:

$$\lambda = \frac{c}{V} \tag{1.63}$$

Where c is the speed of light in the medium. For propagation in a vacuum $c_0 = 2.998 \times 10^8$ m/s. Wavelength (λ) variation for *thermal radiation* in the range of 0.1–100 μm includes a portion of UV and all visible and infrared (IR) pertinent to heat transfer. Thermal radiation emitted by a surface encompasses a range of wavelengths. The magnitude of the radiation varies with wavelength, and the term spectral is used to refer to the nature of this dependence. Emission follows the directional distribution of the emitted radiation with radiation intensity.

1.5.2 Emission from Real Surfaces

A blackbody is an ideal emitter as it can emit maximum radiation compared to any other material at the same temperature. Planck's law describes the spectral density of electromagnetic radiation emitted by a black body in thermal equilibrium at a given temperature *T*. The emissivity and temperature can be correlated as Equation (1.64), where $\sigma = 5.67 \times 10^{-8}$ W/m^2.K^4, known as Stefan–Boltzmann constant:

$$E = \sigma T^4 \tag{1.64}$$

Thus, emissivity may be defined as the ratio of the radiation emitted by the surface to the radiation emitted by a black body at the same temperature. Spectral radiation distribution, as well as directional distribution for real surface emission, are different from Planck distribution of a black body, as shown in Figures 1.19a and 1.19b.

Representative spectral distribution of ε_λ (spectral emissivity) varies for a conductor, nonconductor, and nature of the surface coating (like the presence of oxides), as shown in Figure 1.20a. Total normal emissivity concerning temperature for a different material class is shown in Figure 1.20b [6].

The available data demonstrates that the emissivity of metallic surfaces is generally small, as low as 0.02 for polished metal surfaces. A classical analysis and derivation represent emissivity of a material and is inversely proportional with thermal conductivity, and it can be represented by Equation (1.65) where, k = thermal conductivity:

$$\varepsilon_n = 0.865X10^{-4}\ k^{-1/2} \tag{1.65}$$

At the same time, the oxide layer increases the emissivity of metal surfaces; at elevated temperatures, the stainless steel becomes oxidized, and emissivity increases from 0.3 to 0.7.

In this contrast, the emissivity of nonconductors is comparatively large, generally more than 0.6. The emissivity of conductors increases with increasing temperature; however, depending on the specific material, nonconductors' emissivity may either increase or decrease with increasing temperature. Emissivity strongly depends on the nature of the surface, which can be influenced by the method of fabrication, machining and polishing, thermal cycling, and chemical reaction with its environment.

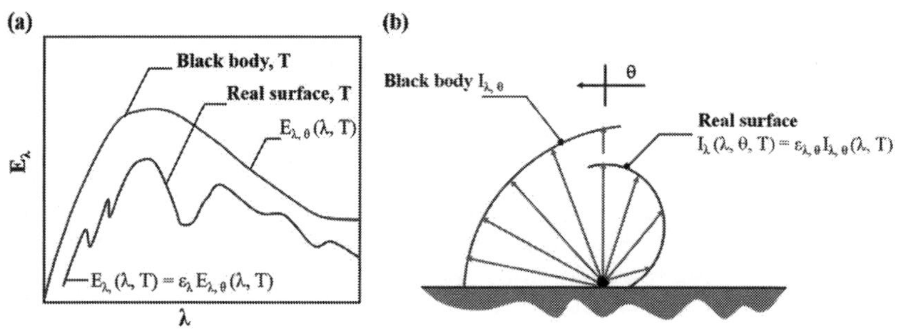

Figure 1.19 Comparison of a black body and real surface emission, (a) spectral distribution, (b) directional distribution.

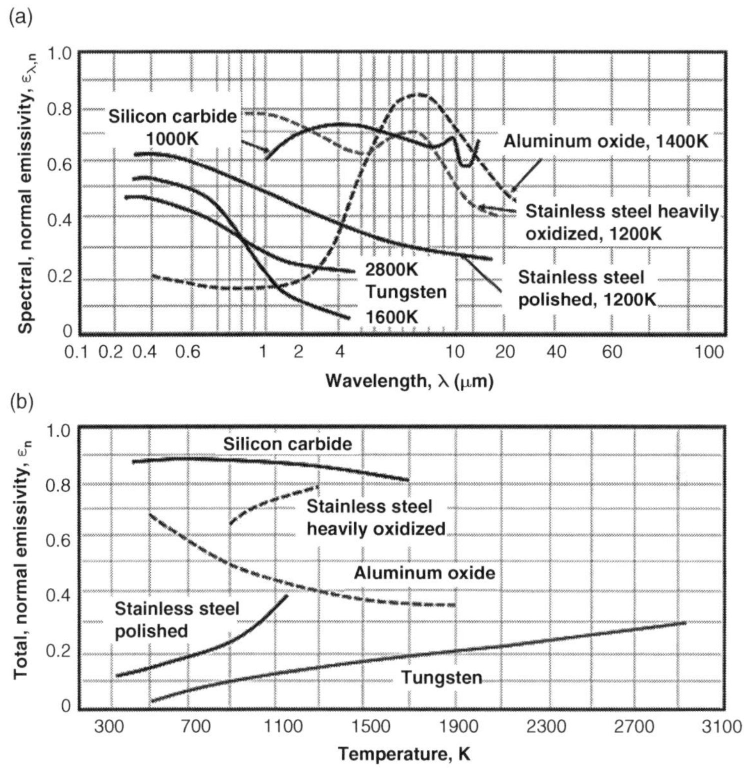

Figure 1.20 (a) Spectral dependence of the spectral, normal emissivity $\varepsilon_{\lambda \cdot n}$, and (b) temperature dependence of the total, normal emissivity ε_n of selected materials.

1.5.3 Absorption, Reflection, and Transmission by Real Surfaces

Many substances, especially metal oxides, are semitransparent in the visible and infrared spectral ranges.

Let's consider how certain irradiation of wavelength (λ) interacts with a semitransparent medium like fused quartz; part of the radiation may be absorbed, reflected, and transmitted represented in Figure 1.21, and can be represented as:

$$G_\lambda = G_{\lambda,\mathrm{ref}} + G_{\lambda,\mathrm{abs}} + G_{\lambda,\mathrm{tran}} \tag{1.66}$$

The $G_{\lambda,tran}$ becomes zero for opaque materials, most engineering material. The remaining absorption and reflection processes may be considered as surface phenomena. However, there is no net effect of the reflection process on the medium, while absorption increases the medium's internal energy. Thus, surface absorption and reflection are responsible for the perception of color. Unless the surface is at a high temperature ($T_s \geq 1000$ K), such that it is incandescent, color is in no way due to emission, which is concentrated in the *IR* region and is hence invisible to the eye.

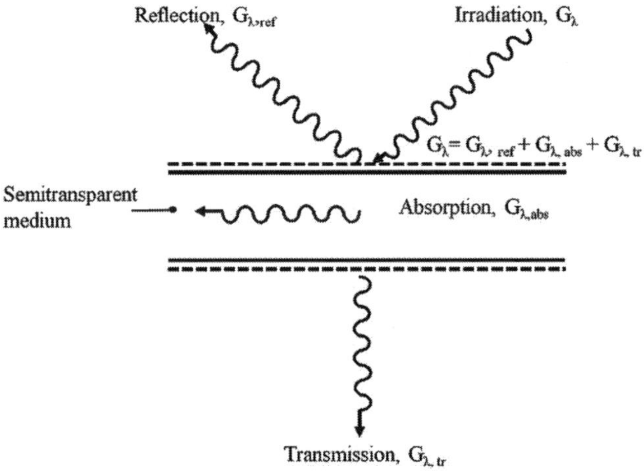

Figure 1.21 Absorption, reflection, and transmission processes associated with a semitransparent medium.

Spectral distributions of the normal reflectivity and absorptivity are plotted in Figure 1.22a for some opaque materials. A material such as glass or water, semitransparent at short wavelengths, becomes opaque at longer wavelengths.

This behavior is shown in Figure 1.22b, representing the spectral transmissivity of several common transparent materials. It is to be noted that inclusion has a significant effect on the resultant transmissions.

1.5.4 Exchange Radiation

Exchange radiation encounters the radiation between two solid surfaces, liquid and polar gases, such as CO_2 and H_2O vapor. The exchange radiation phenomenon depends on the surface geometries and orientations and their radiative properties and temperatures. In an enclosure of two surfaces, the multiple reflection and absorption facilitate radiation exchange between two surfaces.

Consider the enclosure of two surfaces, as shown in Figure 1.23, where radiation is involved within each other. The existence of two such surfaces' net radiant heat must be equal, and in this circumstance, the rate of heat radiation from surface 1 is q_1, equal to the q_2, net rate of radiation transfers to surface 2.

This indicates $q_1 = -q_2 = q_{12}$. The net radiation exchange between surfaces may be expressed as:

$$q_{12} = q_1 = -q_2 = \frac{\sigma\left(T_1^4 - T_2^4\right)}{\dfrac{1-\varepsilon_1}{\varepsilon_1 A_1} + \dfrac{1}{A_1 F_{12}} + \dfrac{1-\varepsilon_2}{\varepsilon_2 A_2}} \tag{1.67}$$

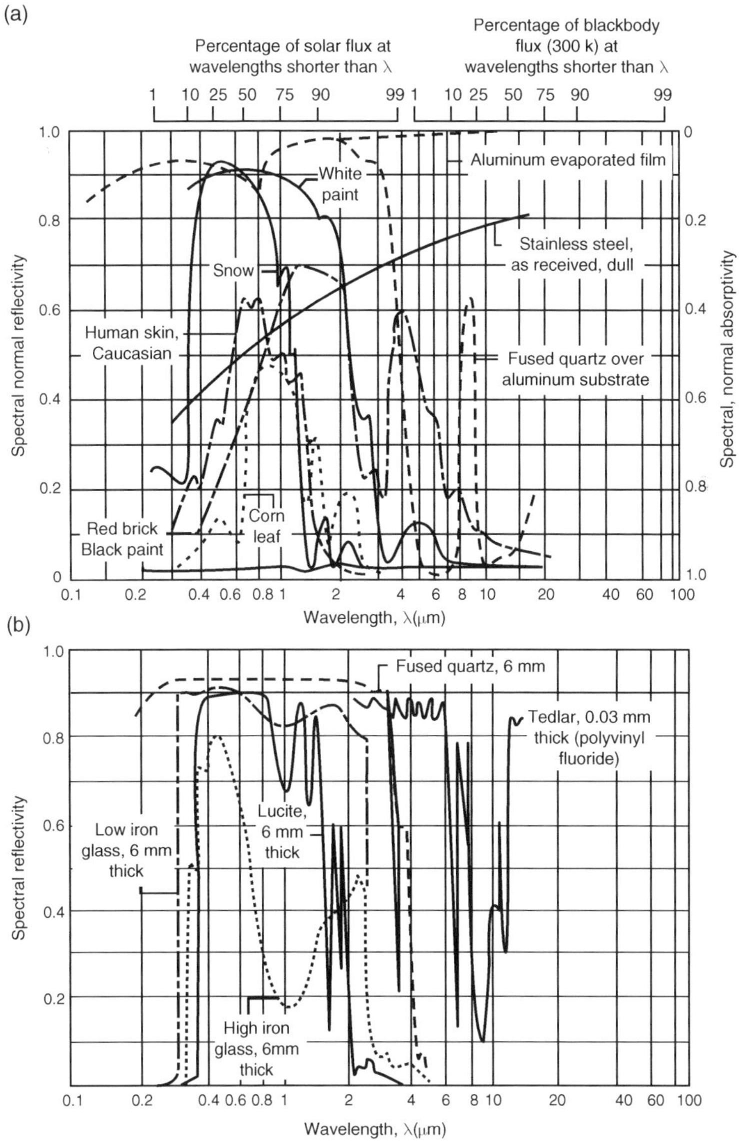

Figure 1.22 Spectral dependence of the spectral, (a) normal absorptivity and reflectivity of opaque materials, (b) transmissivities of semitransparent materials [6] / John wiley & sons.

Where A_1, ε_1, T_1 and A_2, ε_2, T_2 are the surface area, emissivity, and temperature of surfaces, respectively. The view factor F_{12} is the fraction of energy exiting an isothermal, opaque, and diffuse surface 1 (by emission or reflection) that directly impinges on surface 2.

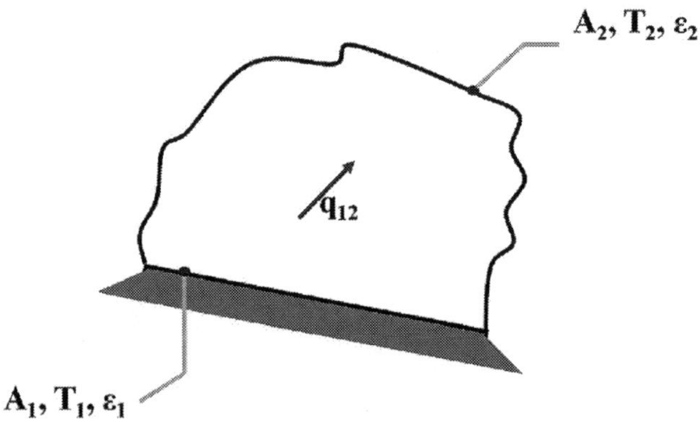

Figure 1.23 Schematic representation of the two surfaces enclosure.

The radiation is not a surface phenomenon for liquid and gaseous, rather more apposite as a volumetric phenomenon. Emission from the gas per unit area of the surface is expressed as Equation (1.68), where E_g is gas emissivity:

$$E_g = \varepsilon_g \sigma T_g^4 \tag{1.68}$$

The rate of radiant heat transfers to a surface due to emission from an adjoining gas is Equation (1.69), where A_s is surface area:

$$q = \varepsilon_g A_s \sigma T_g^4 \tag{1.69}$$

In the same chronology, the net flow rate of thermal radiation energy between a volume of gas and the wall encloses the gas space, where A_v is geometry-dependent absorptance:

$$\dot{Q}_{gw} = A\sigma \frac{\varepsilon_w}{1-(1-\varepsilon_w)(1-A_v)}(\varepsilon_g T_g^4 - A_v T_w^4) \tag{1.70}$$

It is valid only if the temperature, density, and concentration of the gas are constant in space, and its application requires knowledge of the gas emissivity and absorbance.

1.6 Mass Transfer

Modes of mass-transfer are similar to heat transfer. The temperature difference in a medium is the prime driving force for heat transfer; in the same analogous concept, the concentration gradient of chemical species facilitates mass-transfer. Some important aspects related to convection and multiphase mass-transfer are discussed.

1.6.1 Convection Mass Transfer

Some relevant understanding and equations for convection mass-transfer have already been discussed in Section 1.4.1; however, an additional piece of information may provide better insight into the subject. Mobile interface separates two immiscible moving fluids or boundary surface and a moving fluid. In contrast, consider the refractory interaction with molten metal, the convection analysis between the fixed solid surface, and moveable fluid is more important than bulk fluid. Convection mass-transfer is of two types, forced convection, and natural convection. The rate equation for both forced and natural convective mass-transfer is:

$$N_A = k_c \Delta C_A \tag{1.71}$$

Where N_A is the molar-mass flux of species A, ΔC_A is the difference between the boundary surface concentration and the average concentration of the diffusing species in the moving fluid stream, and k_c is the convective mass-transfer coefficient. It defines the mass-transfer rate per unit area per unit driving force and indicates how fast the mass-transfer is by convection. In heat transfer, the ratio of molecular mass-transport resistance to the fluid convective mass-transport resistance is known as the Sherwood number, Sh, and analogous to the Nusselt number Nu. Sherwood number is:

$$Sh = \frac{molecular\ mass\ transport\ resistance}{convective\ mass\ transport\ resistance} = \frac{k_c L}{D_{AB}} \tag{1.72}$$

$Sh = 1$ provides that the diffusion coefficient (mass-diffusivity, D_{AB}) is equal to the product of characteristic length, L (depends on the geometry), and mass-transfer coefficient (k_c). In another way, the dimensionless Lewis (Le) number can be defined as the ratio of thermal diffusivity to mass-diffusivity encountered during the simultaneous convective transfer of mass and energy.

A few significant mass-transfer correlations for different geometries during laminar and turbulent flow are summarized in Table 1.4.

1.6.2 Multiphase Mass Transfer

In multiphase systems, the rate of species mass-transfer is governed not only by the temperature and pressure of the system but also by the mass-transfer conductance, gradients of concentration, kinetics by reaction, activation energy, etc. In a few circumstances, any mass-transfer conductance or kinetics by reaction could control the overall mass-transfer rates; the lowest will be the limiting rate or controlling step. Many theories deal with the mass-transfer among phases, such as the two-film, penetration, and surface renewal approaches. The oldest two-film

Table 1.4 Dimensionless number correlations for mass-transfer in laminar and turbulent flow.

Geometry	Correlations	Conditions
Flat surface	$Sh_L = 0.664\ Re_L^{0.5}\ Sc^{0.33}$ $Sh_L = 0.664\ Re_L^{0.5}\ Sc^{0.33}$	(laminar) $Re_L < 2 \times 10^5$ (turbulent) $Re_L > 2 \times 10^5$
Solid cylinder immersed in water	$kc\ Sc_{0.56}/V_\infty = 0.281\ Re_D^{-0.4}$	(turbulent); $400 < Re_D < 25{,}000$ and $0.6 < Sc < 2.6$
Flow-through pipe	$Sh = 0.032\ Re0.8\ Sc0.33$	$2000 < Re_D < 35000$ and $1000 < Sc < 2260$

theory for gas-liquid mass-transfer assumes the existence thickness of a film (δ) in both the liquid and gas phases divided by an interface. According to the following assumptions: (1) the transfer of mass occurs by molecular diffusion through the film, beyond which the concentration (C_{Ab}) is homogeneous; (2) under the steady-state conditions, the transfer of mass occurs through the film; and (3) the transfer of mass occurs at low concentration when the flux is small.

The concentration profile is thus linear for convective mass-transfer as shown in Figure 1.24, and the coefficient of liquid side mass-transfer is expressed by Equation (1.73):

$$kL = \frac{D_{AB}}{\delta_L} \tag{1.73}$$

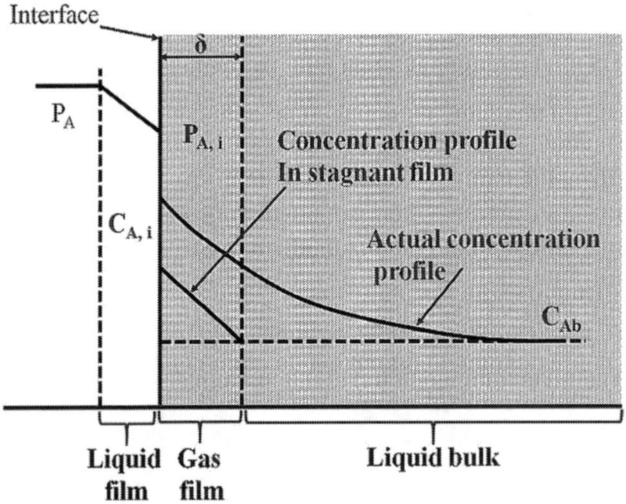

Figure 1.24 Schematic of two-film theory.

Where D_{AB} is the molecular diffusivity of the gas (solute A) into the liquid (solvent B), k_L is proportional to the diffusivity of the power one. The diffusivity primarily depends on the temperature, solvent viscosity, composition, and nature of the solvent. Several investigations established a relation with k_L and D_{AB} from experimental observations in $k_L \infty (D_{AB})^m$, where the ^{m} value varies 0.33–0.66.

1.6.3 Analogy—Heat, Mass, and Momentum Transfer

Dimensionless equations become analogous if both of the same forms govern two or more processes. Analogies between the transfer of mass, heat, and momentum originate in the statistical explanation of the effects or the physical parameters used for the quantitative description. It is known that the mass-diffusion and heat conduction have similar equations to explore these analogies. Specifically, Fick's Law describes the diffusion in one dimension as:

$$J_A = -D_{AB}\frac{dc_A}{dz} \tag{1.74}$$

Similarly, Fourier's law describes heat conduction, where k is the thermal conductivity:

$$q = -k\frac{dT}{dz} \tag{1.75}$$

A similar equation, Newton's law, describes the momentum transfer:

$$\tau = -\mu\frac{du}{dz} \tag{1.76}$$

Where momentum flux (or shear stress) is $\tau_{s,}$ and viscosity of a fluid μ.

It has become conventional at this stage to draw an analogy between the transfer of mass, heat, and momentum. A simple law combined with a balance of mass or energy, or momentum is used by each process. Several analogies are considered among transfer phenomena suggested due to the similarity in their mechanisms in this section. The analogies help to understand the transfer phenomena and as a satisfactory means of assessing the safety of systems for which minimal quantitative data is available. Accordingly, the similarity of the transfer phenomena and the presence of the analogies requires that the five primary following conditions exist within the system.

1) The constant physical properties.
2) There is no reaction chemically within the system due to no mass or energy induced.

3) There is no absorption or emission of radiant energy.
4) There is no dissipation of viscous energy.
5) Velocity profile does not affect the mass-transfer rate; thus mass-transfer rate is low.

Reynolds, known as the Reynolds analogy, established a common analogous mass, heat, and momentum transfer behavior. Although this analogy is limited to gases, a further modification was done by the Chilton–Colburn analogy considering all fluids, gases, and liquids. Reynolds suggested that the mass, momentum, and energy transfer mechanisms are similar and indicate a clear relationship between the various transport phenomena. Correspondingly:

$$\frac{k_c}{u_\infty} = \frac{h}{\rho u_\infty c_\rho} = \frac{f}{2} \tag{1.77}$$

Where h is a heat transfer coefficient, f is the friction factor, u_∞ is the velocity of the free stream. This relation is found to be accurate when Prandtl ($Pr = c_p\mu/k$) and Schmidt ($Sc = \mu/\rho D_{AB}$) are equal to one. The mass-transfer coefficients from the heat transfer coefficient can be calculated during turbulent eddies in gases. In the *Chilton–Colburn analogy*, there are no restrictions on the unit values of Prandtl and Schmidt numbers. These are defined through j factor for mass-transfer as j_D and heat transfer as j_H, and can be written as:

$$j_D = \frac{k_c}{u_\infty} Sc^{2/3} \tag{1.78}$$

$$j_H = StPr^{2/3} \tag{1.79}$$

St. refers to the Stanton number:

$$St = \frac{Nu}{RePr} = \frac{h}{\rho u_\infty c_\rho} \tag{1.80}$$

Exhaustive data for both laminar and turbulent flow regimes established:

$$j_D = j_H = \frac{f}{2} \tag{1.81}$$

This analogy is valid for gases and liquids within the range of $0.6 < Sc < 2500$ and $0.6 < Pr < 100$. The Chilton–Colburn analogy is valid for much different geometry, for example, flow over flat plates (refractory lining), flow in pipes (during casting), and flow around cylinders (submerged monoblock stopper). While comparing the equations relating to heat and momentum transfer with heat and mass-transfer, it can be shown that:

$$\left[\frac{h_c}{\rho c_p h_m}\right] = \left[\frac{\alpha}{D}\right]^{\frac{2}{3}} \tag{1.82}$$

The above analogies are especially useful when implemented, and if the friction factor is known, it is also helpful to determine the heat transfer coefficient. From the information of heat transfer coefficient, the mass-transfer coefficient can be determined.

1.7 Heat Transfer in Refractory Lining

The heat transfer analysis for multilayer refractory lining in both transient and continuous operation is critical to optimize the working and backup lining with knowledge of their physical, thermal, and mechanical properties. An optimum lining provides consistent performance. Insulating refractories are thermal barriers that keep in the heat and save energy. As the cost of energy has increased, the role of insulating refractories has become more important. Determining temperature profile through different layers either in a kiln, a refractory manufacturing unit, or ladle, the steel manufacturing unit is essential for design and operational troubleshooting. From the perspective of structure and energy efficiency, the design of refractory linings is becoming increasingly important for effective operation practice. Strict "fitness for service" (FFS) guidelines for pressure vessels must be maintained to provide the plant's safety, including assets and public health. Knowing accurate thermal property and convection/radiation boundary conditions makes it possible to predict temperature profiles under various conditions with reasonable accuracy. An optimum thickness is necessary to enhance the refractory life. For example, low insulation thickness may facilitate the buckling of steel ladle and hot spot on the outer shell when the temperature is > 370 ^{0}C, but more insulation thickness can overheat the working lining; thus rapid oxidation and corrosion of MgO-C brick in contact with steel. The thumb rule says that just a 50 ^{0}C increment of working lining temperature may boost the refractory's double corrosion rate.

1.7.1 Tunnel Kiln

A well-established tunnel kiln provides continuous high output. Excellent thermal insulation of the walls enhances the efficiency of the modern kiln. They also use lighter support materials on the car floor, reducing the thermal inertia of assembly. The high efficient kiln can be programmed for faster firing cycles without any critical restrictions imposed by the particular features of loaded refractory. The sketch in Figure 1.25a is related to the transversal section of the kiln in the firing zone for effective heat transfer in the entire inner space and minimum heat loss through the kiln [8]. The combustion process produces high-temperature gases that pass through the whole chamber. Such a process is responsible for different degrees of energy transfer processes, and these are listed as follows:

- Radiation and conduction between the loaded refractories;
- Radiation and conduction between the refractories of car and load;
- Radiation between the outside loaded refractories and the kiln structure inner surface;
- Convection inner and outer of the loaded refractories with the gases due to combustion;
- Radiation between the surface of car and base of the kiln;
- Convection between car and kiln car base;
- Effective conduction in between kiln walls;
- Convection between the inner surfaces of the wall and flowing gases inside the tunnel; and
- Radiation and convection between the wall outer surfaces and the environment.

The influence of insulation lining is demonstrated in Figure 1.25b. A typical temperature profile for insulated and non-insulated refractory lining exhibits the load temperature difference considering the same amount of fuel consumption and cumulative refractory thickness.

Non-insulated lining governs lower temperature distribution, and thus a bulk amount of energy is lost to the outside ambient. Indeed, bulk energy will be required to stimulate the firing process, and a smaller amount of energy can be recovered at the dryer in the cooling zone. Systematic heat transfer analysis may assist in designing and selecting the refractory for such kiln lining.

1.7.2 Ladle Lining

Steel ladles, composed of refractories and steel construction components, act as transportation vessels and refining units for the steel-melting shop. Refractory linings act as insulating lining and protect the steel shell and consecutive heat loss through the shell despite the working lining. A well-lined steel ladle offers efficient temperature control of the steel melt and adequate steel quality and productivity. A typical steel ladle lining consisting of metal/slag working zone MgO-C, backup LCC-70, and flow control system is shown in Figure 1.26a. It experiences several heat transfer phenomena during different operational conditions. In the steel-melting shop (SMS) the ladle is continuously used for molten metal transport for casting. Uninterrupted operation without delaying ladle engagement is always a challenging task for SMS [9].

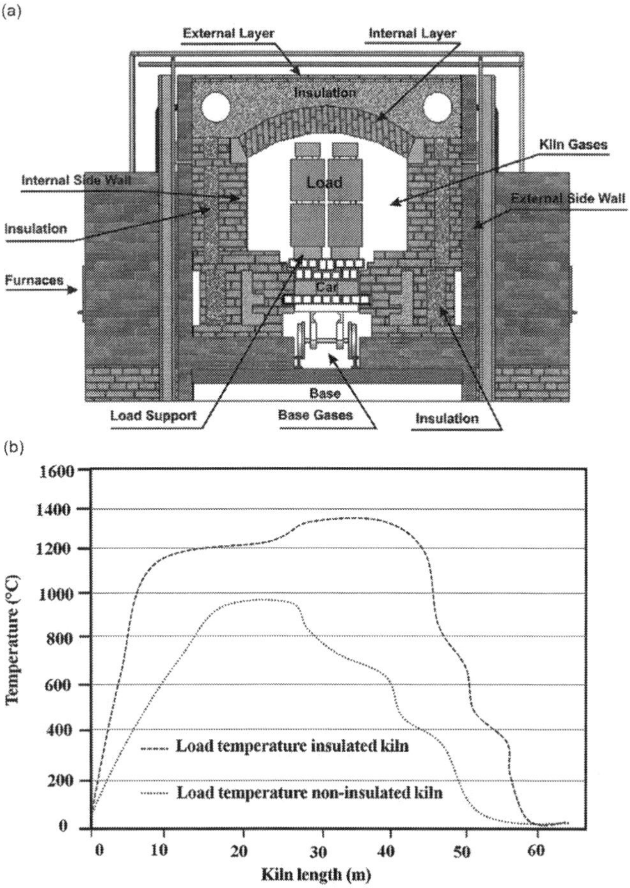

Figure 1.25 A typical representation of (a) tunnel kiln cross-section consists of refractory lining pattern, refractory loading pattern in the car for effective performance, (b) effect of kiln insulation on the central temperature up to maximum 1100 °C [8] / ABCM.

Several unavoidable issues such as delay in ladle lining, different preheating sequence, empty ladle, or idle time may severely affect and delay operation that eventually changes the temperature profile of refractory lining and affects bricks' life. In this context, a typical temperature profile variation for different preheating times and empty ladle times is shown in Figure 1.26b. Consider an empty time of 30 minutes, after which the hot face temperature has fallen significantly, but the temperature profile in the working lining is still high. Thus, immediate preheating is mandatory to increase the hot face temperature by flame temperature. Low heat flux occurs when the temperature gradient becomes small between the lining and flame. As the safety lining and working

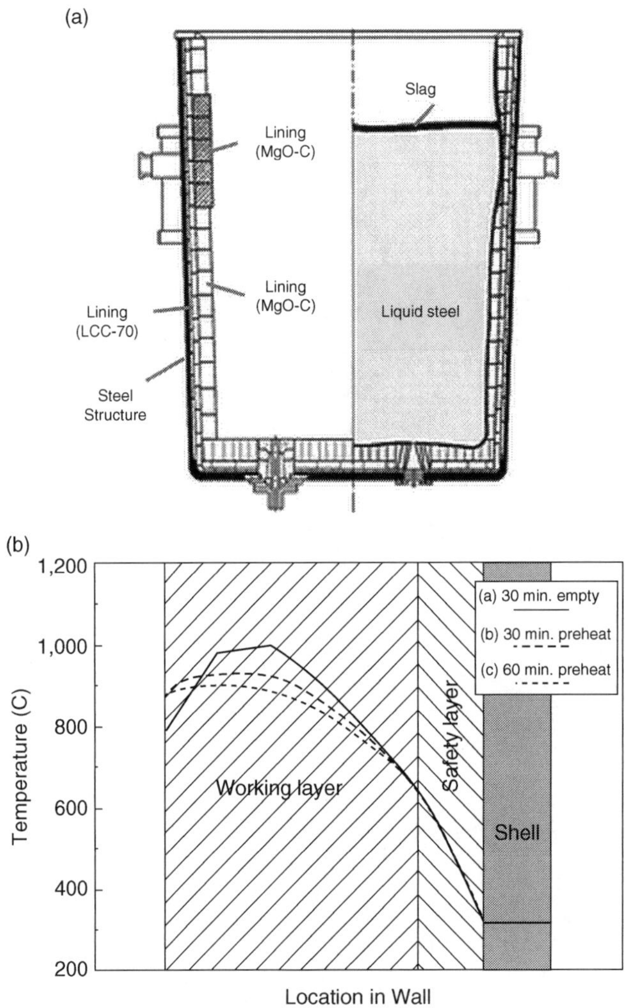

Figure 1.26 A typical schematic representation of (a) steel ladle lining consisting of MgO-C (working), LCC-70 (backup), and different flow control arrangements, (b) temperature profiles in the ladle lining for different preheating and empty times.

heat content remain virtually unaffected, heat loss to the surrounding occurs through the steel shell. Analogous to this incidence and operating conditions, a detailed temperature profile and heat transfer understanding may assist in designing the steelmaking refractories and discussed in Chapter 8.

References

1 Carslaw, H.S. and Jaeger, J.C. (1986). *Conduction of Heat in Solids*, 2e. London: Oxford University Press.

2 Yovanovich, M.M. (1998). Conduction and thermal contact resistances (Conductances). In: *Handbook of Heat Transfer* (ed. W.M. Rohsenow, J.P. Hartnett, and Y.I. Cho), New York: McGraw-Hill.

3 Kakac, S. and Yener, Y. (1993). *Heat Conduction, Taylor & Francis*. Washington, DC.

4 Holman, J.P. (2010). *Heat Transfer*. McGraw Hill.

5 Mills, A.F. (1992). *Heat Transfer*. CRC Press.

6 Incropera, F.P., DeWitt, D.P., Bergman, T.L., and Lavine, A.S. (2006). *Fundamentals of Heat and Mass Transfer*, 6e. Wiley.

7 Shah, R.K. and London, A.L. (1978). *Laminar Flow Forced Convection in Ducts*. New York: Academic Press.

8 de Paulo Nicolau, V. and Dadam, A.P. (2009 October-December). Numerical and experimental thermal analysis of a tunnel kiln used in ceramic production. *J. Braz. Soc. Mech. Sci. & Eng.* XXXI (4): 297.

9 Biswas, S. and Sarkar, D. (2020). Introduction to refractories for iron and steel making. Switzerland: Springer Nature. ISBN 9783030438067.

2

Equilibrium and Nonequilibrium Phases

2.1 Introduction

A phase diagram study is mandated to synchronize material formation at a specific temperature, pressure, and composition. Several experimental data represent the temperature and composition limits of two or more stable constituents at different phases. Efficient heat distribution in a kiln or furnace can facilitate the desired equilibrium phase during refractory manufacturing, eventually fulfilling steel industry demand. However, any nonfavorable conditions may develop nonequilibrium phases that reduce refractory performance. For example, high-temperature refractory–slag interaction expedites the glassy phase during the steelmaking process.

There are multiple binary and tertiary equilibrium phases from refractory compositions and slag chemistry, and significant phases are represented in Figure 2.1. Not all phases are formed in one product or single contact zone; instead, these are often in a wide span of refractory products and steel processing starting from BOF to Tundish operation. Some essential and valuable explanations from the perspective of the ceramic systems are summarized to analyze the failure of refractories. Despite these combinations, there are also possibilities to form several phases in the presence of other oxides or different mixtures of oxides; such incidences are discussed in relevant chapters as required.

2.2 Basics of Phase Diagram

2.2.1 Gibb's Phase Rule

In an equilibrium condition at uniform temperature and pressure, each constituent's vapor pressure or chemical potential remains the same in all phases. If the temperature or pressure is altered, there is a tendency to heat or mass transfer in

Fundamental Design of Steelmaking Refractories, First Edition. Debasish Sarkar.

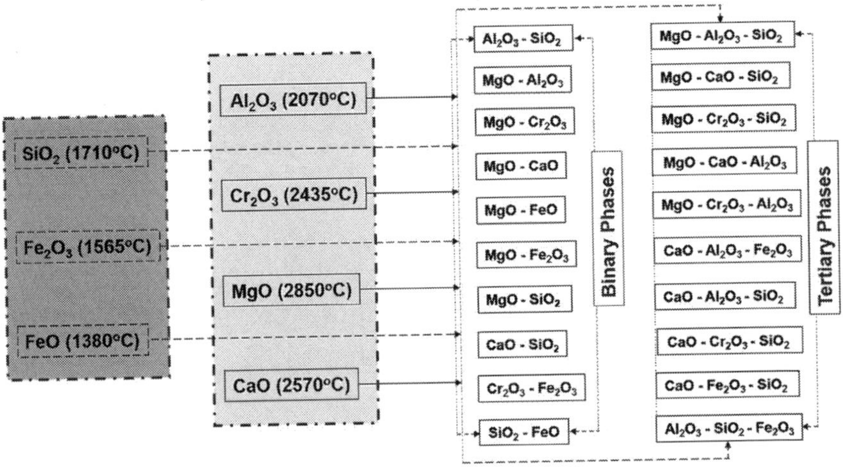

Figure 2.1 Major working lining refractory crystals are in the light shade zone; relatively low melting impurities in the dark shade zone either present in refractory or interact with slag in the steelmaking process. Binary phases either exist in refractory products or are formed in the steelmaking process. The effect of impurities on the refractory is represented as tertiary phases.

the system. In consideration of equilibrium condition, Gibb's phase rule states that the number of phases *P*, number of variance *F*, number of components *C*, and follows a relationship as Equation (2.1), is

$$P + F - C = 2 \tag{2.1}$$

the backbone for the establishment and drawing of any phase diagram. Phase defines as physically homogenous and bounded by a surface that can separate mechanically from any part of the system. Variance indicates the intensive variables (composition, temperature, pressure) that can change separately without a new phase. Component refers to chemically independent constituents, in another way a minimum number of separate species defines the composition of all phases.

For one-component ($C = 1$) system, the largest number of phases can present when $F = 0$, and then maximum $P = 3$. Any change in pressure or temperature ($F = 1$) certain phase disappears, and *P* is no longer maximum, reducing to 2. For a two-component ($C = 2$) system, any arbitrary modification in whichever composition or temperature of one phase is considered $F = 1$ results in $P = 3$. For the three-component ($C = 3$) system, both composition and temperature change ($F = 2$), the maximum number of phases $P = 3$ attain. But the pressure effect on condensed-phase (like ceramic solid or liquid) system is minimal, and most often interested in analyzing at constant pressure, near atmospheric pressure, and Equation (2.1) becomes:

$$P + F - C = 1 \tag{2.2}$$

In isobaric conditions, two components A and B, exhibit the coexistence of their three phases in equilibrium at their eutectic point, where $F = 0$ and $P = 3$ (liquid, crystal of A, crystal of B), but with any change in the composition of the liquid or the temperature, the number of phases becomes 2. Similarly, three components have a maximum number of coexisting phases 4; thus $P = 4$ when $F = 0$, known as independent components.

When three phases are present, each phase's composition and temperature become fixed, following a parallel line with the composition axis. Herein, the three-component system is graphically represented in composition in a triangle, and temperature as a vertical coordinate compared to x-axis composition and y-axis temperature in the two-component system. Several important two-component (binary) and three-component (ternary) phase diagrams may appear if other oxides replace the base refractory crystal in the above examples. These are shown in Figure 2.1; these have significant importance in analyzing the steelmaking refractories.

2.2.2 Binary Phase Diagram and Crystallization

Several classic pieces of literature are available [1, 2]; few fundamental concepts only express the book's content. A basic understanding of a few definitions, including solidus, liquidus, eutectic temperature, eutectic composition, congruent melting, incongruent melting, crystallization, solvus, solid solutions, and exsolution, assists in explaining the binary phase diagrams. Sometimes lack of practice and discontinuity provokes difficulty recalling the phase diagram's utility for the product development and corrosion analysis of refractories.

Let us consider two different materials, pure A and pure B, which have different melting temperatures, and their various compositions are placed in separate platinum capsules at constant pressure, 1 atm. All compositions are put into the furnace to provide a specific high temperature to form equilibrium phases.

The "certain point" is not higher than the melting temperature of pure A (T_{mA}) and pure B (T_{mB}). After that, the platinum capsules are quickly taken off from the furnace and quenched. The quick cooling ensures no further reactions occur, and the high-temperature process can be maintained at ambient atmosphere. Some liquids may have been present at a high glass temperature during the quenching process. The temperature–composition plot is shown in Figure 2.2a, with pure A plotting at 100% A, corresponding to 0% B, and pure B plotting at 100% B, corresponding to 0% A, respectively. A curved line assembles several data points for different phases, and the subsequent phase diagram is called a binary eutectic diagram, as shown in Figure 2.2b. It is an essential and standard two-component phase diagram with crystals of A, pure B, and liquid with compositions varying from pure A to pure B as potential phases.

The curves separating the fields of A + *liquid* and B + *liquid* from *all liquid* are termed liquidus curves. The horizontal line separating the (A + *liquid*) and (B + *liquid*) areas from $A + B$ all solid is termed the solidus, as shown in Figure 2.3.

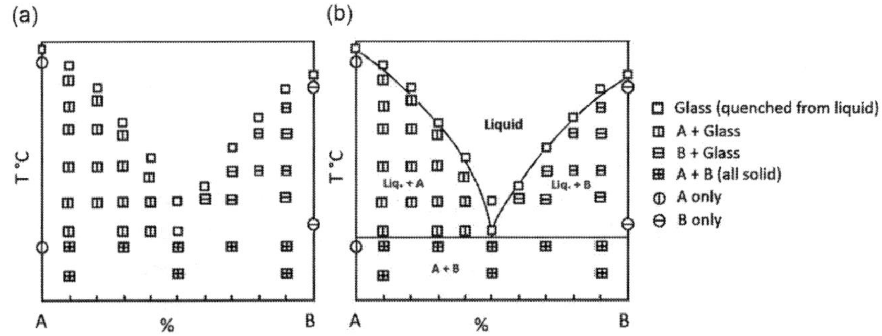

Figure 2.2 (a) Indication of different phase formations concerning the variation of composition at some temperature, and (b) data point assembled with the formation of liquidus, solidus, and eutectic point.

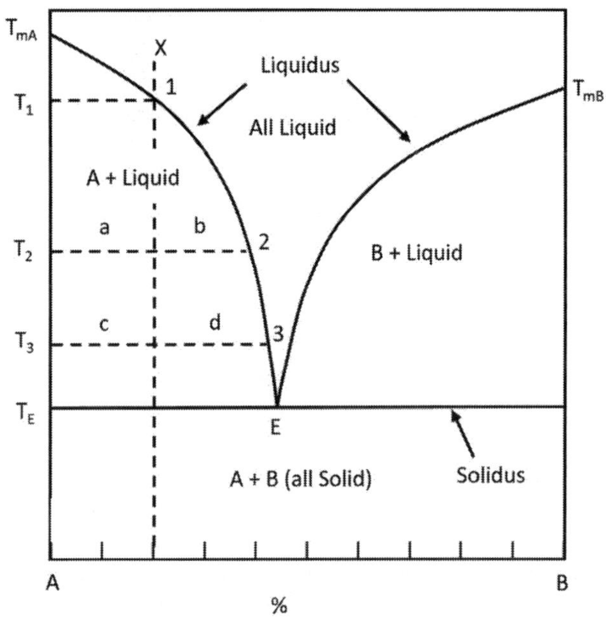

Figure 2.3 Schematic representation of binary phase diagrams consist of the solidus, liquidus, eutectic temperature, eutectic isotherm, and eutectic composition.

The point E is a Eutectic point representing the intersect of liquidus and solidus curves, where a two-component system, all three phases, liquid crystals of A and B, exist in equilibrium. A eutectic is an invariant point. The melting temperature is significantly reduced for all compositions between pure A and B, and melting starts at the eutectic temperature T_E. One of the lattice components is supposed to melt first in a eutectic system, while the mixture's temperature must increase for (all) the other component lattice(s) to melt further.

Crystals of B begin to form at the eutectic temperature, T_E, where crystal A, crystal B, and liquid coexist. T_E must be maintained before one of the phases has vanished. As a result, when the liquid entirely crystallizes, there only remains pure solids of A and B. These two solids phases mixtures are in the proportions of the initial mixture, i.e., 80% A and 20% B. Under this circumstance, the X composition follows, $T > T_1$ (all liquid); $T_1 - T_E$ (liquid + crystal A); at T_E (liquid + crystal A + crystal B); $T < T_E$ (solid crystal A + solid crystal B). As crystallization increases with decreasing time, the left-side portion of the composition line is considered a liquid and the right-side crystallization zone. Thus, Lever rule defines that the liquid percentage at T_2 is $[a/(a + b)] \times 100$, and crystal percentage $[b/(a + b)] \times 100$. The crystal percentage $[d/(c + d)] \times 100$ increases at T_3 compared to T_2. This analysis clearly shows that crystallization is just a reverse melting phenomenon.

Conversely, as a non-eutectic mixture cools down, each mixture's component gradually solidifies (forms its lattice) at a specific temperature until all material is solid. Liquidus temperature is always higher than or equal to solidus, but those need not coincide. The existing gap between liquidus and solidus is the freezing range in which the substance behaves like a slurry (suspension of crystal solid in liquid). The substance is completely liquid above liquidus temperature, the highest temperature at which crystals coexist with the melt in thermodynamic equilibrium. If one waits long enough below the liquidus temperature, depending on the material, more and more crystals can form in the melt. Alternatively, homogeneous glasses can be obtained by allowing the crystallization process to be slowed down sufficiently, i.e., kinetic inhibition. The primary crystalline or primary phase is the crystal phase that forms first when a material is cooled to its liquidus temperature. The primary crystalline phase field is defined as the preliminary phase's composition ranges remaining constant. The liquidus temperature is essential while considering the glassy phase formation during interaction with slag–refractory and frequent incongruent melting that eventually help understand the corrosion and failure behavior of refractory.

Components composed of two or more elements occurs in the system referred to as the intermediate component. In a binary system, these components are composed of various ratios of two-component of the system.

In combination with elements A and B, the component AB is formed, divided into two eutectic binary systems A-AB and B-AB. When a compound is melted, congruent melting happens when the liquid forms the same chemical composition as the solid.

It is most common in two-component systems. Let the two components *A* and *B* in this example and *AB* be the stable solid compound produced by their chemical combination. Since the solid and liquid phases now have the same structure, we can see in Figure 2.4 that the top point *D* of the phase diagram is the *AB* compound congruent melting point. *AC*, *BE*, and *CDE* are three freezing point curves. The congruent melting point indicates an actual temperature, just like the points of the pure components of melting. Component *AB's* congruent melting point *D* lies above the melting points of pure elements *A* and *B* in the phase diagram (Figure 2.4).

The congruent melting point is the temperature at which a phase melts into a liquid of the same chemical composition as the solid. However, incongruent melting occurs when a phase melts to a liquid with a different composition than the solid, resulting in a solid composition than the original solid. While discussing incongruent melting, another critical phenomenon called a peritectic point refers to a phase diagram where a previously precipitated phase reacts with the liquid to form a new solid phase. The temperature must remain constant after this point is reached until the reaction is completed. *A* peritectic point is an invariant point as well.

In incongruent melting, the final composition of the melting in a closed system or equilibrium crystallization is always identical to the initial system composition. *A* simple method demonstrates how to analyze the phase transformation behavior

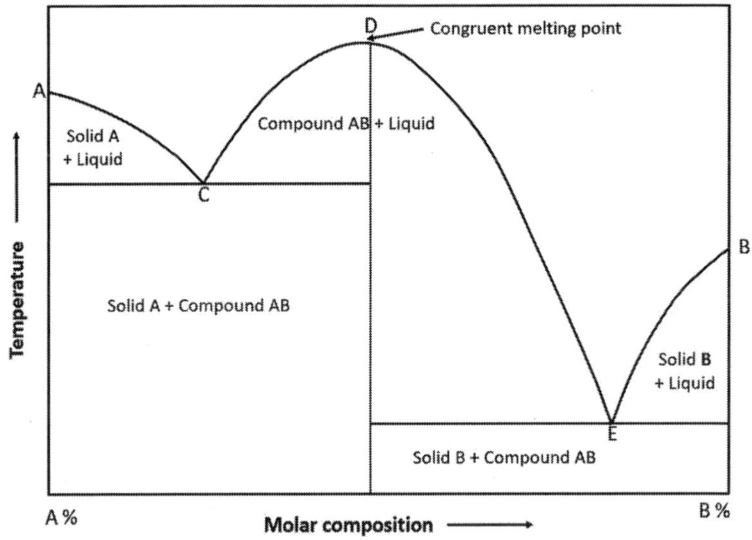

Figure 2.4 Congruent melting of initial phase *A* and *B*, forming intermediate phase *AB* during crystallization.

incongruent melting with having composition X ($A = 87\%$, $B = 13\%$), as indicated in Figure 2.5. Let's consider the incongruent melting of two phases A (Mg_2SiO_4) and B(SiO_2), where an intermediate phase C($MgSiO_3$) is developed during crystallization.

The crystallization behavior of three compositions X, Y, and Z, in which X is between pure A and pure B, Y is equivalent to pure C, and Z is in between C and B. In this situation, the crystallization history of X composition containing A and C crystals only, i.e., no B, is present in the mixture of final crystalline. As a liquid X composition is cooled, A's crystals begin to precipitate below the liquidus temperature of 1800 °C (T_1).

The liquid composition is gradually modified along the liquidus toward the peritectic (P) as the temperature is reduced further, and the forming crystals from the liquid become pure A before P is reached, where three phases are in equilibrium at around 1580 °C (T_5): A, liquid (SiO_2), and C. Some of A's crystals react with the liquid at this temperature to form C ($Mg_2SiO_4 + SiO_2 = 2MgSiO_3$). This reaction eventually absorbs all liquid, leaving only A and C crystals. Applying Lever rule, % of crystal A is $[d/(c + d)] \times 100$, and % of crystal C is $[c/(c + d)] \times 100$ at temperature 1520 °C (T_7). At any intermediate stage the percentage of crystal increases; thus % of crystal A is $[b/(a + b)] \times 100$, and % of liquid is $[a/(a + b)] \times 100$ at 1700 °C (T_2).

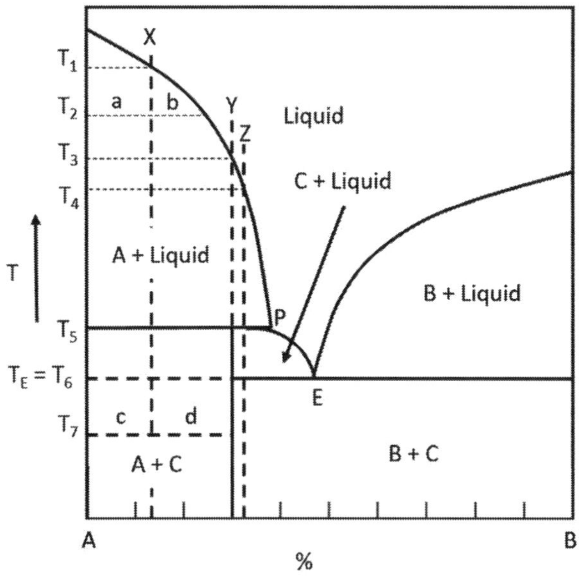

Figure 2.5 Incongruent melting of initial phase A and B, and formation of intermediate phase C during crystallization. $A = Mg_2SiO_4$ (Forsterite), $B = SiO_2$ (Quartz), $C = MgSiO_3$ (Enstatite). The temperature axis indicates T_1 is highest and T_7 is lowest, but not in scale.

If perfect equilibrium is preserved, Y composition is equivalent to pure C, and C only can act in the final crystalline product. As composition Y is cooled from a fully liquid state, it crystallizes around 1650 °C (T_3). At 1650 °C (T_3), A crystals start to precipitate from the liquid, followed by additional quenching before the peritectic temperature is reached. In this time, the liquid composition must become more enriched in B, and it shifts along the liquidus until it comes to the peritectic, P. Herein, the remaining liquid tends to react with precipitated A to develop crystals of C at the peritectic temperature (1580 °C = T_5). The temperature remains constant until the reaction is complete; pure C is the only phase present. As can be shown, $MgSiO_3$ (enstatite) melts in an incongruent manner. As pure enstatite is heated to 1580 °C (T_5), it transforms into A + liquid.

Since composition Z lies between C and B, crystallization occurs between C and B. When the temperature rises above liquidus 1610 °C (T_4), the entire system becomes liquid. A is likely to begin to precipitate below T_4. The liquid composition starts to change along the liquidus toward the peritectic, P. All of the A that had previously precipitated initiate to react with the liquid at P, eventually forming C crystals. A certain amount of liquid still remains after this reaction, and all of the previously precipitated A has been consumed. As the temperature drops, more C crystals form, and the composition of liquid shifts along the liquidus toward the eutectic, E. At E (T_6), B crystals began to form, and the temperature remains steady until all of the liquid is used up, leaving B and C crystals as the final solid. Since composition Z is so similar to pure C, the final crystalline product is mainly C with a small amount of B. A hypothetical vertical line drawn from point P to 100% B shows that all compositions between point P and 100% B behaves the same way as the simple eutectic method discussed previously. Despite equilibrium crystallization, fractional crystallization has a somewhat different cooling and crystallization history, where crystals are isolated from the system and do not react at reaction points like P. Density difference in crystal and liquid facilitate crystal sinking in liquid and gradual separation. Crystals of A and C become a final crystalline product of X under equilibrium conditions. In contrast, the final product included C and B in the fractional crystallization case.

A certain degree of resemblance between two cations aids in forming a solid solution between two solid crystals. A reliable solution is a single step in which one or more solutes are suspended in a solvent. While the addition of the solutes does not alter the crystal structure of the solvent, and the chemical components remain in a single homogeneous form, the multicomponent system is referred to as a solution rather than a compound. Since certain elements have chemical properties identical to other factors, substituting one component for another is common. When such substitutions occur, the process transforms into various compositions, depending on how much substitution occurs. A solid solution is also a strong function of temperature, with the solid solution being preferred at elevated temperatures and unmixing and/or ordering being selected at lower temperatures. Several viable options are listed, including:

- Substitutional solid solution: chemical variation is accomplished by simply substituting one form of an atom for another in the structure, e.g., $MgSi_2O_4$.
- Interstitial solid solution: chemical variation is accomplished by adding atoms or ions to structures not usually occupied, e.g., carbon in iron (steel) is one example of an interstitial solid solution.
- Coupled substitution: equivalent to a substitutional solid solution, but cations with different valences are swapped in a compound. Two coupled cation substitutions are needed to maintain the charge balance, e.g., Albite ($NaAlSi_3O_8$) can change to anorthite ($CaAl_2Si_2O_8$) by having Al^{3+} replaced by Si^{4+}. However, this leaves a negative charge that must be balanced by the (coupled) substitution of Ca^{2+} for Na^+.
- Omission strong solution: chemical variation is accomplished by removing cations from usually occupied cation sites, e.g., Pyrrhotite ($Fe_{1-x}S$).

Thus, essential factors affecting the extent of solid solution provides an insight into it [3]:

1) Atomic/ionic size: When the ionic radii of the atoms or ions in a solid solution are identical, the probability of solid solution formation is always high. If the difference in size is less than around 15%, a comprehensive solid solution is usually possible. Mg^{2+} and Fe^{2+}, for example, have a mismatch size of about 7%, and a complete solid solution between these two elements can be found in a variety of minerals. However, since the difference in size between Ca^{2+} and Mg^{2+} is 32%, we expect minimal Mg substitution in minerals.
2) Temperature: Since elevated temperatures promote the formation of the solid solution, end members that are immiscible at lower temperatures but at high temperature facilitates complete or extensive solid solution with one another. High temperatures encourage more atomic vibration and open structures, which can easily distort locally to accommodate cations of various sizes. Solid solutions have higher entropy than end members because of the increased disorder associated with the randomly distributed cations. The *TS* term in Gibb's free energy ($G = H - TS$) stabilizes the solid solution at high temperatures.
3) Structural flexibility: While cation size is a good indicator of the degree of solid solution between two endmembers, the rest of the structure's ability to bend bonds (rather than stretch or compress them) to accommodate local strains is also essential.
4) Cation charge: At low temperatures, heterovalent substitutions (i.e., those cations containing different charges) rarely result in solid solutions entirely due to complex cation ordering phase transitions and/or phase separation at intermediate compositions. Thus, it is obvious to maintain the local charge balance in the solid solution and satisfy the local strain that drives these processes.

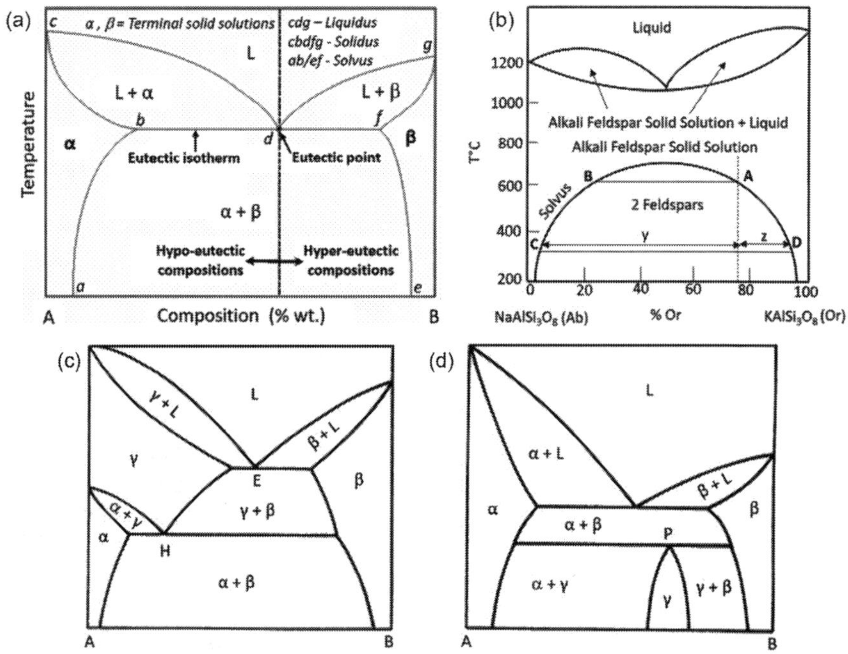

Figure 2.6 (a) Solid solutions of components *A* and *B*, (b) exsolution, (c) eutectoid point *H*, (d) peritectoid point *P*, α + β (cooling) → γ.

Figure 2.6a is a binary phase diagram that depicts the phases of a mixture of two substances *A* and *B*, at various concentrations. It is a preferable incidence of alloys like steel. The area is a solid solution of *A* and *B*, with *B* acting as the solute in an *A* matrix. The area on the other end of the x-axis is also a solid solution, with *A* serving as the solute in a *B* matrix. Instead, microstructure analysis of a mixture in this range shows two terminal phases: solid solution *A* in *B* (α) and solid solution *B* in *A* (β) may form different phases, possibly lamella or grains, with variance depending on the solidification rate.

If an *alloy's* composition isn't eutectic, it's usually either hypereutectic or hypoeutectic. The presence of the alloys to the left of the eutectic composition is called hypoeutectic, whereas those to the right are called hypereutectic. Steels are known as hypoeutectoid (less than 0.77% carbon), eutectoid (at 0.77% carbon), or hypereutectoid (above 0.77% carbon) (above 0.77 wt.% carbon). Solvus lines, denoted by the letters *ab* and *ef* show the maximum solubility (saturated solution) of *B* in *A* or *A* in *B* as a function of temperature, respectively.

Due to a lower temperature, when a solid solution becomes unstable, one or both of the mineral phases must "exsolve" or come out of the solution with the other mineral phase to form a single mineral phase. At the same time, exsolution implies that one or both of the mineral phases in the solution must "exsolve" or come out of the solution with the other mineral phase to form a single mineral phase. The disparity in cation size is the leading cause of this. Cations with significant radii differences are unlikely to substitute readily. As shown in Figure 2.6b, albite $NaAlSi_3O_8$ and orthoclase $KAlSi_3O_8$ are two alkali feldspar minerals. The alkali feldspar solid solution has an X of 70% orthoclase and 30% albite at 750°C. At elevated temperatures, Na^+ and K^+ readily substitute for each other, forming a solid solution; but, at low temperatures, the solid solution becomes unstable, and the albite can only substitute a small amount of K^+, and the orthoclase can only substitute a small amount of Na^+. At 590°C, they like to split into two phases (point A and B), resulting in exsolution. X exsolved further at 300°C, resulting in compositions C and D, where the rich albite solid solution is $[z/(z + y)] \times 100$ and the orthoclase-rich solid solution is $[y/(z + y)] \times 100$. With a lower temperature, the albite and orthoclase in the two solid solutions may exsolve completely, yielding a pure albite and orthoclase step, respectively. In nature, full exsolution does occur only when the temperature is slowly lowered. In this case, the thin white albite layers contrast with usually pink orthoclase, resulting in a perthite texture.

In Figure 2.6c, point E is a eutectic point. Point H is a eutectoid point; three solid phases α, β, γ are in equilibrium. Eutectoid reaction is an isothermal reversible reaction in which a solid phase (usually solid solution) preferentially transforms to two or more intimately mixed solids during colling. The number of solids formed is the same as the number of components in the system. It is similar to a eutectic reaction but occurs from one solid phase to two new solid phases. Pearlite (ferrite + cementite), pearly look, is a eutectoid product when austenite cools below 727°C, occurring in many common steel grades. Hypoeutectoid steel has proeutectoid ferrite and hypereutectoid steel contains proeutectoid cementite [4]. In Figure 2.6d, point P is a peritectoid point. It is an isothermal reversible reaction with two solid phases reacting with each other upon cooling a binary, ternary system to create a wholly different and solid phase. Ferrite is soft and ductile, whereas cementite is hard and brittle; a combination of these two phases assists in synchronizing the steel properties.

2.2.3 Ternary Phase Diagram and Crystallization

A ternary phase diagram depicts the equilibrium between the various phases produced between three components as a function of temperature. Pressure is normally not viable in constructing ternary phase diagrams, keeping constant at 1 atm. For $C = 3$, the maximum number of phases coexist when $F = 0$, so $P = 4$ when $C = 3$ and $F = 0$, and components are independent systems.

This section determines the total composition, the number of phases, the chemical composition of individual phases, the sum of each phase, and the solidification series. The Gibbs triangle can decide whether a point's overall composition

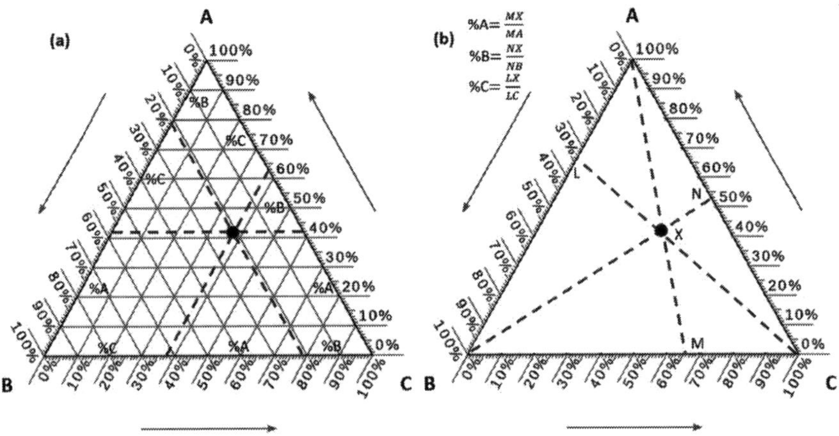

Figure 2.7 (a) Gibbs Triangle and composition determination of an arbitrary point *X* through the interaction of (dotted) parallel lines of each side *A′C′* of *AC*, *A′B′* of *AB*, and *B′C′* of *BC*, (b) composition determination in Inverse Lever Rule.

is expressed in "weight percent" or "molar percent." It is an equilateral triangle in which the pure elements are each other's representation. The parallel lines of each side $A'C'$ of AC, $A'B'$ of AB, and $B'C'$ of BC can intersect a commonly targeted composition, say point *X*. The % composition can be estimated from any side of a triangle as represented in Figure 2.7a. It indicates the wt.% composition of point *X* is, $B = 21$ wt.%, $A = 43$ wt.%, and $C = 36\%$. The sum of the concentration of the three components must add up to 100%.

Similarly, any composition can be determined through the Gibbs Triangle ternary diagram. Figure 2.7b represents another standard method, the inverse Lever rule to determine the composition for any particular point *X*. A simple process defines drawing straight lines from each corner through *X* and measuring the length, resulting in $\%A = MX/MA$, $\%B = NX/NB$, $\%C = LX/LC$, the total percentage should be equal to 100%.

The isomorphous system for ternary phases consists of only one solid phase, and all other components are soluble in the other components. The liquidus surface plots the temperature above a homogenous liquid form for any given overall composition. The solidus surface plots the temperatures below in which a homogenous solid phase forms for any given overall composition. Let us consider three binary systems as shown in Figure 2.8a. While combining all individual binary systems, a ternary isomorphous system is available in Figure 2.8b. A convex contour surface defines the starting of a complete liquid phase, liquidus surface, and concave contour surface represents the solidus surface below which all are solids (Figure 2.8c–d).

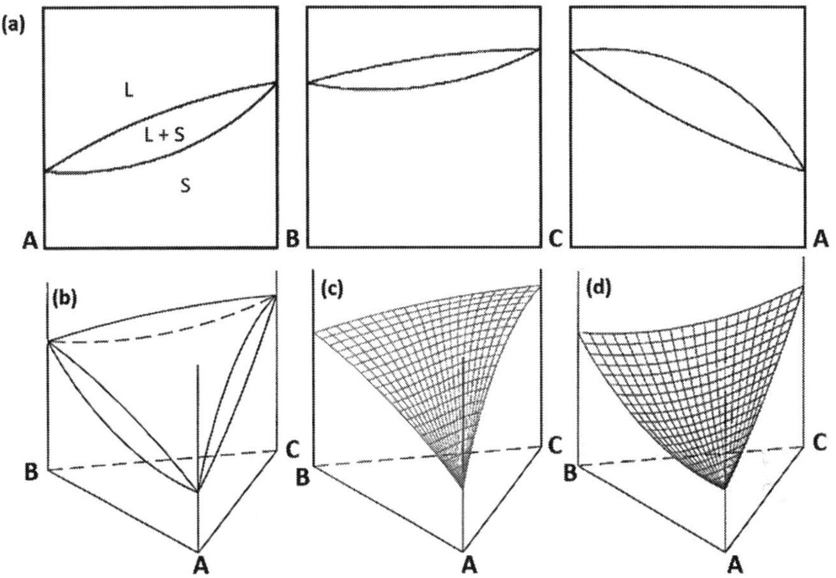

Figure 2.8 (a) Three solid binary solutions (consist of S, L + S, and L), (b) ternary isomorphous system generated from merging of three binary systems, (c) convex contour as liquidus surface, (d) concave contours as solidus surface.

Cutting through the space diagram at a defined temperature, say T_1, produces an isothermal segment of a ternary process. Draw parallel lines of three baselines and develop the triangle at temperature T_1. The intersecting points on the liquidus and solidus surface are marked as 1, 2, 3, and 4, respectively. The isotherm's intersection with the liquidus surface is shown by the line connecting points 1 and 2.

As shown in Figure 2.9a, the line connecting points 3 and 4 shows the intersection of the isotherm with the solidus surface. The confined area *A-B-1–2* represents the homogeneous liquid phase, area *C-3–4* is the homogeneous solid phase, and area *1–2–3–4* is a combination of liquid + solid two-phase region, as shown in Figure 2.9b. A straight tie line is joining liquidus and solidus ternary compositions and assists in determining the fraction of solid and liquid within the solid + liquid zone (hatched zone) by the inverse lever rule.

Analogous to Figure 2.8, we can consider equilibrium crystallization of all binary eutectic systems and provide Figure 2.10a when merging three binary phase diagrams on a triangle, where both binary and ternary eutectic points are indicated in 3D of three-component systems. Temperature (or pressure) is measured vertically, and composition is measured along the sides of the basal triangle. A surface with contours reflecting lines of constant temperature is shown at the top of the figure. Isotherms are the names given to these contours. Every binary system's eutectic points project into ternary structures as lines called boundary curves. Any

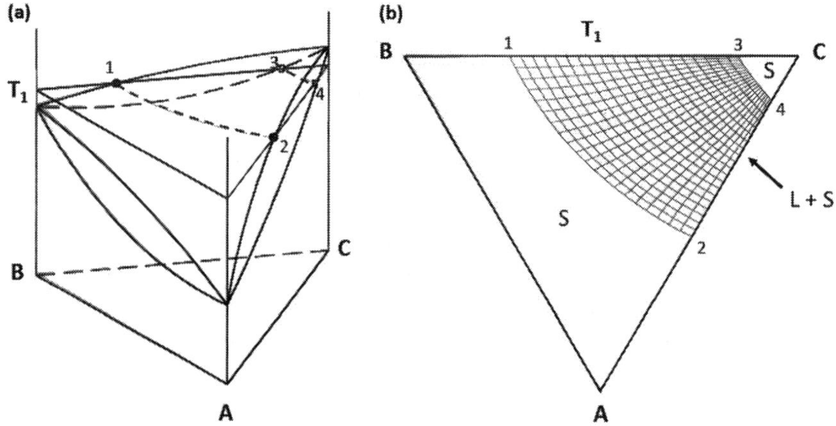

Figure 2.9 Isothermal section at (a) temperature T_1, (b) the distinct difference between liquidus, solidus line, and their intermediate zone at temperature T_1.

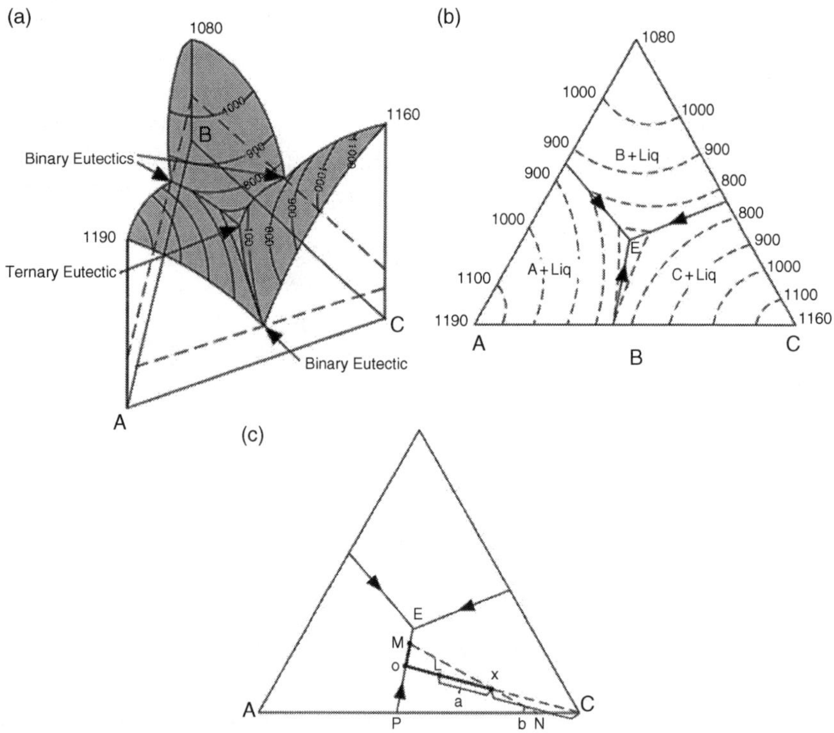

Figure 2.10 (a) 3D representation of the 3 component (ternary) system ABC, (b) 2D representation of 3D, curvature lines represent constant temperature, i.e., isotherms, boundary curves on a basal triangle, (c) similar to without isothermal lines.

composition on one of these curves crystallizes the two phases on either side of the curve. As seen from above, Figure 2.10b depicts the exact figure in two dimensions. The isotherms and boundary curves are also projected onto the basal triangle. It is worth noting that the temperature drops when it gets closer to the middle of the diagram. The crystallization of composition *X* is shown in Figure 2.10c. Figure 2.10c is identical to Figure 2.10b, but the isotherms have been removed for clarity.

The resultant solid must have $A + B + C$ crystals as the initial composition is in the triangle of *ABC*. The liquid of composition *X* intersects the liquidus surface at a temperature of around 980°*C*. It would start to precipitate *C* crystals at this stage. *C* crystals begin to form as the temperature is reduced, and the liquid's composition shifts away from *C* in a straight line. This is due to *C* precipitation, which causes the liquid to become depleted in *C* and enriched in $A + B$. We may calculate the relative proportion of crystals and liquid at a temperature of about 820 °*C*, point L in Figure 2.10c. $a/(a + b) \times 100$ = crystals percent; $b/(a + b) \times 100$ = liquid percent.

The direction of the liquid composition intersects the boundary curve at point *O* as it cools further. The crystals of *A* would then precipitate at the boundary curve. After that, the liquid path follows the boundary curve to point *M*. During this interval, and the solid phase precipitated bulk composition becomes a mixture of $A + C$ in the proportion indicated by point *P*. The bulk composition of the solid phases that have precipitated through the cooling history is at point *M*. In contrast, the solid phases that have yet to precipitate at point *N* are the percentage of distance ratios that can determine an extrapolated part *M*, the percent solids. Herein, the solid at this stage is made up of *A* and *C* crystals. As a result, it is necessary to break down the solid's percentages even more. This is accomplished as follows:

The percentage of the solid *A* is determined by the distance between *C* and *N* concerning the distance between *A* and *C* or by the formula: (distance NC/distance AC) × 100 = % *A* in solid. Similarly, the formula for calculating the percentage of a solid made up of *C* crystals is percent *C* in solid = (distance AN/distance AC) × 100.

At 660°*C*, we can now determine the exact percentage of all phases present in composition *X*, in which liquid composition is at point *M*. The formulas are as follows:

$$\%\,\text{A} = \left(\text{distance NC/distance AC}\right) \times 100 \times \left(\text{distance MX/distanceMN}\right);\ \text{and}$$
$$\%\,\text{C} = \left(\text{distance NA/distance AC}\right) \times 100 \times \text{distance MX/distance MN.}$$

In the same fashion, it is easy to assess all compositions in the system. Analyzing the composition of point *M* of the basal triangle yields the liquid's composition. The solids have a composition of 100%*t A* and 100% *C*, indicating that they are pure solid phases (not mixtures). At a temperature of about 650 °C, the liquid composition shifts to the ternary eutectic, *E*, where *B* crystal precipitates. The temperature remains constant until the completion of liquid consumption. All three $A + B + C$ crystals in the proportions determined by the initial X composition eventually synchronize the final crystalline substance. For all other compositions in the ternary method, crystallization proceeds in the same way.

With all consideration of boundary line, alkemade line, binary, and ternary invariant points, the congruently melting compound *ABC* formed within its primary phase field as shown in Figure 2.11a. However, if *ABC* melts incongruently (Figure 2.11b), the primary phase field of this compound is separated from the position that represents the compound. The ternary subsystems are not accurate ternary systems. Despite some basic understanding of the ternary systems, several incidences such as crystallization of a congruently melting compound, equilibrium, and fractional crystallization of an incongruently melting compound, and a solid binary solution have been discussed for a particular refractory composition.

2.2.4 Alkemade Lines

A binary join is a straight line that connects the points in the ternary system that describes the crystal compositions. An alkemade line links the compositions of the primary crystals of two areas with a common boundary line. The use of alkemade lines makes ternary phase relations easier to analyze and interpret. Alkemade lines and joins can be illustrated from a couple of examples. An intermediate binary compound *BC* with a primary phase field labeled *bc* in the scheme shown in Figure 2.12a.

Alkemade lines include *AC*, *AB*, *B-BC*, and *BC-C*. The line *A-BC* is an alkemade line since it connects the compositions of the primary crystals of two areas separated by a common boundary line, *ED*. Since areas *c* and *b* are not adjacent, line *BC*, despite being a join, does not satisfy the condition of the common boundary line; thus *BC* is

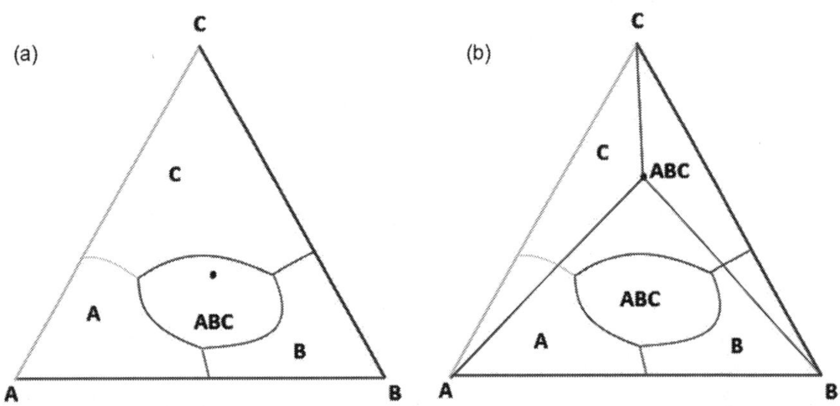

Figure 2.11 (a) *ABC* congruently melting compound, formed within its primary phase field, (b) incongruently melting compound, the position shifted from its primary phase field.

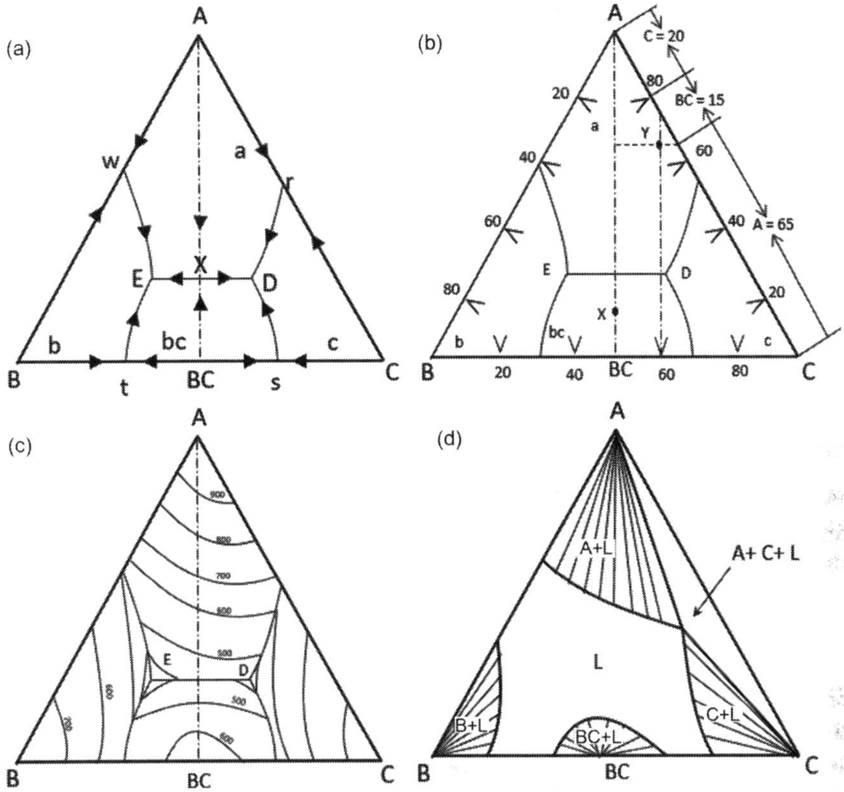

Figure 2.12 (a) Border lines of the liquidus planes and alkemade line of a ternary system with a binary compound *BC*, (b) composition *Y* within the 3-phase equilibrium *A-BC-B* by drawing lines parallel to the legs of the ternary region, (c) isotherms of the "figure b," and (d) isothermal cut of the ternary system at 600°C.

not an alkemade line. The intersection of a boundary line (or boundary line expanded) and its corresponding alkemade line (or alkemade line extended) denotes a temperature maximum on the boundary line and a temperature minimum on the alkemade line, according to the alkemade theorem. The alkemade theorem allows one to evaluate the direction of the slope of boundary lines in terms of temperature. The alkemade theorem was used to calculate the slopes of the boundary lines for the scheme *A-B-C* in Figure 2.12a. Three boundary lines converge at *E* and *D* to form a common junction. These are called eutectic points.

Alkemade lines separate a ternary diagram into compositional triangles, where three alkemade lines generate a compositional triangle. Alkemade lines never cross each other. Upon cooling a melt, a combination of crystals with a composition of the corners of the compositional triangles is formed. The composition *Y* is within the triangle *ABC-C* in Figure 2.12b. The relative amounts are given by the intersections of the lines parallel to the legs of the compositional triangles. eutectic point: *D* (30% *A*, 48% *BC*, 22% *C*). *A-BC* represents a true binary system. If the alkemade line does not cross the boundary line, no actual binary system is formed by the alkemade line. Figure 2.12c shows the isotherm of Figure 2.12b. The areas with the straight lines represent two-phase regions. Figure 2.12d represents the isothermal cut at the temperature of 600 °C.

2.3 One-Component Phase Diagrams

2.3.1 Water

Water is a particular element but has different forms and characteristics in several aspects. One of these unique characteristics is that frozen water (ice) becomes less dense than liquid water, a little above the freezing temperature. Figure 2.13 represents the phase diagram of water, where the pressure is typically plotted with temperature and analyzes the water phase under various pressure and temperature conditions. For instance, we can define the ordinate of operating pressure $P = 1$ atm and analyze the transition phases of water at 0 °C and 100 °C, respectively. Point *A* is the triple point, and *C* represents the typical boiling point of transition phases of water. The water phase diagram shows a negative rather than positive slope between liquid and solid states. Interestingly, the solid phase of water is not dense compared to the liquid form; thus ice floats on liquid water. Hence, there is an opposite effect on these two phases due to pressure change.

If ice is significantly closer to its melting temperature, it can be transformed into liquid water by applying pressure. The water particles are truly nearer self-possessed

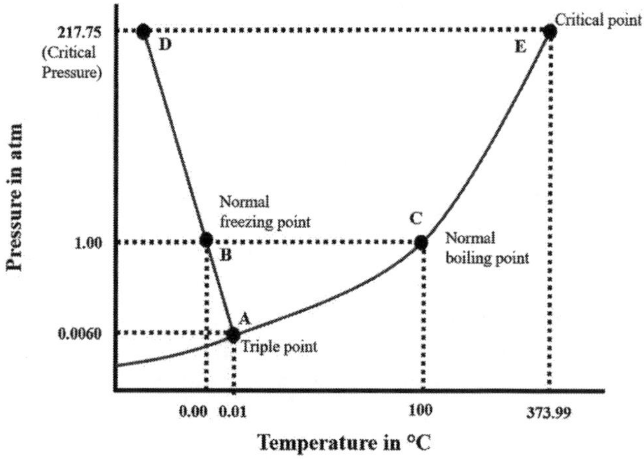

Figure 2.13 Phase diagram of H_2O.

during the liquid phase than they are in the solid phase. Point E represents the critical point at 373.99 ºC, the high rapid movement of the water fragments in the gas phase. The pressure is added to the gas because the gas phase can't be liquefied at any temperature higher than the one. The critical pressure (P_c) is the pressure exerted on the gas at the critical temperature to liquidate. The critical water pressure is 217.75 ºC atmospheric, which is too high [5]. The critical point represents the intersection point of critical temperature and critical pressure. Illustrated on the water phase diagram, the conditions of pressure and temperature of the triple point, at which phases of all three are in equilibrium. The gaseous and liquid phase densities become equivalent when the temperature and pressure conditions are critical, and the system no longer differentiates between the two phases. These critical points are at extremely nonstandard conditions of pressure and temperature.

2.3.2 Quartz

Quartz is a crucial phase and changes to different phases while altering temperature and pressure and has a significant influence on the formation of eutectic phases through a combination of other oxides during refractory application. In equilibrium, five condensed phases exist α-quartz, β-quartz, tridymite, cristobalite, and anhydrous melt. At atmospheric pressure, α-quartz is stable between 0° and 573 °C, β-quartz is stable between 573 °C and 867 °C, and an anhydrous melt forms above 1713 °C, as seen in Figure 2.14.

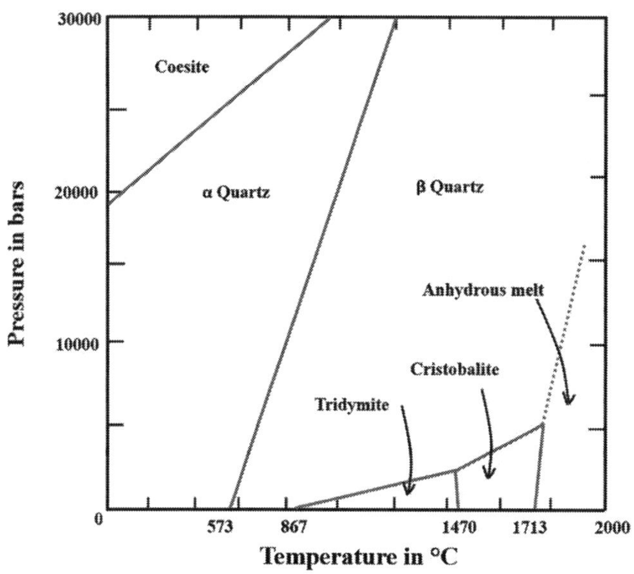

Figure 2.14 Phase diagram of quartz.

The system becomes divariant when one phase-stable ($F = C - P + 2 = 1 - 1 + 2 = 2$), univariant along the boundary curves where two phases can coexist stably ($F = 1 - 2 + 2 = 1$), and invariant at the junctions, known as triple points, where three phases can exist in equilibrium ($F = 1 - 3 + 2 = 0$). The α-quartz persists when both pressure and temperature are the degrees of freedom, the boundary between α-quartz and β-quartz fields coexists in one independent variable. The triple point for the quartz fields of tridymite, cristobalite, and β, where any temperature or pressure change, the number of phases reduce and coexist in equilibrium. The equilibrium phase diagram graphically describes the conditions for minimum free energy in a system. A new phase can be observed at 20–30-kilobar pressure in nature due to meteorite impacts is called coesite. Because of the greater importance of ceramic applications, there is still some debate about the role of minor impurities in the low-pressure phases of SiO_2.

2.4 Two-Component Phase Diagrams

2.4.1 Fe–C

An alloy states physically homogeneous phases, and it may exist in various phases. The existence of precise chemical composition specifies the atomic bonding and arrangements; thus the crystal structure imparts distinct properties to individual phases. A specific application demands particular properties that are eventually controlled by the phases developed in the system. The heat treatment process at a specific temperature produces different phases. Some specific alloys may exist in multiple phases at the same temperature. The Fe–C phase diagram is mainly applicable to numerous phases of cast iron and steel. *A* mixture of carbon and iron results in both the cast iron and steel and a few trace elements also exist in both alloys. The Fe–C phase diagram is plotted with the *C*-concentrations by weight on the *X*-axis and the temperature on the *Y*-axis, as shown in Figure 2.15.

Carbon atoms occupy interstitial sites in iron so that the configurations of any solid solution of different phases (e.g., α, γ, δ, etc.) at constant composition depend solely on the distribution of carbon atoms on the interstitial sublattice. The formation of an alloy may be an FCC (face-centered cubic) lattice or a BCC (body-centered cubic) lattice. The Fe–C phase diagram represents several ferrous alloys depending on the carbon wt.% from 0% to 6.67%. Iron or pure iron may exist at room temperature in the form of α-ferrite when the carbon content is maximum up to 0.008%. With the variation of carbon content from 0.008 wt.% to 2.14 wt.%, different steel grades are obtained, like mild steel (low carbon steel), medium, and high carbon steel. The cast iron stage may exist when the carbon content exceeds 2.14 wt.%. It is rigid cast iron, but there are limited applications due to its severe brittleness and forming methods [6].

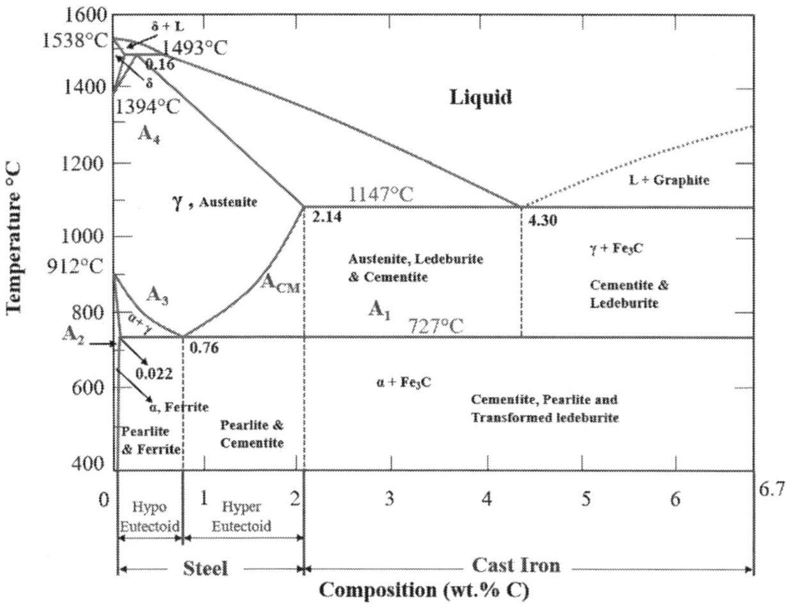

Figure 2.15 Fe–C phase diagram.

A liquid phase freeze into two solid-phase mixtures due to eutectic reactions at these points. The process occurred when a eutectic composition liquid alloy to its eutectic temp undercooling. The formation of alloys at this point represents the eutectic alloys. On neighbor sides, alloys are called hypoeutectic and hypereutectic alloys, respectively ("hypo" in Greek means less than, "hyper" means greater than). In the Fe–C phase diagram, the intersecting boundaries each other and mark the specific regions. Multiple phases or various phases may exist together within the particular area. These regions are distinct field phases in a particular composition and alloy temperature.

While analyzing the Fe–C diagram, there are several lines designated as A_1, A_2, A_3, A_4, and A_{CM}, the abbreviation of letter *A* represents the word "arrest." Temperature variation facilitates phase change incidence more prominently when the temperature reaches the boundary. Usually, an alloy subjected to heating results in increased temperatures. But along these lines (A_1, A_2, A_3, A_4, and A_{CM}), structure realignment into the various phases; thus a complete phase has changed when stopping the further temperature increment in heating. The process represents thermal arrest when there is no change in temperature. The position of boundaries may affect alloy steel elements such as chromium, manganese, molybdenum, and nickel on the phase diagram. Based on the usage of elements, there is a possibility of shifting boundaries in any direction. For example, the nickel addition decreases the boundary value of A_3, as shown in the Fe–C phase diagram. The melting point of multiple phases represents the eutectic point. The lines A_1, A_2, A_3, and ACM melting points represent the eutectic point in the Fe–C phase diagram. These points' formations are coincidental.

Alpha-ferrite with less carbon content exists at low temperature and solid carbon solution of α-ferrite in BCC Fe. At room temperature, it has a stable phase. As shown in the iron-carbon phase diagram has A_2, this phase is magnetic below 768 °C. It has a maximum content of carbon 0.022%, and it transforms at 912 °C to γ-austenite. A γ-austenite phase is a solid carbon solution in FCC Fe with a maximum of 2.14% *C* solubility. It converts into BCC δ-ferrite at 1395°C subjected to further heating. At the temperatures below eutectic (727 °C), unless cooled rapidly, the γ-austenite phase is unstable, this is a nonmagnetic phase. The δ-ferrite phase has a structure similar to α-ferrite but exists only at elevated temperatures and has a melting temperature of 1538 °C.

A metastable process is one with a fixed composition of Fe_3C. At room temperature, it decomposes too slowly into iron and carbon (graphite). This decomposition process can take a long time, much longer than the application's service life at room temperature. Other factors that encourage graphite formation, such as high temperatures and certain alloying elements, can influence this decomposition. Cementite is hard and brittle, making it ideal for steel reinforcement. Its mechanical properties are determined by its microstructure, which is controlled by the presence of a ferrite Fe–C liquid solution. It's labeled *L* on the diagram and found in the diagram's upper region. It's a carbon-in-iron liquid solution, as the name implies.

2.4.2 Two Oxides Phase Diagrams

In reality, there is always a possibility of several binary phases through permutations and combinations of preliminary high refractory oxides and impurities; however, few essential binary phases have been discussed, and others are pointed out throughout the book as required.

2.4.2.1 SiO_2–Al_2O_3 Phase Diagram

A system containing Al_2O_3 and SiO_2 content has significant importance in Al_2O_3–SiO_2 refractories. The composition includes corundum, mullite, high alumina, low alumina, fireclay, semi silica, and silica refractories. These refractories reflect high-temperature transformation, and their relationships are represented in the phase diagram, as shown in Figure 2.16. SiO_2 and Al_2O_3 are the two end members, characterized by their phase diagram, and mullite formed as an intermediate compound $3Al_2O_3{\cdot}2SiO_2$. The system has two eutectic points, and one is obtained between cristobalite and mullite at 1595 °C, and the second one is found between mullite and corundum at 1840 °C, respectively.

The mullite mixture has greater than Al_2O_3 content at above 1595 °C, indicates the occurrence of liquid. The melting point of SiO_2 gradually decreases to 1595 °C

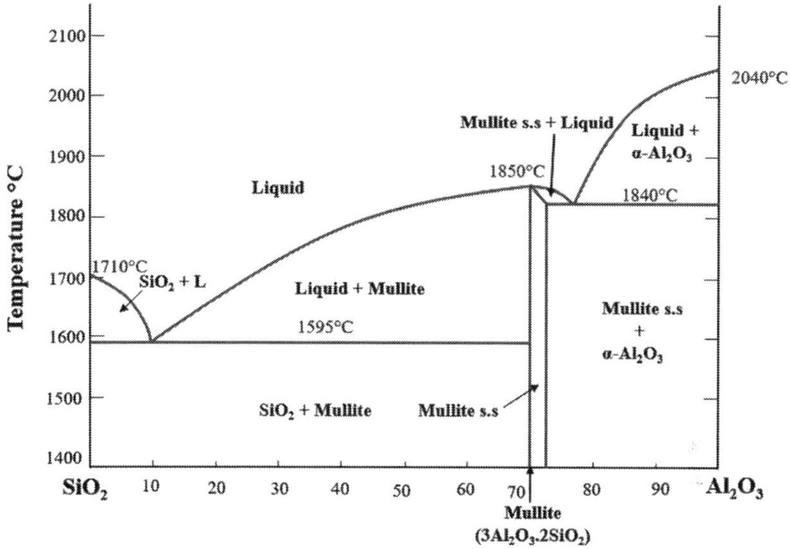

Figure 2.16 SiO_2–Al_2O_3 phase diagram.

when 5 wt.% of Al_2O_3 is added that eventually acts as a low flux for SiO_2. The composition melting point gradually increases with the increment of alumina content from 5 to 72 wt.%. Fireclay (25–45% Al_2O_3) and low alumina (45–65% Al_2O_3) refractory composition can tolerate a certain amount of liquid substance during service because of the viscous characteristic of the melt. The phase diagram predicts that the high alumina composition (72–100% Al_2O_3) based refractory develops liquid phase above 1840 °C. However, certain impurities in specific alkali may reduce the performance temperature of alumina refractory.

2.4.2.2 MgO–Al_2O_3 Phase Diagram

The MgO–Al_2O_3 system encompasses significantly both shaped and unshaped refractories for modern steelmaking. However, certain oxides, including silica, calcia, and iron oxide, develop different tertiary compounds in slag. Most of the high-temperature activity of MgO–Al_2O_3 refractories can be related to phase diagram relationships, as seen in Figure 2.17. MgO and Al_2O_3 are two end members of the phase diagram, and the presence of two eutectics are observed; one between magnesia and spinel at 1996 °C and the other at 1991 °C between spinel and alumina.

The stoichiometric MAS melts at 2135 °C and contains 28.2% MgO and 71.8% Al_2O_3. Even though MgO has a very high melting point, a small amount of MgO in Al_2O_3 decreases its melting point. Few terminal solubilities imply Al_2O_3 dissolves very little MgO, and vice versa. The MgO–Al_2O_3 framework can also be divided into two-step diagrams, with MgO and $MgO.Al_2O_3$ as one part (left side) and $MgO{\cdot}Al_2O_3$ and Al_2O_3 (right side) as the other. $MgAl_2O_4$ spinel has a broad

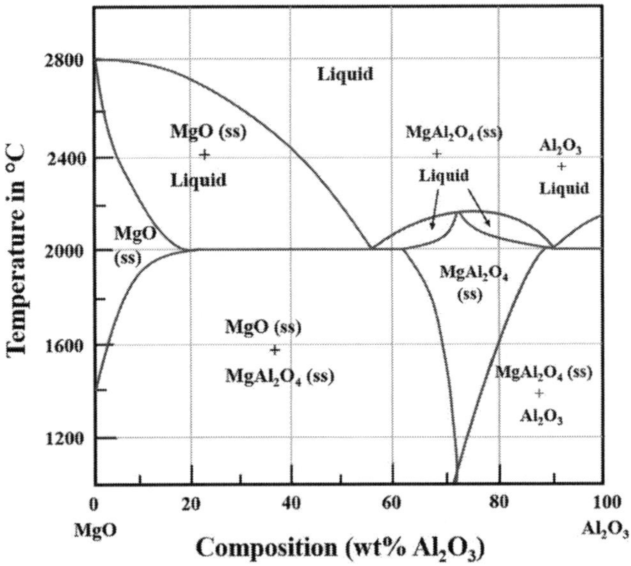

Figure 2.17 MgO–Al_2O_3 phase diagram.

stoichiometry range and preferentially form at temperatures as high as 1600 °C. The MgO and Al_2O_3 solubility of solid in spinel at this temperature is calculated to be 2% and 6%, respectively. The solubility of these two oxides in spinel increases to 3.0% and 10.0%, respectively, when the firing temperature increases to 1700 °C, and eventually solid solution ranges from 40% to 80% at 2000 °C. At a temperature of 1990 °C, the value of n in MgO.$n$$Al_2O_3$ varies from 0.6 to 0.7. However, the pressure acting on spinel has been found to impact its significant high-temperature stability.

2.4.2.3 MgO–CaO Phase Diagram

One of the most common refractory materials is dolomite, a mixture of CaO and MgO. Dolomite refractories bonded with pitch are used in the primary oxygen-furnace method for steel production, and the strong solubility at high temperature forms a high-temperature bond. MgO–CaO binary phase diagram as shown Figure 2.18 is an important aspect in developing refractory and steel manufacturers.

CaO and MgO have the same structure as NaCl, the solution phase formed between them is halite [7]. A miscibility gap in a single solution step due to cation size difference of more than 15% may be used to characterize the minimal, neutral solubility. It can be defined as $(Ca, Mg)_1O_1$ and a regular-solution representation. While considering the effect of impurity like SiO_2, it has a lower solid solubility

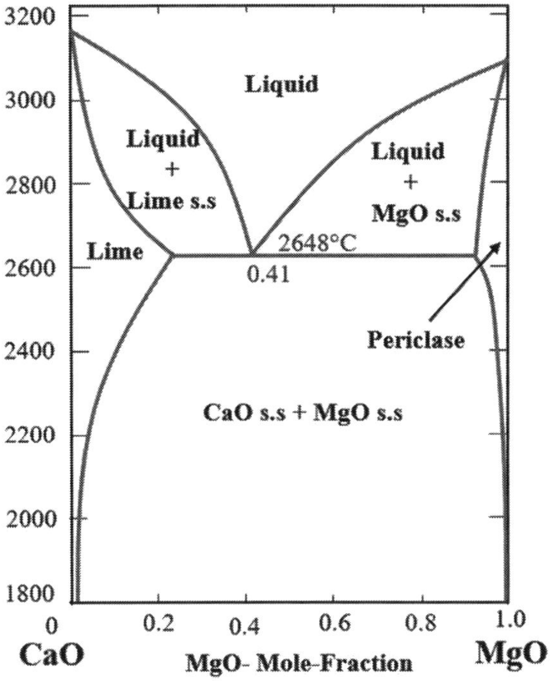

Figure 2.18 MgO–CaO phase diagram.

than CaO in MgO; thus excess CaO is essential to consider avoiding the formation of low-melting intergranular silicates in magnesia refractories. Fusion-cast refractories composed of MgO–CaO appear to have a microstructure (fewer facets) that is more like low-entropy-of-fusion metallic systems than $MgO–MgAl_2O_4$ compositions. The phase diagram of the MgO–CaO system shows a simple eutectic and minimal solid solution of the end member oxides at elevated temperatures.

The microstructure of the compositions near the eutectic can be described as finely disseminated, rounded to rod-like, periclase crystals in a CaO matrix. As the MgO content increases (52.9% MgO, and 47.1% CaO), larger periclase crystals occur that are surrounded by the eutectic structure. With additional MgO (70.6% MgO, and 29.5% CaO), large periclase crystals with a quasi-continuous periclase-periclase bond are surrounded by a CaO matrix are observed. Minor silicate impurities occur as isolated pockets in this system. This may be due to two refractory phases, which decrease the wettability of the silicate phase. Fusion-cast CaO–MgO refractories are especially resistant to corrosion by basic slags (lime/silica ratio greater than 2/1), and they have good high-temperature strength and thermal shock resistance.

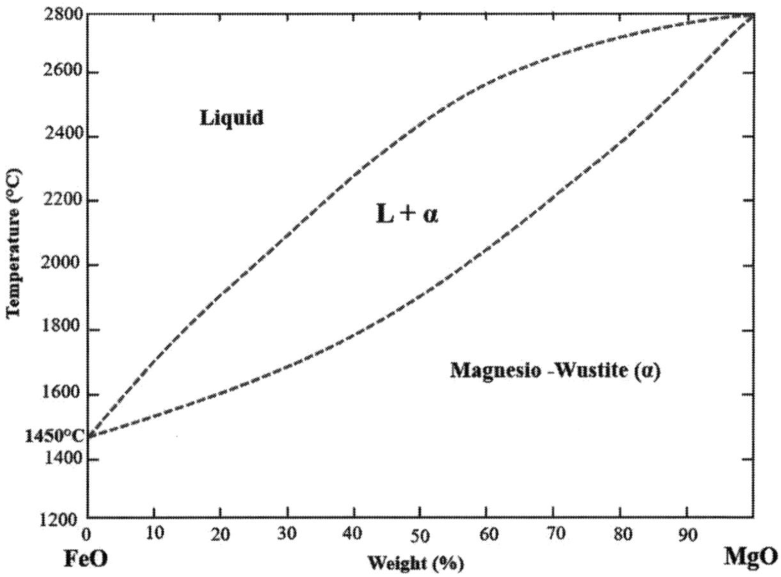

Figure 2.19 MgO–FeO phase diagram.

2.4.2.4 MgO–FeO Phase Diagram

FeO–MgO phase diagram is shown in Figure 2.19, where FeO and MgO have complete miscibility and forms the solid solution. Increasing the MgO content increases the melting point of wüstite. The solid reaction products magnesio-wustite (MgO–FeO) or magnesio-ferrite ($MgO.Fe_2O_3$) form at the refractory/liquid iron oxides interface separating magnesia from the slag, account for MgO's strong resistance to FeO's solvent action. MgO can dissolve up to 70 wt.% FeO without forming a liquid at 1600°C and no liquid forms even after MgO absorbs 100 wt.% Fe_2O_3.

MgO consumes 65 wt.% FeO and 70 wt.% Fe_2O_3 without forming a liquid phase when heated to 1700°C. For high values of iron oxide, the liquidus temperatures increase dramatically under high oxidizing conditions.

2.4.2.5 MgO–SiO_2 Phase Diagram

Mg_2SiO_4 and $MgSiO_3$ are the prime phases in MgO–SiO_2 phase diagram (Figure 2.20), with an immiscibility region in the silica-rich end.

It can be observed from the phase diagram that, MgO (periclase) has the highest melting point, and it decreases with the addition of SiO_2 first phase formed being forsterite (Mg_2SiO_4) and upon further increasing the SiO_2 content enstatite ($MgSiO_3$) is formed. Forsterite has higher temperature stability, compared to that of the enstatite. Both undergo incongruent melting. As we move toward the

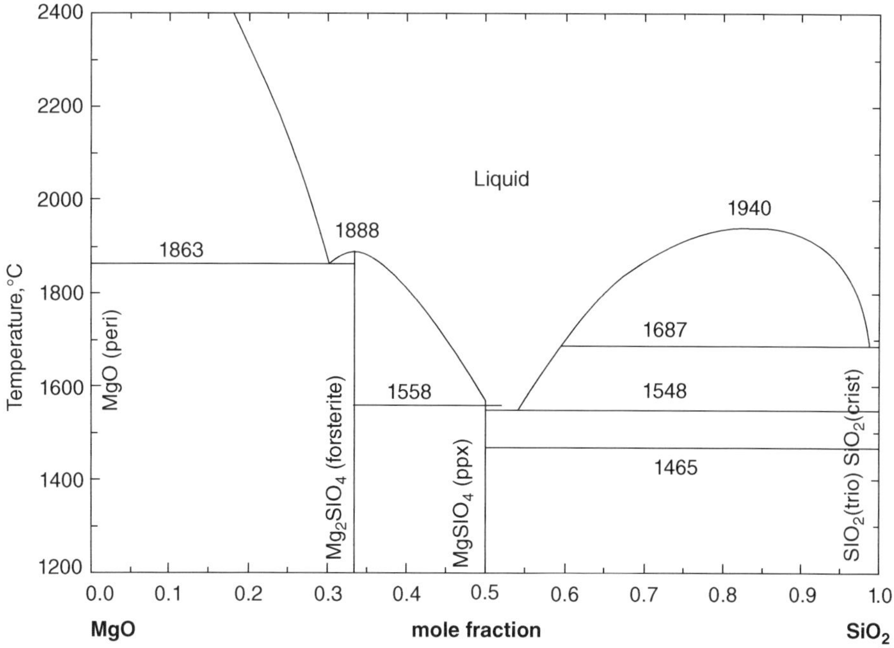

Figure 2.20 MgO–SiO_2 phase diagram.

saturated SiO_2 region, SiO_2 undergoes phase transformation at 1465°C, from low temperature phase tridymite to cristobalite and with the further increase in temperature, during melting above 1687 °C it exists as two immiscible liquids before existing as a single liquid.

2.4.2.6 CaO–SiO_2 Phase Diagram

The prime crystalline phases Ca_3SiO_5, Ca_2SiO_4, and $CaSiO_3$ are included in the CaO–SiO_2 phase diagram as shown in Figure 2.21, which has a small immiscibility field in the silica-rich portion of the binary.

Similar to MgO–SiO_2 system, in Cao–SiO_2 system CaO (lime) has the highest melting temperature, and as we go on increasing the SiO_2 content it forms four intermediate phases i.e., Ca_3SiO_5 (tricalcium silicate, hatrurite), Ca_2SiO_4 (calcium silicate), $Ca_3Si_2O_7$ (rankinite), and $CaSiO_3$ (pseudo-wollastonite). Hatrurite is stable between 1300°C and 1799 °C and it decomposes into CaO and Ca_2SiO_4 beyond that. Rankinite melts peritectically, whereas the other three compounds show eutectic melting. Again, similar to the above system, here SiO_2 undergoes phase transformation from tridymite to cristobalite and then there exists a range of two immiscible liquids.

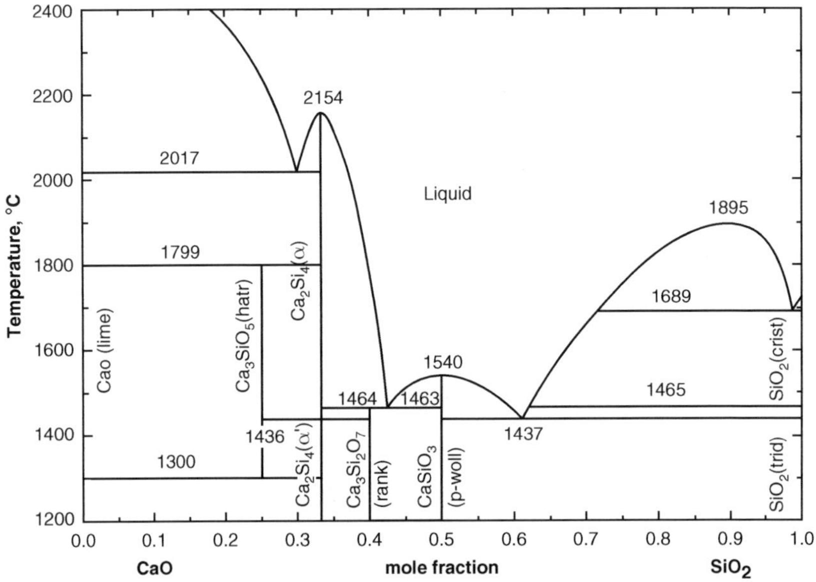

Figure 2.21 CaO–SiO_2 phase diagram.

2.5 Three-Component Phase Diagrams

Ternary diagrams of a different combination of oxides represent the equilibrium between the different phases produced through three components as a function of temperature. Pressure is not usually a viable element in constructing the ternary phase diagram and, is thus kept constant at 1 atm. Few ternary phase diagrams related to steelmaking refractory are represented as the mass or mol functions. Phase equilibrium details can be easily depicted from the ternary phase diagrams.

2.5.1 Three Oxides Phase Diagrams

2.5.1.1 MgO–Al_2O_3–SiO_2

The ternary compound system phase diagram of MgO–Al_2O_3–SiO_2 is shown in Figure 2.22. It consists of binary compounds of MgO–Al_2O_3 (spinel), and 2MgO–SiO_2 (Forsterite) and having a congruent melting temperature 2135 °C and 1900 °C, as well as a binary mixture of MgO–SiO_2 with an incongruent melting temperature, respectively. The ternary system has eutectic and peritectic reactions. The MgO–Al_2O_3–SiO_3 system has a 2MgO–2Al_2O_3–5SiO_2 ternary incongruent melting compound. It has different solid solutions like Al_2O_3 in MgO–Al_2O_3 (spinel) and 3Al_2O_3–2SiO_2 (mullite) in 2MgO–2Al_2O_3–5SiO_2 (cordierite), respectively.

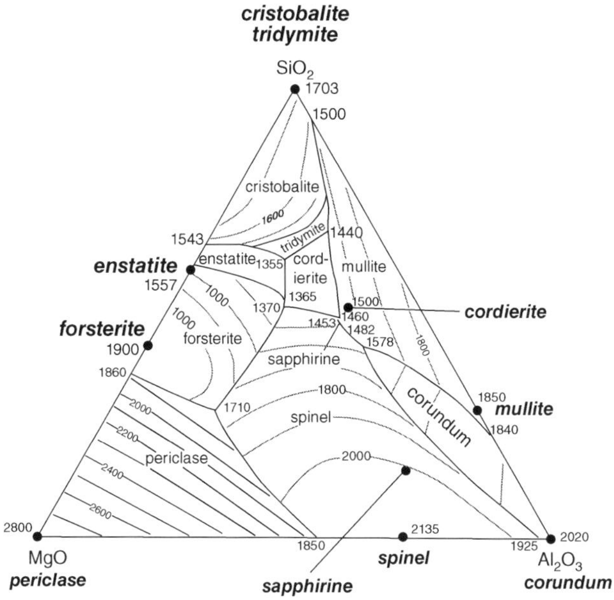

Figure 2.22 Ternary phase diagram for the system of $MgO-Al_2O_3-SiO_2$.

The system $MgO–Al_2O_3–SiO_2$ has a significant prominence in the ceramic industry [7]. Several compositions, including Al_2O_3, SiO_2, mullite, spinel, and forsterite, are used as refractories for furnace linings at elevated temperatures. Moreover, cordierite, SiO_2, Al_2O_3, MgO, and forsterite are the best electrical insulators and are mostly used in electronics industries. Cordierite has good thermal shock resistance and has a low thermal expansion, respectively. The phases such as Cordierite ($2MgO·2Al_2O_3·5SiO_2$ (2:2:5)) and sapphirine ($7MgO·9Al_2O_3·3SiO_2$ (7:9:3)) are ternary solid phases. The crystal structure of these two ternary phases exhibits structural disorderliness at a higher temperature.

2.5.1.2 $MgO-CaO-SiO_2$

The ternary system of $CaO–MgO–SiO_2$ is shown in Figure 2.23, and it indicates the composition mineral behavior of component ratios differently. The ternary phase diagram has pure oxides at their corners of the triangle with congruently melting temperatures of CaO at 2570 °C, MgO at 2800 °C, and SiO_2 at 1710 °C, respectively. Different components with separate regions show the SiO_2-rich part of the phase diagram has two liquid melts. These two liquid melts in the system provide direct evidence of the processes of separation. In the SiO_2-rich portion of the ternary system, liquid–liquid separation occurs. Also, sub liquids phase separation can be predicted in the compositions immediately surrounding them. Crystal phases that can be developed under thermodynamic equilibrium are also indicated in the phase diagram.

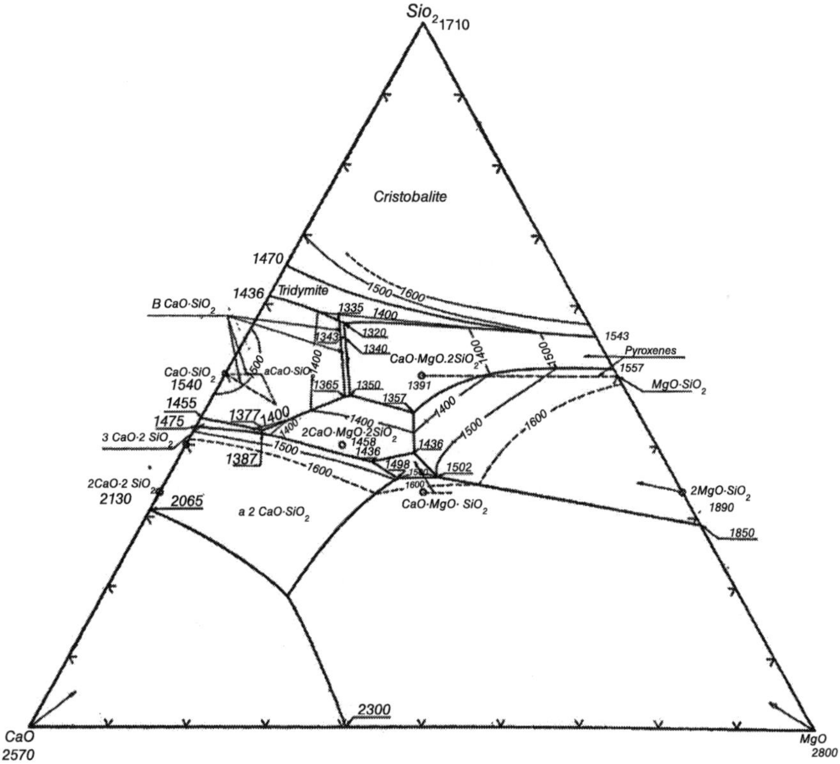

Figure 2.23 Ternary phase diagram for the system of MgO–CaO–SiO_2 system.

The CaO–MgO–SiO_2 system, diopside, $CaMgSi_2O_6$, and hypothetical phase clinoenstatite, $Mg_2Si_2O_6$, are the end-parts of the solid pyroxene solution. There are three other solid solutions of pyroxene with various crystal structures: (1) orthopyroxene, where the end-parts are the imagined phase with the structure of orthoenstatite, $Mg_2Si_2O_6$, $CaMgSi_2O_6$, and orthopyroxene; (2) protopyroxene, where the end-parts are the imagined phase, which has a system of protopyroxene, protoenstatite, $Mg_2Si_2O_6$, and $CaMgSi_2O_6$; (3) low-clinopyroxene, in which the end-parts are the hypothetical phase with a low-clinopyroxene structure, low-clinoenstatite, $Mg_2Si_2O_6$, and $CaMgSi_2O_6$. They are modeled in the same way as the solid solution of clinopyroxene. It has a large region of low melting point minerals such as centered or diopside CaO–MgO–2SiO_2 that must be eliminated. If the CaO/SiO_2 is <1.5, liquid solution forms in a particular region. It was based on this critical selection of raw materials of magnesia for refractory materials [8].

2.5.1.3 MgO–Cr_2O_3–SiO_2

Reactions with a significant bearing on chromite refractories' are demonstrated by considering phase equilibria in the system MgO–Cr_2O_3–SiO_2, shown in Figure 2.24.

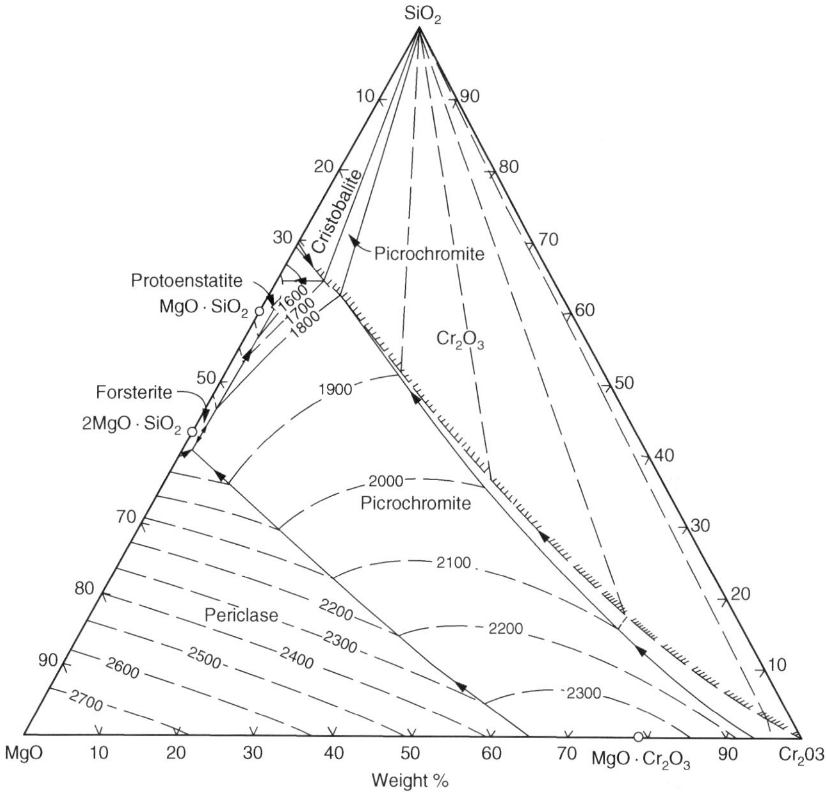

Figure 2.24 Ternary phase diagram for the system $MgO–Cr_2O_3–SiO_2$.

The diagram represents several characteristics of the Cr_2O_3 containing structure and has a good starting point for considering relations of liquidus phase.

The most prominent elements of the diagram are the sweep of the two regions of liquid across the diagram from the $Cr_2O_3–SiO_2$ to the $MgO–SiO_2$ line, the extensive spinel primary phase area (picrochromite, $MgCr_2O_4$), the wedge of the two regions of liquids of this preliminary phase area, and the limited scale of the magnesium silicate preparatory phase areas (protoenst and forsterite) adjacent to the $MgO–SiO_2$. The latter suggests that it needs a comparatively small quantity of Cr_2O_3 to saturate the spinel-phase silicate liquid. The spinel system's movement through the two-liquid region results in a highly complex series of phases of appearance, disappearance, and reappearance across some mixtures' crystallization in the system. Limited information on the chromium content of the silicate crystalline phases (protoenstatite and forsterite) in equilibrium with picrochromite has been obtained for the $MgO–Cr_2O_3–SiO_2$ system.

2.5.1.4 $MgO-Cr_2O_3-Al_2O_3$

The ternary system of the $MgO-Cr_2O_3-Al_2O_3$ phase diagram is shown in Figure 2.25. Magnesia grains initially pick up Cr_2O_3 from the slag to form a solid solution of periclase. Al_2O_3 also dissolves in magnesia, albeit to a much lower degree due to magnesia's lower affinity.

MgO with Cr_2O_3 has higher binding energy than with Al_2O_3; thus the reaction of synthetic spinels ($MgCr_2O_4$) with magnesia at high temperature exhibits very restricted Cr^{3+} diffusion from spinel ($MgCr_2O_4$) to MgO. Hence, dissolved Cr_2O_3 and Al_2O_3 (in periclase grains) emerges from the slag diffusion of Cr_2O_3 and Al_2O_3 in contact with periclase grains, forming spinel.

2.5.1.5 $CaO-Al_2O_3-SiO_2$

A ternary system of $CaO-Al_2O_3-SiO_2$ phase diagram represents the mineral component's mix behavior together in Figure 2.26. One is lower right and left, and one is at the top. In the middle, are all the possible mixture combinations at all temperatures. The ternary phase diagram corners indicate the pure oxides and have their respective melting points alumina at 2054 °C, silica at 1723 °C, and lime at 2572 °C, respectively. The outermost three lines of a triangle show

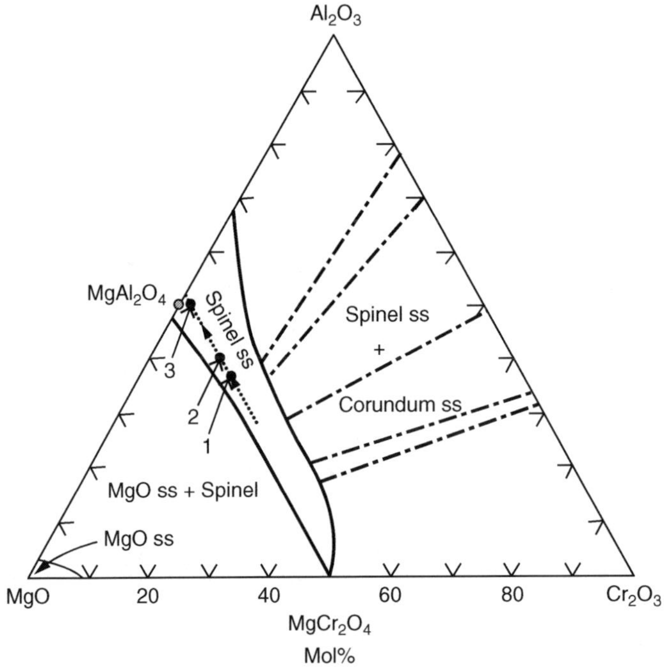

Figure 2.25 Ternary phase diagram of $MgO-Cr_2O_3-Al_2O_3$ system at 1700°C.

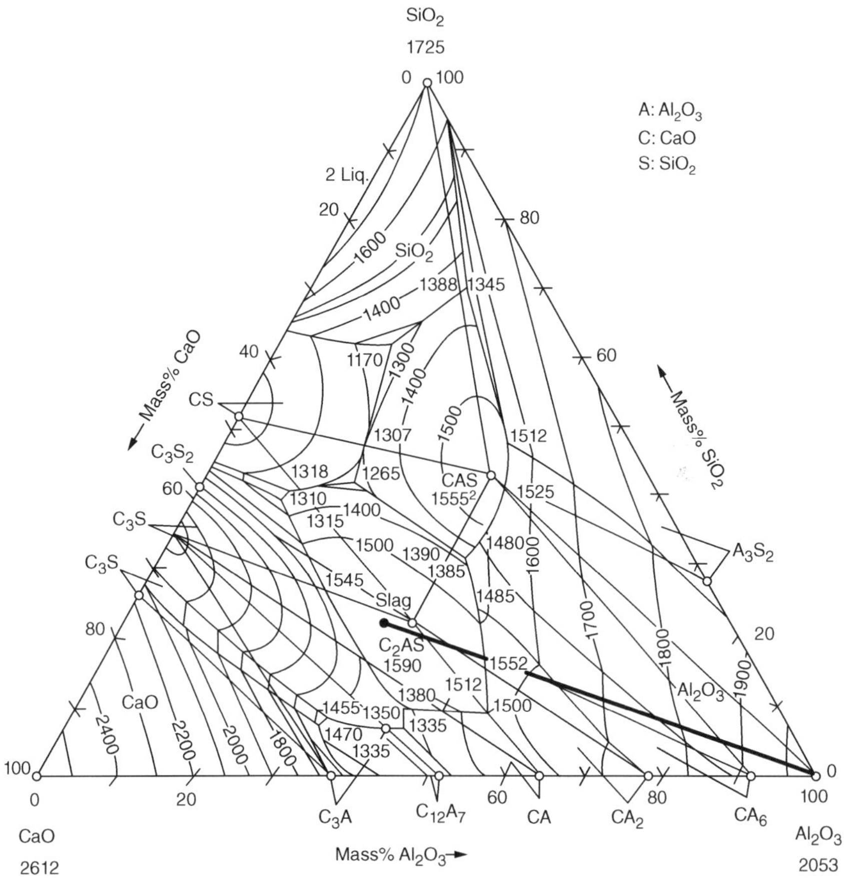

Figure 2.26 Ternary phase diagram of $CaO–Al_2O_3–SiO_2$ system.

the $CaO–SiO_2$, $CaO–Al_2O_3$, and $Al_2O_3–SiO_2$ binary phase diagrams. The composition mixture in the center of the diagram represents the minerals' formation using components concerning the components' distinct ratios.

The specific regions indicate the minerals from the adjacent areas, likely SiO_2 (Cristobalite/Tridymite), CaO (Lime), Al_2O_3 (Corundum), $3Al_2O_3–2SiO_2$ (mullite), $CaO–SiO_2$ (Pseudowollastonite), $3CaO–2SiO_2$ (Rankinite), $CaO–Al_2O_3–2SiO_2$ (Anorthite), and $2CaO–Al_2O_3–SiO_2$, respectively. The first liquid solution is observed at the eutectic temperature of 1170 °C if the composition is in that region. Numerous steep surfaces of liquid are also there, particularly on the silica and lime side regions. Hence, mineral ratios to avoid the early formation of liquid in the refractory design.

2.5.2 $FeO-SiO_2-C$

The ternary system of $FeO–SiO_2–C$ is shown in Figure 2.27. It is evident from the complete phase diagram observation that the region in which the phase of SiC takes part in the state of equilibrium is vast, mainly spreading between SiO_2 and C. It could be associated with relatively good stability at this temperature and decreasing situations of the SiC phase (at the expense of the less steady silicon phase). This assumption promotes its development during the significant decrease of silica.

The $FeO/(FeO+SiO_2)$ ratio ~0.7 provides the maximum bound for the formation of SiC in regions where graphite (solid carbon) exists at the state of equilibrium. The ratio has an excellent value (it might relate to the regular amount of the blast furnace); the phase of SiC does not form the state of equilibrium. Both things combined illustrate fascinating and new ways of interacting and transferring silicon from the organic substances into the molten material (its effect would add to the exchange of Si and C that occurs through the gaseous phase of CO, CO_2, and SiO). Thus, the "silica-rich" segments are the excellent potential for the formation of SiC and the volatility of SiC in areas where liquid iron is present as a stable phase.

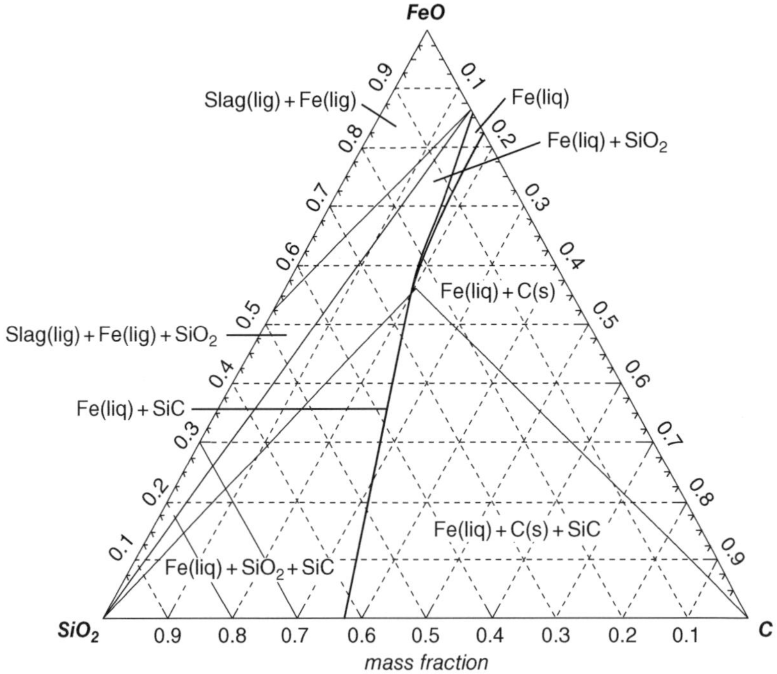

Figure 2.27 Pseudoternary phase diagram of $FeO-SiO_2-C$ system at 1550 °C.

2.6 Nucleation and Crystal Growth

Inevitably, high-temperature contact of foreign particles or molecules facilitates a new phase during refractory application. Such reaction always initiates from the nucleation of new constituents and gradual growth and maturity with time development phases. These may be stoichiometry or non-stoichiometry compounds depending on a molecules' availability from molten slag/metal and atmosphere. Several refractory mixes follow in-situ phase formation at an elevated temperature that regulates refractories' properties despite external resources. Moreover, the resultant phase may be either crystalline or glassy depending on the cooling rate and orientation of atoms during an interaction. This incidence eventually alters the microstructure of the virgin phase through insertion or adherent of secondary phases. When a nucleation and growth process develop a new phase, essentially, it begins as a tiny area and expands in size. Its surface-to-volume ratio is high at first, making it unstable due to its high surface energy that follows different degrees of nucleation. Homogeneous nucleation refers to nucleation from a homogeneous phase; heterogeneous nucleation is from surfaces, grain borders, second-phase particles, and other incoherence in the structures that serve as favorable sites for catalyzing nucleation. In the refractory application atmosphere, the most popular form of nucleation is heterogeneous nucleation.

2.6.1 Homogenous and Heterogeneous Nucleation

The interface between two phases aids the formation of nuclei. The energy interface is less frequently associated with the energy volume reduction until the nuclei form small particles and reach a sufficiently large scale. Hence, the overall shift in free energy on developing the new phase is negative. The fluctuations in a homogeneous system must cause the local increase in free energy, leading to fewer areas of a new process. The formation of free energy of the small field, including its surface energy and the transport atom rate at the interface boundary, are factors in the kinetics of nuclei formation. Consider creating a new phase with spherical geometry and a radius of *r*; the change in free energy is given by Equation (2.3), where is the interface energy and ΔG is the new phase's free-energy change per unit volume. The first part of the equation takes precedence because as an embryo's size grows, so does the amount of free energy needed to shape it.

$$\Delta G_r = 4\pi r^2 + \frac{4}{3}\pi r^3 \Delta G_v \tag{2.3}$$

However, as embryos develop larger in size, the second term takes over, so that once they reach a critical size (Figure 2.28), the second term takes over, and further development leads to lower free energy and a more stable system.

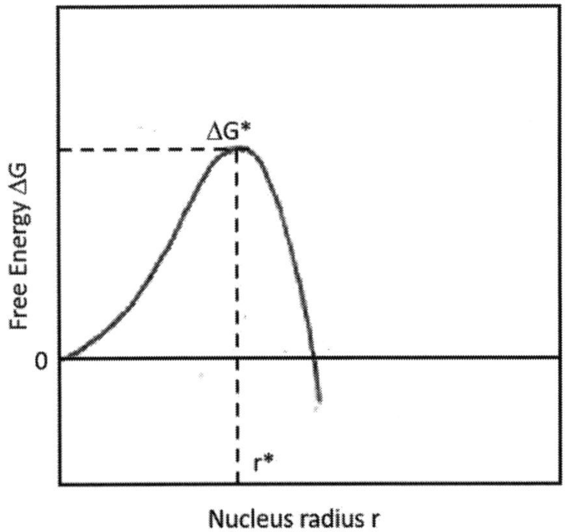

Figure 2.28 Free energy of nucleus as a function of size.

Some critical radii must be exceeded before the nucleus becomes stable. The critical radius r^* and corresponding free energy ΔG^* for the new phase can be represented as:

$$r^* = -\frac{2\gamma}{\Delta G_v} \tag{2.4}$$

$$\Delta G^* = \frac{4}{3}\pi r^2 \gamma = \frac{16\pi\gamma^3}{3(\Delta G_v)^2} \tag{2.5}$$

A new phase with less than critical radius r^* is called subcritical embryos, and a more extensive case is called supercritical nuclei. This nucleus size is much too large to be the product of a single fluctuation, so the nucleation mechanism is thought to involve adding molecules to an embryo one by one. Collision of virgin phase molecule expedites the new phase nucleation, and the frequency and the collision frequency may often be expressed as:

$$\vartheta = \vartheta_o \exp\left(\frac{-\Delta G_m}{kT}\right) \tag{2.6}$$

Where ϑ_o is the molecular jump frequency, and ΔG_m is the activation energy for transport across the nucleus matrix interface. The value of the free-energy shift of the phase transition is determined by the relation when considering increasing supersaturation at a constant temperature:

$$\Delta G_v = -\frac{kT}{V_n} ln \frac{p}{p_o} \tag{2.7}$$

Where p is the actual vapor pressure, p_o is the equilibrium vapor pressure, and V_n is the molecular volume of the new phase. The steady-state rate of homogeneous nucleation in highly viscous liquids is so low that the formation of crystals in the interiors of specimens is often misidentified as heterogeneous or homogeneous. Most liquids display homogeneous nucleation when cooled at $\Delta T^*/\Delta T_o$ between 0.15 and 0.25, where T_o (K) is the melting point, and T^* (K) is the lowest temperature to which the liquid can be supercooled. In reduced temperature units, $\Delta T^*/\Delta T_o$ is the maximum supercooling. In this case, the surface energy is transformed into:

$$\gamma_{Sl} = \beta \Delta H_f \tag{2.8}$$

Where γ_{Sl}, and ΔH_f are the surface energy of crystal-liquid per atom and the heat of fusion per atom, respectively. The constant and $\beta = 1/2$ for metals and $= 1/3$ for nonmetals.

Maximum transformation of phases is initiated heterogeneously on walls of a defective container site, structural flaws, or impurity particles. The general effect of such nucleating substrates is to lower the nucleation barrier expressed by surface energy. The free energy of forming a critical-size cluster with the shape of a spherical cap is given when a new phase nucleus begins on a flat substrate with a contact angle (Figure 2.29).

$$\Delta G_s^* = \Delta G^* f(\theta) \tag{2.9}$$

$$where, f(\theta) = \frac{(2 + cos\theta)(1 - cos\theta)^2}{4} \tag{2.10}$$

ΔG^* is the free energy for homogenous nucleation, given by Equation (2.5). Equation (2.9) shows that the thermodynamic barrier for nucleation on a substrate should decrease with decreasing θ and approach zero as θ approaches zero. Similar atom configurations favor potent nucleation catalysis in the interface planes in the nucleus and substrate [9].

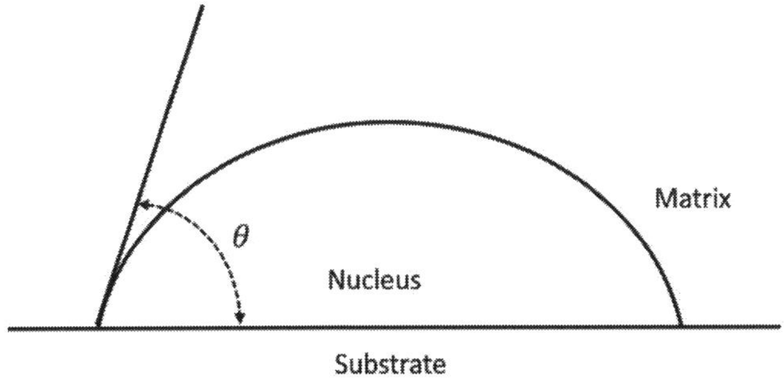

Figure 2.29 Spherical cap model of heterogeneous nucleation.

When cooled to the glassy state and then reheated to a temperature *T* between the glass transition and the melting point, many glass-forming materials exhibit copious crystallization. When a sample of the same substance is cooled to *T* from a temperature above the melting point, the sample persists for a long time without visible crystallization. This disparity in behavior is caused by the nuclei that form during cooling and reheating in the former type of heat treatment.

2.6.2 Crystal Growth Process

The degree of supersaturation and temperature control the growth of the nucleus. The growth rate is determined by the rate at which material reaches the surface and can be built into the crystal structure. It's easier to distinguish melt growth and crystal growth from dilute solutions or the vapor phase. For crystallization processes involving small entropy changes, even the most closely packed interface planes are preferentially rough on an atomic scale. Growth rate anisotropy, or variations in growth rate for different orientations, like to be small. Usually, the most closely packed faces appear smooth, the less closely packed faces have a rough surface, and the growth rate anisotropy must be high for crystallization requiring significant entropy changes. Large entropy changes are necessary for most organic and inorganic compounds, including most silicates and borates, growth from a vapor or dilute solution, and growth from the melt. The development of metals, SiO_2, and GeO_2 from the melt should result in small entropy changes.

Several models have been proposed to explain the crystal growth mechanism. Each is based on a different perspective of the interface and the potential growth sites. The available models are a) Normal (rough surface) growth, b) Screw dislocation growth, and c) surface nucleation growth [10].

a) The growth rate can be expressed as atoms added to or removed from any site on the interface:

$$u = \vartheta a_o \left[1 - \exp\left(\frac{-\Delta G}{kT}\right)\right] \tag{2.11}$$

where u is the interface growth rate per unit area, ϑ is the frequency factor for transport at the crystal–liquid interface, a_o is the distance advanced by the interface in a unit kinetic phase, roughly a molecular diameter, and G is the free-energy shift following crystallization. The frequency factor ϑ can be written as follows:

$$\vartheta = \frac{kT}{3\pi a_o^3} \tag{2.12}$$

where the viscosity is η, for the regular growth model to be accurate, the interface must be rough on an atomic scale and have many phase sites where atoms can be added and removed preferentially.

b) Crystal growth may take place at step sites provided by screw dislocations intersecting the interface. Such dislocations provide a self-perpetuating source of steps as molecules are added to the crystal. Assuming that dislocation ledges are the only places where growth occurs, the fraction *f* of preferred growth sites on the interface in melt growth is approximate:

$$f \approx \frac{\Delta T}{2\pi T_E} \tag{2.13}$$

and the growth rate is:

$$u = f \vartheta a_o \left[1 - \exp\left(\frac{-\Delta G}{kT}\right)\right] \tag{2.14}$$

Undercooling causes the dislocation to wind into a tighter spiral, resulting in a decrease in the separation between the dislocation ledges and a rise in *f*. This model predicts a growth rate that varies as the undercooling or driving force square for minor deviations from equilibrium. For the model to provide a proper change representation, the interface must be smooth on an atomic scale and imperfect, with growth occurring only at screw dislocation steps.

c) Growth occurs at phase sites given by two-dimensional nuclei produced on the interface in this process, and the growth rate can be written:

$$u = A\vartheta \exp\left(\frac{-B}{T\Delta T}\right) \tag{2.15}$$

A is the amplitude factor, and *B* is an exponential constant that varies depending on the model used in the study but is always proportional to the square of the two-dimensional nucleus's edge surface energy. The growth rate predicted by Equation (2.15) should differ exponentially with driving force and be unobservable low for a minor driving force. The interface must be seamless on an atomic scale and ideal for the model to conform to life (free of intersecting screw dislocations).

2.7 Nonequilibrium Phases

One of the most common deviations from equilibrium behavior in ceramic systems is glass, as many silicates can be cooled from their liquid state to form noncrystalline products. The driving force for the liquid-crystal transition must be low, and the process's activation energy must be high. For several silicate structures, both requirements are met. The crystalline phase nucleation rate forming from a liquid is determined by the product of the energy difference between the crystal and the liquid and the mobility of the constituents that form a crystal. These factors shift in silicate systems, favoring the formation of glasses as the silica content rises. The broad network and bridge forming anions are the ones who limit mobility. Even though data for the diffusion coefficient is not commonly available, it is usually inversely proportional to viscosity.

With a variation of temperature, strain, and composition of any system, different nonequilibrium crystalline phases are often observed in ceramic systems. These remain metastable due to the low transition rate caused by the high activation energies needed to convert into more stable phases. Figure 2.30 shows how the energy relationships between three phases of the same composition can be interpreted. If one of these phases has formed, it takes a long time to transition into a more stable phase. This system's transition to the lowest energy state, in particular, takes a long time. When a stable crystal is exposed to a new temperature or pressure range and does not change into a more stable state, metastable crystals form. Second, a residue or transition results in creating a new metastable phase: for example, if phase 1 in Figure 2.30 is cooled into the stability region of phase 3, it can transform into the intermediate phase 2, which can then remain as a metastable crystal.

Nonequilibrium phases in ceramic systems are most often caused by reactions that do not complete in the time available during firing or heat treatment. The most popular forms of incomplete reactions are incomplete solutions, incomplete solid-state reactions, and incomplete resorption or solid–liquid reactions. Both are caused by reaction products that serve as barrier layers, preventing further reactions. Since almost all metals are thermodynamically unstable in the atmosphere but oxidize and corrode slowly, the entire metallurgical industry is perhaps the most striking illustration of incomplete reactions.

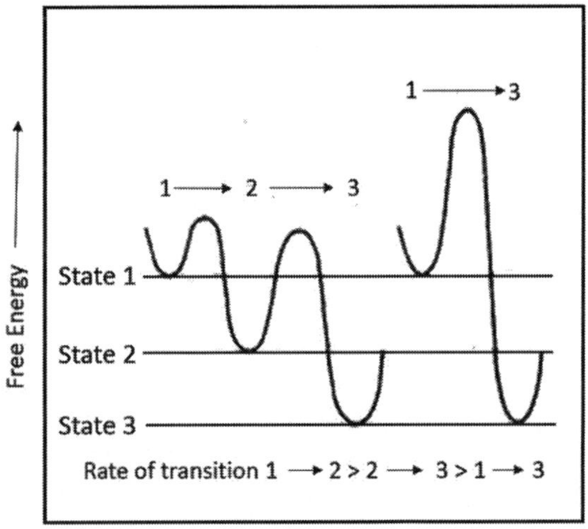

Figure 2.30 Illustration of energy barriers between three different states of a system.

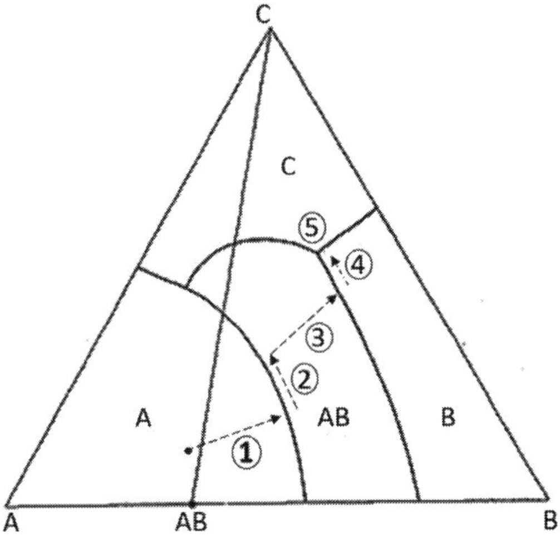

Figure 2.31 Nonequilibrium crystallization path with (1) *Liquid* → *A*, (2) *A* + *liquid* → *AB*, (3) *Liquid* → *AB*, (4) *Liquid* → *AB* + *B*, (5) *Liquid* → *AB* + *B* + *C*.

The exciting example of nonequilibrium conditions critical in understanding phase-equilibrium diagrams is incomplete resorption, which can occur whenever the reaction $A + Liquid = AB$ occurs during crystallization. Such is the case when a primary phase combines with a liquid to create a new compound during cooling. A coating forms on the surface of the initial particle, forming a barrier to further reaction. As the temperature is decreased, the final products are not those expected by the equilibrium diagram. A nonequilibrium crystallization course for incomplete resorption is shown schematically in Figure 2.31.

References

1 Kingery, W.D., Bowen, H.K., and Uhlmann, D.R. (1960). *Introduction to Ceramics.* John Wiley & Sons.

2 Gordon, P. (1968). *Principles of Phase Diagrams in Materials Systems.* New York: McGraw Hill Book Company.

3 *Hummel.* (1984). *Introduction to Phase Equilibria in Ceramic Systems.* CRC Press.

4 Campbell, F.C. (2012). *Phase Diagrams, Understanding the Basics.* ASM International.

5 Carter, G.C. (1977 January). Applications of Phase Diagrams in Metallurgy and Ceramics. *Proceedings of a Workshop, National Bureau of Standards, Gaithersburg,* Maryland. 1.

6 Hillert, M. (2008). *Phase Equilibria, Phase Diagrams and Phase Transformations.* Cambridge University Press.

7 Mao, H., Fabrichnaya, O., Selleby, M., and Sundman, B. (2005). Thermodynamic assessment of the MgO-Al_2O_3-SiO_2 system. *J. Mater. Res.* 20: 975–986.

8 Hudon, P., Jung, I.H., and Baker, D.R. (2005). Experimental investigation and optimization of thermodynamic properties and phase diagrams in the systems CaO–SiO_2, MgO–SiO_2, $CaMgSi_2O_6$–SiO_2 and $CaMgSi_2O_6$–Mg_2SiO_4 to 1.0 GPa. *J. Petrol.* 46: 1859–1880.

9 Sarkar, D. (2018). *Nanostructured Ceramics: Characterization and Analysis.* CRC Press.

10 Scheel, H.J. and Capper, P. (2018). *Crystal Growth Technology: From Fundamentals and Simulation to Large-Scale Production.* Wiley India.

3

Packing, Stress, and Defects in Compaction

3.1 Introduction

Shaping can be defined as the processing step in which a system constituted by isolated particles is transformed into a consolidated body with a specific shape, size, and microstructure. The selection of one or another industrial-scale shaping procedures is based on the required characteristics of the sintered form—shape and size of the material, microstructure—grain size and shape, dispersion of secondary phases, and density—on the number of pieces to be produced and its necessary reliability, and on environmental and economic aspects. The most expensive stage of fabricating pieces of hard materials such as submerged entry nozzle (SEN) is final machining; thus, great effort is devoted to developing shaping procedures to avoid machining.

Fabrication of shaped refractory by dry pressing a wide range of particles (for example, 6 mm to less than 1 μm) is a preferred and most widely used forming protocol to manufacture ceramic components. Lab-scale prototype fabrication to commercial-grade production requires unidirectional die pressing or isostatic pressing to achieve the desired shape and products. Both pressing methods can be automated to a high degree and are used at optimum pressure in the mass production of bricks, slide gate plates, nozzles, and several shapes by the uniaxial press, and shroud, monoblock stopper, and submerged entry nozzle by the isostatic press, respectively. Uniaxial pressing is the most used at an industrial level when it is possible to reach accurate dimensions in large quantities due to the ease of automation and the high production speed.

Uniaxial pressing has limitations and tends to develop a density gradient when the pressing direction exceeds 200 mm. A homogenous isostatic pressing is smart to obtain uniform dense compact from powders or add further densification of isotropic green bodies that cannot be developed in uniaxial pressing. Steel mold with an optimum hardness is typically used for uniaxial pressing, whereas the isostatic pressing molds are made of silicone, elastomers, and polyurethanes.

Fundamental Design of Steelmaking Refractories, First Edition. Debasish Sarkar.

A prerequisite amount of additive is beneficial for dry pressing, allowing less expensive additive removal operations. However, adequate particle grading in refractory mixture facilitates the flowability of the process that frequently experiences deformation (depends on material) and stress–strain during the compaction process. The ability of refractory mixture flow depends on grading, binder content, aging time, and temperature to control the mold feeling and resultant shape. A high content of fine particles provokes extensive elastic relaxation and associated cracking on removing the load, whereas larger particles result in segregation and enhancement of surface porosity or roughness. Binder forms strong bonding in particles and reduces segregation. Due to the lack of plasticity of advanced ceramic powders, organic additives (binders, plasticizers, and lubricants) are usually added to provide plasticity during forming and the green strength of the formed product. Nonuniform matrix and sintering defects initiate because of heterogeneous pressure transmission, leading to the segregated particle arrangements and density variation of the compacts. Although several commercial grade uniaxial press such as friction screw and hydraulic press are predominately used and have limitations to make a long shape like ladle shroud.

Particle surface has different degrees of adsorption capability concerning their size, shape, porosity, angularity, temperature, and pH. Coulomb's force of attraction between the moisture ions and particles is the principal reason for the existence of green body cohesion. The applied pressure is high (~50–200 MPa) when the saturation degree is very low (0.01–0.5). Usually, common practices are starting moisture content less than 7 wt.% for dry pressing and 15 wt.% for semidry and wet pressing. Uniaxial or two directions biaxial pressing executes the applied pressure in one direction; thus the wall friction is the prime source for the density variation of the green body. In this contrast, the applied pressure employed through the entire mass during isostatic pressing avoids the density gradient characteristics of uniaxial pressing. This chapter briefly describes the fundamental aspects of dry pressing, the practice of uniaxial die and isostatic pressing, the preparation of refractory mixtures for the pressing, and their consecutive analysis.

3.2 Refractory Grading and Packing

Shaped refractory demands a wide range of particles to maintain effective particle packing in developing high dense refractories. A wide range of comminution processes is well adapted to obtain a few millimeters to a few micron particles [1]. Crushing and grinding are the basic comminution processes to reduce coarse mineral ore or calcined raw materials by four to nine times through impact by a harder material as compared to feed. A tap density and sieve analysis is an excellent approach to selecting the grading and synchronizing the density. However, this trial should be done before and after adequate particle mixing with binders and other additives. Otherwise, the obtained sieve analysis and particle size distribution may change from virgin raw materials grading. During mixing, fine particles are supposed to adhere to coarse particles through a binder, reduce the fine fraction, and misinterpret the data. Despite these incidences, binders help in denser packing by increasing the ease of particle movement, but the high binder thickness may cause a negative effect on the packing density. The calculation for a simple ternary system consisting of spherical particles can be represented by Equation (3.1):

$$\begin{aligned} \textit{Packing density for ternary mixture}(PF_{mix}) = PF_c + (1 - PF_c)PF_m \\ + (1 - PF_c)(1 - PF_m)PF_f \end{aligned} \tag{3.1}$$

here PF_c, PF_m, and PF_f are the packing fractions of the coarse, medium, and fine particles, respectively, in a ternary mixture, which resembles the discrete model. However, this model may differ somewhat for hard ceramic particles as it has rare sphericity and deformation. Preferentially oxide refractories achieve near 50% relative density after pressing, and high-temperature firing at 1650 °C facilitates obtaining high density during sintering. This transformation depends on particle packing in the green stage, fine fraction content, impurities content, peak temperature, and retention time. On the contrary, the unfired MgO-C does not need any high-temperature firing; rather, tempering up to 300 °C is enough to form the polymerization network and strength. However, carbon-based shaped refractories change to several phases through in-situ phase transformation during application and alter the microstructure that eventually determines the refractory performance. For example, more porosity, low eutectic phases due to impurities, and cracking because of thermal stress open up the passage for the molten metal/slag penetration and failure of refractories. Thus, early stage density design is a significant step, but an optimum density is beneficial for particular applications. It is worth remembering that the consistently high dense refractory may not require maintaining high life and performance. For example, more than 15% porosity of Mag-Chrome brick is within the acceptable limit for AOD; however, 3–5% porosity is allowed for MgO-C brick in the slag zone of LF.

3.2.1 Binary and Ternary System

Obtaining the green density and resultant density of refractories are prime objectives of pressing and sintering, respectively. Characteristics of starting graded particles, including particle size and distribution, particle shape and geometry, and degree of agglomeration of particles, have a profound effect on the sintering kinetics and densification and microstructural development of refractory brick. A system containing a non-agglomerated and high tap density mixer expedites high packing density and eventually experiences preferential mass transport through volume or grain boundary diffusion instead of surface diffusion or vaporization and condensation, ensuring sintering. High dense, coarse particles lead to high green density, but a higher percentage of smaller particle sizes or fines in the starting mixer facilitates a faster densification rate at elevated temperatures during sintering. However, the scenario for fine powder sintering depends on the removal of agglomerates and aggregates. A certain degree of deagglomeration enhances the sinterability and resultant microstructure. Effective mixer and mixing sequence and ingredients have an important influence on it. The presence of too high a percent fine may create some other difficulties, including:

- Nucleation of microcracks in lamination during pressing or ramming that finally extend and expose after final stage drying or sintering.
- Such a dense body may experience relatively low compressive strength, low HMOR, and less load-bearing capacity.

Narrow size distribution is imperative for obtaining a high sintered density. Following postulates can improve the manufacturability of high value added refractory products through adequate particle size distribution and particle–particle interaction. In practice, binary refractory mixture, i.e., combining two different particle sizes, provides better packing density than single particle size as per Equation (3.1).

For binary systems, the apparent volume varies, as shown in Figure 3.1a, in which the minimum apparent volume at about 70% coarse material [2]. The lower straight line results when the size ratio is infinite, whereas an identical size follows the top horizontal line. A dotted curvature behavior is obtained when the diameter ratio is near 10:1. The pore volume decreases with the addition of a third size, and the significant change is shown in the ternary diagram in Figure 3.1b.

Thus, during composition design and raw material processing, i.e., before firing the refractory bricks, the particle or powder characteristics obtained from a green mixture sieve analysis can be considered a set of parameters. To achieve desired properties for field applications and mass production of different refractory bricks, including Mag-Chrome, Dolomite, MgO-C, Alumina-spinel, Slide gate refractories, and Al_2O_3-C, are necessary to synchronize different gradings starting from highest coarse to lowest ultrafine particles. In continuous production, mixer sieve

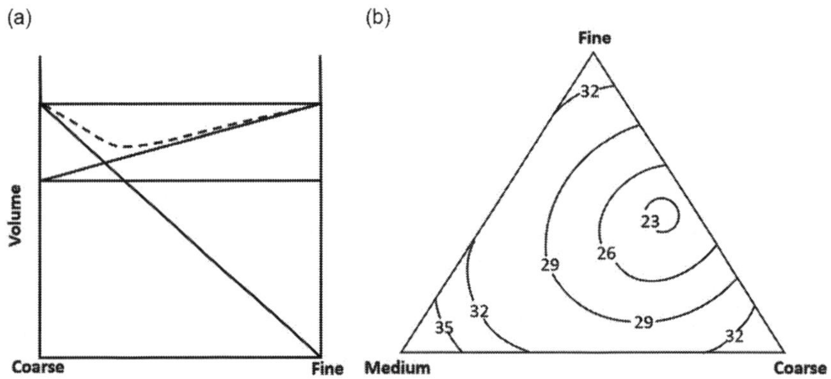

Figure 3.1 Particle packing in (a) binary and (b) ternary refractory mixtures.

analysis with a regular interval is a mandatory check, and adequate precaution is required to ensure the desired fired properties of refractories. The pressing consistency of the mixer for a prolonged time is essential to obtain a smooth surface with desirable densities. Apparently, this behavior depends on binder content, mixing efficiency, and atmosphere. During pressing, the particles experience shear stress, and it is maximum for irregularly shaped particles and minimum for a structured flake particle like graphite. This stress becomes nonuniform when product length is more, which results in surface cracking. Thus, hexagonal structured graphite coating inside the mold enhances the particle motion and reduces the defect, similar to BN coating for graphite mold in hot press.

3.2.2 Particle Morphology and Mechanical Response

Packing depends on particle size ratio, and each size and shape has a phenomenological aspect to dominate the role in the stress, strain, and sintered strength of compact mass. The particle morphology in the compact mass alters the microstructure and is responsible for the variation of strength behavior. When different granular materials are subjected to loading, alteration of particle size and their associated voids govern the load distribution mechanism of the granular mass depending on the individual grains as load transfer particle to particle, and the macroscopic response of granular mass is the result of the individual response of the particles. The stress–strain and strength characteristics of grains are influenced by grain size, distribution, and packing during loading and subsequent deformation. A cumulative microscopic behavior variation may affect the macroscopic behavior of granular materials.

Several processing protocols are required to make different grades of raw materials, and obtained shapes (3D geometry) and sizes are different. While using particles for refractory processing, the void space among the particles alters due to shape, size, and distribution. The particle size variation influences the rolling and sliding resistance and is also responsible for the deviation of the shear strength behavior of the granular media. As a result, when the granular body is subjected to load, the stress, strain, and strength responses also differ with particle size.

While considering the particle interaction phenomenon, it is essential to describe the influence of morphology on the formation of voids and plausible interactive movement under loading to modify the microstructure. Dry contact analysis of particles can establish the effect of particle size on the shear strength and emphasize a few facts. In evidence, the nature of sliding movement alters in the presence of a binder or secondary phase like graphite.

When dry refractory grains touch each other at a finite contacting area, the particulate mass experiences stress by external load or stress during mixing. This external force excites the system and facilitates particle movement. It is apparent that the three-dimensional movement associated with shearing and this deformation under triaxial compression originates along the sliding plane, where maximum develops $(\tau/\sigma)_{max}$, τ and σ represents the shearing stress and normal stress, respectively. The shearing stress and normal stress alter the contact and sliding angles. Eventually, different interactive angles and resultant stresses depend on particle number (N) and particle size (r = average radius), which is further associated with the volume of solid (Vs), the volume of voids (Vv), and void ratio (Vr).

Let the total volume $V = Vs + Vv$, and the total number of particles per unit area be N. Then, the volume of spherical solid particles $(Vs) = N\,4\pi r^3/3$. Again, the total volume $V = Vs\,(1 + e)$, where e = void ratio; thus $Vs = V/(1 + e)$.

The volume per particle $V_1 = 1 \times d^3 = (2r)^3 = 8r^3$, so total volume $V = NV_1 = 8Nr^3$

Thus,

$$Vs = (8Nr^3)/(1+e), \text{ or } N = Vs(1+e)/(8r^3) \tag{3.2}$$

The number of particles and the void ratio is a function of average particle size; herein, particles are spherical. Particle size controls the microscopic properties, and eventually, it plays an important role in the macroscopic response of bulk compacts. Different grades of particles, uniform particles (nearly all the same size), and graded particles (wide range of particle sizes) exhibit different friction and stress in particle interaction. With an increase in particle size, the maximum shear strength and angle of internal friction increase, and the normal load also play an important role. Similarly, with an increase of gradation from smaller to wider, maximum shear strength and angle of internal friction increase. Finally, the response of maximum shear strength is higher in graded particles as compared to uniform particles. The normal stress, shear stress, and the resultant force alter due to a change in particle size, which also satisfies the experimental results.

Despite the spherical morphology, variation of sphericity, roundness, and regularity of crushed 'particles' are apparent and have a significant role in controlling the void ratio, resulting in alteration of friction angle, interactive slippage over another, and strain under external mechanical load. Spherical volume can be measured by a mathematical relation $(4/3)\pi r^3$; r is the radius of the sphere, and this behavior is independent of its size. It can be quantified as the diameter of the largest inscribed (max—in) sphere relative to the diameter of the smallest circumscribed (min—cir) sphere. However, the roundness measures the degree of sharpness of the edges and corners of a particle. It is quantified as the average radius of curvature of surface features relative to the radius of the maximum sphere that can be inscribed in the particle. Two different morphological properties: sphericity predominantly varies if particles elongate, whereas roundness depends on how the sharpness of angular protrusions (convexities) and indentations (concavities) from the object [3]. The regularity is the average of dimensionless values of sphericity and roundness.

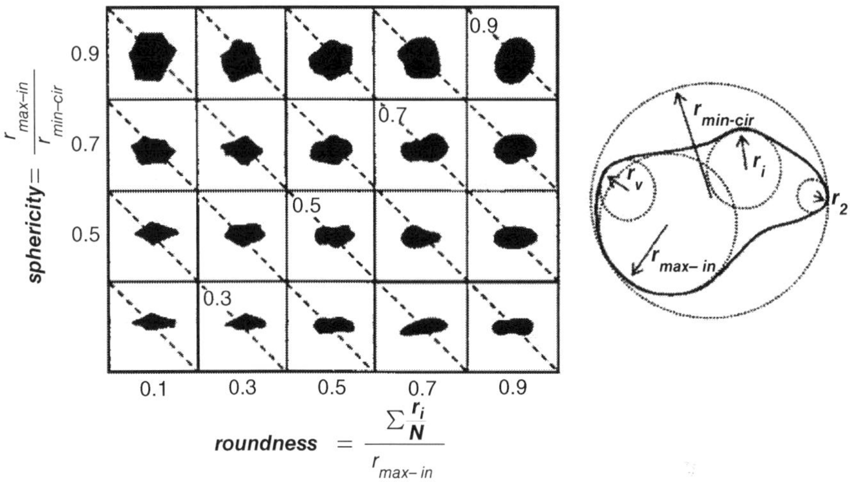

Figure 3.2 Particle shape determination—sphericity *S* and roundness *R* chart. Diagonal dotted lines correspond to constant particle regularity $\rho = (R + S)/2$ [4]. Credit: Krumbein, W. C., and Sloss, L. L. _1963_. Stratigraphy and sedimentation, 2nd Ed., Freeman and Company, San Francisco

Figure 3.2 describes the basic understanding of particle sphericity, roundness, and regularity [4]. Continuous processing enhances the abrasion and alters the morphology; for example, exhaustive milling may enhance the roundness of the particles and eventually change the coordination and contact angle. The particle interlocking and external force alter the void ratio and resultant microstructure.

3.2.3 Nanoscale Particles and Mechanical Response

A perfect combination of coarse-grained and fine particles encompasses the dense microstructure. Wear resistance depends on coarse grain and a continuous complex matrix resulting from wide distribution particles, in which the lowest is the nanoscale particles. In monolithic, even the smallest fraction has a significant role in the particle flowability and densification during high-temperature processing. Compact brick also demands a definite proportionate of smallest particles. Thus, the mechanical response of such particles provides a better insight into designing the refractory composition for the particular zone of interest. Briefly stated, it is evident that the mechanical response of individual nanoscale particles plays an important role in controlling the cumulative densification, microstructure, and performance. The inter-*NP* interaction is normally weak for most nano particle assemblies (*NPAs*), including van der Waals, capillary, and/or electrostatic forces. Several strategies have been developed to reinforce the *NPAs*, in which forming physical inter-NP bonds through sintering and increasing *NP* shape anisotropy to minimize the deformation instability are predominant while bulk ceramic is a concern. However, many *NPAs* have poor mechanical properties, largely hindering their applications. Interestingly, the mechanical properties like *E*, *H*, and shear strength of nanoscale particle varies with diameter [5]. Kendall established an analytical model for the *E* of *NPAs* and is represented by Equation (3.3):

$$E = 17.1(1 - p)^4 (\gamma E_{NP}{}^2)^{1/3} d^{-1/3} \tag{3.3}$$

where E_{NP} is the bulk modulus *NP*, *p* is the porosity, and γ is the *NP's* interface energy. However, this model demanded further modification and developed a more exhaustive mechanical response through the indentation of monodisperse silica nanoparticles as a representative material. An et al. estimated the mechanical response hardness (*H*), elastic modulus (*E*), and shear bonding force (*f*) concerning particle size (d) [6]. Figure 3.3a shows the drastic effect of particle size on the hardness and elastic modulus and follows the power-law correlated with *d*, i.e., $H \propto d^{-n}$ and $E \propto d^{-m}$, the power exponents *n* and *m* decrease with decreasing "*d*." Particularly, for *H*, $n \approx 2$ for $d > 100$ nm, $n \approx 1$ for 50 nm $< d <$ 100 nm, and $n \approx 1/2$ for $d <$ 20 nm. For *E*, $m \approx 1$ for $d >$ 50 nm, and $n \approx 1/3$ for $d <$ 50 nm. Thus, the strength (hardness) and stiffness (modulus) of the *NPAs* are *d*-dependent: the smaller is stronger and stiffer, predominantly due to the higher planar density of *NP* junctions (*NPJs*) at a smaller *d*.

With consideration of Equation (3.2) and three assumptions (1) ordered packing (e.g., *FCC*); (2) smooth surface, and (3) Hertzian contact with attractive force

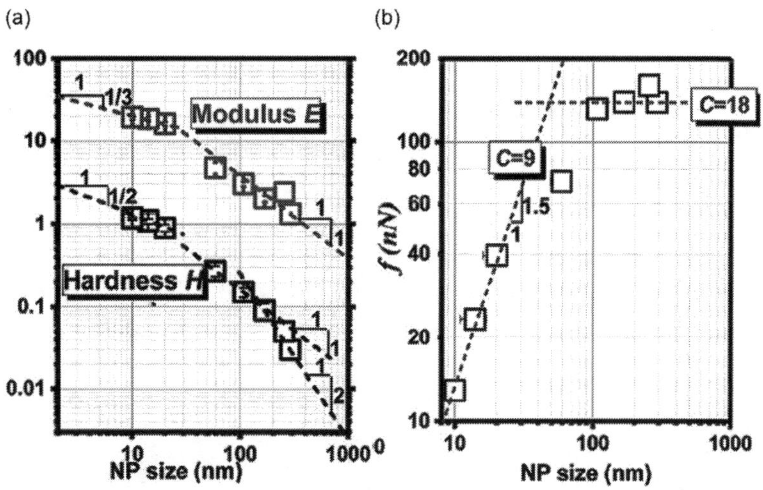

Figure 3.3 (a) *H* and *E* vs. *d*, and (b) estimated shear strength *f* vs. *d* [8] / The Royal Society of Chemistry.

between particles, the effect of d on E can be correlated and preferentially $E \propto \gamma^{1/3} d^{-1/3}$, for FCC packing (i.e., $d >$ ~100 nm). The porosity p would also primarily affect E and H. In fact, moduli of nanosize material could largely deviate (generally decrease) from their bulk values when the size is <5 nm. Thus, the 4 nm *NPA's ENP* may be significantly less than that for larger *NPs*.

In consideration of shear-dislodging the $f = (Hd^2)/C$, where $C = 18$ for specific *FCC* type packing of nanoparticles. Here f is the shear bonding force for an *NP* junction (*NPJ*), i.e., the force to break an *NPJ* by shearing. However, the shear bonding force and particle size relation are associated with three assumptions; (1) each *NP* is rigid, so the *NPA's* strength is correlated only to f, (2) the *NPA* is in *FCC* packing; and (3) three *NPJs* need to be broken for dislodging one *NP* from a {111} plane due to the collective-shear-dislodging on the close-packed plane {111} of *FCC* packing. Without physical bonding around *NPJs*, the particles experience only motion instead of crashing and plasticity of *NPs* themselves. While considering the *FCC* class of particle packing, the assumption is reasonable for $d >$ ~100 nm. In summary, $C = 18$ *for* $d >$ ~100 nm; C is estimated as ~9 *for* $d <$ ~50 nm and transits from 18 to 9 for ~50 nm $< d <$ ~100 nm. Thus, estimated f, as plotted in Figure 3.3b, shows that, for $d <$ 50 nm (in the case of disordered packing), f decreases with decreasing d as intuitively expected, and approximately $f \propto d^{1.5}$. However, for d > ~100 nm (in the case of *FCC* ordered packing), f is fairly constant, ~140 nN. This constant f independent of d is surprising and is attributed to the complex convolution of van der Waals (f_v) and capillary interactions (f_{cap}). Consequently, the trend of f vs. d (Figure 3.3b) implies that, for $d <$ ~100 nm, the d-effect on $f_{cap} \propto d$ may dominate, and for $d >$ ~100 nm, the d-effects on both f_v and f_{cap} contribute. However, in general, the van der Waals (calculated in dry air) and capillary forces cannot be directly added due to the effect of the water medium. The f_v for an *NPA* in dry air ($RH = 0$) is about 10 times the f_v for the same *NPA* completely immersed in water.

In cumulative, the significant particle-size-effect (d-effect) on the hardness (strength), modulus (stiffness), of nanoparticle assemblies (*NPAs*) for particle-size d in the range of 10–300 nm. In fact, f is shown to be fairly constant (~140 nN) for the ordered *NPAs* with $d =$ 100–300 nm. Then, f decreases with decreasing d from ~100 nm to ~10 nm, and $f \approx$ 13 nN at d = 10 nm.

3.2.4 Binder and Mixing on Packing

Organic binders are typically long-chain polymers that serve the primary function of strength attainability of the green body through bridging and network formation between particles. The selection of binders and mixing protocols are essential processing components to obtain a homogenous mixture of refractory aggregates that eventually reduce the product defect after the final stage of high-temperature processing. While considering a binder, one has to consider the following characteristics to select the best binder for a particular system:

- Facilitate homogenous mixing with refractory aggregates
- Avoid reaction with the refractory mix at any temperature
- A minimum amount of unwanted residue is expected after firing
- Easy burnout at an elevated temperature
- It should not affect the bulk density and apparent porosity during processing
- Green and dry bricks are capable of withstanding high mechanical strength
- It is capable of providing flowability during pressing to attain a consistent density
- Permanent linear change of compact is not affected by it
- Need to avoid sticking the refractory mix with mold because of binder
- Avoid toxicity and air pollution after burnout of the binder
- Refractoriness of the final product does not change because of it
- Inexpensive is expected concerning economics

However, it is difficult for all features to exist in a single composition binder. Thus, selecting an adequate binder is a key parameter to enhancing the production yield and minimizing operation costs for refractory manufacturers. While a single entity is insufficient, a binder made of multicomponent molecules exhibits a unique range of properties to improve the process and product quality. Definite quality of specific additives is mandated for manufacturing hydraulic pressed bricks, shaped bricks by other processes, and other unshaped refractory products. Usually, the natural products have a neutral CO_2 footprint, are environmentally friendly and nonhazardous, and provide several advantages in a particular process. An expert opinion and field test may provide confidence for selecting a binder for a specific process.

A particle size variation facilitates high dense refractory, and their adequate mixing sequence and procedure in the presence of binder results in a nonsegregated mixer and final product. Moreover, rotational motion and subsequent shearing force are common in mixing refractory particles. Thus, proper design, speed, filling, and other parameters of the mixer machine are essential to maintain a quality product.

As a consequence of mixing, there is an equilibrium limit where further mixing is balanced by segregation and other de-mixing processes. This mixing limit and the rate at which it can be reached typically depend on the design and operation of the mixing devices. Mixing homogeneity is significantly affected by impeller geometry, operational conditions of the mixer, and particle properties to be mixed. Particle properties are preferentially size, density, and geometry. The particle properties of the mixture can also have a significant effect on the mixing behavior, the most significant apart from the size difference being the density difference. An increase in size or density difference results in increased segregation tendencies in mixers and improper packing of refractory products.

Generally, a binary mixture of particles with uniform density but different sizes segregate so that large particles prefer to collect at the periphery of the rotating drum. On the other hand, a mixture of particles with uniform size but different densities further segregates so that the denser particles collect at the center of the rotating drum. Interestingly, a binary mixture where large particles are denser shows different behavior. Here, both the size and density segregation effects compete against each other resulting in good mixing. Similarly, the powder mixture can be improved under certain size and density combinations in the case of rotating drums. In particular, the mixing improves when the size ratio $r_s = d_{heavy}/d_{light}$ is substantially larger than the density ratio $r_d = \rho_{heavy}/\rho_{light}$; here, d and ρ represent the particle diameters and densities of a binary mixture, respectively.

Table 3.1 represents the typical composition of MgO-C metal zone brick. Significant ingredients are MgO ($\rho = 3.58$ gm/cc) and graphite ($\rho = 2.26$ gm/cc), where different size fraction of MgO is required to maintain higher r_s than r_d, and effectively produce a homogenous mixture for continuous commercial production.

Several classic pieces of literature address segregation phenomena for granular media, the mechanisms being described as mainly due to percolation and the Brazilian nut effect, which occur at different particle sizes [7]. In the Brazilian nut effect phenomenon, larger particles move to the top of the container when a container with two different types of particles are under shaking conditions. A reverse effect may also occur depending on the density of the particles; namely, the larger particles can sink to the bottom. In addition, it has been shown that percolation and buoyancy can oppose each other if large-heavy particles and light-small particles are used in a rotating drum. Although the Brazil nut effect has been described as due to buoyancy, studies that illustrate the direct involvement of forces acting on the particles in the segregation process are rare. The transition between states of mixing and segregation due to size and density differences at a given volume fraction can be predicted for rotating drums based on an expression [8]. The same incidence is promising to quantitatively predict the mixing states of

Table 3.1 Typical composition of metal zone MgO-C refractory.

Raw materials	Size (mm)	Weight %
MgO	2–4	20
	1–2	30
	0–1	25
	–0.044	15
Graphite flake	0.1–0.3	10
Liquid resin	–	4 parts

dry and wet particles at low cohesion when the particle flow is in the continuous flow regime. The cohesive particles can be divided into two groups, one with dry particles where particle sizes are generally less than 10 μm, and the cohesion is due to the van der Waals force, and another with wet particles where the cohesion originated from capillary forces. Mixing processes involve dry fine solid particles that play an important role in manufacturing processes used in many industries and found that some cohesion can improve the mixture quality.

With the improvement of refractory material production technology, the requirement of mixing technology is also increasingly improved. Thus, intensive high shear, high energy, and extremely efficient mixer applied perfect structure, and advanced mixing technology is well developed to produce various kinds of regular and special refractory material.

3.3 Stress–Strain during Compaction

The significant volume of refractories is manufactured by the compaction method, either uniaxial or isostatic pressing. Compression of an unsaturated plastic, ceramic body or mixture of particles may cause permanent consolidation through:

Stage I: Sliding, reorientation, rearrangement, and packing of particles
Stage II: The increase of the contact surface areas between the particles by elastic compression of particles, binders, lubricants, liquid, and the fragmentation of the brittle ceramic particles
Stage III: The compression or dissolving of gas in the liquid; particle orientation; the migration of liquid and gas from interstices; and porosity reduction.

Some refractory mixes may expedite plastic deformation, depending on secondary phases like graphite, metal powder, etc. The compressibility may be determined under uniaxial or isostatic confined compression conditions, as shown in Figure 3.4.

In uniaxial compression, the specimen is confined in a smooth, rigid die; compression is the dominant source of strain, but die wall friction causes shear stresses, as indicated on the *q*-*p* diagram. The ratio of horizontal stress σ_h to vertical stress σ_v, $K_{h/v} = \sigma_h/\sigma_v$, may be determined as a function of pressure by using a pressure transducer in the die wall. Permeable pistons may be substituted to determine the drainage effect of the pore fluid on the mechanical behavior. In the isostatic compression test, the specimen is enclosed in a thin, impermeable membrane and compressed by a fluid.

Proper isostatic compression causes a decrease in volume, but, as indicated on the *q*-*p* diagram, no shear ($\sigma_1\sigma_2 = 0$) and no distortion happens during compaction. During isostatic or uniaxial compression of loosely packed powders, the rolling, sliding, and rearrangement of particles may cause large volumetric compaction. Although the average shear stress is zero during isostatic compression,

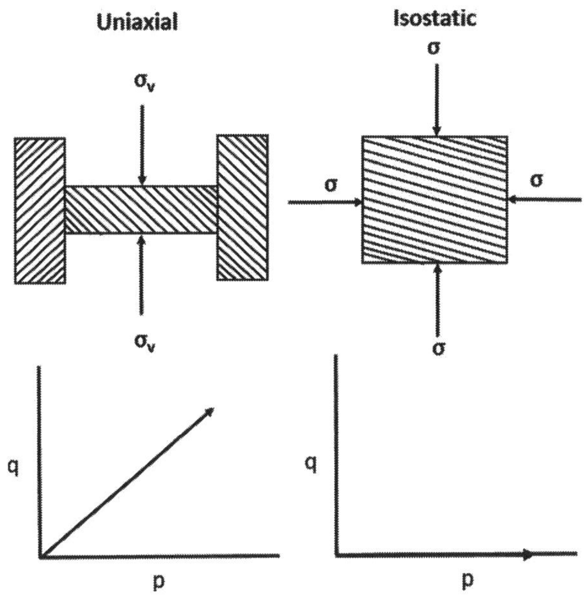

Figure 3.4 Loading and stress path during (a) uniaxial and (b) isostatic pressing.

shear forces at individual particle contacts may be quite large. The stress ratio $K_{h/v}$ is higher for a saturated but highly porous body, and $K_{h/v}$ approaches 1 when the friction angle ϕ_f approaches 0°. For confined compression and elastic behavior:

$$K_{h/v} = n/(1- n) \tag{3.4}$$

where ν is the effective Poisson's modulus of the porous compact.

While compaction provides compaction, the compact density as a function of the applied pressure is commonly used to characterize the compaction behavior.

3.4 Agglomeration and Compaction

Van der Waals forces or electrostatic attraction expedites agglomerates of dry and nonmagnetic powder. Electrostatic forces occur because of surface-adsorbed ions or in the transfer of electrons between particles in contact regions [9]. For nonconductors particles, charge mobility is limited to the surface, and charges persist until contact with an electrically conducting material occurs. An adsorbed film may also diminish electrostatic charges, enhancing the charged mobility. Electrostatic and Van der Waals forces produce relatively fragile agglomerates. However, these weak agglomerates can impede the flow and packing of dry powder and colloids.

Agglomeration is also caused by the capillary force of a wetting liquid at contacts between particles.

The DPS is a measure of pore saturation and is defined as the ratio of the volume of the suspending liquid to pore volume. Agglomeration is also caused by the capillary force of a wetting liquid at contacts between particles. The effect of capillary forces on the agglomerate strength using a model of packed spheres of equal size containing a liquid distributed, as shown in Figure 3.5. The mean number of contacts per unit area N_A is $(aNc)/3$ where, "a" is sphere diameter, and Nc is defined by Equation (3.5). For a random packing, $CN = \pi/(1 - PF)$, and; CN—coordination number, PF—packing fraction [10].

$$N_c = \frac{3(PF)(CN)}{\pi a^3} \tag{3.5}$$

$$N_A = PF / \left[(1 - PF)a^2\right] \tag{3.6}$$

The tensile strength (S_t) is the product of the number of contacts (bonds) per unit area and the capillary force per contact "b":

$$S_t = \left[PF / (1 - PF)a^2\right] b \tag{3.7}$$

For randomly packed uniform spheres separated by an adsorbed liquid, the nearly identical equation by other means ($\phi = 1 - PF$):

$$S_t = 9/8\left[(1 - \phi) / \phi a^2\right] b \tag{3.8}$$

The capillary force b is equal to the force produced by surface tension and the capillary pressure multiplied by the area of the liquid in the neck. The force b increases as the liquid contents are reduced. When the neck radius is very small compared to the radius of the circular liquid in the plane of contact, the wetting angle is 0°. Equation (3.8) indicates that the tensile strength from capillary liquid bonding varies directly with the surface tension and packing density and inversely

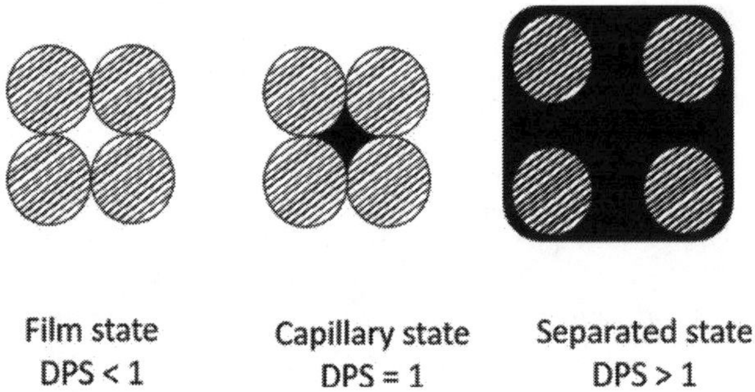

Figure 3.5 Liquid distribution in film, capillary, and separated particle states. The degree of pore saturation (DPS) is <1 for powder, granules, and plastic; DPS = 1 for paste, and DPS > 1 for slurry and suspension.

with the particle size. A hard, brittle mass formed from a soft mass on drying from the capillary state to the film state is also commonly observed. Capillary liquid bonding can produce a tensile strength ranging up to about 350 kPa for water in an oxide powder of about 1 μm particle size. When pores are saturated, internal capillary liquid bonding disappears. Agglomerates in the capillary state are usually of lower strength and more plastic. Agglomeration affects the particle size distribution, more prominent when particle specific surface area is very high. This phenomenon may alter the grain size distribution during compaction.

During uniaxial die-pressing of ceramics, density gradients are developed in the compact. Density gradients are density differences over a distance or area. It impacts the physical and mechanical properties and can cause cracking, warping, or distortion after compaction or sintering. Large density gradients must be avoided to improve the quality of uniaxially pressed products. In a compact, density gradients and the density distribution are caused by the stress distribution induced by the applied compaction pressure. The stress distribution develops due to a balance of forces among the external compaction pressure, interparticle friction, and die-wall friction. With the increase of pressure during compaction, inter-particle friction and die wall friction occur afterward. Geometrically, the friction can be expressed as axial and radial friction, superposing internal friction. An increasing angle of the internal friction produces larger stress and density gradients.

Figure 3.6 shows typical density distributions of uniaxially die-pressed compacts [11]. For one-sided pressed compacts, the typical form shows high density regions at the upper edges and in the middle of the bottom for common height-to-diameter-ratios h/d below ≈ 1. If two-sided die-pressed, high density regions are in the upper and lower edges and the center of the compact. Generally, the arising density distribution, final mean density, and dimensional accuracy depend on various factors.

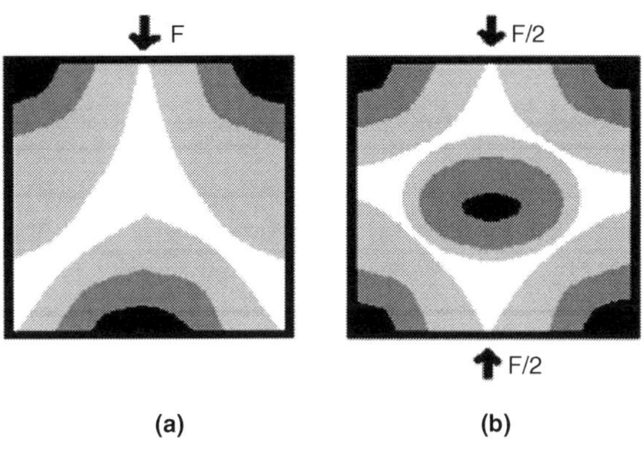

Figure 3.6 Typical density distribution of uniaxially die-pressed compacts (higher density regions are darker), (a) one-sided compaction, (b) two-sided compaction [11] / with permission of Elsevier.

Such factors include the used equipment, like the press mold with the die-play, and the clearance or gap between punches and the press mold. The mold's material type, filling behavior, and material distribution also affect the density distribution. Furthermore, the developing properties depend on operational factors such as compaction pressure, compaction speed, and the type of de-airing. The final compaction steps like type and speed of pressure relief and de-molding also have an impact on the density distribution and dimensional accuracy.

Different technological measures can improve the density distribution and uniformity of the compact. In general, a decreasing height-to-diameter-ratio h/d leads to smaller gradients. Furthermore, a mold roughness with asperities smaller than the particle size reduces the occurring friction. Decreasing pressure and stress lead to decreasing axial and radial friction. Two-sided pressing reduces the density gradients by positioning the neutral plane in the middle of the compact height. If further improved uniformity is required, isostatic compaction can be applied. Usually, the effect of different lubricating additives on the compaction of constant compositions with a maximum grain size of 6 mm does not influence the density gradients. However, the particle size distribution plays a key role in the developing density distribution in coarse grained ceramics. The influence of particle size distributions with a maximum grain size of 1 mm on properties of two-sided pressed compact ensures the fines have a lubricating effect between the coarse particles or aggregates. The shrinkage for increased amounts of fines (<200 μm) also increased, resulting in fired bulk densities correlating with the unfired strength and shrinkage. The cold crushing strength (CCS) can also be improved by an increased number of inter-particle contacts and, therefore, a higher density. Furthermore, an optimum combination of coarse and fines led to stronger sintering and ceramic bonding for refractory applications.

3.5 Uniaxial Pressing

Uniaxial pressing is the most common method for the fabrication of ceramic bodies. During uniaxial pressing, the powder is compacted in a rigid die by applying pressure along a single axial or double direction using a punch or piston. The process is easily automated; thus the method is especially suitable for mass production. Uniaxial pressing is primarily suitable for low L/D ratio and definite shapes with having compact mass.

Pressing high bodies usually results in the inhomogeneous packing of ceramic particles due to the friction of the die (wall and particle). A sprinkle of graphite powder in carbon-based refractory and lubricant oil in oxide refractories are used to reduce the pressing by friction screw or hydraulic press. Figure 3.7 shows the changes in packing pressure for different die geometries [12]. The contour plot for

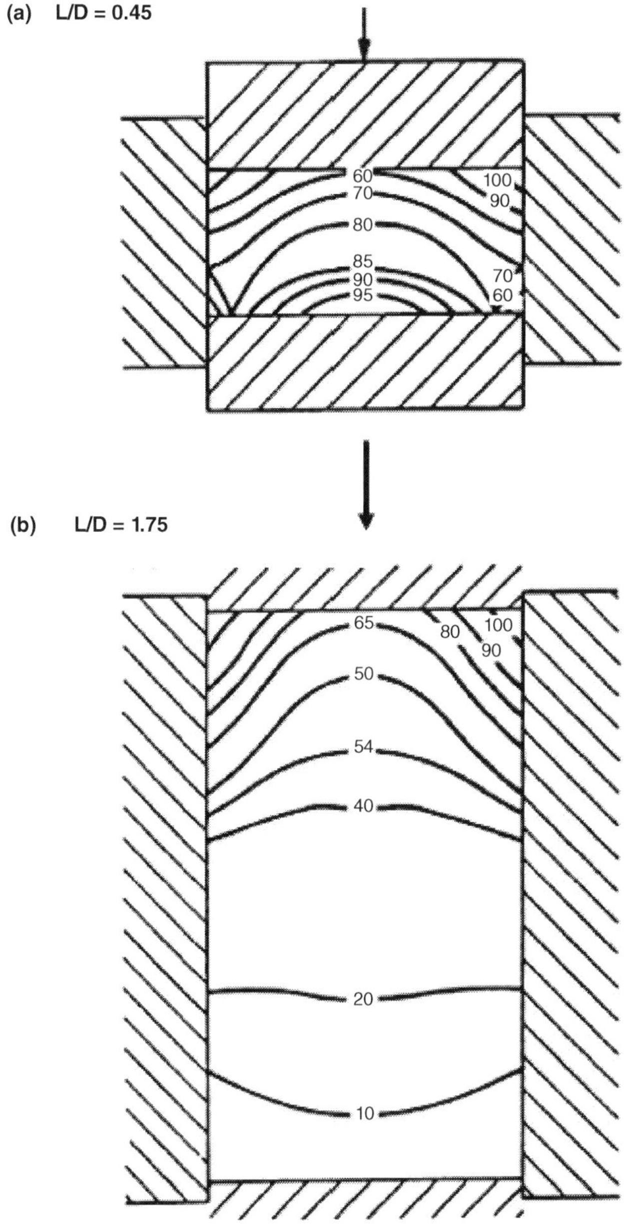

Figure 3.7 Pressure variations in the powder compact during uniaxial pressing for two ratios of height-to-diameter of the compact [12] / Taylor & Francis.

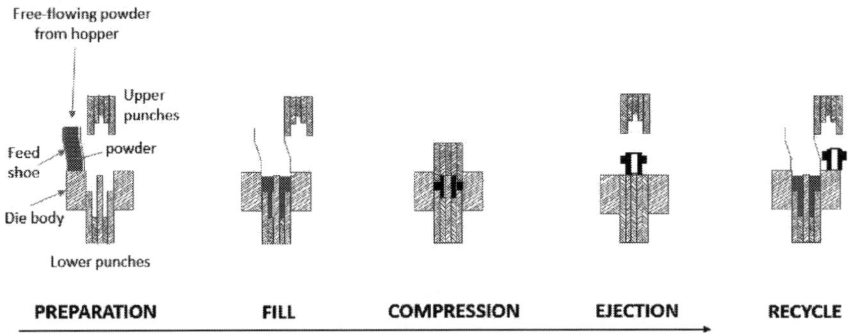

Figure 3.8 Schematic presentation of uniaxial pressing with segmented punches.

the material region indicates more discrete pressure distribution when L/D ratio is high; thus one should fix the mold design to obtain the desired product. During sintering, the lower-density areas either are not completely densified or shrink more than the surrounding areas.

Both can result in deformations or even cracks in the sintered body. While considering the pressing of refractory bricks, a few parameters are essential to synchronize, such as adequate packing of materials in mold, proper de-airing, movement rate of die punch, retention of pressing load, and number of strokes. There are several automatic presses taking care of such operations.

However, the density variation can decrease if the pressure is applied from both sides in double-acting presses. A schematic example of pressing with segmented punches is shown in Figure 3.8. The homogeneity of particle packing in complex-shaped powder compacts or compacts with irregular height can be improved by pressing with segmented punches. Further improvement can be obtained if the pressure of each punch segment is independently controlled. However, this process demands further modification when L/D ratio is very high and needs homogenous strength, including the wall of a hollow tube. A common technique is employing cold isostatic pressing (CIP) to achieve large components like a shroud, monoblock stopper, and subentry nozzle.

3.6 Cold Isostatic Pressing

The isostatic pressing follows either a cold or hot environment. The refractory industry adopted cold isostatic pressing to manufacture several components, especially black refractories for continuous casting. Simply, it is a process by which the refractory mix or particle mixes make homogenous compact before machining or sintering. Usually, the powder or graded particles are subjected

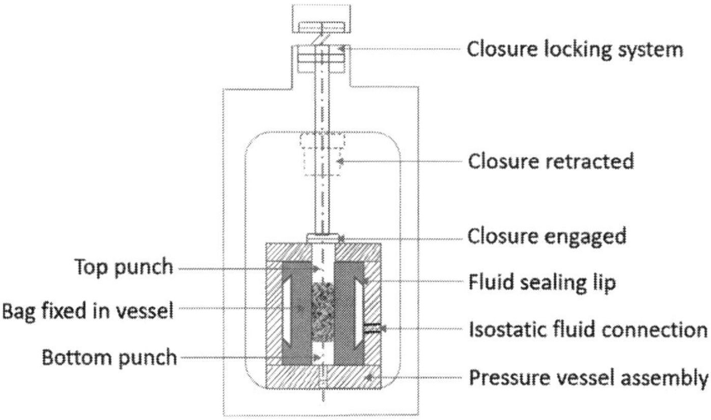

Figure 3.9 Schematic representation of dry bag CIP process.

to equal pressure from all directions; that is impossible to maintain homogenous density in the uniaxial pressing process. The pressing mechanism refers to hydrostatic pressing capable of producing high integrity and large cylindrical components or performs that exhibits little distortion or cracking when fired. A typical schematic representation of dry bag CIP is given in Figure 3.9.

Further, the CIP process can be classified as a wet bag and dry bag process, and their process and equipment arrangement are also different. The resulting compact in the wet bag process has a uniform density with low entrapped stress, making it machined in green condition (unfired). Uniform pressing facilities develop uniform stress distribution and eventually reduce or eliminate the distortion and machining that demand expensive diamond tools and difficult processes. A method sometimes used to reduce the cracking effect of entrapped air is to evacuate the tool and seal it before pressing.

The dry bag process lends itself to automation and is therefore suited for pressing relatively long runs of compacts at high production rates (compared with wet pressing). Dry CIP is preferentially an effective tool to make balls, hollow balls, rods, tubes (open and closed ends), nozzles, grinding wheels, spark plug insulators, fuse tubes, sodium-sulfur battery electrolyte, oxygen sensors, seals, crucibles, sodium lighting tubes, pump shafts, and dinnerware. The main advantages over wet bag pressing are high production rates and increased compact accuracy and consistency.

High production rates are achieved because minimal time is required to fill, pressurize, and eject a compact. Since the bag remains within the pressure vessel, it is necessary only to open the clamp mechanism sufficiently to fill and eject. The volume of isostatic fluid required to produce a compact in the dry bag system is much less than that for a wet bag because utilization of the pressure vessel is much more significant—typically 95% as against approximately 20% for a good wet bag system. Consequently, pressurization time and volume are reduced considerably; thus single-shot, hydraulically driven intensifiers are used. It provides the advantages of pulse-free controllable pressures without needing control valves on the high-pressure side.

Accuracy is improved in dry bag pressing because precious cast or molded bags are used, alignment of punches and mandrels is machine-controlled and accurate. The filling is at a consistent rate and constant volume, the pressurization cycle is repeatable, and the bags can have cavities "corrected" to improve the external shape of the compact. Another critical advantage of dry bag pressing is that it can be easily linked to other automatic processes.

Starting powder or material quality and tooling design imparts successful exploration of the process. When the method is used to its full capabilities, it can produce compacts requiring no further machining. Graded particles or powders mixed with no or low wax are filled in the impervious flexible and moldable bag and placed in a pressure chamber consisting of a water-based liquid medium with an isostatic pressure at room temperature. Near-net shapes are obtained after pressing, consisting of bag and core material. The obtained component may not have scope for tight tolerance limits like uniaxial pressing; thus critical size and shape eventually made by computer numerical control (CNC) machining either in green stage or pre-sintered or after final stage sintering.

Some flow control black refractories produced through CIP is shown in Figure 3.10. This process has several Pros and Cons.

Pros:

- Any simple and complicated shapes can be achieved through uniform hydrostatic pressure on all surfaces; thus maintaining consistent material density.
- It can avoid the die friction phenomenon as an elastic material is used for mold and applied pressure in all directions rather than uni-direction.
- Most green (unfired) compacts can be machined.
- Parts can be fired without drying.
- Shapes with high L/D ratios (greater than 10) of length to characteristic diameter can be densified and handled.
- Dual characteristics refractory particle mixers can press in two or more layers for complicated shapes.
- Parts with re-entrant and complex 3D geometries can be fabricated.
- An exact product dimension such as cylindrical bores or outside diameters can be controlled easily through elastomer machining by metal tooling inserts.
- Thin-walled sections and parts can be pressed.
- Elastomer tooling costs are low.

Figure 3.10 Flow control refractories. Credit: Krosaki Harima Corporation/ ATS Metalurji

Cons:

- Direct compaction without machining makes it challenging to obtain precise dimensions as there is no presence of any reference surface like a uniaxial press.
- Sometimes, it is difficult to pinpoint the presence of entrapped pore and their distribution during filling.
- Relatively rough surface finish compared to uniaxial die pressing, except certain zones that directly contact reference surface in hybrid type molds.
- Very low production rate compare to die pressing.
- Refractory materials have high hardness compared to elastomer mold materials; thus abrasion reduces the mold life.
- Mold tearing and poor sealing are an immense operational issue as it allows leakage in CIP vessel and material contamination with working fluid and material loss.
- Mold leakage may allow water penetration and hazardous reaction with material and quality deterioration.

3.7 Defects in Shaped Refractories

Composition, particle packing, and temperature variation may produce a different class of defects in both fired and unfired shaped refractories. While considering the fired refractories, large scale production depends on continuous tunnel kiln firing atmosphere and firing profile. In actuality, several controlling parameters are responsible for making defect-free fired refractories in tunnel kiln. The predominant defects are underfiring, dimensional intolerance, overfiring, mute bricks, cracks, black core and embossing, white bricks, bloating, black bricks, efflorescence, and limestone burst.

- Underfiring

Appropriate heating–isothermal treatment-cooling cycle is critical to attaining defect-free fired refractory bricks. Thus, one can pick up the temperature from the phase diagram and subsequent lab scale and pilot scale trial before production. Unusual low firing temperature selection or less retention time at peak temperature may result in immature underfiring bricks. Low temperature and the rapid heating cycle may not be sufficient to initiate or complete the phase transformation or liquid phase formation to develop bonding in the entire matrix, resulting in a dull appearance with poor strength. Such defects can be eliminated when firing at a high-temperature with adequate retention time.

- Dimension intolerance

Mold design and machining is the first step to achieving the desired dimension. It is worth mentioning that it is mandatory to consider the firing shrinkage or expansion behavior before mold fabrication. Usually, the dimension defects are caused by missing edges, corners, under/overfiring, and particle removal after sintering. The plausible reason for such defects is attributed to

- Insufficient matrix strength because of bonding systems
- Uneven mixing of refractory particles, additives, and binders
- Unexpected high external force during transportation
- Segregated particles after drying of mixes
- Insufficient mold pressure and dwell time
- Low firing temperature
- Improper kiln loading

The mold material possesses an optimum hardness to avoid expansion during pressing despite the perfect mold dimension that results in distortion. Irregular dimension may affect the level and lining life performance through hot metal/slag penetration or bulging.

- Overfiring

High heating temperature, high holding time, and fuel feeding may enhance the kiln temperature, resulting in unexpected liquid phase formation and a high degree of shrinkage. Such bricks have high density and strength but are unqualified for application. The size variation may severely affect the completion of the lining. The overfired brick may have an unexpected phase that may facilitate the formation of a weak zone and reduce the performance temperature in the presence of foreign particles during application. High precision peak temperature control and proper cooling schedule if the holding time is too long can avoid such defects.

- Mute bricks

Mute bricks are found throughout the kiln, which may happen due to operational error. It is essential to check, analyze, and reject such bricks to avoid premature lining failure. The following points can be considered to avoid such defects:

  - Store dried green bricks in a dry environment and regulate the kiln flue gas discharging in time to avoid reappearance
  - Proper temperature selection and heating schedule of high moisture content brick without alteration of the environment
  - Reasonable cooling rate, not very fast or very slow

- Cracks

There are two types of cracks, and the causes are different.

- Hairline crack

Rapid cooling may severely affect and initiate such defects through an equilibrium temperature distribution on the outer and core surface temperatures. The thermal gradient expedites the surface to become cool and shrink, whereas the core is still hot and expands, resulting in fine hair cracks due to internal stress. Strictly controlling the car and air movements, keeping stable firing temperatures, and optimum cooling rates can eliminate hairline cracks.

- Burst

The rapid preheating expedites the fast moisture evaporation in a short period, and bursting occurs when the vapor pressure exceeds the strength of the brick at that particular processing temperature. A control water vapor removal at an optimum preheating rate and extended time with a slow temperature rise can eliminate this defect.

- Black core and embossing

The black core formation and embossing occasionally happen when bricks are fired in a confined zone. During the firing process, sometimes green bodies experience insufficient air; thus reducing atmosphere, which enables the reduction of impurities like high valence *Fe* into low valence and bricks form spot green or blue or black in color. Sometimes high impurity content may change to the black color of the fired brick, called black brick. Low valence iron oxide has solid fluxing action; thus forming a more liquid phase and high shrinkage. Usually, specific considerations can assist in reducing such defects:

- Minimization of the brick stack contact and overlapping area
- Extension of holding time and burn out the available fuel in the cavity, if any
- Low temperature curing with a controlled heating rate can minimize the early stage of vitrification
- Ensure the presence of enough air for the complete combustion of fuel
- Enhancing the ventilation rate in a kiln and adequate firing of bricks homogenously

• White brick

The presence of sulfides, either from refractory raw materials or low-quality fuels, are prime resources for forming sulfur dioxide by combustion reaction. With time, this dioxide transforms into trioxide in an oxidizing atmosphere. Sulfur-enriched refractory surface absorbs gas during preheating and further transforms to sulfates after high-temperature firing, resulting in a white refractory surface. Quality fuel, reducing moisture content in green bodies, lowering the stack, and increasing ventilation space can reduce this defect.

• Bloating

Overfired bricks are prone to develop a liquid phase early, resulting in vitrification, but the gas formation in core is still active and cannot escape the gaseous phase through a high viscous molten phase. A gradual accumulation of evolved gas and sudden escaping of the gas pressure results in bloating. Control over the internal combustion substances in brick, firing schedule, long time holding at low temperature, the slow firing process, and calculated excess air may reduce such defects in shaped refractories.

• Efflorescence

Recrystallization of soluble salt solutions on the brick surface is commonly known as efflorescence. Extensive efflorescence defects may appear as scaly peeling, affecting the durability of the brick. Taking specific measures such as high fineness of particles and extended firing time can reduce or eliminate such defects.

• Limestone burst

Limestone burst is not a common phenomenon for all refractory bricks, instead, it can be found when a system has a certain amount of lime as primary constituents or impurities. After refractory firing completion in a kiln, the lime is highly prone to hydration in the presence of vapor in the atmosphere. Transformation of such reaction accelerates volume expansion ~2.5 times of particles and develops enormous tensile stress that results in bursting with the appearance of flaking and crushing. Such defects may be eliminated by adopting particle size usually under 1 mm, and enhancing the firing temperature to promote the reaction with silica and the formation of silicates. Eventually, this measure can help minimize the defect and improve the mechanical response of the brick. In this context, a detailed understanding of firing and hence sintering is necessary and discussed in the Chapter 4.

References

1 Sarkar, D. (2019). *Ceramic Processing: Industrial Practices*. CRC Press.

2 Kingery, W.D. *Ceramic Fabrication Process, Part IV*. New York: Technology Press, Cambridge, and John Willey & Sons. 1976.

3 Cruz-Matías, I., Ayala, D., Hiller, D., Gutsch, S., Zacharias, M., Estradé, S., and Peiró, F. (2019). Sphericity and roundness computation for particles using the extreme vertices model. *J. Comput. Sci.* 30: 28–40.

4 Krumbein, W.C. and Sloss, L.L. (1963). *Stratigraphy and Sedimentation*, 2e. San Francisco: Freeman and Company.

5 Kendall, K., Alford, N.M., and Birchall, J.D. (1987). *Nature* 325: 794.

6 An, L., Zhang, D., Zhang, L., and Feng, G. (2019). Effect of nanoparticle size on the mechanical properties of nanoparticle assemblies. *Nanoscale* 11: 9563.

7 Bridgwater, J., Foo, W.S., and Stephens, D.J. (1985). Particle mixing and segregation in failure zones–theory and experiment. *Powder Technol.* 41: 147–158.

8 Alonso, M., Satoh, M., and Miyanami, K. (1991). Optimum combination of size ratio, density ratio, and concentration to minimize free surface segregation. *Powder Technol.* 68: 145–152.

9 Rahaman, M.N. *Ceramic Processing and Sintering*. Mercel Dekker. 2003

10 Reed, J.S. (1995). *Principles of Ceramics Processing*. John Wiley & Sons, Inc. (US).

11 Fruhstorfer, J., and Aneziris, C.G. (2017). Influence of particle size distributions on the density and density gradients in uniaxial compacts. *Ceram. Int.* 43 (16): 13175–13184.

4

Degree of Ceramic Bonding

4.1 Introduction

When a particle compact transforms into a densified body with the help of thermal energy, the process is known as sintering. Sintering expedites the formation of ceramic bond within grains and the matrix; however, the degree of such bonds are influenced by sintering characteristics that strictly depends on several parameters, including starting particle size, temperature, time, atmosphere, etc. Sintering consists of densification and grain growth. Sintering is often used to synthesize or process basic elements of materials in science and engineering. Depending upon the shaping techniques applied to the green product, the sintering conditions need to be adjusted accordingly, which may vary the sintering techniques. Various techniques and processing variables can be altered in the sintering step to achieve desired microstructure and properties. Microstructure control means managing grain size, sintered density size, and distribution of other phases, including pores.

There are two processes into which it is divided, solid-state and liquid-state sintering. There are many pros and cons to liquid phase sintering. Some of the pros are easy to control microstructure by controlling the composition and the amount of liquid phase, one of the methods that preferably has lower processing costs, and can be used for materials that need properties of grain boundary phase. Cons are mostly a reduction in the mechanical properties of the material. Removing isolated pores takes more time and follows almost all the sintering time.

In brief, the driving force of the sintering is the reduction of total interfacial energy that actually needs to encounter refractory composition design.

Total interfacial energy of powder compact = γA
γ = specific surface (interface energy)
A = total surface (interface) area

Fundamental Design of Steelmaking Refractories, First Edition. Debasish Sarkar.

Reduction of total energy expressed as:

$$\Delta(\gamma A) = \Delta\gamma A + \gamma\Delta A \tag{4.1}$$

For the above terms sintering in general means

$\Delta\gamma$ = represents the reduction in internal energy due to densification

ΔA = represents the reduction in interfacial area due to grain coarsening

In solid-state sintering, $\Delta\gamma$ is related to replacing solid vapor interfaces (surface) with a solid–solid interface. For the precursors or powder, the particle size is between 0.1 and 100 μm, and the corresponding total surface energy is near 500–0.5 J/mole. This energy is small relative to the energy during the oxide formation, so to utilize this (low energy change) for the desired sinter microstructure, we need to further explain controlling the sintering variables. The sintering variables are divided into material variables and process variables, which are summarized in Table 4.1 as given below:

Table 4.1 Variables affecting sintering.

Variables	Type
Raw material • Physical properties	Particle shape, size distribution, agglomeration behavior, probability interaction and mixing, mixing sequence, etc.
• Chemical properties	Material composition, percentage impurities, defects and nonstoichiometry, elemental homogeneity, etc.
Process	Temperature, heating and cooling rate, holding time, pressure, atmosphere, etc.

Homogeneity and distribution of different particles are essential in obtaining the desired microstructure during firing or sintering.

4.2 Importance of Heating Compartment

4.2.1 Loading and Heating

A continuous tunnel kiln is always expected to achieve steady-state operating temperature regimes that eventually help continuous production and gradual conversion from pre-fired to fired brick with substantial ceramic bonding. Usually, the tunnel kiln is heated up after immediate installation and keeps operating year after year. The loaded car displacement is simply succeeding one after another with a fixed moving rate throughout the kiln. The temperature zone length is critical to provide satisfactory temperature and holding time and results in desired phase and removal of residual energy in the cooling zone. The entire structure

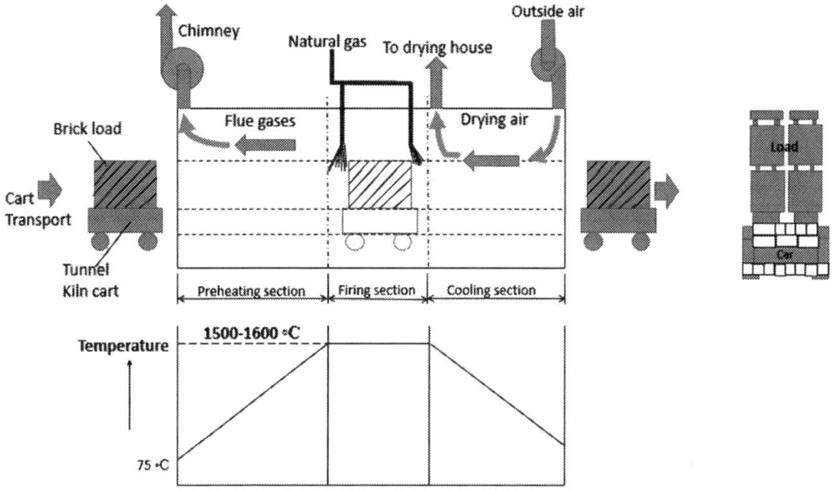

Figure 4.1 A typical loading, movement, and heating schedule in a tunnel kiln.

remains hot and prerequisite insulation thickness can allow minimum heat loss through the side and top wall.

Figure 4.1 represents the schematic view of the tunnel kiln along with the probable heating and cooling profile. The front view of a typical car loading that enters from one side consists of green brick, passes through different hot zones, and leaves sintered bricks from another end. Usually, prime three zones are designed to perform thermal treatment and are referred to as preheating, firing, and cooling zone. The scheme also shows the probable flows of gases after combustion and air circulation, heating, cooling, car loading, and movement. A quick review of dimensional features of the kiln corresponding to the refractory plant is shown in Table 4.2.

Table 4.2 Dimensional characteristics of the tunnel kiln [m].

Useful width 2.5
Useful height 3.2
Kiln length 70.20
Length of the preheating zone 15.20
Length of the firing zone 24.80
Length of fast cooling zone 4.50
Length of normal cooling zone 25.70
Wall thickness in the preheating zone 0.72
Wall thickness in the firing zone 0.91
Wall thickness in the cooling zone 0.52

4.2.2 Heat Distribution

Modern tunnel kilns have stable refractory walls and use lighter support materials to reduce thermal inertia. A kiln is mainly divided into three components:

1) Preheating zone
2) Firing zone
3) Cooling zone

Fuel combustion with air gradually attains steady state thermal equilibrium in the furnace, and this thermal energy has four critical movements during the completion of the firing process:

i) The load and the car utilize the energy to remove the humidity from the green body in the preheating zone and are also used to mediate the probable physical and chemical changes in raw material.
ii) Energy in the flue gases released from the chimneys.
iii) The energy in the form of heat is lost through the kiln envelope by conduction, followed by convection and radiation from the walls, ceilings, and conduction from the base.
iv) The energy associated with the cooling.

A series of energy transfers occur within a tunnel kiln, mainly conduction, convection, and radiation from different elements present inside the kiln. For these elements, equations on the energy conservation law equation are:

$$q''_k . A_k + q''_c . A_c + q''_r \cdot A_r + E_g + E_{reactions} = 0 \tag{4.2}$$

However, more details are available in Chapter 1.

The first three terms represent conduction, convection, and radiation (here each heat flux is multiplied by the area dependent on the interface). The conduction in the longitudinal direction is neglected as it is small compared to other directions. The fourth term represents the energy transfer due to the mass displacement occurring at the inlet and outlet of the tunnel. The fifth and last terms represent the energy generation rate from physical and chemical reactions. In a recent article, an analysis represents the temperature variations in the load as the kiln car traversed through the tunnel kiln [1]. In operation, there is always the possibility of obtaining different production rates with different degrees of defects. It can be overcome through a predefined firing profile of the kiln.

4.2.3 Temperature Conformity

Attainment of predefined temperature in different zones, from preheating to firing to a cooling zone, is the most critical parameter that eventually helps maintain temperature conformity and defect-free products. While temperature increment of the loaded car from the heating resource is a concern, convection and radiation are predominant, but conduction and convection facilitate temperature conformity of loaded material on the car. The kiln efficiency depends on several parameters, and the specific energy consumption depends on it. Some important parameters and their effect on energy variation are both pointed out.

- Density: High dense material provides a higher number of molecules in a unit volume and helps increase thermal conductivity and heat transfer through conduction. With increasing temperature, both the number of free electrons and lattice vibrations increase; thus, the material's thermal conductivity is expected to increase.
- Thermal conductivity: High thermal conductivity of a kiln car facilitates a high heat transfer rate and more energy loss.
- Dimension: Sufficient stack height results in lower energy loss because of the smaller temperature difference.
- Car emissivity: It depends on the refractory quality used to construct the kiln car. Usually, fireclay, alumina, and magnesite have a favorable and low emissivity; thus these are common materials to construct the car, as shown in Figure 4.2. However, SiC is rarely employed because of high emissivity in any temperature horizon.
- Heat transfer coefficient: As the mass flow rate of the gas is low, the heat transfer coefficient varies in the small range. It is due to changes in Reynolds number and Prandtl number, which are the combustion gas properties that change every time.
- Axial heat flux: In the firing zone, the temperature of the gases is always high during the operation; thus the radiation flux dominates the convective flux. However, the convective flux dominates during the cooling period because the emitting gases are absent here, and no combustion occurs.

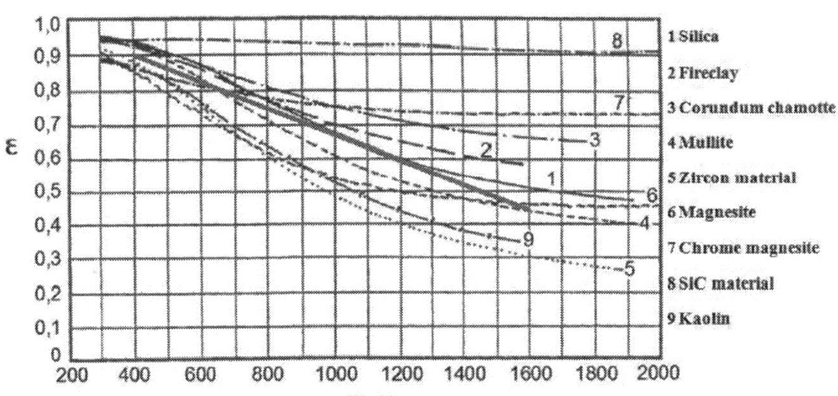

Figure 4.2 Mean emissivity of ceramics [2] / SAGE Publications / Public domain CC BY 3.0.

It is worth noting that effective heat exchange assists in maintaining the high surface temperature of the kiln car compared to the bottom of the kiln, which is attributed to the car's low conductivity. In a stipulated time, the car consumes sufficient surface temperature similar to the gas temperature for the desired sintering, phase transformation, densification, etc. Material sintering is a classic study, and well-established facts are documented in the literature. Few fundamental concepts provide insight into the gradual conversion process during the firing or sintering of refractories.

4.3 Initial Stage Sintering

Solid-state sintering consists of three overlapping stages (initial, intermediate, and final), which can be explained using a different model. Two particle models usually exhibit shrinkage up to 3% for the initial sintering stage. In contrast, the maximum densification is attributed to the minimization of channel pores during an intermediate sintering stage. The isolated pore model explains the final stage.

4.3.1 Sintering Mechanisms of Two-particle Model

Suppose it is assumed that powder compacts are made of spherical particles of similar size. In that case, the driving force of sintering in the initial stage is mainly driven by the neck geometry and size of two spherical particles. In two-particle model theory, two consecutive phenomena can be encountered:

i) In the beginning, starting material becomes interactive in which the spheres have just begun to stick together, and there is no possibility of shrinkage.
ii) In the second stage, just the later step of the initial stage of sintering, shrinkage is found to be mass transferred from the particles to the neck region.

$$r \approx \frac{x^2}{2a} \tag{4.3}$$

(Without shrinkage, Figure 4.3a)

$$A \approx 2\pi x 2r = \frac{2\pi x^3}{a} \tag{4.4}$$

$$V = \int A dx = \frac{\pi x^4}{2a} \tag{4.5}$$

Where a is the particle radius, x is the neck radius, r is the radius of the neck curvature, A is the neck area, and V is neck volume for particles that do not change size during the sintering and have 180° dihedral angle between them.

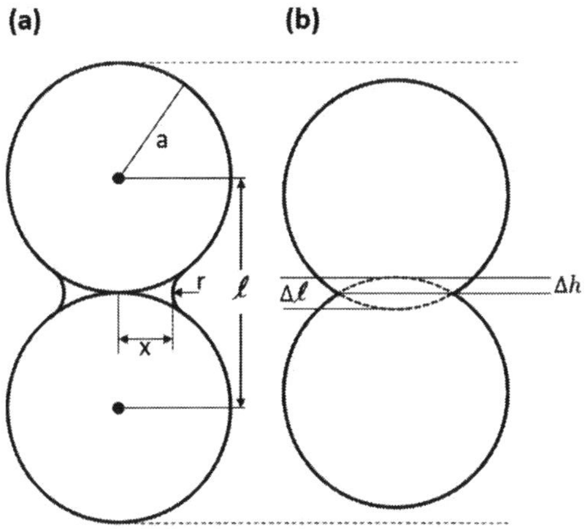

Figure 4.3 Two particles (a) without shrinkage, (b) with shrinkage.

$$r \approx \frac{x^2}{4a} \tag{4.6}$$

(With shrinkage, Figure 4.3b)

$$A \approx \frac{\pi x^3}{a} \tag{4.7}$$

$$V \approx \frac{\pi x^2 2r}{2} = \frac{\pi x^4}{4a} \tag{4.8}$$

In Equations (4.5) and (4.8), shrinkage exhibits half volume compared to the non-shrinkage condition. The driving force for the material transport depends on the neck geometry; this geometry is due to differences in bulk pressure, vacancy concentration, and vapor pressure. Such material transport occurs under the influence of many parallel actions of various mechanisms. These mechanisms are mentioned below in Table 4.3.

Only some processes contribute toward densification and shrinkage among the mechanisms mentioned above in Table 4.3. Refractory materials are mostly crystalline; thus the material flows in bulk from the grain boundary by atomic movement during the densification of crystalline compacts. The interparticle contact does not reduce as densification during shrinkage, but the neck area increases as the materials are transported from the particle surface to the neck. The grain boundary sintering perception becomes predominant when the pores at the grain boundary shrink, but the pores inside the grains remain intact.

Table 4.3 Material transport mechanism during sintering.

Material transport mechanism	Material source	Material sink	Related parameter
Lattice diffusion	Grain boundary	Neck	Lattice diffusivity, D_l
Grain boundary diffusion	Grain boundary	Neck	Grain boundary diffusivity, D_b
Viscous flow	Bulk grain	Neck	Viscosity, η
Surface diffusion	Grain surface	Neck	Surface diffusivity, D_s
Lattice diffusion	Grain surface	Neck	Lattice diffusivity, D_l
Gas phase transport Evaporation/Condensation Gas diffusion	Grain surface	Neck Neck	Vapor pressure difference, D_p Lattice diffusivity, D_g

4.3.2 Atomic Diffusion

The driving force for the sintering is attributed to the difference in the bulk pressure, vapor pressure, and vacancy concentration. These differences help in the diffusion, which means the movement of atoms due to the concentration gradient. The atomic diffusion can be of two types,

Case A: Atom movement due to vacancy diffusion under a vacancy concentration gradient.
Case B: The movement of atoms under stress gradient

The equations related to *vacancy movement* are:

$$J_{vac} = -D_v \nabla C_v = -\frac{D_v C_{v,\infty} V'_m}{RT} (\Delta P) \frac{1}{L} \tag{4.9}$$

Where,

J_{vac} = the vacancy flux
V'_m = molar volume of vacancies
D_v = the vacancy diffusion coefficient
C_v = vacancy concentration per unit volume
$C_{v,\infty}$ = the equilibrium vacancy concentration in the material with a flat surface
L = diffusion distance, R = Gas Constant, ΔP = pressure difference

When *atom movement* occurs:

$$J_{atom} = -C_a B_a \nabla(\mu_a - \mu_v) \tag{4.10}$$

C_a = atom concentration per unit volume
B_a = atom mobility
μ_a = chemical potential of the atoms
μ_v = chemical potential of the vacancy

The local equilibrium vacancy concentration does not facilitate the vacancy movement if their formation and destruction occur freely. In this situation, the vacancy movement can be represented as:

$$J_{atom} = -C_a B_a \nabla \mu_a = -C_a \frac{D_a}{RT} \nabla \sigma V_m = -\frac{D_a}{RT} \nabla \sigma = -\frac{D_a}{RT} (\Delta \mathrm{P}) \frac{1}{L} \tag{4.11}$$

Where, J_{atom} is atom flux, D_a is the atom diffusion coefficient, V_m is the molar volume of the solid, σ is pressure, and the rest of the symbols represent the same as the previous equations.

The above equation describes the kinetics of sintering due to diffusion. Usually, the diffusion gradient is consistent; therefore, the required time to attain equilibrium is negligible. The sintering mechanism preferentially is controlled by atomic movement and not by the reaction kinetics at the interface. However, the stress is well distributed around the grain boundaries, so the atomic movement from grain boundaries to the neck region becomes easy. In the sintering process, no additional energy is required for the attachment and detachment of atoms when the source is the grain boundary, and the sink is the particle surface. But such a hypothesis does not match a real system where extensive faceted surfaces exist in grain boundaries. Imperfect sources demand enormous energy to remove self atoms or secondary phase particles from grain boundaries. The driving force for atomic movement alters if the dislocation at the grain boundary affects the atom or vacancy movement.

4.3.3 Sintering Kinetics

Usually, the neck size depends on the sintering time. Certain assumptions are essential to explain the neck growth during the initial stage of sintering; these are:

- The neck size should be smaller than the particle size ($x/a < \sim 0.2$), where, a is the particle radius, x is the neck radius.
- The angle between the neck center and the neck surface should be less than 1 ($\theta \ll 1$).

The undercutting process occurs when the material moves from the particle surface to the neck, so the surface near the neck changes its shape. The driving force for sintering is reduced when this undercutting occurs. The different processes where material transport occurs are:

i) Lattice diffusion from grain boundary

In this mechanism, the grain boundary is a site of vacancy annihilation. The role of the grain boundary is similar to Nabarro Herring Creep. The atom movement from one grain boundary to another occurs, analogous to the creep where the grain boundary is under compressive stress, and the other is under tensile stress. In lattice diffusion, the atoms move from grain boundary to neck, where the neck region experiences tensile stress and grain boundary under compressive stress, so it is evident to the formation of a stress gradient between the neck center and neck surfaces. Thus, shrinkage and densification occur along with the neck growth and center-to-center approach between two particles.

For neck growth:

$$dV / dt = JAV_m \tag{4.12}$$

or,

$$\frac{dV}{dt} = \frac{\pi x^3}{a}\frac{dx}{dt} = \frac{D_l}{RT}\nabla\sigma A V_m \approx \frac{D_l}{RT}\left(\frac{\gamma_s}{r} - \frac{1}{x}\right)\frac{\pi x^3}{a} V_m \tag{4.13}$$

$$x^4 = \frac{16\, D_l\, \gamma_s\, V_m a}{RT} t \tag{4.14}$$

For shrinkage:

$$\frac{\Delta l}{l} = \frac{r}{a} = \frac{x^2}{4a^2} = \left(\frac{D_l \gamma_s V_m}{RTa^3}\right)^{1/2} t^{1/2} \tag{4.15}$$

D_l = Lattice diffusion coefficient; l = sample size

Shrinkage occurs when there is material movement from grain to neck surface regions.

ii) Grain boundary diffusion from grain boundary

It is similar to Coble creep, where grain boundary diffusion contributes to diffusion creep. The neck growth and shrinkage equations represent this phenomenon:

For neck growth:

$$\frac{dV}{dt} = \frac{\pi x^3}{a}\frac{dx}{dt} = \frac{D_b}{RT}\left(\frac{\gamma_s}{r} - \frac{1}{x}\right) 2\pi x \delta_b V_m \tag{4.16}$$

$$x^6 = \frac{48\, D_b\, \delta_b \gamma_s\, V_m\, a^2}{RT} t \tag{4.17}$$

For shrinkage:

$$\frac{\Delta l}{l} = \frac{r}{a} = \left(\frac{3D_b \delta_b \gamma_s V_m}{4RTa^4} \right)^{1/3} t^{1/3} \tag{4.18}$$

D_b = grain boundary coefficient; δ_b = diffusion thickness of grain boundary diffusion.

Grain boundary diffusion-assisted material transfer to the neck surface may facilitate grain growth if the material is not sufficiently redistributed rapidly; thus fast redirection can control the grain growth.

iii) Viscous flow

The viscous flow mechanism, as stated by Frenkel, can be used for sintering of viscous materials like glass, so if the material has a Newtonian fluid-like character, then the shrinkage and neck growth is described as:

Equations for neck growth and shrinkage are:

Neck growth:

$$\varepsilon^{\cdot} = \frac{1}{h}\frac{dh}{dt} = \frac{1}{\eta}\Delta\sigma = \frac{1}{\eta}\frac{\gamma_s}{r} \tag{4.19}$$

$$x^2 = \frac{4\gamma_s a}{\eta} t \tag{4.20}$$

For shrinkage:

$$\frac{\Delta l}{l} = \frac{h}{a} \approx \frac{\gamma_s}{\eta a} t \tag{4.21}$$

Where η represents the viscosity of the material, h is the penetration depth of one particle into another, which is very close to r.

iv) Surface diffusion from particle surface

Here, diffusion occurs through the surface via atomic movement from the sphere surface to neck surface; no shrinkage occurs. It is assumed that a stress gradient is derived from the capillary pressure on the neck surface and is found up to a distance equal to the neck curvature radius.

$$\frac{dV}{dt} = \frac{2\pi x^3}{a}\frac{dx}{dt} = JAV_m = \frac{D_s}{RT}\frac{\gamma_s}{r}\frac{1}{r}\left(2\pi x 2\delta_s\right)V_m \tag{4.22}$$

$$x^7 = \frac{56 D_s \gamma_s \delta_s V_m a^3}{RT} t \tag{4.23}$$

Where D_s represents the surface diffusion coefficient, and δ_s represents the diffusion thickness of the surface diffusion.

v) Lattice diffusion from particle surface

The neck growth occurs by lattice diffusion, and the material is transferred from the particle surface without any shrinkage. The equation for neck growth is as follows:

$$\frac{dV}{dt} = \frac{2\pi x^3}{a}\frac{dx}{dt} = \frac{D_l}{RT}\nabla\sigma A V_m \tag{4.24}$$

$$x^5 = \frac{20 D_l \gamma_s V_m a^2}{RT} t \tag{4.25}$$

vi) Evaporation/Condensation

The atoms which are evaporated from the surface of the sphere were condensed in the neck region. The mechanism for gas-phase transport depends upon the distance between the evaporated and the condensed region. If the distance is smaller than the mean free path of gas atoms, the evaporation/condensation overpowers. In contrast, gas diffusion is the major part of the slower gas reactions at the interface, only. Gas-phase transport is a mechanism for material transport and not for sintering mechanisms. Neck growth depends on gas adsorption, Langmuir equation says:

$$\frac{dx}{dt} = \frac{m}{d} = \frac{\left(p_\infty \frac{\gamma_s}{r}\frac{V_m}{RT}\right)\left(\frac{M}{2\pi RT}\right)^{1/2}}{d} \tag{4.26}$$

$$x^3 = \sqrt{\frac{18}{\pi}\frac{p_\infty \gamma_s}{d^2}}\left(\frac{M}{RT}\right)^{3/2} at \tag{4.27}$$

Where $d\ (= M/V_m)$ is the material density.

vii) Gas diffusion

The neck growth is controlled by gas diffusion because it is the slower step with respect to the interface reaction.

$$A\frac{dx}{dt} = A D_g \nabla C V_m \tag{4.28}$$

$$\frac{dx}{dt} = D_g \frac{V_m}{(RT)^2}\frac{\gamma_s}{r^2} p_\infty V_m \tag{4.29}$$

$$x^5 = 20p_\infty D_g \gamma_s \left(\frac{V_m}{RT}\right)^2 a^2 t \tag{4.30}$$

Where D_g is the diffusivity of the gas atoms, p is the vapor pressure of the solid.

4.3.4 Sintering Variables

Smaller particle sizes and higher sintering temperatures may lead to higher densification or sintering rates. A wide range of refractory compositions, including oxides, non-oxides, metal powders, and other chemical entities, have a multidirection effect on obtaining the desired density through sintering.

i) Particle size

While considering the effect of particle size, it is assumed that the resulting microstructure is similar for all the powders of different sizes as a similar sintering mechanism is applied throughout the sintering process. Powders have similar shapes of different sizes and are also subjected to following the same sintering schedule and mechanism. Thus, Henry's scaling law predicts the sintering time required for a different set of particles of similar shapes to reach the same degree of sintering. For example, if we consider similarly sized particles of two different sizes a_1 and a_2 (xa_1), so the required sintering time t_1 and t_2, respectively, will be interrelated as:

$$t_2 = (x)^\alpha t_1 \tag{4.31}$$

Where α is an exponent, for lattice diffusion α is 3, which means that for the same degree of sintering, the sintering times are related as x^3. The general expression for sintering is:

$$t = \frac{V}{JAV_m} = \frac{L^3}{((D_l / RT)(2\gamma_s / L)(1 / L))L^2 V_m} \propto L^3 \tag{4.32}$$

Here, (x/a) is constant to have a consistent sintering degree, a^{m-n} t must be constant, so, $\alpha = m - n$.

ii) Temperature

Temperature and some properties like viscosity, diffusivity, etc., are exponentially related. These variables are mostly considered because sintering is a process that is activated thermally, and these variables are sensitive to temperature. The sintering time (t) is mostly related to temperature (T):

$$\ln t \,\alpha\, (1/T) \tag{4.33}$$

but, for different sintering mechanisms, the relationship varies.

iii) Pressure

The densification rate primarily enhances with increased pressure (including external pressure) from hot pressing, gas pressure sintering, etc. The driving force of sintering is the difference in capillary pressure due to curvature difference (internal pressure), so when external pressure is applied, the total pressure (internal and external) increases.

$$P_t = \frac{\gamma_s}{r} + P_{appl.} f\left(\rho, geo\right) \tag{4.34}$$

Where $f(\rho, geo)$ represents a function of relative density and particle geometry.

iv) Chemical composition

The sintering is diffusion controlled process, so the sintering rate can be increased by increasing the vacancy concentration. In the case of ionic compounds, the increase in vacancy concentration can be regulated by adjusting, comparing, and calculating dopant addition. The atomic size of sintering species has also influenced the degree of sintering.

4.3.5 Limitations of Initial Stage of Sintering

A local equilibrium of atom with capillary pressure is maintained everywhere, and a dihedral angle of 180° is common consideration for the sintering equations. However, some limitations include; earlier sintering mechanisms being determined by measuring neck growth and shrinkage kinetics. However, results were not appropriate as the experimental condition favored simultaneous operative two or three mechanisms, so the exponent (n) would vary, and concluded that experiments could not determine n exactly. Thus, it is expected that the experiment design must be accounted to regulate only one operating mechanism during diffusion assisted sintering.

4.4 Intermediate and Final Stage Sintering

4.4.1 Intermediate Stage Model

Coble's microstructure model is the predominant consideration in which bcc-packed tetrakaidecahedral grains create cylindrical pores along the grain edges and follow equal shrinkage of pores in radial directions. This model is limited in terms of real-life sintering, but it allows the evaluation of the effect of sintering variables on sintering kinetics. If the edge effect is not considered, then the atom flux toward the cylindrical pore is similar to the thermal flux toward the pores under the influence of heated wire with a diameter equal to grain boundary diameter. If λ = heat conducting, and J_{heat} = thermal flux per unit length:

$$J_{heat} = -\lambda(dT / dx) \tag{4.35}$$

On solving we get:

$$J_{heat} = 4\pi\lambda\nabla T$$

J_{atom} = atom flux per unit

$$= -\left(\frac{D}{RT}\right)\left(\frac{d\sigma}{dx}\right)$$

$$J_{atom} = 4\pi\left(\frac{D}{RT}\right)\Delta\sigma \tag{4.36}$$

For this J_{atom} two mechanisms are available:

i) Lattice diffusion

$$\begin{aligned}\frac{dV_p}{dt} &= -\frac{14}{2}2rJ_{atom}V_m \\ &= -14r4\pi\frac{D_l}{RT}\left(\frac{\gamma_s}{r}\right)V_m\end{aligned} \tag{4.37}$$

$$\frac{dP_v}{dt} = \frac{\frac{dV_p}{dt}}{l^3 8\sqrt{2}} = -\frac{d\rho}{dt}$$

$$\frac{d\rho}{dt} = \frac{7\pi D_l\gamma_s V_m}{l^3 RT\sqrt{2}} = \frac{336D_l\gamma_s V_m}{RTG^3} \tag{4.38}$$

Where l is the length of a grain edge, ρ is the relative density, G is the grain diameter.

ii) Grain boundary diffusion

$$\frac{dV_p}{dt} = -\frac{14}{2}4\pi\frac{D_b}{RT}\frac{\gamma_s}{r}\delta_b V_m \tag{4.39}$$

$$\frac{dP_v}{dt} = \frac{\frac{dV_p}{dt}}{l^3 8\sqrt{2}} = -\frac{d\rho}{dt}$$

$$\begin{aligned}&= -28\pi\frac{D_b\gamma_s\delta_b V_m}{RT}\frac{1}{l^3 8\sqrt{2}}\frac{1}{\sqrt{P_v}}\sqrt{\frac{12\pi}{l^2 8\sqrt{2}}} \\ &= -\frac{854\left(D_b\gamma_s\delta_b V_m\right)}{RTG^4}\left(\frac{1}{P_v}\right)^{1/2}\end{aligned} \tag{4.40}$$

Where δ_b represents the diffusion thickness.

4.4.2 Final Stage Model

In order to explain the final stage model, Coble considered tetrakaidecahedral grains with spherical pores of radius r_1 at each grain corner and concentric sphere diffusion of atoms from a distance r_2 to the pore surface, as shown in Figure 4.4. The atom flux passing through any concentric sphere surface with a radius of r, J_{total}, is constant.

$$J_{total} = const. = -4\pi r^2 \frac{D_l}{RT}\frac{d\sigma}{dr} \tag{4.41}$$

$$J_{total} = -4\pi \frac{D_l}{RT} \Delta\sigma \frac{r_1 r_2}{r_1 - r_2} \tag{4.42}$$

If $r_1 \ll r_2$, then

$$\frac{d\rho}{dt} = -\frac{24}{4} J_{total} \frac{V_m}{\frac{1}{6}\pi G^3} = \frac{288 D_l \gamma_s V_m}{RTG^3} \tag{4.43}$$

$$\frac{d\rho}{dt} = \frac{441 \gamma_s V_m}{RTG^3} (1-\rho)^{\frac{1}{3}} \tag{4.44}$$

$$\frac{d\rho}{dt} = \frac{735\left(D_b \gamma_s \delta_b V_m\right)}{RTG^4} \tag{4.45}$$

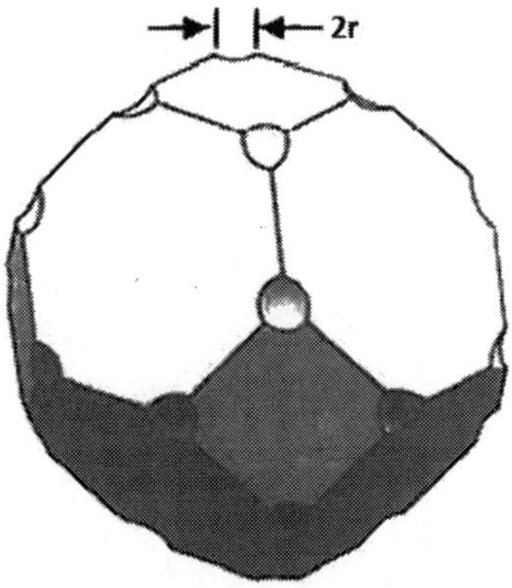

Figure 4.4 Coble model for final stage sintering.

Equations become similar to initial stage sintering, where particle size is the main factor on which the growth and shrinkage depend. The above equations assumed that if $r_1 \ll r_2$, only constant material flux from grain boundary to the surface is possible irrespective of pore size. Herring's scaling law can also be used to derive the sintering kinetics, where the effect of the surface of pores on the material flux from grain boundary to the pore is considered. Kang and Jung derived the densification rate for volume diffusion and lattice diffusion, where lattice diffusion Equation (4.44) is similar to that of Coble's model, and Equation (4.45) for grain boundary diffusion is found to be similar to that of the initial model [3].

Densification, as well as grain growth, occurs simultaneously in the final stage of sintering. The densification in the final stage of the sintering is mainly denoted by the shrinkage of isolated pores present in the corners and the grain junctions if the lattice diffusion governs the final stage of sintering. Usually, the lattice diffusion mechanism does not encounter grain boundary as the resource of atom source for the diffusion and densification. Thus, irrespective of the pore size, the constant material flux from the grain boundary to the pore surface is not necessary to encounter. However, in the case of grain boundary diffusion, densification is dependent on the grain size, which is similar to the initial stage model. The grain size is mainly considered for predicting densification kinetics. Harmer and coworkers confirm that the grain boundary diffusion governs densification at the final stage of the sintering [4].

4.4.3 Influence of Entrapped Gases

The densification depends on the sintering atmosphere as the pores contain atmospherically entrapped gases, which can be easily removed under fast diffusing conditions. However, densification becomes difficult if the entrapped gas has slow diffusion; and thus needs high external pressure to remove it. The final density and densification kinetics depend on initial pore radius r_i, surface energy γ_s, etc. [5].

It is found that the maximum density is reached when the pressure of the entrapped gases becomes equal to the capillary pressure of the pore.

Equation (4.46) provides the condition to attain the maximum attainable densities:

$$P_i\left[(r_i / r_f)^3 - 1\right] = 2\,\gamma_s / r_f \tag{4.46}$$

P_i is the initial pressure inside the pore before the isolation of the pore. Initial stage pore radius and their distribution play a significant role in achieving resultant density. Content of smaller pores expedites high final density at optimum pressure, but larger pores experience more internal gas pressure and difficulty attaining the same density at the same pressure where smaller pores disappear. However, initial pore size does not significantly impact the resultant density during pore isolation or avoid pore coalescence in the matrix. In this context, an experiment revealed that pore isolation by two-stage sintering or sinter-HIP facilitates density increment irrespective of pore size; eventually, the overall densification time reduced as the densification rate increased [6].

4.5 Microstructure Alteration

The usual ceramic processing procedure involves the compaction of crystalline and amorphous powders followed by firing at a higher temperature to obtain desired densification and properties. In the initial stage, phase transformations and decompositions are observed, while in the later stages, the grains start to grow, and the pores change their shape and are reduced in number for lower porosity. These stages can be further studied in detail through various processes such as recrystallization, grain growth phenomena, and the densification of single phase systems. The combination of new phase formation, alteration of pore size and distribution, recrystallization, and grain growth expedites the continuous microstructure alteration of the compact.

4.5.1 Recrystallization and Grain Growth

There are three predominant processes, primary recrystallization, grain growth, and secondary crystallization. In the process of primary recrystallization, the nucleation and growth of strain-free grains occur in the matrix, which has been plastically deformed. The driving force for the primary recrystallization is mainly due to the increased energy of the matrix from the deformed matrix. Even though it is smaller in value than the heat of fusion, it still has an optimum amount of energy to affect the grain boundary movement and size. Constant grain growth is usually observed after the induction stage for isothermal treatment. The induction period is the time required to grow the unstable embryos and attain a stable nucleus. A nucleus is said to be stable when the reduction in the free energy is equal to the increase in the surface energy, which is found when the nucleus grows larger than the critical diameter and attains a size that would be stable.

$$d = U(t - t_0) \tag{4.47}$$

Where U = growth rate (cm/sec), t = time, t_0 = induction period, g = grain size.

As mentioned in the above Equation (4.47), the growth rate is found to be constant until the grains impinge on each other. The size of the grains at the final stage is determined by the number of nuclei formed during the induction period; these nuclei formed during the nucleation period would grow by the atomistic process where an atom jumps from one part of the grain boundary to the other.

Both the nucleation rate and the growth rate are temperature dependent, thus the recrystallization also depends on the temperature. Some of the reasons for grain growth could be that when the temperature is increased, the grains are gradually larger because the rate of growth is rapid concerning the rate of nucleation, or if the temperature is high, then the nucleation process is completed earlier. The stable nuclei then start to grow, and at that time, the temperature is still available to them. As the driving force is mainly due to the plastic deformations in the matrix, thus the growth rate would be higher in the presence of high deformations, but the grain size would decrease. So the points to consider are:

- Minimum deformation must be present to ease out the process of nucleation so that there is no problem during recrystallization.
- The degree of deformation must be small, and the temperature must be high for recrystallization so that the embryos become stable.
- Greater annealing time can reduce the recrystallization temperature, so relatively small sized grains can be attained.
- The final grain size depends on various factors like the degree of deformation, the grain size from which the growth starts, and the crystallization temperature, continuous exposure to a higher temperature can result in larger grain growth.

Primary crystallization is commonly not found in ceramic materials; they seldom undergo plastic deformations during the processing, so it is common among metals. However, the graphite-containing matrix may experience deformation and an effect during microstructure alteration.

4.5.2 Grain Growth: Normal and Abnormal

In the process of grain growth, a continuous increase in the average size of the stress free grains with increasing temperature is evident. Grain growth follows two types; normal and abnormal grain growth. Normal grain growth where the distribution is invariable and the grains having average sizes grow in equal proportions. In abnormal grain growth, a bimodal distribution is found where some of the grains become exceptionally large relative to other smaller grains.

4.5.2.1 Normal Grain Growth

If fully dense polycrystalline materials are considered, the grain movement may vary from grain to grain depending on grain orientation. The grain boundary energy might vary, which would further result in the growth and mobility of the grain. The atoms present across the grain boundary experience different pressures, and for the atoms to be in local equilibrium, the pressure difference $(\Delta P)2\gamma_b / R_o$. As the growth rate of average-sized grains must be proportional to the average velocity of grain boundary movement, hence the Equation (4.48) justifies it:

$$\begin{aligned}\frac{\mathrm{d}\bar{G}}{\mathrm{d}t} &= \alpha\bar{v}_b = \alpha J\Omega \\ &= \alpha\frac{D_b^{\perp}}{kT}(\nabla P)\Omega \\ &= \alpha\frac{D_b^{\perp}}{RT}\frac{2\gamma_b}{\bar{R}_o}\frac{V_m}{\omega}\end{aligned} \tag{4.48}$$

Where, R_o = is grain boundary's radius of curvature

α = proportionality constant

J = atom flux

Ω = atom volume

$D\frac{1}{b}$ = atom diffusion coefficient across the grain boundary

ω = grain boundary thickness

k = Boltzmann constant

As the average radius of curvature is proportional to the average grain size so the above equation could be written as Equation (4.49):

$$\bar{G}_t^2 - \bar{G}_{t_o}^2 = \frac{4D_b^{\perp}\gamma_b V_m}{\beta RT\omega}t \tag{4.49}$$

Thus, the grain growth is proportional to the square root of the annealing time (t) for bulk polycrystalline structures. If two-dimensional structures are thin plate-like, the reaction kinetics reduces by half due to the reduction of pressure difference. This phenomenon is explained by the classical theory of grain boundary movement. In classical theory, it is assumed that the driving force for the grain boundary is dependent only on the radius of curvature and that the grain growth is proportional to the grain boundary movement. Such incidents can be proved from the microstructure of the grains having normal grain growth or consisting of any abnormal grain growth.

With an increase in the annealing time, the grain grows, reaching a maximum point after which it cannot grow further. Various experimental studies reported that kinetics rarely satisfy the Equation (4.49) and that the exponent is generally greater than 2. In some cases, it is also shown that the small grains that are supposed to have six-sided hexagon-like growth structures are liable to disappear due to instability after reaching an approximate average number of edges ~4.5 of the grains. Some impurities might also get incorporated, and the annealing temperature might also be low, which is conflicting and unrelated according to Equation (4.49). Refractories are a mixture of heterogenous oxides, and sintering is done frequently in the presence of impurities in the kiln or operating vessels.

Sometimes the presence of second-phase particles at the grain boundary can hinder the grain growth, which is explained by the Zener effect. The drag force of the particle against the boundary movement is defined by the equation, on an assumption of radii of secondary phase particles to be constant r and the number of particles attached to the unit area of grain boundary be f_v, the drag force by Zener can be written as the drag force is F_d, which is given by the equation:

$$\begin{aligned} F_d &= \gamma_b \sin\theta \times 2\pi r \cos\theta \\ &= \pi r \gamma_b \sin 2\theta \end{aligned} \tag{4.50}$$

The maximum drag force per unit area of grain boundary F_d^{σ}, would be:

$$F_d^{\sigma} = \frac{3f_v\gamma_b}{2r} \quad (4.51)$$

So the work needed for 1 mole of atoms to pass through grain boundary, would be:

$$W = \frac{3f_v\gamma_b V_m}{2r} \quad (4.52)$$

There is a limiting grain size, which decreases with an increase in the volume fraction of the secondary phase, but in reality, it seems that the second phase particle, which is present before the moving grain boundary, helps it to grain boundary movement. On the occasion of grain growth, secondary particles may grow by grain boundary diffusion or lattice diffusion, but in reality, it does not because the assumption of the distribution of second phase particles in a random manner does not occur during the grain growth. The free energy difference between the atoms present on the two sides of the grain boundary is the driving force that makes the boundary move toward the center of curvature. The rate of grain boundary movement is proportional to the rate at which atoms jump across the boundary and its curvature.

If all the grains have similar grain boundary energy, then they meet at an angle of 120° and have six sides, so it can be inferred that the grains, which have less than six sides become small and disappear, and the ones with a size greater than six is likely to grow. So, for the grains, as the radius of curvature is proportional to the grain diameter, the grain growth rate is inversely proportional to the grain size. For the grain boundary movement, an equation for the mechanism is mentioned:

$$B_i = \left(\frac{D_b}{kT}\right)\left(\frac{\Omega}{Sw}\right) \quad (4.53)$$

D_b is grain boundary diffusion coefficient, Ω is Atomic volume, S is boundary area, and w is boundary width. However, it cannot be completely pure; it has lattice defects associated with boundaries. This lattice defect sharply reduces the grain boundary movement. It reduces the grain boundary velocity, and these defects increase as the grain size increases, the segregate concentration of the solute increases, and the average boundary curvature decreases. Grains stop growing once they reach a size equal to the sample size. For example, grain in a rod sample would grow until it reaches the size of its diameter. When inclusions are added, they hinder the grain growth as they increase the energy; thus making it difficult to grow. So if the inclusions are found in greater amounts in the grain boundary, they make it difficult for the grain boundary to move. It can be explained by Equation (4.54) mentioned below:

$$d_l \approx \frac{d_i}{f_{d_i}} \quad (4.54)$$

Where d_l represents limiting grain size, d_i represents particle size of the inclusion, and f_{di} is the volume fraction of inclusion. When a grain boundary passes through an inclusion, it gets attached to it first and then subsequently moves away from it, or the grain boundary may approach in such a manner that it might drag the secondary phase particle along with it. In this case, the inclusion velocity must be similar to the grain boundary to drag and material transport across the particle. The driving force controls this grain boundary velocity on the boundary, and it is represented by Equation (4.55), where B_p is the particle mobility, v_b is the particle velocity, p represents the number of inclusions per, and F_b represents a force for the grain boundary.

$$v_b = \frac{B_p F_b}{p} \tag{4.55}$$

So, three conditions arise when inclusion is encountered:

- It moves with the grain boundary without offering any resistance
- It moves with the grain boundary but in such a manner that it limits the velocity of the grain boundary
- It does not move, and the grain boundary just passes through it

The inclusion particles that move with the grain boundary are collected at the boundary intersections and further agglomerated as the grain growth proceeds.

The existence of a pore alters the original properties of the matrix, and such inclusion is considered as a secondary phase that tends to agglomerate and develop intergranular or intragranular pores with time. The driving force and curvature of grains are high during the initial sintering stage, and the grain boundary rapidly moves across the pores and remains in the same place. In the later stages, the driving force is lower when the grains are larger, and the pores get dragged along with the boundary, and the grain growth is sluggish. The presence of the liquid phase can further affect grain growth. A minimal amount of liquid can help secondary recrystallization, but more significant amounts can lead to complexities.

4.5.2.2 Abnormal Grain Growth (Secondary Recrystallization)

Secondary recrystallization is also referred to as discontinuous or abnormal grain growth. When a grain grows larger at the expense of the small grains nearby, the uniform grain matrix alters and forms a nonuniform matrix, which results in inconsistent properties. It happens due to the presence of secondary inclusions. Once a grain grows large, it has more sides than the small grains with fewer sides, which grow more rapidly. In the case of secondary recrystallization, the only grain boundaries that can move are the larger grains, while the other grains remain uniform in size. After the grain grows to the matrix diameter, its growth is indirectly proportional to d_m (matrix diameter). So an induction period corresponding to the higher growth rate is required at the expense of uniform grains in the matrix. Thus, the grain growth is constant until the diameter of the matrix is unchanged.

Secondary recrystallization mainly occurs for oxides, where the secondary phase inclusions inhibit the grain growth. The secondary recrystallization magnitude of polycrystalline body depends on the initial particle size. If initial particles are larger than the matrix diameter, in that case, they can serve as the embryos for secondary recrystallization. In contrast, smaller particles experience difficulty growing. Only a few grains can grow enough to overcome the matrix diameter, which is an essential condition that must be met for the secondary recrystallization to start. Thus, the availability of the embryos becomes difficult, hence the secondary recrystallization gets affected. These systems are found where the impurity concentration is low, and the driving force is mainly the difference in the low surface energy of the larger grains and the high surface energy of the smaller grains. Under these conditions, transfer can only occur if there is a boundary phase present between the small and larger grains, usually the secondary phases tend to be present at the grain boundary of the larger grains, so it grows once the growth is initiated. If the boundary phase goes on increasing, then the grain growth and the secondary crystallization are inhibited. Secondary recrystallization can affect the properties of the resultant products; thus composition and firing schedules need to be monitored properly. Excessive grain growth could affect mechanical and thermomechanical properties. Electrical and magnetic properties also depend on the grain size, and their properties might vary depending on their use in the specific field.

4.5.3 Pores and Secondary Crystallization

The variables for sintering that are generally easy to control are temperature (which could indirectly control the vapor pressure) and the initial size of the particles. However, refractory manufacturing demands as high as 4–6 mm particle size to attain wear, corrosion resistance, and mechanical response. A proper combination of course size to fine up to a few micron particles facilitates adequate compaction and desired green and fired density, as seen in the Chapter 3. As the sintering rate steadily reduces with time, sintering for a longer period to gain advanced properties is not a critical variable for refractories. The sintering rate is indirectly proportional to initial particle size; thus it must be taken into account seriously. Different models already explained the initial, intermediate, and final stage sintering mechanisms in the previous section, where final sintering is the only process synchronized by boundary diffusion and lattice diffusion from sources on the grain boundary. When pores are present in the boundary, the diffusion process becomes easier as the mean diffusion distance decreases, so material transport becomes easier than with less or no pores. In the initial sintering stage, grain growth and pore growth are found in the initial raw materials when particle sizes are very small preferentially. It attributes to the presence of agglomerations of the finer particles and rapid growth of the grain boundary, resulting in the agglomeration of pores at the grains interface.

Fine particle expedites agglomeration and the intragranular pores after the final stage of sintering. Thus, the breaking of lumps by raw material processing accessories and desired fine particles assist in obtaining better packing efficiency and high density during a high sintering rate. Packing can reflect in the intermediate stage, where pores of different sizes are formed. The presence of local agglomerates and various factors like die-wall friction and heterogeneous mixing can affect the pore concentration at the later stages due to initial particle size variations in the starting material, as shown in Figure 4.5.

The pore concentration may vary because of the agglomerates formed. The rapid pore elimination can also be due to pores near the surface, which can be easily removed because of the temperature gradient. Thus, the kinetics of pore elimination and thermodynamically metastable equilibrium pore configuration is important because some parts shrink due to pore elimination while others do not follow the same degree of shrinkage. Thus, effective pore elimination is required for a stable structure. In Figure 4.5, a spherical pore model is considered, but the shapes of pores may vary depending on the dihedral angle and the surrounding grains. If r/p becomes zero, the interface is somewhat flat. The r is the radius of the circumscribed sphere around a polyhedral pore surrounded by grains, and p is the radius of curvature of the pore surface. When r/p is negative, the pore tends to grow. The conditions for pore stabilization are described in Figure 4.6.

A significant difference between grain and pore sizes is not always a necessity for the stability of pores. The position or site of the pore and size relative to the grain boundary affects the diffusion distance and the driving force. When there is the presence of many pores in the initial stages, the grain growth is inhibited, but if the porosity decreases to such a value that secondary growth prefers, then secondary crystallization is formed at a higher temperature.

Many pores get separated and isolated from the grain boundary once the grain growth occurs, so the diffusion distance also reduces between them, and the

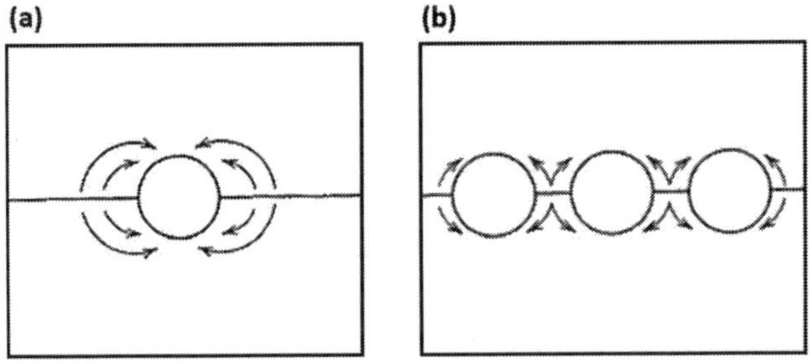

Figure 4.5 The mean diffusion distance for material transport in two cases (a) with the existence of only one pore (b) the presence of multipores.

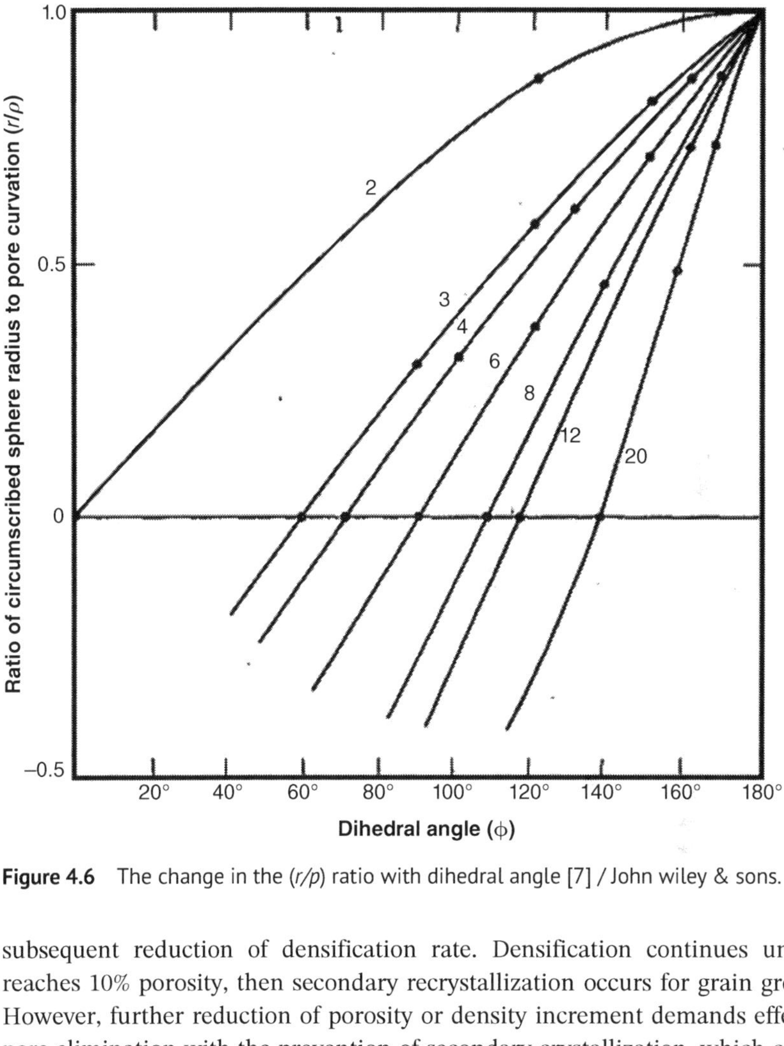

Figure 4.6 The change in the (*r/p*) ratio with dihedral angle [7] / John wiley & sons.

subsequent reduction of densification rate. Densification continues until it reaches 10% porosity, then secondary recrystallization occurs for grain growth. However, further reduction of porosity or density increment demands effective pore elimination with the prevention of secondary crystallization, which can be done by introducing additives that slow down boundary movement.

4.6 Sintering with Low Melting Constituents

Alkaline earth silicate is a typical low melting assisted binder for ceramic systems like carbide, oxides, etc. In this system, solid particles or phases show certain solubility in liquid phases during the high temperature of sintering. After solubility in the solution, reprecipitation also occurs to facilitate grain growth. Under this circumstance, the densification must follow three fundamental points:

- Wetting of solid by a liquid layer
- Presence of an adequate amount of liquid layer
- The ability of the solid to be soluble in the liquid phase

The driving force is the capillary pressure, which is found in between two solid particles. This densification occurs by various processes, and these are:

- After liquid formation, the particles rearrange among themselves to get effective packing. If the interstices are filled perfectly by the liquid phase, then the densification would occur ideally.
- After the rearrangement of the particles, the contact points in the particles undergo plastic deformation and creep.
- During the sintering process, the smaller grains dissolve, and the larger grains grow by reprecipitation; this entire process is known as Ostwald ripening.
- The contact point pressure between the particles increases when liquid penetrates between particles. Thus, the material transfers from the contact point toward the outer surface. It helps to combine nearby particles, and shrinkage occurs as the material transfers from the contact point.
- Beyond this point, the densification slows down, and the process stops.

4.7 Bonding Below 1000 °C

Most ceramic bonding develops beyond 1000 °C; however, a different class of bonding arrangements is used to maintain the initial stage of green binding that eventually facilitates intermediate temperature binding performance when required. The binders are selected based on:

- Binder burnout characteristics
- Molecular weight
- Glass transition temperature
- Compatibility with the dispersant
- Effect on the viscosity of the solvent
- Solubility in the solvent
- Cost

Binders can increase the viscosity and hence flow characteristics, and this property can be effective in the case of extrusions, where it could provide strength. Depending on how they increase the solution viscosity of the binder, they are classified into low, medium, and high viscosity. Binder grade relies on the structure of the polymer chain. Low viscosity grade polymers have smaller chains that exert less drag or resistance on the liquid molecules, so the increase in viscosity with the concentration is also less. The nature of the polymer side groups decides which solution becomes soluble. The solubility of the binder is enhanced when the binders have similar functional groups, i.e., the groups with similar polarity. Binders are polymer molecules, coagulated colloidal particles that help bind the particles together by adsorbing on their surface and interlinking among themselves. The primary purpose of the binder is to enhance and improve the strength for handling the green body before the firing step for further densification. A few examples of binders are explained below.

4.7.1 Organic Binder

Polymers with high molecular weight are usually considered as both binders and surfactants. Adsorption of polymers on particles occurs, after which the suspension is stabilized by electrostatic repulsion due to the adsorbed surfactant molecules. PVA (poly vinyl alcohol), each basic structural unit is known as *mer*" (a repeat unit). The repeating units are denoted by the subscript n, which indicates the degree of polymerization. With an increase in polymerization, the molecular weight also increases. The vinyl backbone is made up of C-C linkage, and the H and OH groups attached are known as side groups, where the OH side is hydrophilic, promoting wetting and dissolving of PVA in polar liquids. The hydrogen bonding of the adhesion to the OH group by the intermolecular attraction is caused by dipolar OH groups. PVA has a strong attraction toward the oxide particles, which are dispersed in water.

The cellulose derivatives are considered a class of naturally occurring binders. Cellulose are natural carbohydrates that contain a cellulose backbone, but they are not as flexible as the PVA binder. The substitution is described by the two parameters: the degree of substitution (DS) and molar substitution (MS). DS represents the number of anhydroglucose unit (AGU) that has reacted or the number of sites on which the modification is made on the monomer. For cellulose to be water-soluble, DS needs to be about 1.62 for effective use as a binder. In hydroxymethyl cellulose, the MS defines the average number of CH_2CH_2O ethylene oxide molecules that have reacted with each anhydroglucose unit.

Some binders can also contain side groups that can be ionized, like sodium alginate. It can also change the behavior of adsorption and flocculation. These are found to be charged molecules that are effectively longer and less twisted. Starch is another organic binder that preferentially breaks down when heated and agitated and is insoluble in cold water. Refined starch is a polysaccharide and is mostly spray-dried particles that are used as binders for the fabrication of the body.

4.7.2 Inorganic Binder

The utility of an inorganic binder is different than solid sintering additives for refractories. Inorganic binders provide green strength to high-temperature properties in an optimum content. Sodium silicate is a popular and common inorganic binder for refractories that does not affect the refractory properties in the presence of sodium and silicon residue. In the solution below the pH value of 7, hydrolysis reaction forms monosilicic or silicic acid [$Si(OH)_4$] consisting of silanol groups Si-OH. Silicic acid polymerizes into particles that aggregate to form chains and networks. The process of the polymerization occurs in three steps:

- Polymerization of monomers for the formation of the particles
- Growth of the so formed particles
- Linking of particles into chains, which converts into gel

Polymerization of silicic acid means condensing the silanol groups resulting in higher molecular weights of silica. Condensation facilitates the formation of Si-O-Si bond that minimizes the number of terminal OH groups.

Condensation results formation of nuclei that further grow with time through Ostwald ripening. It states that smaller particles dissolve and precipitate on the larger particles; thus increasing the average particle size but decreasing the number of particles. The transformation of sol to gel network expedites the stronger network through reprecipitation of silica on the necks. The degree of polymerization varies with the pH of the solution. Usually, three regions, pH < ~2, pH 2~7, and pH > 7, follow different polymerization states.

Figure 4.7 describes that both PZC (point of zero charge: where the surface charge is zero) and IEP (isoelectric point indicates the electrophoretic mobility) are zero at pH −2. In-between pH 2–7, herewith increasing pH exhibits a reduction in the gelation time. Above IEP, the gelation rate becomes proportional to OH^-. The Equations (4.56) and (4.57) describe the base catalyzed polymerization nucleophilic reaction:

$$\equiv Si - OH + H_3O^+ \rightarrow \equiv Si^+ + 2H_2O \tag{4.56}$$

$$\equiv Si^+ + HO - Si \equiv \rightarrow \equiv Si - O - Si \equiv + H^+ \tag{4.57}$$

The surface silanols are deprotonated in accordance with their acidity. The deprotonation is a relatively faster process (as in Equation 4.56); the formation of dimer (as mentioned in Equation 4.57) is slow, but once the monomer form, they react with the dimers to ultimately form tetramer, and cyclization is rapid at this pH range due to the strain resulting from the formation of cyclic trimers. With the reduced Si-O-Si bond angle, such structures are much less, but these are quite stable at higher pH values. As the growth occurs near IEP, there is no electrostatic

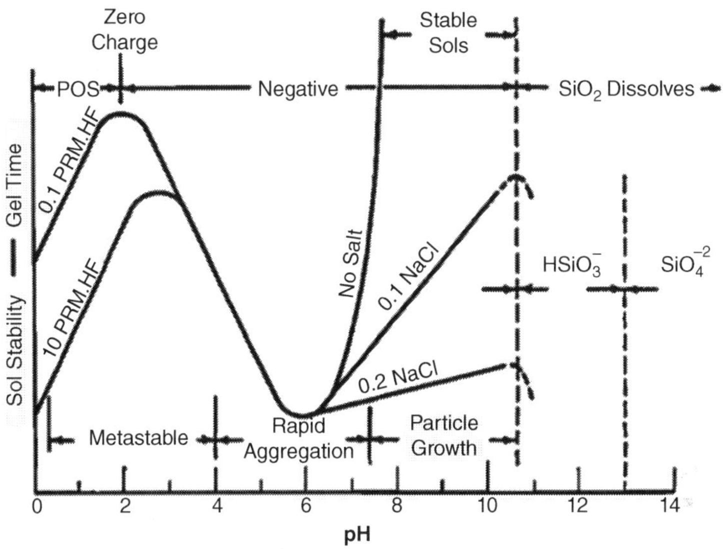

Figure 4.7 Effect of pH in a colloidal system containing silica [8] / John wiley & sons.

repulsion, so aggregation and growth occur freely. In this pH range (2–7), the solubility is low, so the gel network consists of a chain that is made of very fine particles.

At pH levels greater than 7, polymerization follows the same mechanism as mentioned in (Equations 4.56 and 4.57). However, the condensed species become ionized and repulsive. Thus, growth occurs by adding monomers to the condensed species, forming the cyclic trimers. They are stable because the planar-type cyclic configuration allows the separation to be larger in-between the deprotonated sites. The silica solubility is high at this pH (>7), so the grains grow by Ostwald ripening, and the growth rate is dependent on the particle size distribution.

The gelation time is reduced with the decreasing pH, specifically at pH levels below 2. In this condition, the polymerization rate is proportional to H^+. The solubility of silica is low, and the formation and aggregation of the particles are relatively easy, but Ostwald ripening does not contribute much. So the gel network exists only as very fine particles.

4.7.3 Carbonaceous Binder

Polymeric thermosetting resins and coal-tar pitch are usually added as prime carbonaceous binders to form a bond in a refractory matrix, especially for carbon-based refractory like MgO-C, Al_2O_3-C, and many more. The obtained mixers are easily moldable at the initial stages: eventually, they harden and form a structural bond at elevated temperatures. These binders have a low softening point (80–120 °C) and produce graphitized carbon after pyrolysis at 800 °C under a certain reducing atmosphere. Usually, this carbon is a hybrid mixture of highly crystalline graphite-like structures and an amorphous phase. Coal-tar pitch-bonded refractory products release more toxic (such as fumes, carbon monoxide, and benzopyrene) compounds than thermosetting resins. Adequate molecular structures, curing conditions, and additives expedite carbon-containing refractories' properties and microstructural evolution after pyrolysis and in-situ sintering.

Shaped refractories are usually cured in the range of 200 °C to remove volatiles and moisture and induce chemical reactions for crosslinking reactions. The initial stage of tempering up to 400 °C follows low heating rates and temperature thresholds to allow the release of residual monomers and solvents and prevent cracks caused by the large thermal expansion of the polymeric phase. Pyrolysis temperature up to 1000 °C facilitates the rupture of bonds. Most carbons rearrange to generate new structures and release gases that eventually precipitate carbon-based solids over the ceramic grains. Such structure develops a combination of graphitic and non-graphitic carbon (NGC). However, the intermediate temperature in the range of 1500–1700 °C allows for the development of a higher specific surface area, pores, and low crystallinity of partially graphitized carbon. A significant graphitization can be observed after heating samples above 2000 °C for several hours.

References

1 Nicolau, V.D.P. and Dadam, A.P. (2009). Numerical and experimental thermal analysis of a tunnel kiln used in ceramic production. *J. of the Braz. Soc. of Mech. Sci. & Eng*. Copyright, XXXI (4): 297.

2 Specht, E. (2014). *Waerme- und Stoffuebertragung in der Thermoprozess-technik*. Essen: Vulkan Verlag.

3 Coble, R.L. (1961). Sintering Crystalline Solids. *I. Intermediate and Final State Diffusion Models J. Appl. Phys.* 32: 787.

4 Bedu-Amissah, K., Rickman, J.M., Chan, H.M., and Harmer, M.P. (2005). Impact of microstructure on grain-boundary diffusion in polycrystals. *J. Appl. Phys.* 98: 063511. doi: https://doi.org/10.1063/1.2043257.

5 Kang, S.-J.L. (2005). *Sintering, Densification, Grain Growth, and Microstructure*. Elsevier.

6 Davis, R.F. (1991). *Hot Isostatic Pressing, Concise Encyclopedia of Advanced Ceramic Materials*. Elsevier. 210–215.

7 Kingery, W.D., Bowen, H.K., and Uhlmann, D.R. (1976). *Introduction to Ceramics*, 2e. John Wiley & Sons.

8 Iler, R.K. (1979). *The Chemistry of Silica, Solubility, Polymerization, Colloid and Surface Properties and Biochemistry*. New York: John Wiley. 866p.

5

Thermal and Mechanical Behavior

5.1 Introduction

Ceramic matrix has high hardness, compressive strength, elastic modulus, high-temperature resistance, and high stiffness. However, the lack of crack propagation resistance under stress makes it a weak and catastrophic failure. Thus exhaustive microstructure modification can provide all of these together. Refractory is a *classic bulk ceramic composite* that enormously experiences mechanical, thermal, and thermomechanical stresses, which eventually determine the performance of complete lining or component performance. Thus, a critical understanding of fracture behavior may help to design an adequate matrix.

These mechanical properties are among the crucial criteria (crack initiation and propagation being the most critical) for evaluating the efficiency of ceramics in terms of their applications. After briefly discussing some early brittleness theories suggested, Griffith's theory of brittle fracture in brittle solids may address this incidence. It illustrates how to apply the physics of fracture behavior under external stress. The fracture toughness is defined from the definition of stress concentration at the crack edge [1]. The variability in strength properties is presented after exploring fracture theories. Addressing the microstructure inconsistency, spontaneous microcracking results from accumulating residual stresses and the prediction of crack growth usually emanating from a preexisting defect produced during processing or surface finishing.

Thermal behavior in specific thermal conductivity is an important property of ceramics, along with their chemical inertness and oxidation resistance, which makes them uniquely suited as a class of materials to play an extremely challenging and essential role in metal smelting and refining.

As materials are heated, their size and volume increase in a process known as thermal expansion, in small increments depending on the substance being heated; due to this characteristic, the expansion values may differ. The thermal

Fundamental Design of Steelmaking Refractories, First Edition. Debasish Sarkar.

expansion coefficient ratio shows how much a material expands per degree centigrade temperature increase. Usually, advanced ceramics (or fine ceramics) have low thermal expansion coefficients that are less than half those of stainless steel. In this respect, how strong bonds result in solids with low thermal expansion coefficients is essential to consider. Ceramics are prone to thermal shock because of their brittleness and low thermal conductivity; it can crack when exposed to wide thermal gradients. Because of these temperature gradients in ceramics, due to the restrictive effect of adjacent body components, stresses may develop. When the thermal stress is high enough, it can cause crack propagation, which can be either stable or unstable based on the temperature distribution. Furthermore, the thermomechanical behavior of ceramics, such as refractory under load, hot modulus of rupture, and ceramic creep behavior of competitive ceramics, is also addressed.

Wear seems to be the surface injury or loss of material from either or both of two solid surfaces in a sliding, rolling, or impact motion relative to each other. In most situations, wear occurs at asperity by surface interactions. It is carried out by mechanical and/or chemical methods and is typically stimulated by thermal or frictional heating. Thus, different forms of wear mechanisms caused for extracting material from the rubbing surface are carefully highlighted to justify the refractory performance.

Finally, this chapter presents the various significant mechanical properties of ceramics and cracking that is critical for understanding the application, thermal and thermomechanical behavior for high-temperature applications, and different types of wear mechanisms.

5.2 Mechanical Properties

5.2.1 Elastic Modulus

Elastic modulus is the property of ceramic material, which is important for structural applications. It describes the stiffness of a material, which means that with increasing the elastic modulus, the stiffness of the material also increases [2].

5.2.1.1 Experimental Determination of Elastic Modulus

A rectangular sample is prepared and impacted with a rod close to the high frequency sensor. During elastic modulus determination by the dynamic elastic property analyzer, the software estimates the values by using the formulae:

$$E = 0.9465\left(\frac{mf_f^2}{b}\right)\left(\frac{L^3}{t^3}\right)T \tag{5.1}$$

where

$$T = 1 + 6.858\left(\frac{t}{L}\right) \tag{5.2}$$

m is mass, b is width, f_f is natural frequency in flexural mode, t is thickness, L is length, E is Elastic modulus, and T is correction factor.

5.2.1.2 Impulse Excitation Method

The impulse excitation equipment measures the flexural vibration frequency, as illustrated in Figure 5.1, and calculates the Young's modulus using the mass and dimensions of the sample according to the different measurements and standards (ASTM E1876-15, ISO 12680–1, EN 843–2).

5.2.1.3 Ultrasonic Method

The ultrasonic method is a nondestructive testing method, where high frequency sound waves are propagated through materials, and the elastic modulus can be evaluated by this method. Ultrasonic velocity is obtained by the travel time of ultrasonic waves through the known thickness of the sample material. The equation used to calculate the Elastic modulus is:

$$E = \frac{(1+\mu)(1-2\mu)}{(1-\mu)}(\rho C_L) \tag{5.3}$$

Where C_L = velocity of longitudinal wave(m/s), μ = Poisson's ratio, E = elastic modulus (GPa):

ρ = density of specimen (g/cm^3).

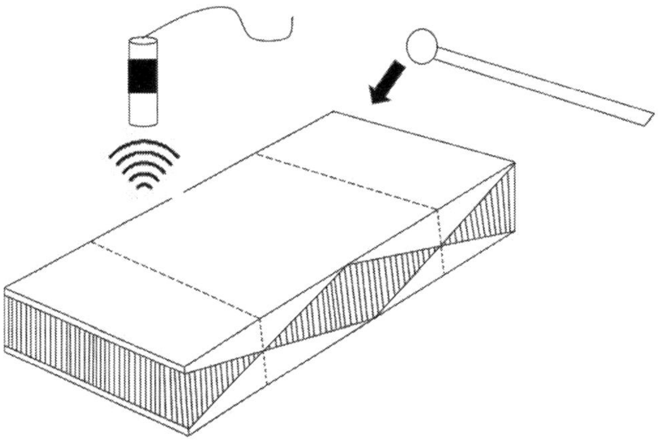

Figure 5.1 Figure of impulse excitation technique.

5.2.2 Hardness

Hardness defines the ability of a material to resist plastic deformation during the indentation of an object. It is a surface phenomenon where hardness is measured by making an indentation on a polished surface. Among several hardness measurement protocols, Vickers hardness is the most effective and accepted method for bulk ceramics.

5.2.2.1 Vickers Indentation Method

Vickers hardness measures the hardness of a material by diamond indenter force on the surface of the material and the area between the surface and four faces of the pyramid indenter. The diamond indenter is most widely used for measuring the hardness of ceramic materials. Formulae used for hardness measurement in Vickers indentation is:

$$H_v = 1.854\left(\frac{P}{d^2}\right) \tag{5.4}$$

Here, H_V is Vickers hardness, P is load applied, and d is the average length of two diagonals.

The following factors must be taken into account to achieve consistent and real hardness values for engineering ceramics:

a) No cracking of the indent corners or ribs caused by the applied indentation load is acceptable during hardness determination. A stable load will allow free indentation of damage.
b) It is expected to calculate the consistency of the formation of an indented assisted flaw in a definite direction by different loads of indents. Eventually, this flaw results in any "indentation size effect," and a conservative "true hardness" estimation.
c) In order to obtain an exact hardness value, the preference is to use high resolution electron microscopy to measure the indent diagonal or length scale in the order of micrometers. Any minimal diagonal measurement errors can reflect the errors in measured or reported data.

5.2.2.2 Instrumented Indentation Measurement

The penetration depth calculation to assess the hardness has the benefit of continuously measuring the penetration depth to create a load, versus displacement curve, as shown in Figure 5.2. Such a measurement is referred to as *instrumented hardness*, which is the measurement that nano-indentation instruments provide. Nevertheless, Vickers and Knoop instrumented indenters still operate. Details of hardness measurements for coating and their data analysis can be found elsewhere [2].

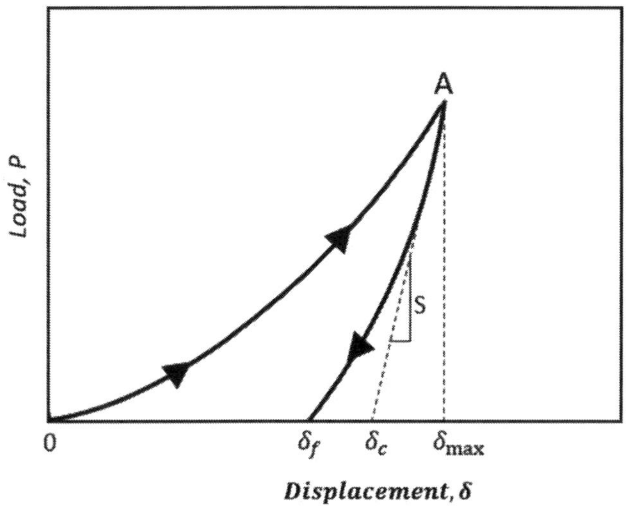

Figure 5.2 Load–displacement curve for instrumented hardness.

5.2.3 Fracture Toughness

With regard to toughness assessment, it is noted that the toughness of brittle material depends on the testing methodology, which is commonly divided into *long crack* and *short crack* methods. The single edge notched beam (SENB) and single edge V notched beam (SEVNB) techniques provide long crack methods, the most accepted method for ceramics. Short crack techniques include measuring the crack lengths (radial median) around hardness indentation, from which different recorded models can be used to estimate the toughness details. A considerable volume of literature is available on the mechanistic and phenomenological explanation of the indentation cracking of brittle materials such as ceramic and glass-based materials, and much of the discussion in this section is based, as described below, on such literature reports as well as other reports. Although the research in indentation induced damage behavior has relevance to the wear resistance property, the indentation technique is also used to test the properties of ceramics for strength or crack growth resistance.

5.2.3.1 Long Crack Method

In performing classical fracture toughness testing on ceramics, the main challenge with long crack techniques is introducing and preparing a sharp crack in the front of the notch, and the cumulative flaw length is c as shown in Figure 5.3.

In particular, for precracking, two regulated methods are used:

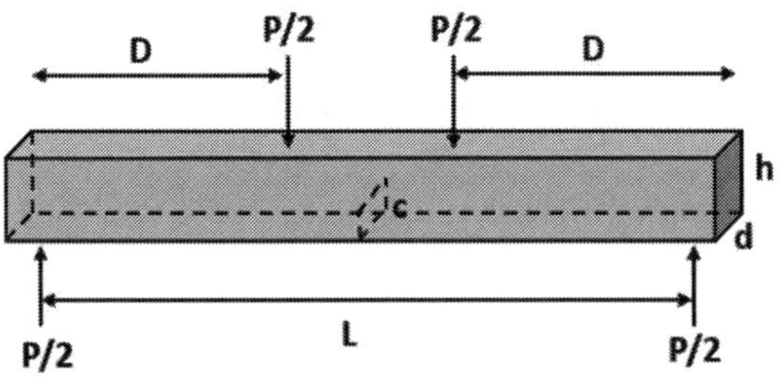

Figure 5.3 SENB tests for a type of geometry and loading configuration to determine fracture resistance.

5.2.3.2 SENB Method

In this scenario, on the SENB specimen, cyclic compressive loads with a cut notch are applied. It leads to the accumulation of damage resulting in crack growth when the sample is loaded in flexure mode in the region upstream of the notch. The above schematic shows the geometry of the testing arrangements of ceramics. SENB's classical expression evaluation of the critical stress intensity factor (K_{IC}) of mode I is followed by:

$$K_{IC} = Y\left(\frac{3Pc}{h^2 d}\right)c^{\frac{1}{2}} \tag{5.5}$$

Where

$$Y = 1.99 - 2.47\left(\frac{c}{h}\right) + 12.97\left(\frac{c}{h}\right)^2 - 23.17\left(\frac{c}{h}\right)^3 + 24.8\left(\frac{c}{h}\right)^4 \tag{5.6}$$

5.2.3.3 SEVNB Method

It is an advanced version of the SENB method in which a sharp V notch is created by sharpening the notch root with a sharp blade infused with diamond paste. The SEVNB method uses rectangular samples prepared to assess the fracture toughness of the materials. A V-*shaped notch* is essential to introduce at the height of the sample using a diamond saw and sharp knife with diamond abrasive paste on a commercial V-notch preparation unit. The samples are then fractured using a four-point bending setup on a UTM.

Figure 5.4 indicates the design of a standard notch. The fracture load is registered, and fracture toughness can be calculated from the following equation:

$$K_{IC} = \frac{P_f\left(L_o - L_i\right)}{BW^{3/2}}\,\frac{3\propto^{1/2}}{2\left(1-\propto\right)^{3/2}}\,f(\propto) \tag{5.7}$$

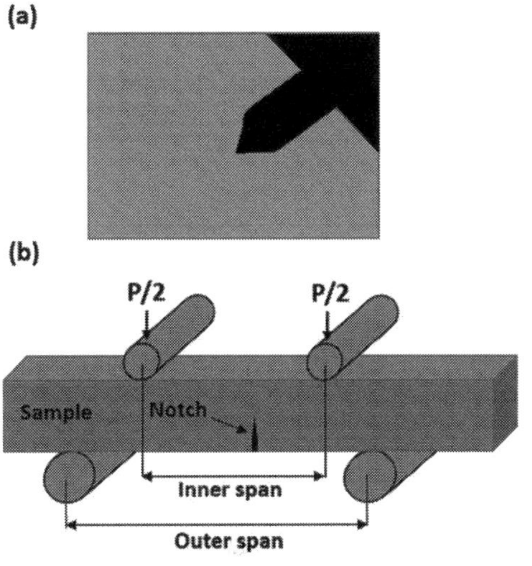

Figure 5.4 (a) Schematic representation of the standard V-shaped notch on the ceramic sample, (b) four point bending configuration for fracture toughness measurement.

Where P_f, L_o, L_i, B, W are fracture load, external length, internal length, specimen width, and thickness of specimen accordingly. $\propto$ is a precrack size and $f(\propto)$ is

$$f(\propto) = 1.9887 - 1.326 \propto - \frac{\propto(1-\propto)(3.49 - 0.68 \propto + 1.35 \propto^2)}{(1+\propto)^2} \tag{5.8}$$

5.2.4 Strength

One of the main issues concerning ceramics is the reliability of strength, and it is a critical factor that restricts the wider use of brittle materials in various structural applications. Different distribution functions, a gamma or lognormal distribution, have been demonstrated to describe the strength properties of certain brittle solids.

5.2.4.1 Strength Distribution Functions

In general, for representing strength data, several distributions can be used. There are two commonly used distributions, the normal or Gaussian and the Weibull distribution.

5.2.4.2 Gaussian, or Normal, Distribution

Gaussian or natural distribution is possibly the most frequently used to describe experimental results. The normal distribution is written in terms of the mean σ_s and variance s^2,

$$P(\sigma) = \frac{1}{s(2\pi)^{\frac{1}{2}}} \exp\left[-\frac{(\sigma - \sigma_S)^2}{2s^2}\right] \tag{5.9}$$

Notice that P (σ) is symmetric to the mean σ_s. It is useful to describe the strength distribution data for values not too far from the mean. Still, it struggles with the reality that it gives a finite likelihood of negative strength occurring for small values of strength. That is, for strengths well below the average; thus the data is not accepted. Fortunately, the Weibull distribution is more suited to smaller strengths and is more effective to use mathematically.

5.2.4.3 Weibull Distribution

The strength distribution of ceramic can be represented in several ways. The most frequently used today is the Weibull distribution.

$$P_f(\sigma) = 1 - \exp\left[-\left(\frac{\sigma}{\sigma_0}\right)^m\right] \tag{5.10}$$

Here, *m* is the Weibull shape parameter or Weibull modulus. It is an inverse calculation of the width of distribution; i.e., a greater value of *m* equals a narrow distribution and supports more reliability of the specimen, σ is failure stress and σ_0 is a normalizing parameter (63 percentile) that is a probability of failure at or below a stress σ_0 is 0.63, or 63%.

5.2.4.4 Determining the Weibull Parameter

The simplest approach is to use the strengths of *n* number of specimens measured in a tensile (ductile) or flexural (brittle) test to determine the Weibull parameter *m*. It is crucial to have high *m's* from a design point of view. Note *m* should not be mistaken for strength since a weak solid with a high 'm' is possible, and vice versa. For illustration, a solid may be weak with large defects, all of which are equal in size, but it may exhibit large m. A *homogeneous microstructure*, including grain size, defects, and inclusion, expedites the large *m* values. Usually, two methods are accountable: (1) A linearized distribution with the least square fitting, or (2) the maximum probability estimation form. The principle of least squares is in common use and is discussed. For Weibull distribution equation, natural logarithms are taken twice:

$$\ln\ln\left(\frac{1}{1-P_f}\right) = m\ln\sigma - m\ln\sigma_0 \tag{5.11}$$

Followed by, it is possible to evaluate the Weibull parameters by fitting a straight line to ln ln $[1/1\text{-}P_f]$ vs ln σ as shown in Figure 5.5.

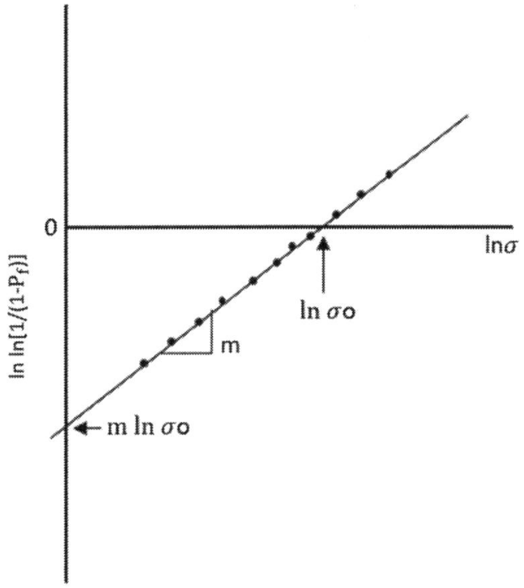

Figure 5.5 Weibull schematic showing a connection between linear regression and Weibull parameter.

5.2.4.5 Compressive Strength of Ceramics

Though ceramics are relatively poor in tension, they exhibit excellent compression properties. The disparity in microstructural resistance to crack development and the mode of crack propagation spread are due to the differences under two separate loads. Figure 5.6 shows a typical plot of stress–strain reaction under tension and compression. While ceramics act up to fracture like a perfect linear elastic material, upon reaching a peak load (greater than that in tension) in compression, the same material class acts nonlinearly beyond a certain load. As compared to tensile crack growth, the cracks stretch vertically along the loading direction of compression. The serration in the compression stress–strain response is basically due to a limited spalling of test volume, as when the growing cracks either go along with each other or enter a surface of free or uncontrolled material. Obviously, in compression, a slow fracture activity is recognized and, usually, compressive strength is about eight times greater than tensile strength. Compressive strength can be calculated by using a universal test machine. In refractory testing, this compressive strength mode is usually known as cold crushing strength (CCS) when performing at room temperature.

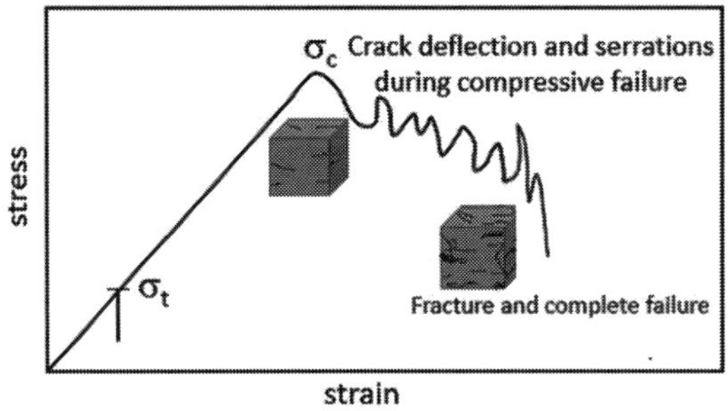

Figure 5.6 A brittle ceramic's stress–strain behavior in compression, with σ_c reflects the compressive strength. The tensile stress–strain graph, for comparison, is overlaid to demonstrate about 8 times greater compressive strength than a brittle ceramic's tensile strength (σ_t). It also reveals the compression failure mechanisms.

A cylindrical test sample with a height-to-diameter ratio of 1.0 or greater is typically used for this purpose. The samples are mounted between two parallel plates of the device and force is applied to the properly positioned samples with a constant crosshead speed (typically about 0.05 mm/s) on. During the entire compression test, the load-displacement response can be recorded using a computer connected to the UTM. The compressive strength (σ_{cs}) of the samples could be calculated from the fracture load and the sample dimension using the basic formula.

$$\sigma_{cs} = \frac{P}{A} \tag{5.12}$$

Where P is ultimate load (fracture load) and A is area of cross-section.

5.2.4.6 Flexural Strength of Ceramics

In the flexure mode, the strength of ceramics is determined either by a three-point or a four-point bend configuration by considering the difficulty of dog bone sample preparation and performing tensile tests on ceramics. For this reason, samples having a bar geometry and either a rectangular or circular cross-section are mounted in a bending fixture.

In the three-point setup, either a focused load applied or a distributed load is applied at two different positions in the four-point setup (see Figures 5.7a and 5.7b). The flexural strength is then determined with the help of the fracture load measured and test sample dimension. Usually, the cold modulus of rupture (CMOR) and hot modulus of rupture (HMOR) of refractory are measured through this protocol. The fracture strength can be determined according to the following expression for three-point loading:

$$\sigma_f = \frac{3PL}{2bd^2} \tag{5.13}$$

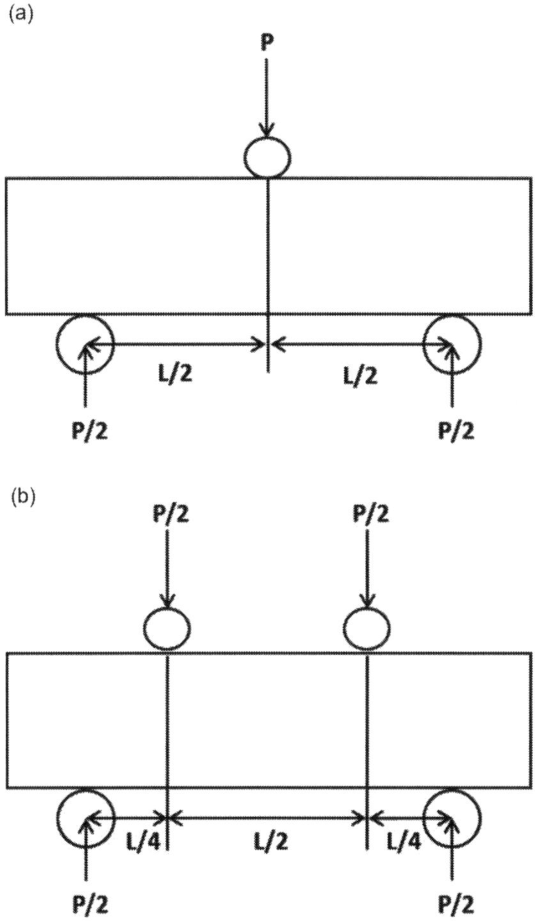

Figure 5.7 (a) Three-point and (b) four-point flexural strength illustration measurement of the rectangular cross-sectional ceramic bar.

If the materials' flexural strength is σ_f, P is fracture load, L is length of the range, b is sample width, and d is specimen thickness. Similarly, it is possible to get the flexural strength for the four-point bend setup from the expression:

$$\sigma_f = \frac{3PL}{4bd^2} \tag{5.14}$$

The loading surface is placed in compression during the flexural testing while the opposite surface is placed under tensile stress. The stress value also decreases along the thickness of the sample (z direction) linearly. In the case of four-point loading, in comparison to the three-point flexural mode, maximal tensile stress is extended over a wider region of the sample, and a lower and conservative strength estimate is thus obtained in four-point flexural tests.

5.2.5 Fatigue

Usually, common experience manifests that ceramics did not experience fatigue at or near room temperature. This assumption is obviously based on reality, where metal fatigue is associated with plastic deformation and presumption. Such appreciable plastic deformation to induce fatigue does not exist in most ceramics at room temperature. However, fatigue at room temperature has now been discovered in many ceramics, even though mechanisms tend to include microstructure weakening through toughening elements rather than plastic deformation.

By contrasting static fatigue curves with S-N (Wöhler curve) curves on same material, fatigue is demonstrated in several ceramics (including zirconia, graphite, alumina, silicon nitride, silica glass ceramics, LAS–Si$\mathbf{C_f}$, $\mathbf{Al_2O_3}$–SiC, and laminated graphite–pyrolytic carbon composites) [2]. In the static fatigue experiment, time to failure against constant stress is determined as a stress function.

5.2.5.1 Fatigue Mechanism of Ceramics

Ceramics have no serious plastic deformation and thus do not have well-established fatigue mechanisms. However, there are two categories of possible mechanisms intrinsic and extrinsic. The intrinsic mechanism facilitates fatigue-damaged microstructure in front of the crack tip. Herein, the possible mechanisms are alternating crack tip blunting and resharpening or, for whisker-reinforced composites, whisker breakage.

The extrinsic mechanism need not require a modification for the propagation of crack during fatigue loading conditions. Rather, unloading adjusts the value of the stress intensity factor for a crack tip by decreasing the crack tip shielding under the next maximum load. Thus, fatigue may reduce the impact of transformation toughening or crack tip bridging on the crack tip shielding effect.

5.3 Cracking

The strength of brittle solids is constrained by the existence of preexisting defects and their development under the application of external stress. Therefore, understanding the nature of cracking and how cracks grow during external loading is important. The cracking lies on the kind of loading carried out in the application. For instance, in the situation of blunt or distributed loading under spherical indents, cone cracks are formed in ceramics, as seen in Figure 5.8. The entire crack pattern creates a conical shape in unloading condition, as shown in the instance of transparent soda lime glass (above Figure 5.8c).

A distinct cracking pattern in brittle solids such as ceramics and glasses is identified during concentrated loading through a sharp indenter. There are two separate crack patterns that are prominent: *radial-median cracks* and *lateral cracks*. During initial loading, the radial-median cracks emerge from the strongly deformed

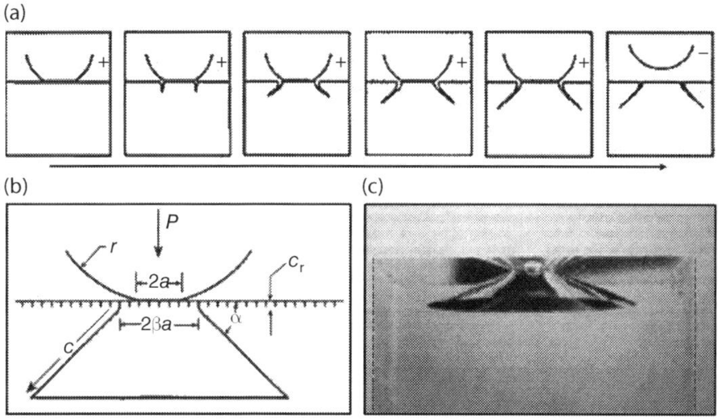

Figure 5.8 Schematic showing the (a) creation of a cone crack during loading–unloading of a spherical blunt indenter against a horizontal plane of a brittle material left, (b) the geometric parameter associated with the configuration of a cone crack (extreme right of figure (a), and (c) cone crack formation is also demonstrated in soda lime glass Adapted from [3].

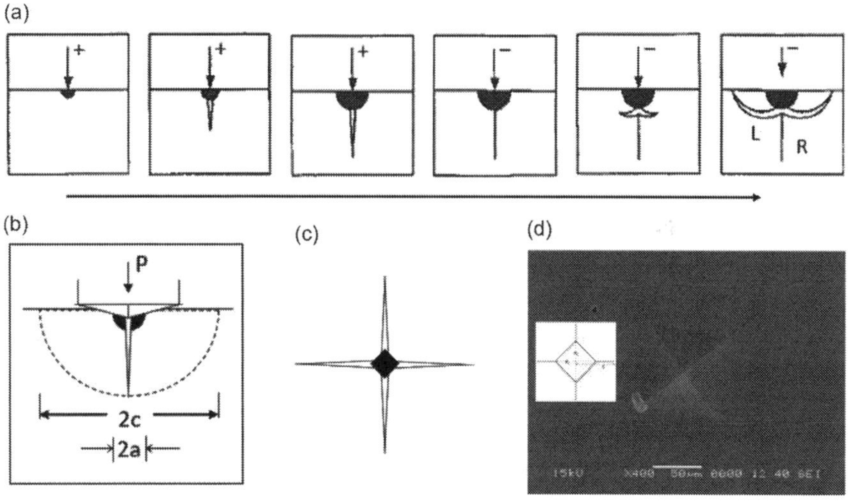

Figure 5.9 (a) The diagram shows the development of a radial-median (marked as R) and lateral crack (marked as L) when the sharp indenter is loaded and unloaded against the flat surface of the brittle material (left), (b) a geometric parameter showing the dimension of the highly deformed zone and the length of the crack, and (c) an indented image on sintered zirconia [4] / with permission of Elsevier.

zone below the indentations, and the whole loading process, where cracks expand with time. A lateral crack pattern forms during unloading, with cracks beginning to migrate in the transverse direction. The largely developed cracks meet the brittle solid surface upon complete unloading as represented in Figure 5.9a. A typical indentation assisted crack is shown in Figures 5.9b–5.9d.

5.3.1 Theory of Brittle Fracture

The brittle behavior of ceramic materials is widely recognized. Ceramics are probably best known for their brittleness instead of any of their excellent properties, including the retention of strength by certain ceramics' (SiC, Si_3N_4) at high temperatures and the extremely strong biocompatibility or bioactivity of others. A few of the significant theories that explain the brittle fracture of ceramic and glass materials are described in this section.

i) Theoretical Cohesive Strength

The theoretical cohesive strength can be calculated from the maxima consideration from the total interatomic force under stress on the basis of first-principal calculations. The initial theory of brittle fracture is established based on the theory of interatomic breakage or rupture of the bond, by forming two additional surfaces. Often in bond rupture leading to failure, a linear elastic stress–strain relationship is encountered and confirmed. With such assumptions, the following expression of theoretical cohesive strength (σ_{th}) can be achieved:

$$\sigma_{th} = \left(\frac{E\gamma}{a_0}\right)^{1/2} \tag{5.15}$$

Where E is the elastic module, surface energy is γ, and a_0 is interatomic distance of equilibrium under unstrained conditions.

ii) Griffith's Theory

Griffith's theory, the most commonly known brittle fracture theory, is a complete change of the potential energy of brittle solids through crack propagation under external tensile loading. The change in total energy for the system can be interpreted as taking into account a rectangular plate with a central elliptical hole loaded in tension by thickness:

$$\Delta u = -\Delta u_{el} + \Delta u_s \tag{5.16}$$

Where elastic strain energy exerted across the elliptical hole is Δu_{el} and surface energy change is Δu_s, while the cavity stretches perpendicular to the direction of tensile stress. Considering that plate does have a thickness t which is much less than width w, $t \ll w$, the terminology for Δu_{el} and Δu_s can also be found by taking into account that in a brittle ceramic with crack length *c*, a sharp elliptical hole of major axis length c is relatively similar to a *real* crack as shown in Figure 5.10a. Griffith also recommended releasing elastic strain energy over an elliptical volume around the crack, with the major axis twice the length of the crack and the minor axis becoming the crack length.

Taking into account the Hooke's law for elasticity in obtaining the difference in elastic strain energy (negative amount) and the associated positive outcome of

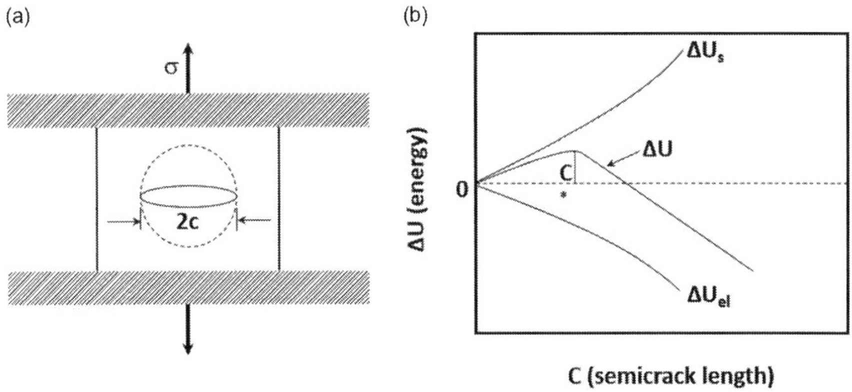

Figure 5.10 (a) Diagram showing a rectangular plate comprising an elliptically shaped cavity through thickness, and (b) energetics engaged in Griffith's brittle fracture theory.

surface energy change attributed to the existence of dual crack surfaces, absolute alteration of the potential energy of the cracked body articulated as:

$$\Delta u = \frac{-a^2}{2E}\pi \times 2c \times c \times t + 2c \times t \times 2\gamma \tag{5.17}$$

Consequently, total energy change with respect to unit thickness can be expressed as:

$$\Delta u = \frac{-\sigma^2}{2E}\ \pi \times 2c^2 + 4c\gamma \tag{5.18}$$

The variance of multiple energy quantities with a length of the crack is schematically shown in Figure 5.10b. The first differential of Δu with regard to c becomes 0 (zero), at limit of Δu-c curve and critical fracture stress or equivalent critical crack length is expressed as:

$$\sigma_c = \sqrt{\frac{2\gamma E}{\pi c^*}} \tag{5.19}$$

$$c^* = \frac{2\gamma E}{\pi \sigma^2} \tag{5.20}$$

The fracture could happen if either of the following requirements is met based on this theory. Under this circumstance, the

a) External stress $\sigma \geq \sigma_c$
b) Intrinsic flaw $c \geq c^*$

It can also be said that, at a given external stress, cracks smaller than c* are not enough to develop a further extension, whereas every extremely small crack propagation contributes to an expected rise in the energy of the system. If $c \geq c^*$, a negligibly slight increase in crack length along its downhill of the Δu-c curve results in an overall decrease in energy. The Δu-c curve is likely to shift to the left with an increase in external stress; thus c* decreases gradually. It indicates that essential cracks of a smaller size expand at higher stress, resulting in early fracture of brittle solids.

One aspect needs to be mentioned at this juncture: cracks with sizes shorter than the critical crack size ($c < c^*$) in glasses and certain ceramics (e.g., Si_3N_4) can become unstable at a defined stress level in a moist or humid environment, resulting in fracture. This is referred to as subcritical crack growth and is due to contact with the environment, contributing to a chemical attack at the crack tip, likely to lead to rapid bond breakage.

5.3.2 Physics of Fracture

The variation in the strength of ceramics is largely due to the intense sensitivity of various sizes to the presence of cracks. Yield strength and the fracture or failure strength of polycrystalline metals are deterministic and independent of volume when for the metallic samples examined, the characteristic microstructural function (grain size) remains the same. However, according to Griffith's theory, the fracture strength of a brittle material is measured by the critical crack length by Equation 5.21.

$$\sigma_f = \frac{K_{IC}}{\sqrt{\pi c}} \tag{5.21}$$

K_{IC} is the critical stress intensity factor (a measurement of fracture toughness) during mode I loading (tensile), where σ_f is failure or fracture strength, and c is half critical or largest crack size. Thus, the longer the critical crack length, the lower the strength values would be. The multitude of ceramic sample experimental findings could be explained simply:

a) Brittle materials contain flaws or cracks, and such cracks prevent their potential strength from expected or targeted demand.
b) The deviation of testing results is due to the flaw size dispersion throughout the developed component.
c) The mean strength, as calculated by similar tests, depends on the size of the test specimens and the design of the loading process.
d) Strength may vary for identical specimens.
e) Larger volume specimens have a high probability of a critical crack length and thus lower the value of strength.
f) The mean strength may decrease as ceramic specimen volume is larger.

These results suggest that (1) there has to be a direct association between the probability of failure of brittle materials and the degree of stress under which they are exposed, and (2) brittle materials still do not have a volume dependent deterministic strength value.

5.3.3 Spontaneous Microcracking

It is known that both single- and multiphase ceramics experience spontaneous microcracks after cooling. The phenomenon is unavoidable regardless of the sintering or firing rate at which the processing temperature is changed. Spontaneous microcracking results from the accumulation of residual stresses that can be triggered by one or more of the three causes:

1) Anisotropy of thermal expansion in single phase materials.
2) Mismatches of thermal expansion in multiphase materials.
3) Phase changes and subsequent changes in volume in single or multiphase materials.

5.3.3.1 Spontaneous Microcracking Due to Thermal Expansion Anisotropy

Usually, the noncubic ceramics with high thermal expansion anisotropy expedite spontaneous microcrack upon cooling. With decreasing grain size, the cracking along the grain boundaries becomes progressively less intense, and it is no longer observed under a certain *critical* grain size [1].

Considering a series of noncubic grains of identical material (a, b, and c are the same) with different thermal expansion coefficients along their x and y axes, as shown in Figure 5.11, where $\alpha_1 < \alpha_2$, is responsible for the internal stress. During high-temperature processing, the mismatch thermal expansion behavior ensures the accumulation of stresses at the grain boundaries and is accountable for failure [5].

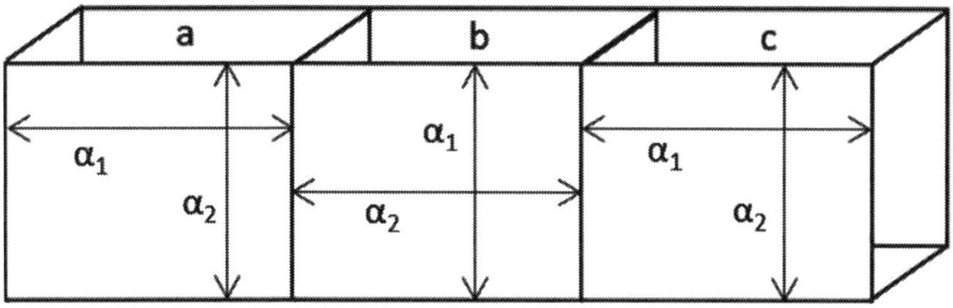

Figure 5.11 Anisotropic thermal expansion characteristic expedites internal stress after sintering. Grain a, b, and c are the same material but have anisotropic thermal expansion characteristics.

5.3.3.2 Spontaneous Microcracking Due to Thermal Expansion Mismatches in Multiphase Materials

Consider Figure 5.11, in which another material replaces grain b with a different degree of thermal expansion coefficient. The existence of a second phase with different CTE facilitates microcracking in its surroundings. Thus, the selection of materials and their further interaction phenomena are a big concern during refractory processing.

5.3.3.3 Spontaneous Microcracking Due to Phase-Transformation-induced Residual Stresses

Here, residual stresses do not evolve due to mismatches in thermal expansion or rapid temperature fluctuations but because of phase transformations. So, these transformations include atomic rearrangements, and they are often correlated with a shift in volume. Conceptually, at this point, the explanation of why such a volume change should give rise to residual stresses is obvious. However, instead of using $\Delta\alpha$, the resulting stresses typically scale to $\Delta v/v_0$, where Δ v is the change in volume associated with the transformation. The stress approximated by:

$$\sigma \approx \frac{Y}{3(1-2v)} \frac{\Delta v}{v_0} \tag{5.22}$$

can be relatively high.

5.4 Thermal Properties

5.4.1 Stress Anisotropy and Magnitude

During polycrystalline material (or two-phase) cooling from the sintering temperature, internal stresses are produced because of anisotropic thermal expansion (various thermal expansion coefficients in different crystallographic directions) or just a two-phase ceramic with different coefficients of thermal expansion values for each phase. The internal stresses are tensile in some volume elements in a given direction, leading to localized microcracking when sufficiently large in the presence of a potential flaw. The microcrack origin is always a tiny pore and is significantly shorter than the grain boundary length, located preferentially at the triple point of the grain boundary. The inside stress typically extends the length of the grain boundary, and the standard microcrack extends the length of the grain. If the local stress intensity factor is in subcritical for a specific flaw due to anisotropic thermal expansion stresses, the microcrack can develop while stress is applied, and eventually, the fracture strength becomes lower in such incidence.

Materials with anisotropic CTE often have variable elastic constants that provoke lower failure stresses through local stress concentrations. For both toughening behavior and strength, these phenomena are critical. If the material is externally loaded, elastic anisotropy only causes internal stresses. The more significant load is supported when one grain is stiffer than its neighboring one. If one grain is stiffer than the adjacent, then the greater load is supported by it. The stresses can be concentrated in the crystal over small volume elements; thus improving stress intensity factors on the initiation site of the local crack. Microcracking from thermal expansion anisotropy occurs spontaneously beyond an optimum grain size, d_c, a common feature of every material. The analysis of complex stress distributions and statistical effects is needed for in-depth analysis. In order to deal with fundamental physics, several simplified models have been proposed [6].

Inside a matrix with a mean coefficient of thermal expansion including all three directions of $< \alpha >$, Evans termed a two-dimensional grain with an anisotropy in thermal expansion cooled from the sintering temperature [7]. This basic approach is used for illustrative purposes, which is inaccurate because the individual stresses around grain boundary also rely on adjacent grain orientation. The crystallographic axis of c has a greater thermal expansion, α_c, than the crystallographic axis of a, is α_a. Here $\Delta\alpha_c = \langle\alpha\rangle - \alpha_c$, $\Delta\alpha_a = \langle\alpha\rangle - \alpha_a$

ΔT is the difference between the temperature of the room and the temperature at which polycrystal is rigid, T_{sr}, such that, where the stresses are no longer relieved by creep. The stress is approximated by the

$$\sigma_a = -\frac{\langle E\rangle\, \Delta T\left(\Delta\alpha_a - \vartheta\, \Delta\alpha_c\right)}{1-\vartheta^2} \tag{5.23}$$

$$\sigma_c = -\frac{\langle E\rangle\, \Delta T\left(\Delta\alpha_c - \vartheta\, \Delta\alpha_a\right)}{1-\vartheta^2} \tag{5.24}$$

While considering the above equations, there is no existence of grain size; however spontaneous microcracking of polycrystalline ceramics is known to be dependent on grain size, occurring only above a particular grain size d_C. In addition, d_c becomes lower with having high anisotropy.

5.4.1.1 Magnitude of Thermal Stresses

An example is a bar or cylinder restricted at the ends and heated in the middle may be considered to understand the magnitude of thermal stress. Analogous to the theoretical concept, a furnace tube clamped in a water-cooled metal housing at both ends and heated in the middle may approximate this. The resulting stress magnitude demands some flexibility in practice to allow the tube ends to be moved. If the heating length is L, the temperature is T, and the temperature at which the ends are rigidly clamped is T_0, the length difference is $\Delta L = \alpha(T - T_0)L$, where α is the mean coefficient of thermal expansion for the temperature range involved. This is a strain of $\varepsilon = \alpha(T - T_0)$ and induces $E\,\alpha(T - T_0)$ stress. Therefore, allowing thermal stresses in ceramic design is essential in devices that encounter thermal gradients. As in the case of steady-state heat flow, such gradients can be constant in time or shift rapidly over time, such as when a furnace or heat engine is shut down. Failure arising from the above condition is usually called thermal shock failure.

5.4.2 Thermal Conductivity

As a consequence of temperature gradients, heat conduction across solids occurs. The relationship between heat flux and temperature gradients $\frac{\partial T}{\partial x}$, analogous to Fick's first law.

$$\frac{\partial Q}{\partial t} = k_{th} A \frac{\partial T}{\partial x} \tag{5.25}$$

Where dQ/dt is the heat transferred per unit time over an area plane, *A* normal to the thermal energy flow, and k_{th} is the material property (analogous to diffusivity) that defines the heat transport capability of the material. The units are J/(s.m.K) or W/(mK).

5.4.2.1 Thermal Conduction Mechanisms

In brief, lattice vibrations usually transport thermal energy in solids, i.e., phonons, free electrons, and radiation. Since the concentration of free electrons in ceramics remains low and most ceramics are not transparent, phonon mechanisms are dominant and are the only ones discussed below.

Imagine a small area of a heated solid. Atoms can have high vibration amplitudes in that region and vibrate violently around their average positions. Since these atoms are connected to other atoms, their motion must also oscillate with their neighbors. As a result, in a wavelike way, the disturbance induced by the heat application propagates outward. Such waves may be scattered through imperfections, grain boundaries, and pores in absolute analogy with the electromagnetic waves or even reflected in other internal waves. In other words, the disturbance can change the direction of its propagation every so often. Before being scattered, the average distance that the disturbance travels is equivalent to the mean free path (λ_{th}), usually defined as the average travel distance by a gas molecule.

Assuming, N_{th} is the number of thermal energy carriers, and v_{th} is the average velocity, the k_{th} is the thermal conductivity and can be expressed as,

$$k_{th} = (\text{const})(N_{th} \lambda_{th} v_{th}) \tag{5.26}$$

In general, open, highly ordered structures consisting of atoms or ions of similar size and mass tend to reduce the scattering of phonons, resulting in increased k_{th} values. The diamond, which has one of the highest thermal conductivity values of any known substance, is an outstanding example. S_iC, B_eO, and $A1_N$ are other good examples. More complex structures, such as spinel, and structures where the mass difference between ions is significant, such as UO_2 and Z_rO_2, appear to have lower k_{th} values. Similar claims imply that the thermal conductivity of a solid is diminished by adding a second component to the resultant solid solution. For instance, the addition of NiO to MgO or Cr_2O_3 to $A1_2O_3$ is well-known.

In addition, in amorphous ceramics, the lack of long-range order results in more phonon scattering than in crystalline solids and consequently contributes to lower k_{th} values. Finally, it is important to note the porosity effect. Since air thermal conductivity is negligible compared to solid phases, adding large volume fractions of pores (>25%) can significantly reduce k_{th}.

5.4.2.2 Measuring Thermal Conductivity

Several techniques are employed to measure thermal conductivity. The laser flash technique is one approach that has recently gained popularity. In general, on one side of the specimen, the method aims to determine the progression of the temperature over time, while a laser pulse heats the other side very quickly. The signal alters in two ways as it moves through the solid. There is a time difference between the solid being pulsed and the maximum response. This gap in time is directly proportional to the thermal diffusivity of the material, D_{th}. A drop in the temperature spike is directly related to the heat capacity, c_p, and is the second consequence. The solid one with thermal capacity, thermal diffusivity, and density and thermal conductivity is related to:

$$k_{th} = \rho\, c_p D_{th} \tag{5.27}$$

Therefore, k_{th} can be determined if the solid density is known along with D_{th} and c_p.

A series of thermal conductivity of oxide and non-oxide ceramics with respect to temperature is shown in Figure 5.12. Such ceramics are an essential component for steelmaking refractories. While considering different refractory systems, a more or less consistent thermal conductivity with the temperature of insulating brick is noticed, whereas a drastic fall in MgO system and follows a theory of thermal conductivity of solids. It has inverse-temperature functionality above the Debye temperature (highest normal mode of vibration of crystals), where electron flow restricts due to lattice distortion. Depending on the composition and crystal structure of the secondary phase, the thermal conductivity may enhance beyond a certain temperature while free electrons are available with an increment of temperature.

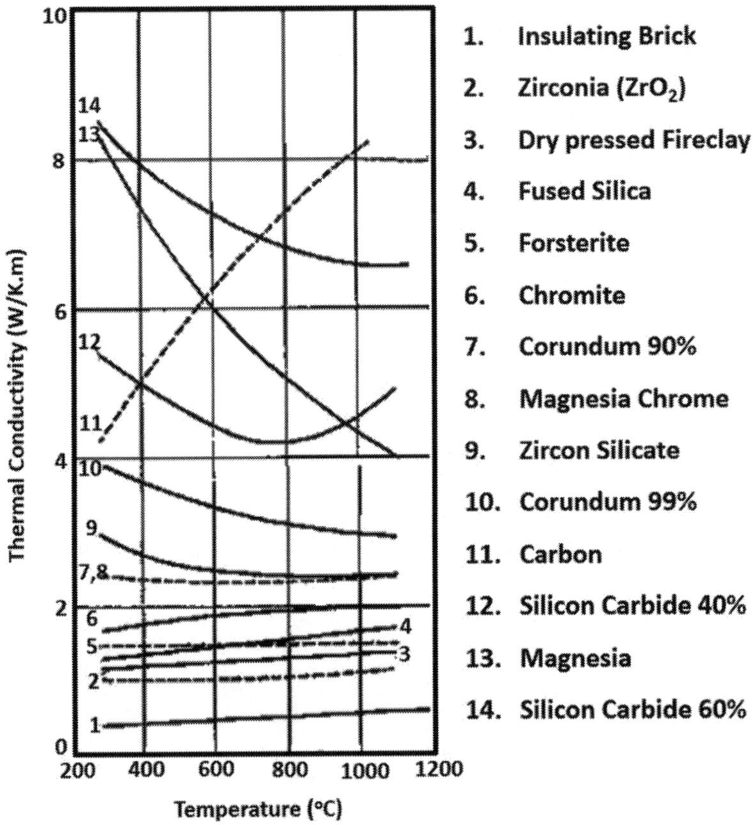

Figure 5.12 Temperature-dependent Thermal Conductivity of oxide and nonoxide ceramics.

5.4.3 Thermal Expansion

It is well-known that upon heating, solids expand. The degree of the expansion is characterized by a linear expansion coefficient, defined as a fractional change in length at constant pressure with a change in temperature, or

$$\alpha = \frac{1}{l_0}\left(\frac{\partial l}{\partial T}\right)_p \tag{5.28}$$

Where l_0 is original length, δl and δT are the change in length and temperature, respectively.

The source of thermal expansion can be directly traced to the anharmonicity or asymmetry of the energy distance curve described in Figure 5.13. The curve

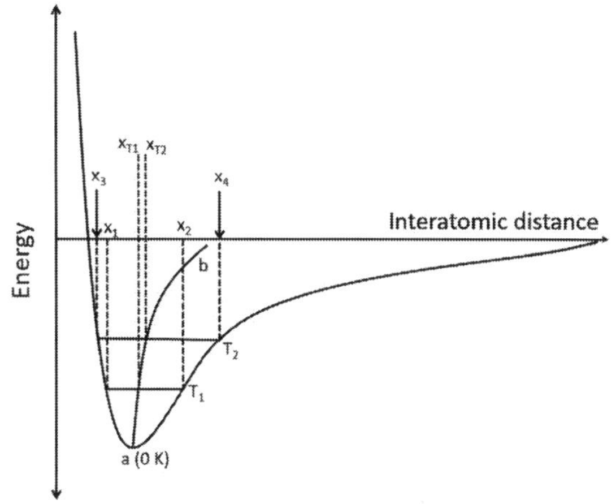

Figure 5.13 The effect of heat on the interatomic spacing of atoms. Be mindful that the thermal expansion is due to asymmetry of the well. The mean location of the atoms in a perfectly symmetric well doesn't change with temperature.

asymmetry expresses the idea that pulling two atoms apart is simpler than bringing them closer. At 0 K, the total energy of atoms is potential energy, and atoms rest at the bottom of the well. The average energy of the system increases correspondingly as the temperature is increased to, say T_1. Atoms vibrate among x_1 and x_2 positions and their energy fluctuates between the pure potential at x_1 and x_2 (i.e., zero kinetic energy) and accelerates somewhere between this.

In other words, atoms act as if they are coupled by springs. Consequently, the average position of the atoms at T_1 would be between x_1 and x_2, i.e., at x_{T_1}. If the temperature is raised to, say T_2, the average position of the atom changes to x_{T_2}, etc. Thus, it follows the average atom position moves along line ab, shown in Figure 5.13 with increasing temperature. Furthermore, the dimensions of the crystal also increase. In general, with declining bond strength, the asymmetry of the energy well increases, and thus the thermal expansion of a solid varies inversely with its bond strength or melting point. The values of α for most ceramics lies between 3 and $10 \times 10^{-6\,0}C^{-1}$.

Refractory is a classic composite material as it is manufactured either from natural raw materials or synthetic materials or a combination of both processed materials. Several phases exist during processing and frequently develop different phases in the application. Thus, different crystal structures experience different degrees of expansion at a particular temperature.

For example, manufacturing magnesia (MgO) refractory with the addition of zircon ($Z_rS_iO_4$) is essential to analyze, as MgO has thermal expansion ~1.9% and zircon has only 0.6% at 1400 °C, as shown in Figure 5.14. Such expansion difference results in cracking after manufacturing that is not recognize during green processing.

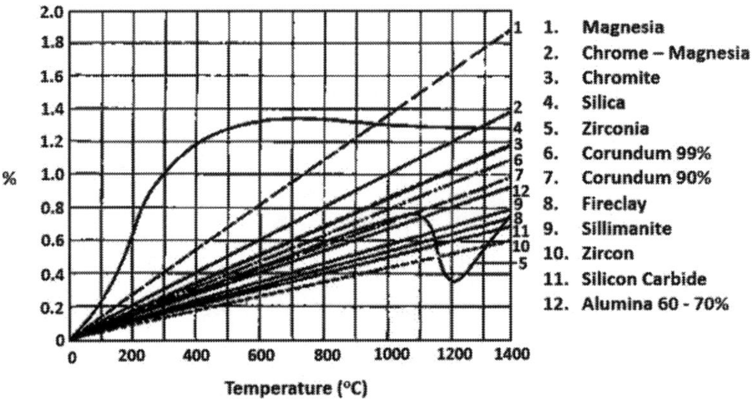

Figure 5.14 Thermal expansion behaviour of different refractories.

5.4.4 Thermal Shock

Ceramics are vulnerable to thermal shock because of their brittleness and their poor thermal conductivity; i.e., when exposed to wide thermal gradients, they can break. It is also not necessarily desirable to pour a very hot liquid into a cold glass container or cold water into a hot ceramic furnace tube, since the rapidly cooled surface tries to contract, but the bulk of the body prevents it from doing so, causing stress to grow. The ceramic cracks if the developed stresses are high enough.

Thermal stresses are generally to be avoided as they can considerably weaken a part. During cooling, a component spontaneously crumbles in extreme cases. It is well known that ceramics are prone to failure during rapid heating or cooling. Such failure is typically known as thermal shock and occurs when thermal gradients and subsequent thermal stresses exceed the strength of the component. For example, when a component is rapidly cooled from a temperature T to T_0, the surface begins to contract, but the bulk of the component is still at temperature T, and in this circumstance, surface tensile stresses are counterbalanced by compressive stress in bulk.

5.4.5 Thermal Stress Distribution

Ceramic with temperature gradients causes thermal stresses due to the restricting action of neighboring elements of the body. It causes crack propagation if the thermal stress is high enough, whether stable or unstable, on consideration of the distribution of temperature. When the stable crack is propagated, a body with larger cracks is the outcome, and the strength is gradually decreased as the magnitude of the shock is raised. Here, the body experiences a discontinuous change in strength if the crack propagation is originally unstable and stops when the strain is released.

5.4.5.1 Influence of Geometry

In a recent article, the influence of geometry and material has been considered and illustrated damaging effects, including crack failure and elastic instability (local or general) [8]. A special combination of material properties exists in an expression for each combination of geometry and failure criterion for the maximum permissible difference in temperature under the stated conditions. This combination is known as a figure of merit (a representative numerical expression to determine the efficiency) for the material. Such analysis is very useful and essential when considering different geometry like the tube, plate, solid, and much more geometry for refractories of different compositions. The following are summarized examples provided by Hasselman.

1) Externally constrained flat plate to avoid thermal expansion in the plane and thickening necessary to avoid thermoelastic instability and ΔT adjustment concerning change in uniform temperature. The maximum permissible difference in temperature is:

$$\Delta T_{max} = \frac{\sigma_f(1-\upsilon)}{\alpha E} \tag{5.29}$$

where σ_f is the strength in tension (σ_{ft}) when $\Delta T < 0$ and is the strength in compression (σ_{fc}) for $\Delta T > 0$; thus, for the maximum permissible temperature difference one requires maximum strength, minimum thermal expansion, and minimum elastic modulus.

2) Infinitely long hollow concentric cylinders undergoing heat flow radially outward

$$\Delta T_{max} = \frac{\sigma_f(1-\upsilon)}{\alpha E}\left[2\ln\left(\frac{b}{a}\right)\right] \times \left[1 - \frac{2a^2}{b^2-a^2}\ln\left(\frac{b}{a}\right)\right]^{-1} \tag{5.30}$$

Where the inner and outer radii of the tube are a and b. The figure of merit for materials with a maximum difference in temperature is $\sigma_f(1-\upsilon)/\alpha E$. Herein when the tube can be used as a heat exchanger, the maximum quantity of heat flux per unit length without thermal shock failure is:

$$Q = \frac{4\pi(1-\upsilon)K}{\alpha E}\left[1 - \frac{2a^2}{b^2-a^2}\ln\left(\frac{b}{a}\right)\right]^{-1} \tag{5.31}$$

Where the thermal conductivity is K. Figure of merit of material in a heat exchanger for maximum heat flux is $\sigma_f(1-\upsilon)K / \alpha E$.

3) Initially solid circular cylinder of radius *b* subjected to an instantaneous temperature decrease in uniform temperature. It is assumed that heat transfer is linearly convective with a heat transfer coefficient (h):

$$\Delta T_{max} = \frac{1.45\ \sigma_{ft}(1-\upsilon)}{\alpha E}\left(1+\frac{3.41K}{bh}\right) \tag{5.32}$$

When conduction is associated with high convective heat transfer rates, the condition is (i.e., $3.41\ K/bh \ll 1$).

$$\Delta T_{max} \approx \frac{1.45\ \sigma_{ft}(1-\upsilon)}{\alpha E} \tag{5.33}$$

High strength, limited thermal expansion, and narrow Young's modulus are ideal for this condition. For better conduction compared with convection (i.e., $3.41\ K/bh \gg 1$):

$$\Delta T_{max} \approx \frac{4.94\ \sigma_{ft}(1-\upsilon)K}{\alpha Ebh} \tag{5.34}$$

This makes it desirable to combine high strength and thermal conduction with low values of thermal expansion and Young's modules.

5.5 Thermomechanical Response

A judicious furnace lining selection can only be made with a detailed knowledge of refractory material properties and probable stresses on the materials during service. The relationship between different thermal operational stresses in industrial furnaces and the important properties of refractory bricks during service forms the basis for classifying the properties of refractory materials and the test methods. These methods are also important for quality control and developing new products.

Thermomechanical properties are determined using high-temperature test methods with external forces causing stresses on the tested material. The stress–strain behaviour of refractories at high temperatures is very complicated because reversible elastic strain and nonreversible nonelastic time-dependent deformations occur. Therefore, the thermomechanical behavior of refractories must be considered as the interrelation of four variables: stress, strain, temperature, and time.

Testing can be performed to client-defined parameters, including variable cross-head speeds, modulus of rupture configurations, loading rates, temperatures up to 1550 °C with variable ramp rates, and various load cycling regimes.

5.5.1 Refractoriness under Load

The temperature at which the material softens indicates refractoriness. Refractory under load (RUL) is a function of the *deformation behavior* of ceramic refractory products that are subject to *increasing temperature with constant load*. RUL is a crucial property for refractory bricks that reflects the service temperature of the shaped and unshaped refractories. It indicates the temperature at which the applied refractory (bricks or other shapes) would collapse in service conditions with a similar load. Generally, the refractoriness is represented by PCE (Pyrometric cone equivalent), and it must be higher than the application temperatures, as the refractoriness decreases under load. Thus, RUL is more crucial than refractoriness while designing refractory composition. A typical schematic representation of RUL setup is shown in Figure 5.15, which used a standard cylindrical shape 50 mm diameter and 50 mm height specimen. During the experiment, it measures softening temperature as refractory is a composite material, and a higher value compared to application temperature confirms the consistent performance of refractory.

Figure 5.16 represents a typical RUL characteristic of alumina containing refractory; the specimen is subjected to initiate deform (0.5%) beyond 1400 °C, and completely collapses or breaks at a temperature of 1480 °C. Thus, deciding on a safer service temperature limit in a particular situation is essential.

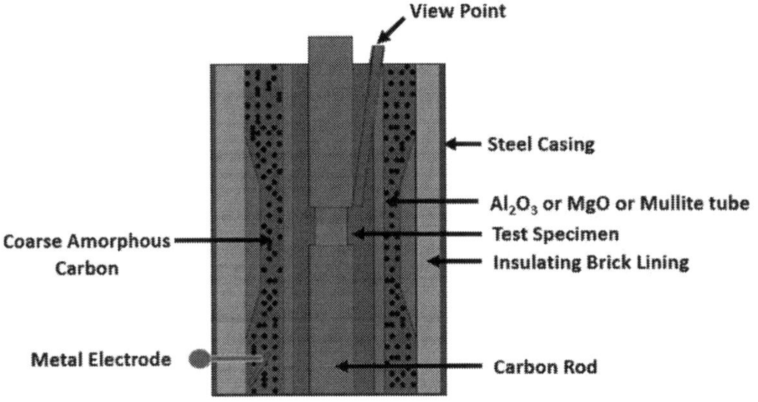

Figure 5.15 A typical RUL testing set up for refractory specimens.

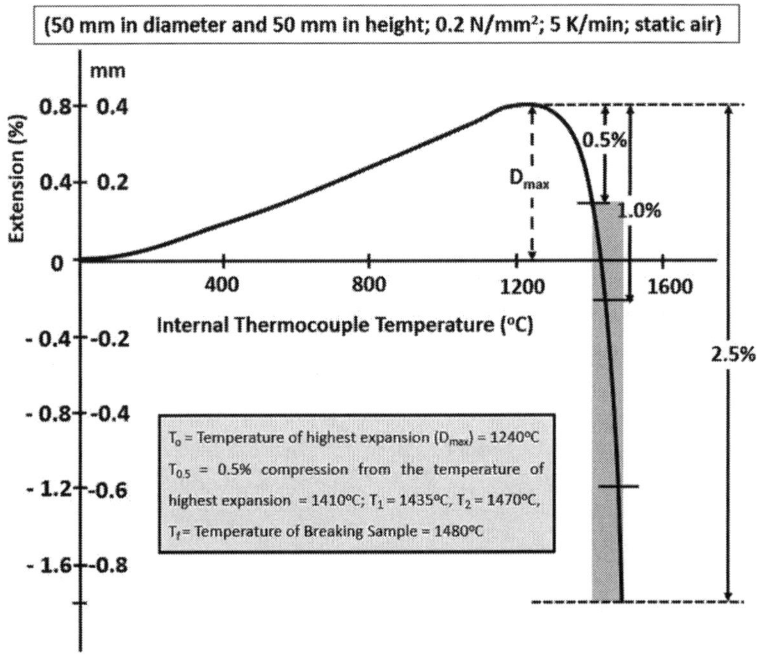

Figure 5.16 A typical RUL behavior at fixed load 0.2 N/mm^2.

Certain philosophies are essential to consider during design or to achieve high RUL, a) minimization of formation of glass phase under load and temperature, b) particle packing to reduce the porosity, and c) high compressive strength of starting raw materials (grains). However, contributing factors to the increased resistance of brick under load during application are a) more thorough distribution of liquid phase rather than localized accumulation, b) the growth of crystals through the influence of heat, and c) crystallization of a portion of the liquid during cooling.

Fireclay brick abruptly deforms, specimen height decreases beyond 1400 °C, and complete failure occurs 1650 °C with a size reduction of 14% from its original size, as shown in Figure 5.17, whether the sillimanite brick can resist more deformation at identical temperatures. It is interesting to note that S_iO_2 retains its strength up to close to its fusion point before sudden failure as it has very low liquid phase content, whereas fireclay softens over a large temperature range. Interestingly, chrome magnesia and magnesia chrome have a different degree of deformation resistivity at identical temperature regions, rather, MgO-Cr_2O_3 exhibits a bit of expansion behavior up to 1650 °C followed by shrinkage and deformation under load. Furthermore, silica for crown and coke ovens has different RUL behavior; thus, RUL data is essential to select the proper composition for designing the refractories.

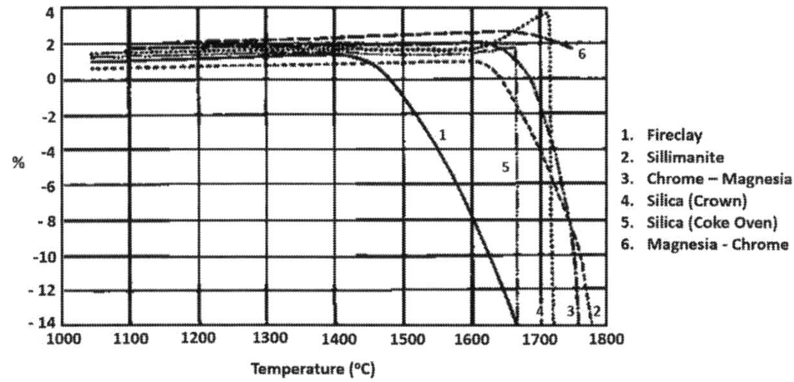

Figure 5.17 RUL behavior of different refractories upto 1800 °C.

5.5.2 Creep in Compression (CIC)

The phenomenon of continuing plastic deformation under constant stress can most clearly be defined as a creep. Creeping in ceramics usually occurs at high temperatures under moderate stress and low strain levels. Creep is often presented as a different condition than plastic deformation; the latter is viewed as it happens when stress is applied. Most frequently, creep is measured under constant stress. Creep strain ε , arising from constant stress application, is also described as:

$$\varepsilon = \varepsilon_0 + \beta\, t^{\frac{1}{3}} + kt \tag{5.35}$$

Where ε_0 is elastic strain and β and k are constants at a specific temperature for a defined material with a defined microstructure (defect structure also). The 2nd and 3rd terms in Equation 5.35 are a primary and steady-state creep, respectively.

5.5.2.1 Nabarro-Herring Creep

Nabarro demonstrates that one direction (say z direction) tension on a small cube of the material results in pure shear stress and sufficient x and y direction compression to produce resultant zero pressure [9]. In this circumstance, the vacancies flow through bulk via a face-to-face tension in compression, causing a shear of material to extend to the z-direction and reduce it to the x and y direction. If the result of Herring and the relation of simple tensile stress with the stress as mentioned in the above system are considered to provide simple shear, the resultant strain rate of the full grain boundary sliding is:

$$\dot{\varepsilon} = \frac{13.3 D_1 \Omega \sigma}{d^2 kT} \tag{5.37}$$

In a basic tensile test, where the tensile stress is σ. D_l is self-diffusion coefficient, Ω is atomic volume, d is average grain diameter, T is temperature, k is constant. The significant features of the equation are that $\dot{\varepsilon}$ is directly commensurate to stress and reversely proportionate to the squared grain diameter.

5.5.2.2 Coble Creep

While material transport moves across the grain boundaries, known as Coble creep. Usually, bulk diffusion prefers at high temperature where bulk diffusion is quicker than grain boundary diffusion. At lower temperatures; however, or for very fine-grained solids, the faster path may be grain boundary diffusion, in which case the expression for the creep rate, known as Coble creep, becomes:

$$\dot{\varepsilon} = \frac{150\ D_{gb}\ \Omega\ \sigma\ \delta}{\pi\, d^3 kT} \tag{5.38}$$

Where the constant of diffusion for transport within grain boundaries is D_{gb} and width of the grain boundary is d.

Coble creep is preferred over Nabarro Herring creep at lower temperatures because the activation energy for grain boundary diffusion is less than that for diffusion through the grains. Coble creep is also preferred over Nabarro-Herring creep for very small grain sizes due to the inverse cubic dependency on grain size. A typical schematic in Figure 5.18 represents the diffusion path for Nabarro-Herring (lattice diffusion) and Coble (grain boundary diffusion).

5.5.2.3 Measuring Creep

The creep test aims to precisely measure the rate at which secondary or steady-state creep occurs. Increasing the stress or temperature increases the *slope of the*

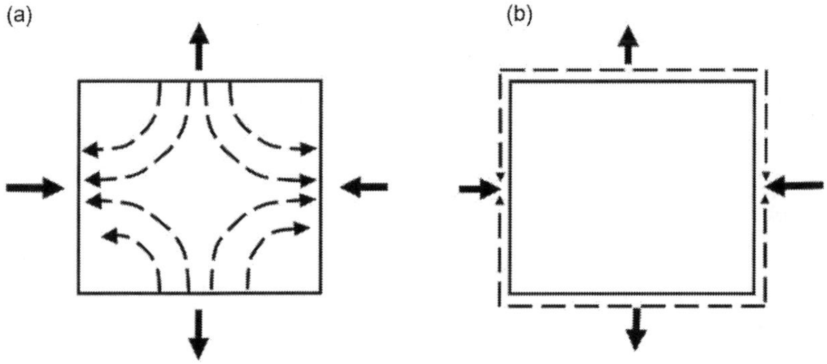

Figure 5.18 a) Nabarro-Herring creep through lattice diffusion. (b) Coble creep through diffusion at grain boundaries.

line, i.e., the amount of deformation in a given time increases. Several types of creep tests are available to assess high-temperature deformation of materials. Typical examples are tensile creep, compressive creep, flexural creep, indentation creep, etc. Creep in Compression (CIC) is most effective for refractories as these experience compressive stress under load and temperature during application. Under uniaxial compression, creep occurs not only in the axial direction but also in the lateral or normal directions. It is referred to as lateral creep. By analogy to elastic strains, lateral creep to axial creep is termed creep Poisson's ratio.

The compressive creep test preferentially required samples with a typical diameter of 35 mm and a height of 70 mm and heated within 5 K/min to the targeted testing temperature. After reaching the testing temperature, a certain dwell time, say one hour, is required to maintain a homogenized temperature. During fixing and heating up, a preload is to be applied, say 100 N, to fix the sample. Afterward, the load application can start at a 0.3 mm/min rate to the defined load. The testing time, i.e., the time from load application to fracture of the sample, depends on the creep resistance of materials, the testing temperature, and the applied load.

The creep response of a solid is usually discovered by calculating the strain rate as a function of the load applied. It can be achieved most effectively by applying a load to a sample, heating it, and calculating its deformation as a function of time. The resulting strain, as shown in Figure 5.19a, is plotted against time. Herein, three regions can be classified as,

Stage I: There is an initial response, almost instantaneous, followed by a decreasing rate of strain increase over time. This area is known as the region of primary creep.

Stage II: There is an area where the strain linearly increases with time. It is known as the steady-state or secondary creep stage, the most significant stage from a practical point of view.

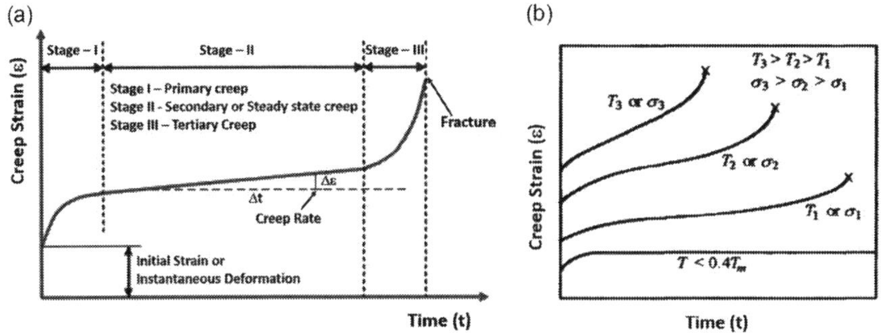

Figure 5.19 (a) Standard curves of strain versus time creep. Not all three regions are often experimentally observed, (b) the slope of the line increases with increasing stress and/or temperature.

Stage III: There is an area called the tertiary creep stage, where the strain rate increases rapidly over time, occurring just before the specimen fails.

The creep rate decreases initially with time (primary creep, its second derivative is negative), subsequently remains constant for a prolonged period (secondary creep, its second derivative is zero), and finally increases (tertiary creep, its second derivative is positive) before sample rupture, considering Equation 5.35 [10]. Increasing temperature and stress (Figure 5.19b) enhances the creep rates of both instantaneous strain and steady-state strain, which eventually reduces the time to failure. Data, as illustrated in the figure, can be further decreased by recording the steady-state creep rate $\dot{\varepsilon}_{ss}$ relative to the log of the stress applied at constant T. Such curves normally generate straight lines, indicating that:

$$\dot{\varepsilon}_{ss} = \frac{d\varepsilon}{dt} = \mathbf{A}\,\sigma^{p} \tag{5.36}$$

Where A is a temperature dependent constant, and p is the exponent of the creep law and generally lies between 1 and 8. This creep is frequently referred to as power law creep for $p > 1$.

Refractoriness under load (RUL, according to ISO 1893) is a measure of the deformation behavior of refractory ceramic products exposed to constant compressive stress (0.2 NN/mm^2) and increasing temperature up to a dwell time. The temperature varieties in which the softening occurs are not identical to the melting range of the pure raw material. Creep in compression (CIC, according to ISO 3187) refers to the percent of shrinkage of a refractory specimen under a continuous load and exposed to a constant high temperature over a long period of time. One drawback is that the creep starts during the heating up period and is superimposed by thermal expansion. Hence, the beginning of creep cannot be exactly defined. Furthermore, the stress level in the test is much lower than those in typical application cases [11]. A regular creep in compression behavior at 1400 °C is shown in Figure 5.20 and illustrates the measuring creep rate for 10 hours.

Creep data set at 1500 °C for different grade alumina brick, magnesia chrome brick, and low iron content magnesia brick is shown in Figure 5.21. While comparing alumina (65%) and corundum (95%) bricks, the secondary phase and impurity content reduce the creep resistance. Similarly, low iron containing MgO brick exhibit high creep resistance. Thus, raw material and process design is an important phenomenon to achieve an optimum dense refractory for a particular application.

5.5.3 Hot Modulus of Rupture

The strength of the refractory material at room temperature is given by the modulus of rupture (MOR)/cold modulus of rupture. It provides the optimum load carrying capacity in the bending phase at high temperatures. The mode of testing

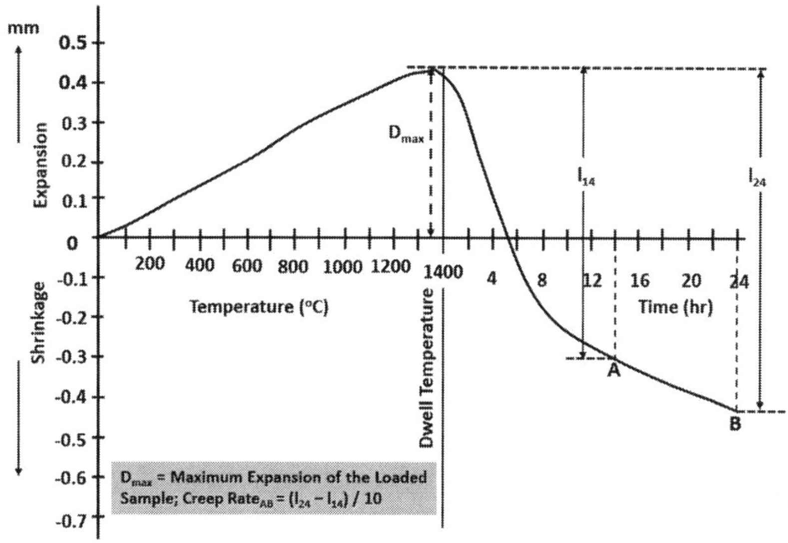

Figure 5.20 A typical creep behavior at fixed temperature 1400 °C.

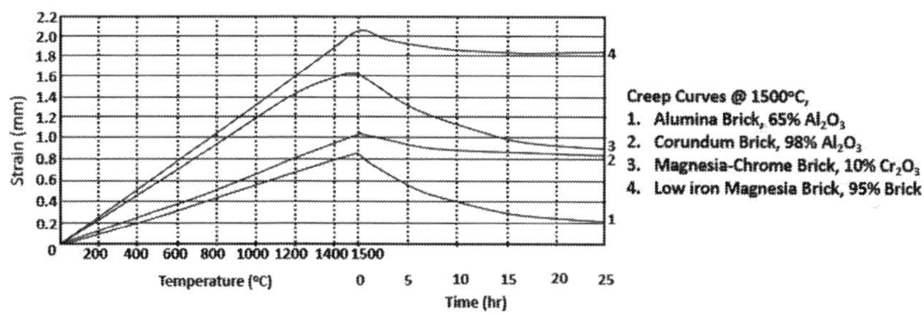

Figure 5.21 Creep behavior of different refractories at fixed temperature of 1500 °C.

is already discussed in Figure 5.7. The three-point bending test is used to detect the MOR of the sample. The regular dimension is 150 × 25 × 25 mm and the span length is usually 125 mm (ASTM C133). The unit of MOR is kg/cm^2. Despite high temperatures, the hot rupture modulus is analogous to the cold rupture modulus. During HMOR measurement, the specimen is preserved in the stated location in the apparatus and gradually fired at 1400 °C with a heating rate of 5 °C/min in the air atmosphere and kept for a dwelling time of 30 min, followed by performing a 3-point bending operation. A typical HMOR testing arrangement is shown in Figures 5.22a-b exhibits the refractory HMOR testing arrangement, and Figure 5.22b represents the broken specimen at an elevated temperature.

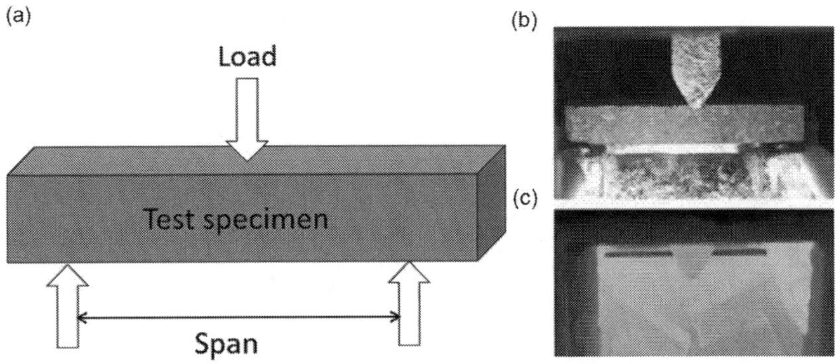

Figure 5.22 (a) Schematic representation of the test for 3-point bending, (b) sample position during testing, and (c) breaking of specimen at experiment temperature.

The HMOR measurement formula is $\frac{3PL}{2bd^2}$ unit of HMOR is kg/cm^2.
P = maximum load(kg) that breaks the specimen.
L = span length (mm), b = sample width (mm).
d = the specimen's depth (mm).

5.6 Wear

5.6.1 System-dependent Phenomena

Usually, the direct interaction and sliding against two refractory surfaces are rare in steelmaking refractories. However, slide gate plate movement can pick up as an example, which experiences relative motion, and in this aspect, a brief adhesive and abrasive wear are discussed. Furthermore, particle impingement on the refractory lining is a common incidence during material charging; thus a fundamental concept of erosion wear is also encountered. Furthermore, oxidative wear is important for carbon-based refractories. Usually, the wear phenomenon is typically measured by the wear rate. The wear rate per unit time or per unit sliding distance is the volume or mass of material removed. Other forms may be dimensionless, such as the material depth per unit sliding distance or the volume removed per apparent contact area and per unit sliding distance. Typically, the wear rate is not stable. The wear rate is, in general, a complex function of time. The wear rate can start off slowly and rise, or vice versa. The wear rate remains steady for a while and can alter when the change from one mechanism to another occurs throughout a wear test.

In the first region, the wear rate variation is referred to as the period of run-in or break-in, as shown in Figure 5.23. Wear during run-in relies on the structure

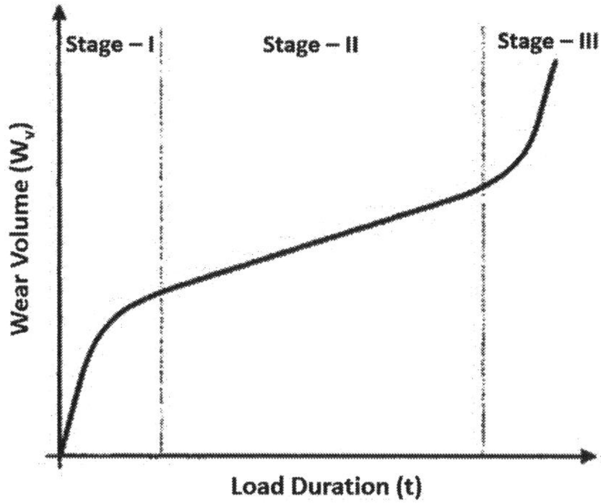

Figure 5.23 Three hypothetical wear volume situations as a feature of sliding distance displaying run-in, steady-state, and failure regions.

and properties of the initial material and surface conditions, such as surface finishing and the nature of any films present. During this transition phase, plastic deformation changes the surface roughness to a steady-state condition. Initial factors influence the damage and its duration during the transition phase. Eventually, the wear volume dramatically increases with time.

The wear rate of a material, like friction, depends on the counter face or mating material, surface preparation, and operation conditions. Wear rate is generally explained by means of the nondimensional wear coefficient (k) for a material pair. The wear rate is also known as the wear volume with respect to normal load applied and sliding distance (mm^3/Nm).

5.6.1.1 Wear of Ceramics

Refractory is a classic example of a mix of either ceramic oxides, non-oxides, and carbonaceous materials or a combination of all the above. Ceramics possess good mechanical strength and can never sacrifice much, except tensile strength. It can withstand high temperature (despite non-oxide ceramics and carbonaceous materials) and a corrosive environment. Effective mechanical properties with a smooth surface (low R_a) expedite minimum contact area and are less prone to damage under friction. In clean environments, friction coefficients and wear rates for ceramic tribocouples do not exceed very high values usually found in clean metals, particularly in ultra-high vacuum conditions or in the lack of oxygen. Ceramic materials exhibit limited plastic flow and even less ductility than metals at room temperature. The fracture toughness of ceramics is an important property in ceramic wear, usually low but can be upgraded through different toughening mechanisms. Operating conditions significantly influence wear modes and wear rates: normal load, sliding velocity and environment [12].

Tribo-chemical reactions of ceramics with liquid or gaseous environments govern the wear and friction of ceramics. Based on the chemical response between ceramic and the atmosphere, wear and friction can decrease or increase, altering the wear mechanisms and wear rates. These chemical interactions often lead to surface composition modifications and a reduction in the purely chemical form of wear through dissolution in a liquid environment. Still, it may experience chemical fractures, which enhance resultant wear rates.

As suggested for metals, no clear wear mechanism operates over a wide range of ceramic conditions. Typically, the wear residue is finely broken and can differ chemically from the bulk material. However, brittle wear mechanisms dominated by intergranular fractures lead to severe wear with higher wear rates and rough surfaces. Generally, the wear debris is angular and not chemically distinct from the substrate. Mild wear occurs when low pressures and velocities are combined, whereas extreme wear occurs when high pressures and velocities become effective.

5.6.2 Adhesive

Adhesive wear occurs when two nominally flat solid bodies are in sliding contact with each other (e.g., relative motion of slide plates), whether lubricated or not. At the interface, adhesion (or bonding) takes place, but those contacts are interrupted by rubbing. That may lead to the removal of a fragment from one surface and attachment to another surface. The relocated fragments come off the surface from which they are moved and loosely relocated back to the initial surface as the sliding continues or else form loose wear particles. During repeated loading and unloading behavior, some are fractured by a fatigue mechanism, resulting in loose particle formation.

Various mechanisms are suggested for the detachment of a fragment from the surface. These are proposed that shearing may happen at the initial interface or the weakest point of one of the two contact bodies. In most situations, the interfacial adhesion strength is assumed to be small compared to the breaking strength of the corresponding local regions; hence, the fracture during shearing takes place on the surface (path 1) in most contacts, and no wear occurs during the sliding period. A break in one of the two bodies may occur at a small fraction of the contacts (path 2), and a small fragment (shaded area in Figure 5.24) may be connected to another surface. These transfer particles are irregularly shaped and blocky. Galling, scuffing, welding, or smearing are sometimes considered extreme forms of adhesive wear; however these words are also used generally to define other types of wear.

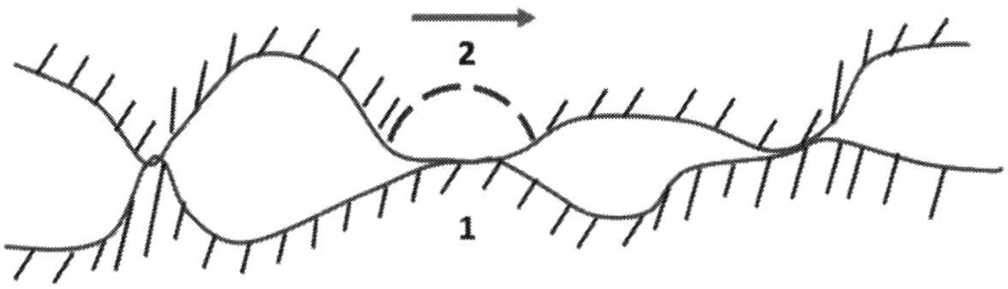

Figure 5.24 Schematic showing two break possibilities (1 and 2) during interface shearing.

5.6.3 Abrasive

Abrasive wear arises when rough, hard surface or hard particles slide over a softer surface and damage the interface through plastic deformation or fracture. In the case of ductile materials with high fracture toughness (e.g., metals and alloys), the plastic flow of the softer material results from hard asperities or hard particles. During sliding, most metallic and ceramic surfaces show strong signs of plastic flow, even some of the brittle ceramic materials also. Contacting metal asperities deform at the lightest loads plastically. In the case of brittle materials with poor fracture toughness, brittle fracture causes wear. In these cases, significant cracking takes place in the worn zone.

There are two general abrasive wear situations (Figure 5.25). In the first case, for example, in mechanical operations such as grinding, cutting, and machining,

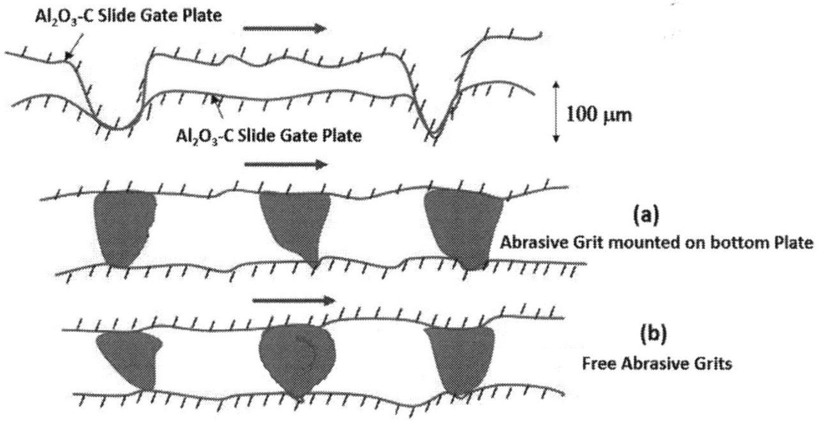

Figure 5.25 (a) Rough, hard surface or surface placed on a softer surface with abrasive grits, and (b) free abrasive grits collected among surfaces with at least one surface that is softer than the abrasive grits.

the hard surface is the hardest of two rubbing surfaces (two-body abrasion); and in the second case, the hard surface is the third body, usually a tiny particle of abrasive, trapped between the other two surfaces and sufficiently hard to detach either one or both of the mating surfaces (three-body abrasion), for example in free abrasive lapping and polishing.

The wear mechanism is adhesive at the beginning. In some situations, it creates wear debris stuck at the interaction resulting in three-bodies abrasive wear. Scratching (also the softer surface) in most abrasive wear situations is seen as a pattern of chords parallel to the direction of sliding (plowing). Depending on the degree of severity, scraping, grading, or gouging are other words often loosely used for abrasive wear.

5.6.4 Erosive

Erosion may occur through jets and streams of solid particles, liquid droplets, and the implosion of bubbles produced in the fluid. Erosion of solid particles occurs by impingement of solid particles (as seen in Figure 5.26a). It is a type of abrasion that is normally handled somewhat differently since when it meets a surface, the *contact stress occurs from the kinetic energy* of particles moving in an air or liquid stream. The kinetic energy of impinging particles is given by the particle velocity and impact angle combined with the size of the abrasive, that is, the square of the velocity. As a result of repetitive impacts, the surface becomes eroded and wear debris is produced during erosion, as seen in Figure 5.26b.

Erosive wear is initiated by plastic deformation and/or brittle fracture, depending on the material being eroded away and operational conditions, as in the case

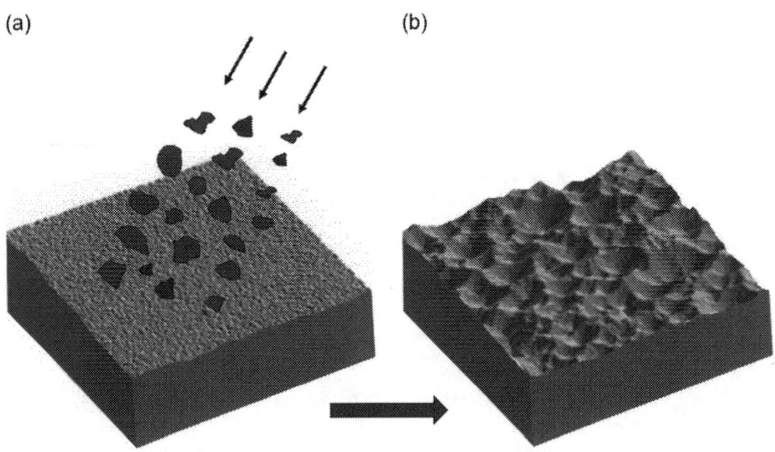

Figure 5.26 A diagram of a jet of abrasive particles striking a surface with high speed.

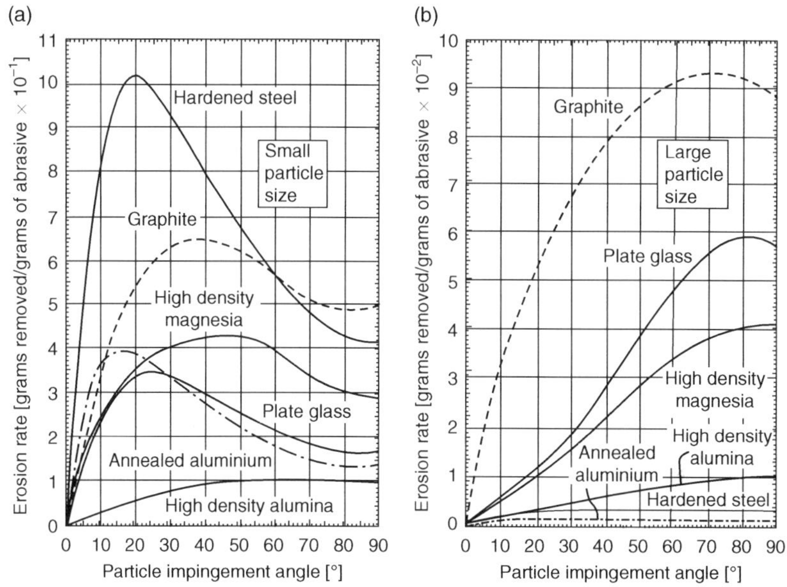

Figure 5.27 Erosive wear rate, as a function of attacking angle of impinging particles (with respect to the material plane) [13] / with permission of Elsevier.

of abrasive wear. The dependence of the wear rate upon impact angle for ductile and brittle materials is different, as shown in Figure 5.27 [13]. Moreover, the erosive particle size influences and changes the highest erosion rate in particle impingement angle. A plastic deformation process through which the material is separated by the displacing or cutting action of the eroded particle makes ductile materials wear. On the other hand, inside a brittle material, the formation and intersection of cracks that radiate from the impact point of the eroded particle remove material.

The size and morphology of abrasive particles determine the pattern of plastic deformation over each indentation; thus displacing the amount of material from each impact. In the case of brittle materials, the shape of abrasive particles will influence the degree and intensity of cracking. Compared to the more rounded particles, sharper particles result in more localized deformation and subsequent wear.

5.6.5 Oxidative

Chemical or corrosive wear happens in a corrosive atmosphere when sliding takes place. In the air, oxygen is the most dominant corrosive medium. Chemical wear in the air is thus commonly referred to as oxidative wear. The chemical corrosion products (e.g., oxides) usually form a film less than a micrometer thick on the surfaces without sliding, which eventually helps slow down or even avoid the corrosion. Still, the sliding motion wears away the chemical film so that the virgin expose further, and the chemical attack proceeds repeatedly. Thus, both chemical reactions (corrosion) and rubbing facilitate chemical wear. Despite oxidative wear, high-temperature-assisted corrosion is the most dangerous wear process that deteriorates the refractory performance and is extensively discussed in Chapter 6.

References

1. Barsoum, M. Fundamentals of Ceramics.
2. Sarkar, D. (2018). *Nanostructured Ceramics: Characterizations and Analysis*. crc press.
3. Stachurski, Z.H. (2006). Testing materials' hardness science, technology & application. *Mater. Forum*. 30.
4. Sarkar, D. and Maiti, H.S. Zirconia. MSEC.
5. Wachtman, J.B., Cannon, W.R., and Matthewson, M.J. (2009). *Mechanical Properties of Ceramics*. Wiley.
6. Evans, A.G. and Blumenthal, W. (1983). High temperature failure in ceramics. In: *Fracture Mechanics of Ceramics*, 6 (ed. R.C. Bradt, A.G. Evans, D.P.H. Hasselman, and F.F. Lange), 423–448. New York: Plenum.
7. Evans, A.G. (1978b). Microfracture from thermal expansion anisotropy—I. single phase systems. *Acta. Metall.* 26: 1845–1853.
8. Hasselman, D.P.H. (1985). Thermal stress resistance of engineering ceramics. *Mater. Sci. Eng.* 71: 251–264.
9. Nebero (1948).
10. Stueckelschweiger, M., Gruber, D., Jin, S., and Harmuth, H. (2019 1 June). Creep testing of carbon containing refractories under reducing conditions. *Ceram. Int.* 45 (8): 9776–9781.
11. Gruber, D. and Harmuth, H. (2014). Thermomechanical behaviour of steel ladle linings and the influence of insulations. *Steel Res. Int.* 85 (4): 512–518. doi: https://doi.org/10.1002/srin.201300129.
12. Bhusan, B. Introduction to tribology.
13. Stachowiak, G.W. and Batchelor, A.W. (2014). *Engineering Tribology*, 4e. [In] Abrasive, Erosive and Cavitation Wear. Elsevier Butterworth-Heinemann. 525–576.

6

High Temperature Refractory Corrosion

6.1 Introduction

Refractory materials erode during service in harsh environments due to thermodynamic and kinetic variables, including high temperature, high pressure, molten metal, slag, and glass, oxidizing and reducing atmosphere. Corrosion of refractories not only reduces their service life but can also lead to catastrophic failure, resulting in unplanned shutdowns of manufacturing/processing units. Manufacturing and installing new refractories consume energy and natural resources, in addition to economic and working-hour losses and safety risks. Solid-state processes, as well as interactions with liquid phases and gas species, can cause refractory corrosion. Liquid phases consist of molten metals and slags that are particular to each application. Water vapor, reducing gases, alkali vapor, and halogens are examples of corrosive gas species.

Failure results from various events and causes, and it is worthwhile to investigate from the ground up to comprehend and provide suitable slag design for future refractory improvement. It is necessary to correlate critical parameters and slag–refractory interaction phenomena to design refractories. High temperature corrosion of refractories is defined as the wear of the exposed refractory face expedited through chemical assault by a corroding fluid in a process in which the refractory and the corroding fluid interact in the zone of contact. Corrosion reactions progress in a direction toward chemical equilibrium at a limited level. As described in Chapter 2, phase equilibrium diagrams are important for understanding corrosion circumstances and anticipating chemical solutions to reduce corrosion and wear rates. Such a concept provides two possibilities for those interested in refractory corrosion. The first is the *phenomenological approach*, which views corrosion as a chemical and physical process without detailed phase equilibrium diagrams. The second is to employ phase equilibrium diagrams with

Fundamental Design of Steelmaking Refractories, First Edition. Debasish Sarkar.

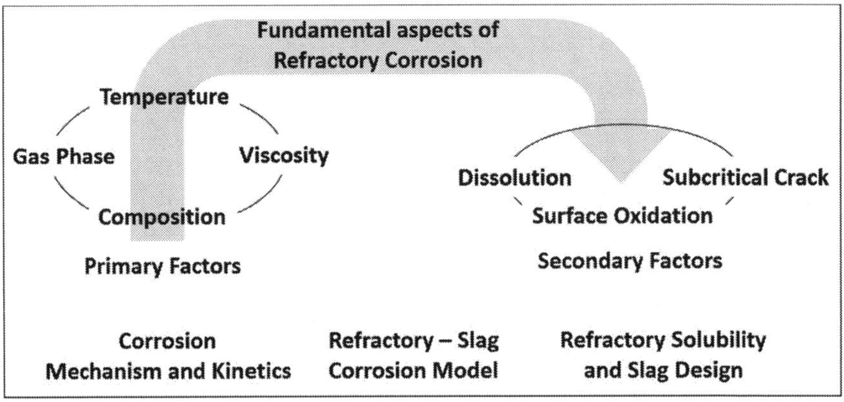

Figure 6.1 A quick scheme on the basics of the high temperature corrosion of refractories.

the information from the phenomenological technique. For a complete grasp of refractory corrosion, the second choice is essential

Figure 6.1 represents the scheme of the entire discussion, where critical parameters are essentially temperature, gas phase, composition, and viscosity of interactive materials that eventually expedite the secondary factors comprising dissolution, subcritical crack growth, and surface oxidation. In applying steelmaking refractories, the environment provokes to tolerate high temperature, mechanical stresses, and resist attack by liquids such as molten metal, slags, fluxes, and corrosive atmospheres. Attack of refractories is a complex phenomenon involving chemical wear (corrosion) and physical/mechanical wear (such as erosion/abrasion) processes that may act synergistically. Thus, a cumulative approach on the mode of penetration, corrosion mechanism, and kinetics and their simulation with an established model provides the failure analysis of refractories. Eventually, the basic philosophy to design the slag chemistry and viscosity with an understanding of refractory solubility is considered and discussed as sometimes such phenomena are mandated during steel processing and operation.

6.2 Thermodynamic Perceptions

Metal corrosion in contact with an aqueous environment follows different degrees of chemical reaction with the variation of solutes in aqueous media, temperature ($<100\,^{\circ}$C), pH, etc. On the other hand, the corrosion of carbon-based refractory, a predominant component for steel processing, is initiated by the oxidation process of solid–gas or solid–liquid interactions. Graphitized carbon surfaces are preferentially non-wetting with molten metal and slag but dissolved oxygen in molten liquid may initiate the oxidation of carbon and other elements like Al, Si, and B that eventually transform to their oxides and interact with refractory matrix and frequently form low eutectic phases. The corrosion reaction system is influenced by temperature (>1000 °C), pressure, concentration and mobility, solid, liquid, and gaseous reactants, solvent concentrating, molten phase reactivity, viscosity, and capillary effects of the molten reactants and slags.

The complexity of the oxidation phenomena generally requires the graphical representation of a thermochemical analysis, which is most often employed in oxidation, to be called a diagram by Ellingham. It describes the standard free energy of formation versus temperature diagrams, which allow the thermodynamic data for oxides to be represented in a compact form. Despite this ΔG–T plot, the vapor species diagram allows the vapor pressures of compounds to be presented as a function of conventional variables such as partial pressure of a gaseous component. The plot of Ellingham is a handy way of representing the amount of 1 mol of oxidizing species (M), and is the standard equation:

$$\frac{2a}{b}M(s)+\ O_2(g)\rightarrow\frac{2}{b}M_aO_b(s) \tag{6.1}$$

The free energy change (ΔG) can be calculated by using the Equation (6.2):

$$\Delta G_T=\Delta G_T^o+RT\ln\frac{a_{M_aO_b}^{2/b}}{a_M^{2a/b}\,.\,P_{O_2}} \tag{6.2}$$

Where, $\Delta G^o{}_T$ = standard free energy of formation; R = gas constant; T = absolute temperature; a = activity of solid; P_{O2} = oxygen partial pressure. The obtained ΔG value predicts the process; if the calculated $\Delta G < 0$ is oxidation, $\Delta G = 0$ is an equilibrium state; and $\Delta G > 0$ is reduction process.

The standard free energy $\Delta G^o{}_T$ of the process at a particular temperature T is:

$$\Delta G_T^o=\Delta H_T^o-T\Delta S_T^o \tag{6.3}$$

Where, ΔH_T^o and ΔS_T^o are conventional reaction enthalpy and entropy, respectively. In balance ($\Delta G = 0$), from Equation (6.2),

$$\Delta G_T^o=-RT\ln\frac{a_{M_aO_b}^{2/b}}{a_M^{2a/b}\cdot P_{O_2}} \tag{6.4}$$

Where the oxide is stoichiometric, and there is no miscibility between metal and oxide, the metal and oxide activities are considered uniform, and the P_{O2} is hence the partial oxygen pressure with which metal and oxide are balanced. The partial pressure of reacting gas is given by:

$$P_{O_2} = \exp\left(\frac{\Delta G^o}{RT}\right) \tag{6.5}$$

The oxide dissociation pressure is defined where both metal and oxide coexist at a definite oxygen partial pressure. Thus, if the oxygen partial pressure of the atmosphere is greater than the dissociation pressure, a metal is oxidized, and if it is lower than the dissociation pressure, the oxide is reduced to metal.

As shown in Figure 6.2, ΔG^oT (or RT ln P_{O2}) is plotted versus temperature for the oxides in the Ellingham diagram. Although ΔS and ΔH vary with temperature, the changes are so small as to be negligible, and so the lines are approximately straight. It indicates that they are of the form $y = mx + c$. The standard enthalpy and entropy of the formation of the compounds are also considered to be independent of temperature over large temperature ranges, an Ellingham approximation. In the diagram, $\Delta G^o{}_T$ values are expressed in kJ.mol^{-1} O_2 indicates straight lines and the change in the slope corresponding to the phase changes of the metal or the compound, which alters the entropy change. The lower position of the line diagram represents more stabilities of oxides.

The values of p_{O2} may be obtained directly from the oxygen monograph on the Ellingham diagram by placing a straight line from the origin marked O ($T = 0$, $\Delta G = 0$) through the free energy line at the temperature of interest and estimating the oxygen partial pressure from their meeting point at the right side p_{O2}. Similarly, it is possible to obtain the oxygen potential of the environment in terms of p_{O2}, p_{H2}/p_{H2O}, p_{CO}/p_{CO2}, and the oxygen partial pressure in equilibrium with the oxide of interest to determine whether or not the oxide is likely to form thermodynamically. The most stable oxides have the largest negative values of $\Delta G^\circ{}_T$, or the lowest values of p_{O2}, or the highest values of p_{H2}/p_{H2O} and p_{CO}/p_{CO2}.

In addition to air, industrial gaseous atmospheres may consist of complicated gases (CO_2, H_2O, SO_2, SO_3, etc.), multiple gas combinations, or contain gaseous corrosion products or heat breakdown products from corrosion. Vapor species that form in any given high temperature corrosion situation strongly influence the attack rate. The rate is generally accelerated when volatile corrosion products are essential in the kinetics of high temperature oxidation of carbon, silicon, aluminum, etc. [1]. Preferentially, the vapor pressure data in oxide systems are log p_{MxOy} versus log p_{O2} at a constant temperature, and Arrhenius diagrams of log p_{MxOy} versus $1/T$ a constant oxygen pressure provide the characteristic behavior oxides [2].

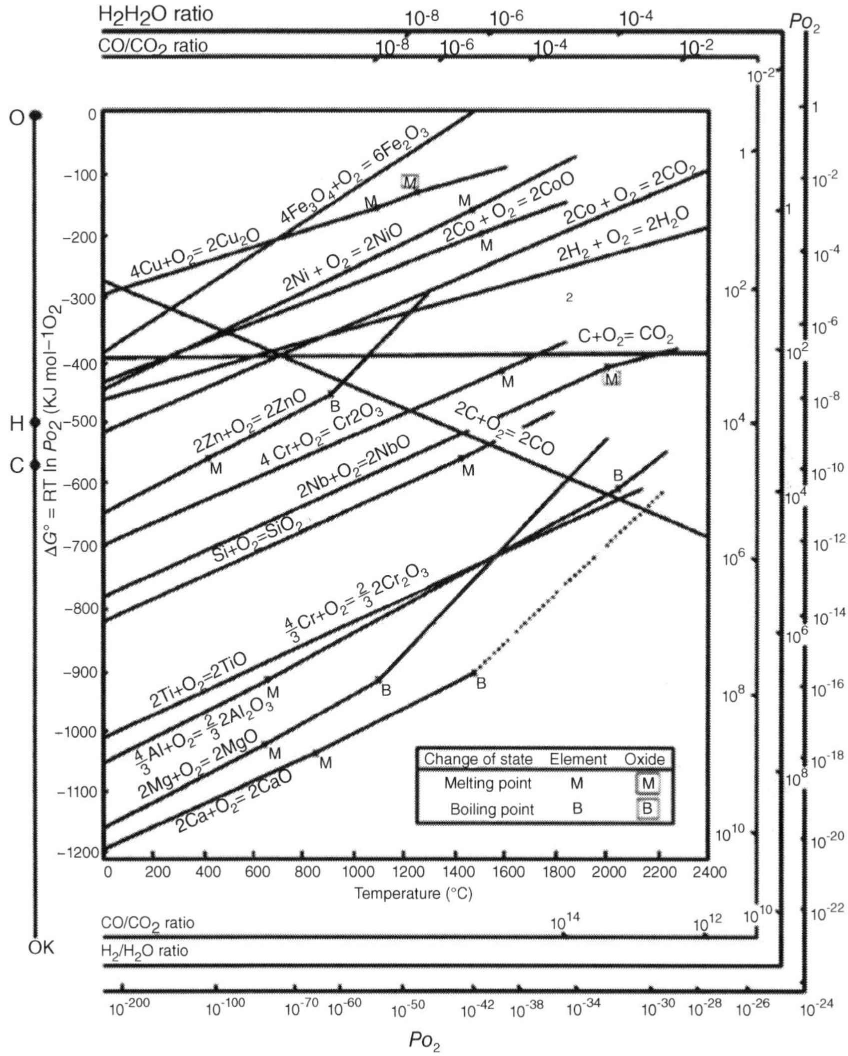

Figure 6.2 Standard free energy of the formation of oxides as a function of temperature.

6.3 Effect of Temperature and Water Vapor

Temperature, viscosity, composition, and atmosphere significantly influence imitating the corrosion of refractories. Temperature expedites the reaction rate, and the low viscosity of molten material facilitates penetration, whereas the molten acidic/basic oxides and atmosphere are responsible for developing in-situ formation of multi oxide phases. A brief on the effect of temperature and water vapor is discussed, and other critical parameters such as the influence of viscosity and molten materials are analyzed later.

The rate of most dissolution reactions grow exponentially with temperature, as shown by the Arrhenius equation:

$$Rate = A_o e^{\frac{-E_a}{RT}} \tag{6.6}$$

The pre-exponential factor, A_o, has a theoretical complexity and generally is considered empirical. Activation energy E is used to measure the barrier height, which must be overcome during the reaction. Diffusion processes offer low energy barriers, and E acquires a few kJ a mole values. Surface reactions typically have a higher 80 kJ/mole barrier, while recrystallization processes have several hundred kJ/mole activatory energies [3]. Experimental value E, the rate dependence, rate constant, diffusiveness, or solution ion uptake dependent on temperature can be estimated. These data can be used to demonstrate the sort of mechanism for dissolution. Since the values of activation energy for the dissolution of reaction are usually the same, regardless of the chemistry of dissolution, the temperature dependency for many different types of reactions seem reasonably similar.

Like refractory classification, the gaseous environment can also be categorized as acidic, basic, and neutral, and will be discussed later when required. Herein, the major focus is given to the influence of water vapor on the few critical oxide-based refractories such as SiO_2, MgO, and Al_2O_3, which are frequently used for shaped and unshaped refractories. If the pressure of the generated volatile species is below 10^{-6} atm, the oxides are likely to become stable. The volatile species otherwise evaporate and cause the refractions to recess [4]. The corrosion mechanism is significantly more complex and depends on the refractory and environmental elements as well as the evenly located temperature gradient of the refracting lines. Mechanisms dominate refractory corrosion in a gas environment:

- Volatile production and evaporation of species resulting in refractory linings regression
- Corrosion from the deposition of the contaminants in the low-temperature zone induced by condensation and deposition
- Dissociation of refractories

Low cement castable uses a substantial amount of Al_2O_3- and SiO_2-based raw materials, whereas carbon-based MgO and Al_2O_3 refractory use Al and Si metal powder that eventually oxidizes and forms different carbide and oxide phases further. In considering the importance of such refractories, the effect of water vapor on three major oxides, SiO_2, Al_2O_3, and MgO are discussed.

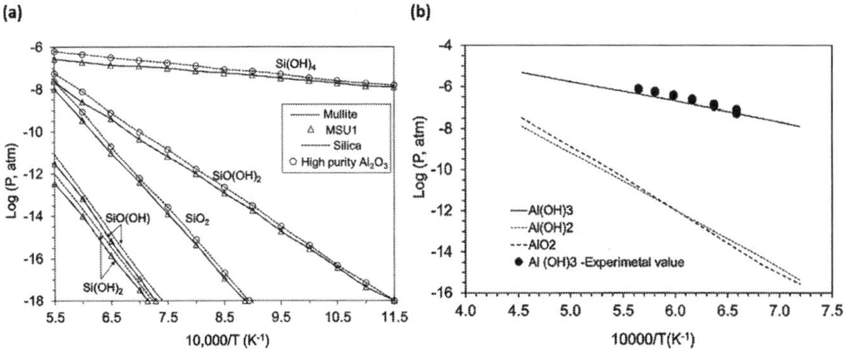

Figure 6.3 Equilibrium partial pressures of (a) Si vapor species form mullite ($3Al_2O_3.2SiO_2$), MSU1 (3.7% $3Al_2O_3.2SiO_2$, 96.3% Al_2O_3), silica (SiO_2), high purity alumina (99.96% Al_2O_3, 0.04%SiO_2) in humidified air (O_2: 13.217%, H_2O: 36.909%, Ar: 0.589%, CO_2: 0.019%, N_2: 49.267%). P(total) = 1 atm [5]. (b) Al vapor species form Al_2O_3 in 0.5 atm H_2O and 0.5 atm O_2 mixture Adapted from [6].

While considering alumina-containing SiO_2 (e.g., for the LCC-70), various volatile silicon vapors such as SiO(OH), $SiO(OH)_2$, $Si_2O(OH)_6$, $Si(OH)_4$ develops due to SiO_2 interactions in the air vapor. The main vapor species are $Si(OH)_4$ and $SiO(OH)_2$, since their partial pressure is considerably higher than the other vapor species, as seen in Figure 6.3a [5].

$$SiO_2 + 2H_2O = Si(OH)_4 \tag{6.7}$$

$$2SiO_2 + H_2O = 2SiO(OH) + ½\,O_2 \tag{6.8}$$

$$SiO_2 + H_2O = SiO(OH)_2 \tag{6.9}$$

$$2SiO_2 + 2H_2O = Si_2O(OH)_6 \tag{6.10}$$

Various researchers have experimentally established the evaporation at 900 °C–1000 °C of $Si(OH)_4$ and $SiO(OH)_2$ at 1100 °C–1500 °C [7]. The partial pressure calculated for these types is between 10^{-6} and 10^{-8} atm.

According to reactions in Equations (6.11–6.14) between Al_2O_3 and water vapor, $Al(OH)_3$, $Al(OH)_2$, AlO(OH), AlO, and AlO_2, are the common volatile species of aluminum [6].

$$½\,Al_2O_3 + 3/2\,H_2O = Al(OH)_3 \tag{6.11}$$

$$½\,Al_2O_3 + ½\,H_2O = Al(OH) + ½\,O_2 \tag{6.12}$$

$$\frac{1}{2}\,Al_2O_3 + \frac{1}{2}\,H_2O = AlO(OH) \tag{6.13}$$

$$\frac{1}{2}\,Al_2O_3 + H_2O = Al(OH)_2 + 1/4\,O_2 \tag{6.14}$$

$Al(OH)_3$, followed by $Al(OH)_2$ and Al_2O_3, is the predominant species while other species' partial pressure is insignificant. The estimated values of these species' partial pressures are closely matched to the experimental values derived by transpiration, as illustrated in Figure 6.3b. However, high temperature facilitates the evaporation rate of volatile species.

MgO fabricated by carbonate calcination at 1000°C followed by sintering at 1800°C, and their hydration was studied in classic literature to understand the influence of temperature, vapor pressure, and eventually calculated the activation energy with hydration from MgO to $Mg(OH)_2$ mechanism [8]. Figure 6.4a represents the linear relation between both fraction reacted x, and $\alpha(x)$ with time, where

$$\alpha(x) = \left[1-(1-x)^{\frac{1}{3}}\right] = \left(\frac{v}{r_0\rho}\right)t \tag{6.15}$$

v = mass rate of reaction per unit area, ρ = spherical particle density, t = time, r_0 = radius of unreacted material at time t = 0. Equation (6.15) reveals that the

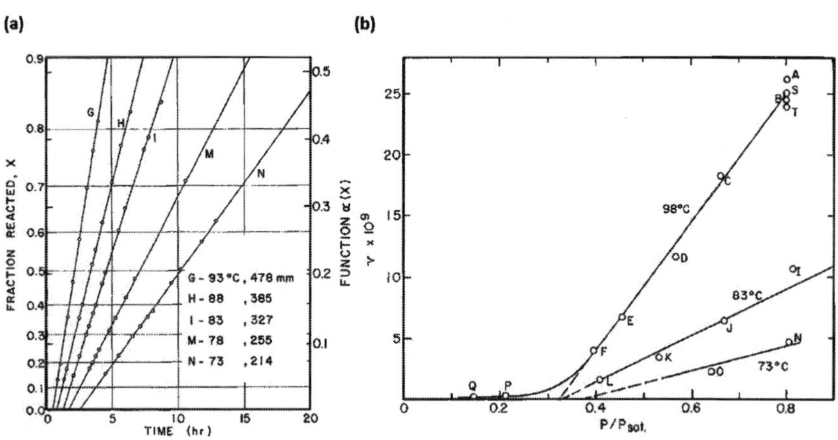

Figure 6.4 (a) Fraction of x of MgO reacted as a function of time for various temperature-conditions of pressure. The ordinates of x correspond to equal increments of the function set to the right scale. The direct line relationship verifies that the value of a value of $\alpha(x)$ varies linearly with time, and (b) reaction rate v by water vapor pressure for samples of 98°, 83°, and 73°C of MgO powder hydrated.

reaction efficiency depends on particle size; smaller particles have higher surface energy and hydration than larger particles.

Figure 6.4b displays an image of the reaction v versus (p/p_{sat}) at 98°, 83°, and 73°C, where the hydrating reaction is carried out by a vapor pressure-dependent mechanism that exceeds an adsorbent equilibrium pressure p^* to the maximum of 0.3 p_{sat}. At low pressures, due to insufficient adsorption, the reaction proceeds relatively slowly.

Nuclear creation is extremely slow at low temperatures and vapor pressures, but the reaction process controls the supply of liquid surface water film, which eventually speeds the kinetics and the creation of hydroxide via surface micro cracks and fractures. According to the Arrhenius type equation, the reaction rate consists of a temperature of 16,100 cal per mole.

6.4 Slag–Refractory Interactions

More than two-thirds of crude steel output worldwide comes from iron ore-based steel. Crude steel is made in two ways in the primary process. By combining the iron ore with coal/coke and celestone in a blasting oven (BF) the first approach reduces to iron and generates BF slag (BFS). The second step is a basic oxygen furnace to transform iron into steel (BOF). The BOF slag (BOFS) is manufactured utilizing the transported iron from the BF in the steel production process. In the second process, the scrap-based steel industry is produced slags. Scrap-based steel is used in one-third of worldwide steel production. In an electric arc furnace (EAF), the technique recycles steel with electrical energy to produce an electric arc that smashes the scrap. The EAF Slag (EAFS) and the second stage are carried out in the initial phase of the manufacturing of scrap-based steel. Other sources, such as metallic iron, are also used in some plants. After the BOF and EAF, the steel is put through a ladle furnace (LF) to refine the molten steel to obtain different steel qualities. The LF slag (LFS) is the product of steel refinement and is usually a heavy metal carrier such as chrome, zinc, etc.

Thus, starting from primary to refining steelmaking process has several scopes to form different slags. Refractory composition design is needed to synchronize steel composition, slag composition, and operation parameters that eventually enhance the vessel life. Steel slags are generally categorized by the type of furnace in which they are manufactured. The slag's qualities rely on the type of process for producing crude steel, the slag's cooling conditions, and the recovery procedure. Slag preparation with a prerequisite composition is an essential technology to produce the desired composition of steel and nonferrous metal production. It eventually improves the production process and products. Table 6.1 displays the main slag kinds of the steel industry, and their probable processing for application and properties [9] / with permission of Elsevier. The fundamental aspects of the slag–refractory interaction during the steelmaking process are cataloged as follows.

Table 6.1 Oxide composition of main types of iron and steelmaking slags [9].

		Chemical composition													
		CaO	SiO_2	Al_2O_3	MgO	Fe(Total)	SO_3	MnO	TiO_2	P_2O_3	Free CaO	Density (g/cm^3)	Processing	Application	Mainproperties
Blast furnace	ABS*	30–45	30–40	10–15	5–15	0–1	0–2	0–1	-	0–0.1	-	2.5	Slow cooling (air cooled), crushing, sieving	Concrete and road aggregate	Crystalline, density, strength
	GBS*	30–45	30–40	10–15	5–15	0–1	0–2	0–1	-	0–0.1	-		Rapid cooling (water cooled), grinding (<100 μm)	Cement, concrete addition	Vitreous, hydraulic
Basic oxygen furnace	BOFS*	30–50	10–15	1–3	5–15	10–30	0.1–0.3	1–5	0.5–2	1–3	2–10	3.3	Slow cooling, crushing, sieving	Concrete and road aggregate	Crystalline, strength, high skid resistance
Electric arc furnace	EAFS-S*	40–50	25–30	2–10	3–10	0–2	0–2	1–3	0–1	-	-		Slow cooling, crushing, sieving	Cement addition and concrete	Crystalline, strength, high skid resistance
	EAFS-C*	20–40	10–20	2–10	5–15	20–40	0.1–0.5	2.5–5	0.5–1	0.5–2	0.5	3.5	Slow cooling, moistened, crushing, sieving	Concrete and road aggregate	Crystalline, strength, dense, high skid resistance
Ladle furnace	LFS*	35–60	5–50	1–30	4–18	0–5	0–3	0–2	0–3	0–0.5	0.5–20	2.6	Slow cooling, moistened, crushing, sieving	pH stabilizer, roads, cement	Crystalline

ABS: Air-cooled Slags; GBS: Granular Blast Furnace Slag; BOFS: Basic oxygen Furnace Slag; EAFS-S: Electric Arc Furnace Slag (stainless steel); EAFS-C: Electric Arc Furnace Slag (crude steel); LFS: Ladle Furnace Slag.

6.4.1 Diffusion in Solids

Diffusion is a fundamental and ubiquitous phenomenon in nature, as a process for mitigation and mixing caused by irregular particle movement (atoms, ions, molecules). Therefore, it is a process where particles intermingle due to their spontaneous thermal agitation movement. It is manifested in the solid state by the migration of "solute" atoms through a defective structure of the system. The diffusion process at elevated temperatures is improved as the defect creation rate increases with the atoms' vibratory energy. However, a solid's inherently rigid structure leads to a slower diffusion process than liquids and gases. Vacancies and interstitial atoms are the primary points defects inside a crystal mesh and are, in general, caused by and combined of these diffusions in the solid state. Various atomic diffusion mechanisms in crystals through interstitial, vacancy, divacancy, interstitial, and interstitial-substitutional exchange mechanisms are the prime resource of atomic diffusion, as shown in Figure 6.5. The solute atoms in the interstitial mechanism are usually integrated into the interstitial sites and a solid-interstitial solution,

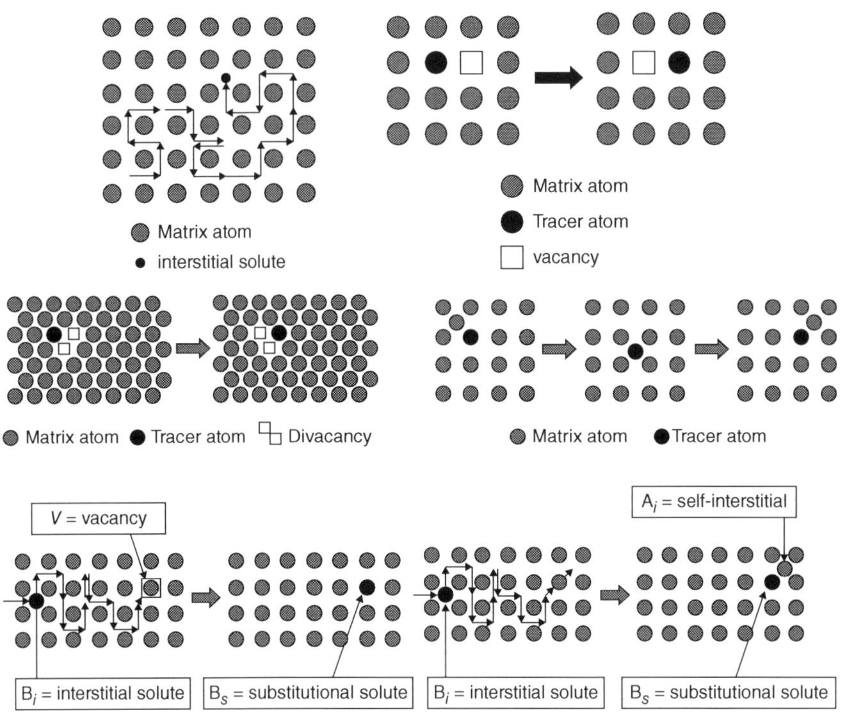

Figure 6.5 (a) Direct interstitial mechanism, (b) vacancy mechanism, (c) divacancy mechanism, (d) interstitial mechanism, (e) dissociative mechanism in interstitial-substitutional exchange mechanism, and (f) kickout mechanism in interstitial-substitutional exchange mechanism.

much smaller than solvent (lattice) atoms (i.e., hydrogen, carbon, nitrogen, and oxygen). In the octahedral or tetrahedral positions, interstitial solutes are commonly used. The vacancy mechanism defines the transmission into a neighboring vacant site of the self-atoms or substitutional solution atoms.

Self-atoms or replacement solute atoms can also be diffused via bonded vacancy pairs, known as vacancies and vacancy pairs. In the interstitial (or indirect interstitials) mechanism, self-interstitials act as diffusion vehicles. A self-interstitial replaces a substitutional atom which then, in turn, replaces a neighboring lattice atom. This mechanism is the counterpart of the vacancy mechanism since the self-interstitial is the anti-detect of the vacancy. Some foreign atoms B may be dissolved on interstitial (B_i) and substitutional sites (B_s). Such atoms may diffuse via the dissociative mechanism or the kickout mechanism. The two mechanisms have in common that the diffusivity of foreign atoms is much higher when they are located in interstitial sites than when they are located in substitutional positions, whereas the opposite is true for the solubilities. Under such conditions, the rapid diffusion of B_i and the subsequent conversion to B_s may lead to the incorporation of B-atoms.

There are two types of interstitial substitution processes that can be identified. The transition includes vacancies (V) according to the dissociative mechanism:

$$B_i + V \, f \, B_s \tag{6.16}$$

This mechanism is also denoted as Frank–Turnbull mechanism or the Longini mechanism. If the changeover involves self-interstitials (Ai), according to:

$$B_i \, f \, B_s + A_i \tag{6.17}$$

the mechanism is denoted as kickout mechanism.

The transport of any charge, mass, or momentum is related to the flux J of the diffusing species and is linearly proportional to the concentration gradient dc/dx of that species at any point is:

$$J_A^B\left(\frac{mol}{m^2 s}\right) = -D_A^B\left(\frac{\partial C_A}{\partial x}\right)\left(\frac{m^2}{s}.\frac{mol}{m^3}.\frac{1}{m}\right) \tag{6.18}$$

Where D_A^B is the chemical diffusion coefficient of species A in matrix B. The self-diffusivity D of an atom or ion measures the case and frequency with which that atom or ion jumps around in a crystal lattice without external forces, i.e., in a random fashion. The diffusivity depends on the stoichiometry and purity level of ceramics. Experimentally, it has long been appreciated that D is thermally activated and could be expressed as

$$D = D_0 \exp\left(-\frac{Q}{kT}\right) \tag{6.19}$$

Where Q is the diffusion activation energy, D_0 and Q vary for both interstitial diffusion (Equation 6.20) and vacancy diffusion (Equation 6.21):

$$D_0 = \left(\frac{\lambda^2 \nu_0 z}{6}\right) \exp\left(\frac{\Delta S_m^i}{k}\right) \tag{6.20}$$

$$D_0 = \lambda^2 \nu_0 z \exp\left(\frac{\Delta S_\nu - \Delta S_m}{k}\right) \tag{6.21}$$

Where λ is average jump distance of atom during diffusion, ν_0 is the vibrational frequency (the atomic frequency), z is the coordination number (i.e., the number of interstitial sites that an atom can make in a single jump), ΔS_m^i, the entropy for interstitial migration. The subscripts v and m refer to vacancy formation and vacancy migration, respectively. k represents the frequency of attempts to jump, also known as "the frequency factor" or "the pre-exponential factor."

An interstitial defect is random, no interstitial formation energy is required, and no defect concentration enters the diffusion equation. However, the activation energy (Q) for the vacancy migration is the sum of the activation energies for vacancy formation, Q_v, and migration (E_m).

The randomness of interstitial diffusion is faster than vacancy diffusion as the surrounding atom in the former is weaker and prefers to jump rapidly. Although oxide self-diffusion is an issue, the rate is often well below the sulfides of the metal oxides, as illustrated in Figure 6.6a. However, as demonstrated in Figure 6.6b, chemical diffusion coefficients are comparable in metal oxides and sulfides. In the bulk of transitional metal sulfides, the much-increased auto diffusion rates are essentially the product of much higher levels of defects and not mobility.

6.4.2 Oxidation

The probability of oxidation of ceramic oxides is negligible, but the diffusion of self or foreign atoms is a common incidence and is a function of temperature. In the steelmaking process, a major share refractory is made of carbon, and their different degree of oxidation is expected during different steel processing steps. Despite discussion on the oxidation of metal or formation of oxides through the Ellingham diagram, an argument on the fundamental understanding helps to calculate the cumulative decarburization time of MgO-C refractory during application in the ladle.

The crude steel from the melting aggregates is partially treated in the ladle, followed by elemental synchronization. Such process is known as secondary metallurgy, includes degassing, decarburization, desulfurization, dephosphorization and removal of trace components, fine allocation content adjustment, deoxidation, and inclusion transformation, improved pureness and solidifying

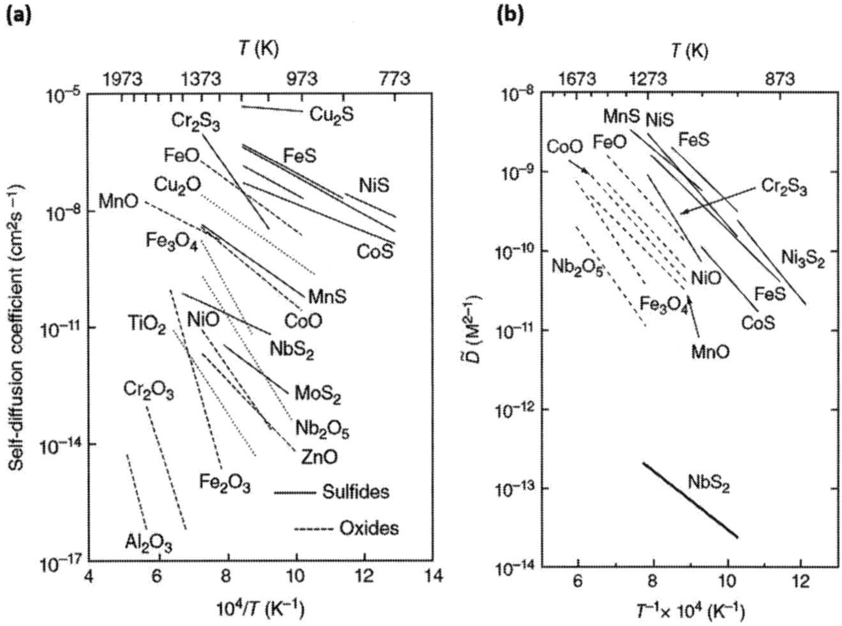

Figure 6.6 a) Comparison of self-diffusion coefficients in several metal sulfides and oxides, b) Chemical diffusion coefficient in niobium sulfide on the background of analogous data obtained for several metal sulfides and oxides [10] / John wiley & sons.

structure, temperature, and chemical adjustment and homogenization. The cold ladle fitted with MgO–C bricks is preheated at elevated temperature in the steel shop for two to four hours to decrease thermal stresses and probable wear due to the real temperature gradient during the filler with steel (T up to 1800 °C). There is no newly walled available in the ladle with a slag layer that protects the stems of the binding carbon from oxidation (graphite 5–15 wt.%). Thus, the graphite oxidizes immediately as a consequence of this preheating technique. The carbon bonding and beneficial characteristics such as the good thermal shock and weather resistance are destroyed by slag. A huge pore volume is also opened at the same time (approximately 35% volume). Steel and slag can easily penetrate and lead to "inner" disintegration in the microstructure. The time advance of the reaction:

$$2C + O_2 = 2CO \tag{6.22}$$

The following differential equation can be represented in a MgO–C brick:

$$\frac{dx_i}{dt} = -2V_c D_p \frac{n_0 - n_i}{x_0 - x_i} \tag{6.23}$$

A sharp change between outside and core is expected due to decarburization, and diffusion controls such a topo-chemical process. The decisive factor is the D_P coefficient of porous diffusion:

$$D_p = D_{O_2} \gamma p_c \quad \left(\text{cm}^2\text{h}^{-1}\right) \tag{6.24}$$

where,

$$D_{O_2} = 2.7 \times 10^{-2} T^{1.79} \quad \left(\text{cm}^2\text{h}^{-1}\right) \tag{6.25}$$

and

$$\gamma = 1.84 \times \sqrt{p_c} - 0.3 \tag{6.26}$$

where D_{O2} is the oxygen diffuser, T is the labyrinth factor (~1373 K), and the p_c is the open porosity of the decarbonized area (~0.35). The result is D_P = 3081 cm^2 h^{-1} for the input of these variables. The answer for the differential Equation (6.23) is provided by [11] for an infinite plate:

$$(1-\phi)^2 = \tau \tag{6.27}$$

where,

$$\phi = \frac{x_i}{x_0} (x_i = x_0 \text{ at } t = 0) \tag{6.28}$$

Where x_i is the position of the reaction front in the plate (cm), and x_o is the total thickness of the plate (cm):

$$\tau = \frac{t}{t_0} \tag{6.29}$$

When t is time (h) and t_0 is time for total decarburization of a plate with x_0 and:

$$t_0 = \frac{x_0^2}{2zV_c D_p \left(n_0 - n_i\right)} (h) \tag{6.30}$$

where V_C is the mole volume of graphite (5 cm^3 mol^{-1}), z, the geometry factor (infinite plate: $z = 2$); n_0, the oxygen content at the plate surface (20 vol%; 2×10^{-6} mol cm^{-3}); and n_i is the oxygen content at the reaction front ($n_i \ll n_0$, $ni \rightarrow 0$). A MgO–C brick with x_0 = 10 cm would be totally decarburized within t_0 = 1014 h (near to 42 days). The time t required to decarburize 1 cm (x_i = 9 cm) of the brick results as follows:

$$\tau = \left(1 - \frac{9}{10}\right)^2 = 0.01 \tag{6.31}$$

With $t = \tau\, t_0 = 0.01 \times 1014h = 10.14h$

Numerous investigations have shown that the process is actually controlled by diffusion, and that some additional additives such as 10 wt.% ZrB_2 are widely prevented. Aluminum (3 wt.%) has a limited antioxidant effect if refractory has high porosity, and it reacts with retained carbon when $T \geq 800\ ^{o}C$ and forms Al_4C_3 and probable $MgO\text{-}Al_2O_3$ spinel at $T \geq 1200\ ^{o}C$.

6.4.3 Infiltration

The degree of contact of the molten substance with refractory is the initiation point of penetration, alteration of microstructure that eventually expedites the dissolution. In this assumption, the wetting theory describes the contact of pure oxide and graphite with fluid or refractory slag material in a thermodynamic balance. The vectors, which act at a time of communication between all three stages, have mechanical equilibrium in all probable surface tensions of refractory–slag systems:

$$\sigma_R - \sigma_S cos\theta - \sigma_{R/S} = 0 \tag{6.32}$$

Where σ_R is the refractory material's surface tension (mNm^{-1}), S is the steel or slag's surface tension (mNm^{-1}), R/S is the interfacial tension between refractory material/steel or refractory material/slag (mNm^{-1}), and θ is the wetting angle. The wetting angle of contact characterizes the interactions of the three phases involved:

$$cos\theta = \frac{\sigma_R - \sigma_{R/S}}{\sigma_S} \tag{6.33}$$

Where,

$$\theta > 90^{\circ},\ if \sigma_R - \sigma_{R/S} < 0\ ,\ \text{(non-wettability)}$$

$$\theta = 90^{\circ},\ if \sigma_R - \sigma_{R/S} = 0$$

$$\theta \langle 90^{\circ},\ if \sigma_R - \sigma_{R/S} \rangle 0\ \ \text{(wettability)}$$

The surface tensions and the wetting angle of contact are the critical parameters for steel or slag infiltration of a capillary or pore, and can be evaluated experimentally. Capillary suction pressure or the pressure difference between the two ends of the capillary, ΔP ($N.m^{-2}$) as shown in Figure 6.7a and can be represented as:

$$\Delta P = \Delta P_c + \Delta P_s = \frac{2\sigma_s cos\theta}{r} + \rho_S g h \tag{6.34}$$

ΔP_c is the capillary pressure, ΔP_s is the hydrostatic pressure caused by the penetrated slag ρ_s density of steel/slag (kgm^{-3}), r is pore radius (m), g is the acceleration of gravity ($m\ s^{-2}$), and h is the penetration height (m) from the pore entry or slag surface. The hydrostatic pressure ΔP_s may be neglected compared to ΔP_c because of $\Delta P_s \ll \Delta P_c$ in the initial stage of penetration when contact angle is very low (e.g., MgO–molten slag contact).

Thus, Equation (6.34) becomes Equation (6.35), and the effective contact angle is $\theta < 90^o$.

$$\Delta P = \frac{2\sigma_s cos\theta}{r} \tag{6.35}$$

When the contact angle is greater than 90° for a non-wetting surface, an external (ferrostatic) pressure is required to allow the liquid steel to permeate into the pore or capillary system and calculate the penetration depth (h). The same is true for a slag that does not wet the graphite phase of the refractory oxides.

It is evident that the dynamic viscosity (η, Pa.s) of the molten steel/slag plays a significant role to estimate the penetration length along with pore dimension and pressure, and the penetration rate dh/dt of slag into a capillary can be represented by Poiseulle's law:

$$\frac{dh}{dt} = \frac{r^2 \Delta P}{8h} \tag{6.36}$$

It is worth noting that the brick temperature gradient has a significant effect on the penetration depth through its impact on the η. Combination of Equations (6.35) and (6.36):

$$\frac{dh}{dt} = \frac{r\sigma_s cos\theta}{4l} \tag{6.37}$$

Integrating and rearranging both sides of Equation (6.37) gives:

$$h = \sqrt{\left(\frac{r\sigma_s cos\theta}{2}\right)t} \tag{6.38}$$

According to the above discussion and relationship, slag penetration can be reduced by increasing viscosity or contact angle or lowering surface tension. A schematic representation as shown in Figure 6.7b can illustrate this phenomenon. This penetration gradually decreases when molten slag transfers from hot refractory face to cold backup lining because of high viscosity at the cold end. From Equation (6.38), a simple relation in penetration height and time is:

$$h = k_c t^{0.5} \tag{6.39}$$

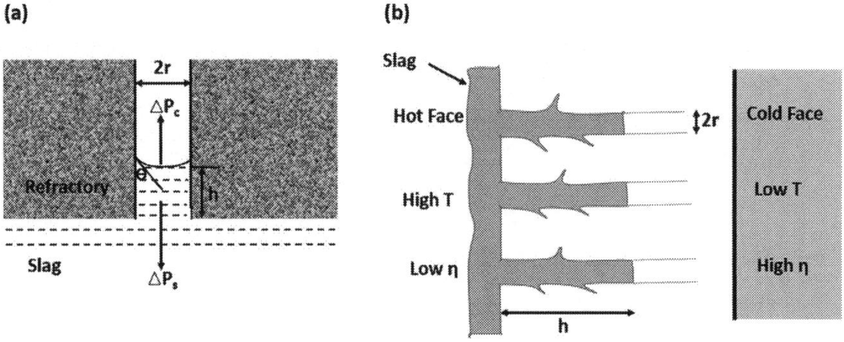

Figure 6.7 Schematic illustration of slag penetration (a) model into a vertical capillary, (b) in the typical refractory lining in furnace.

k_c can be regarded as the penetration coefficient, which may vary to wetting and mode of penetration. The diffusive mass transit rate inside the liquid penetration into the refractory is affected by viscosity, associated with Stokes-Einstein relationship:

$$D = \frac{kT}{6\pi r} \tag{6.40}$$

The ionic diffusivity D in the slag is inversely proportional to the viscosity; therefore changes in composition (e.g., due to solid dissolution) that reduce viscosity also boost transport in the slag, speeding degrading processes depending on diffusive mass transfer. As a result, the liquid and solid react during penetration, especially at high temperatures, altering the matrix phases and microstructure. The reaction products, on the other hand, can be solid, liquid, or gas (or a combination of these), and they can stay attached to the solid (typically solid or viscous liquid), be fugitive (usually fluid liquid or gas), or be a combination of both. While considering the alteration or evolution of refractory matrix microstructure in contact of penetrating slag, the grain boundary (γ_{gb}) and solid/liquid (γ_{sl}) interfacial energies and can be represented as,

$$\gamma_{gb} = 2\,\gamma_{sl} \cos\left(\frac{\varphi}{2}\right) \tag{6.41}$$

Where ϕ is the dihedral angle. If the ratio $\gamma_{gb} / \gamma_{sl} \geq 2$ or ϕ is zero liquid can penetrate into grain boundaries at equilibrium. If $\gamma_{gb} / \gamma_{sl} > \sqrt{3}$, ϕ is $< 60^0$ and the liquid can appear along all three-grain edges as a continuously connected phase.

If γ_{gb} / γ is between 1 and $\sqrt{3}$, then ϕ is between 60^0 and 120^0, and liquid can only partially penetrate along grain boundaries, whereas $\gamma_{gb} / \gamma_{sl} < 1$, when

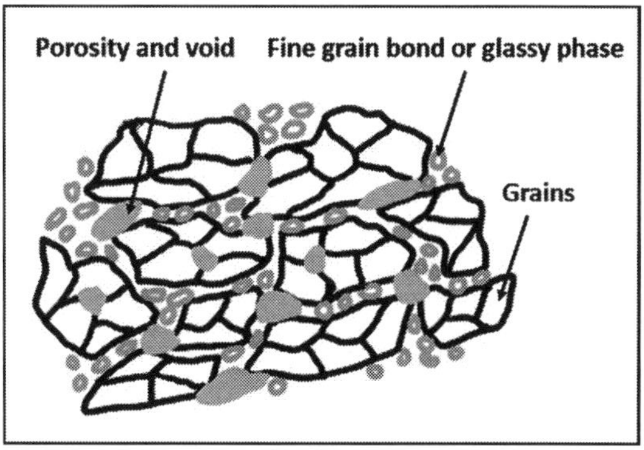

Figure 6.8 Schematic diagram of fired refractory microstructure.

$\phi > 120^{0}$ and no penetration will occur. A schematic of refractory microstructure as shown in Figure 6.8, consists of different phases, including inclusion that eventually accelerates the penetration through pores and grain boundaries.

Through changes in crystallite shape and crystallite development processes, penetration modifies the microstructure of dense refractory. As the system seeks to minimize the overall interface area while maintaining the volume of the solid crystallite constant, the solid crystallites, a unit of polycrystalline aggregate, tend to revert to their equilibrium shape after penetration. It necessitates the dissolution and reprecipitation of the material. Isotropic interfacial energy for MgO, for example, tries to form spheres, whereas anisotropic interfacial energy for Al_2O_3 tries to form equilibrium shapes given by Wulff construction [12].

6.4.4 Dissolution

The following three distinct stages govern dissolution at refractory/melt interfaces:

a) transport (or diffusion) of reactants from melt to the refractory/melt interface,
b) chemical reaction (or solution) at the refractory/melt interface, and
c) transport of products from the refractory/melt interface to the melt.

The rate determining step is used to define the different types of dissolution, which are referred to as transport types of dissolution or interface reaction regulated for step (a) or (c) and diffusion controlled for step (a) or (c) (b). The process is known as *mixed controlled* when neither reaction nor diffusion is the rate-controlling element. Dissimilar chemical phases tend to react at high temperatures; therefore, the refractory and liquid in contact should be of similar chemical composition to restrict the extent of dissolution. Thus, acidic SiO_2 (B_2O_3, P_2O_5, and

V_2O_5) refractories, for example, are utilized with acid liquids (coal gasifiers and glass melts). MgO and CaO are basic in nature and prefer to be used in basic melts (e.g., basic oxygen steelmaking, cement production). Some refractories, such as $MgCr_2O_4$ and carbon, are neutral and can withstand acid and basic slags. Chemical reaction rates are aided in the iron and steelmaking process by congruent or homogeneous dissolution atoms from the solid dissolve straight into the liquid melt. When the diffusivity of reaction products is quicker than the rate of chemical reaction at the interface, direct dissolution can be a reaction or interface control or transport or diffusion control if it is slow. As a result, the transport process typically controls the rate of refractory dissolution in the slag. A reaction of first order concerning a reactant species, whose beginning rate can be described, can directly regulate the dissolution process:

$$J = K\left(\frac{A_C}{A_o}\right)C_m \tag{6.42}$$

Where J is the dissolution rate (g.cm^{-1}.s^{-1}), K is the rate constant, A_c is the true area of refractory (cm^2), A_o is the apparent area of refractory (cm^2), C_m is the concentration of reactant in the melt (g.cm^{-3}). Surface flaws like grooves and porosity, which in the preceding equation raise the ratio A_c/A_o, have a major impact on another microstructural and crystal structural effect. The atoms are diffused from the interface by direct dissolution proportionately to $t^{1/2}$, since in the absence of liquid convection or agitation, reactants are depleted, and dissolved species accumulate.

If the removal rate of reaction products by diffusion is lower than the rate of chemical reaction, a solute-rich frontier layer is built up with a product-saturated interface with the refractory. Concentration gradient provides the driving force for dissolution reactions. The dissolution process is then regulated by the spread of reactants over the boundary layer or outside the products. The so-called indirect, incongruous, or heterogeneous resolution when the border layer results in a solid interface being formed. The Nernst equation can express the rate of corrosion:

$$J = D\frac{\left(C_s - C_m\right)}{\delta} \tag{6.43}$$

Where D is the diffusion coefficient (cm^2.s^{-1}), C_s is the saturation concentration of refractory in the melt (g. cm^{-3}), C_m is the concentration of a reactant in the melt (g. cm^{-3}).

$$\delta = \frac{\left(C_s - C_m\right)}{dc/dx} \tag{6.44}$$

Where dc/dx is the interfacial concentration gradient. In equilibrium, the saturation concentration of refractory composition in the melt and melt saturation in the refractory is balanced. Figure 6.9 represents the effective boundary layer thickness in the concentration gradient.

However, under natural convection alter the dissolution rate; thus J becomes [13] V. G. Levich [1962], Pearson Education,

$$J = 0.5D(C_s - C_m).(\Delta\rho g / D)^{\frac{1}{4}}\, x^{-\frac{1}{4}} \tag{6.45}$$

Where $\Delta\rho$ is the difference in density between saturated and bulk melt (g. cm^{-3}), η the dynamic viscosity (poise), g is the gravitational constant (cm.s^{-2}), and x is the distance from the leading edge of the solid surface (cm). Despite natural convection, forced convection is a common phenomenon during melt agitation; thus the velocity of fluid U replaces the $\Delta\rho$ as a variable. The dissolution rate for such agitation can be rearranged from the above Equation (6.45), and turned into a more useful Equation (6.46):

$$J = 0.5D\frac{(C_s - C_m)}{3.09\left(\frac{\vartheta x}{U}\right)^{\frac{1}{2}}\left(\frac{D}{\vartheta}\right)^{\frac{1}{3}}}. \tag{6.46}$$

Where, U is the fluid bulk velocity (cm.s^{-1}) and ϑ its kinematic viscosity (cm^2.s^{-1}). Suppose the refractory oxide in the bulk slag has been saturated under these conditions, J = 0. As a result, $C_s - C_m$ must be decreased to reduce the dissolution rate. For example, increased MgO content in slag decreases the corrosion of

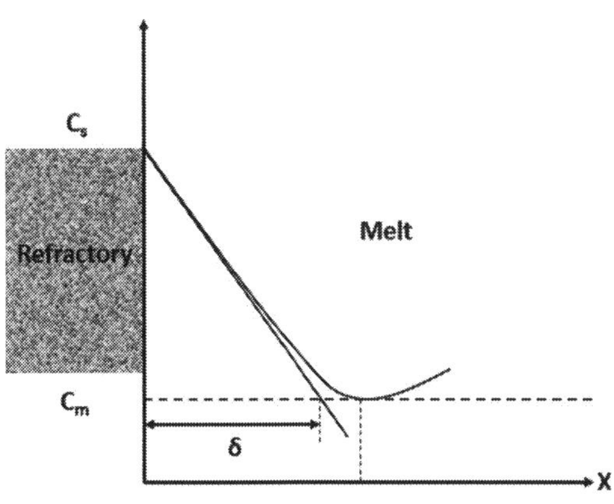

Figure 6.9 The concentration distribution at the interface of a static melt and solid refractory.

periclase, the primary phase in many basic refractories. If $C_m = 0$, then the value of $C_s - C_m$ reaches a maximum; thus so does the dissolution rate. As a result, the solubility of refractory oxides in molten slag and the saturation concentration at the refractory–slag interface is critical in refractory corrosion. In practice, dissolution is frequently under mixed control, making the distinction between first-order reaction and diffusion control unclear.

6.4.5 Crystallite Alteration

During refractory corrosion, not only are solid phases dissolved, but nearby liquids change the composition and become saturated with particular compounds, eventually precipitating and growing out of the saturated solution. It is the opposite of corrosion and very crucial in indirect dissolution. The cooling path on an equilibrium phase diagram must be analyzed to determine if phases originate from the liquid at temperature or during cooling. However, these are frequently unavailable, or systems involve more than the three or four phases that can be (relatively) simply interpreted on a diagram. For each system, thorough thermodynamic calculations utilizing Gibb's energy minimization modules are required to understand phase evolution [14].

The components of a solid with the highest solubility are the first to dissolve during dissolution. Quite often, a smaller radius of curvature persists for smaller particles, and angularly formed protuberances have a larger specific surface area and, as a result, a faster dissolving rate. Pressure at particle/particle connections (necks) boosts solubility, making these junctions dissolve quickly as well.

Because the solid dissolved redeposits on larger particles or in concave places, the removal of such features causes an overall rounding of the microstructure, which is commonly highlighted (shown schematically in Figure 6.10).

The morphology of the crystals developing can be inferred by evaluating their interfacial energies with the liquid phase, as discussed above in the section *Slag Penetration into Refractories*. If the crystals are isotropic, these strive to form

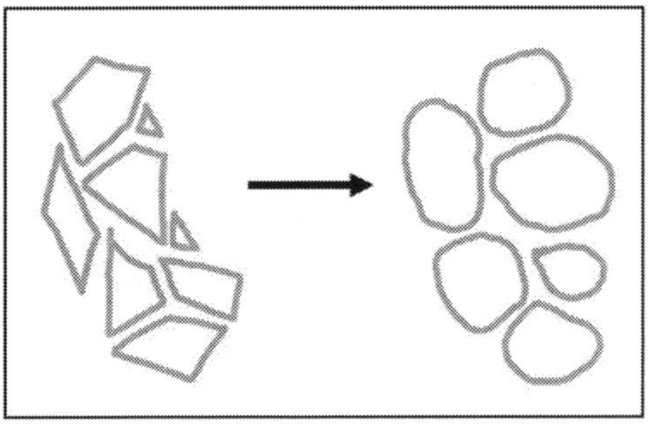

Figure 6.10 Dissolution and redeposition leads to general rounding of angular crystals.

spheres, but if the interfacial energies are anisotropic, then try to create the Wulff construction's equilibrium forms. MgO (spheres), Al_2O_3 (prismatic), mullite (needlelike or cuboidal), calcium hexaluminate (elongated tabular), and other minerals have this equilibrium structure. Local composition changes, such as phase separation or impurity segregation to or from a growing surface, can cause changes in the equilibrium form of the crystal morphology. Hence, these shapes may not always occur for these phases in complicated refractory–slag systems.

Cr_2O_3 from a $CaO.MgO.SiO_2$ (CMS) composition liquid was dissolved in solid MgO, resulting in high crystallographic anisotropy of surface energies and angular ($MgO–Cr_2O_3$) crystal morphologies. The growth of Al_2O_3 is known to be controlled by the rate of reaction at the liquid/alumina interface, whereas MgO growth is controlled by mass transport diffusion via the border liquid phase. Slag penetration into a refractory can result in solid-state diffusion of slag species into a grain phase, followed by exsolution (precipitation) as the refractory cools. This behavior occurs because the diffusing species are supersaturated at lower temperatures than the granular phase. Cr_2O_3 in MgO grains in $MgO–Cr_2O_3$ bricks and (Mg,Fe,Mn)O in MgO grains in doloma are two examples of resolved precipitates seen in refractories [15, 16].

6.4.6 Endell, Fehling, and Kley Model

In classic literature, a hydrodynamic model has been proposed by Endell, Fehling, and Kley to simulate the slag–refractory interactions [17]. The above findings eventually concluded that slag attack is mainly determined by the pace of slag supply and the viscosity of the slag, with only a minor influence from the degree of inclination of the refractory wall. The thickness of the interaction film was estimated based on experimental data, and it is significantly impacted by (a) slag feed or supply, or the amount of slag deposited on the refractory each hour, and (b) slag removal or the speed at which it drains down the refractory. Considering hydrodynamic principles, Fehling proposed alternative equations for a velocity of flow (W_o) and film thickness (d).

$$W_0 = 0.85Re^{0.05}\sqrt[3]{\frac{4.g.sin\alpha}{\nu}\left(\frac{V}{b}\right)^2} \tag{6.47}$$

$$d = 1.31Re^{0.02}\sqrt[3]{\frac{2.\nu}{g.sin\alpha}\left(\frac{V}{b}\right)} \tag{6.48}$$

$$\text{Where } R_e = Reynolds\ number, \frac{W_0.d}{\nu}$$

where η= dynamic viscosity (poises); ϑ = kinematic viscosity ($cm^2/sec(=\eta/\rho)$); α = angle of path of flow to horizontal; ρ= density (gm/cm^3); g = acceleration of

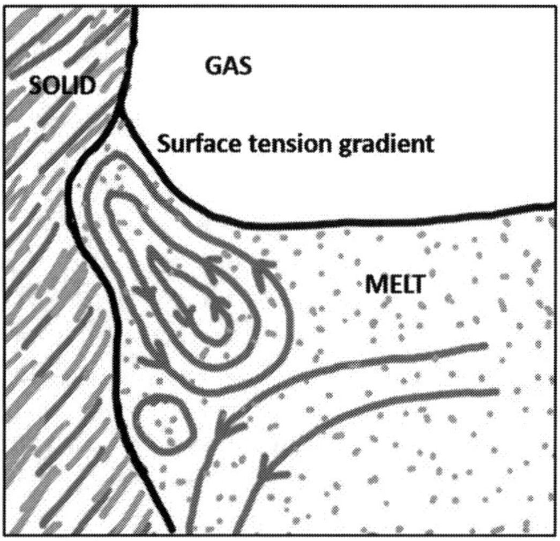

Figure 6.11 Illustration of the motion of the slag film caused by the surface tension phenomena between the refractory and slag.

gravity (cm/sec^2); V = quantity of flow (cm^3/sec); and b = width of path of flow (cm). However, the Reynold number has less influence on the experimental data. The film is thicker at channel wall than middle; thus the flow velocity is larger in the center of the film than at the channel wall. The constants 0.85 and 1.31 are included in Equations (6.47) and (6.48).

Equation (6.38) anticipates that raising viscosity or contact angle, or lowering surface tension, reduces the slag penetration. Thus, the motion of the slag film or film thickness gradient primarily caused by surface tension (wettability) between the refractory and slag essentially causes the local corrosion of refractories at the slag surface, as shown in Figure 6.11. The slag film motion accelerates the dissolution rate of the refractory and induces the abrasion of some refractories.

6.5 Phenomenological Approach and Slag Design

High temperature refractory dissolution or corrosion depends on the acidity and basicity of oxide phases in the molten state. Compatible mineral phases are not reacted at high temperatures, according to a rigorous definition of compatibility. In most cases, the term *more compatible* means less reactive. It means mineral phases *stand beside* or coexist in equilibrium on a microscopic level without reacting to generate new chemicals. However, acidic–basic oxide mixture lowers the mixture temperature and expedites the fluidity resulting in rapid penetration through a pore in refractory. Refractory is a mixture of several acidic and basic oxides, and several metals are added during alloying or processing of steel. The term *V* ratio is used to simplify the examination of slag during steel refining, with a ratio greater than 1.0 implying basic chemistry (typically a basic slag chemistry) and a ratio less than 1.0 indicating acid chemistry. The *V-ratio* can be represented in terms like these in formulations of this type:

$$V = \frac{CaO+MgO+FeO+MnO+\ldots}{SiO_2+P_2O_5+Al_2O_3+Fe_2O_3+Mn_2O_3+\ldots} \tag{6.49}$$

In many cases, three-component phase equilibrium diagrams help analyze corrosion problems. The primary refractory component can be seen as one of the triangle's apexes (corners), with CaO and SiO_2 as the other two components (apices). These calculations make it easier to utilize the simple CaO/SiO_2 ratio. Using the V-ratio to determine *equilibrium* conditions in more sophisticated investigations employing advanced thermodynamic tools may be more accurate.

As discussed earlier, infiltration depends on pore diameter, surface tension, and slag viscosity. Thus, the minimization of open pores during refractory processing is an essential feature. This porosity might be open holes that can be penetrated by fluid media (i.e., apparent porosity) or closed pores that are difficult to penetrate by liquid media (i.e., *hidden* porosity). If there is no porosity in a refractory (or no brick joints, expansion joints, or construction joints), the corrosion reaction is limited to the face exposed to the corrosive media (called the hot face). According to most researchers, slag corrosion rates rise linearly with apparent porosity within the refractory. Because corrosive fluids can enter the refractory, they usually cause disruption and even destruction of its matrix. The matrix is the area where sintered fines hold bonded bricks and monolithic refractories together. The matrix usually has more contaminants and porosity than the aggregate particles. As a result, corrosion damages the refractory's weakest component faster than denser or pure particles. However, slag–matrix interaction in a severe environment facilitates the rapid erosion of coarse aggregates.

A more realistic schematic of slag–refractory interaction is shown in Figure 6.12a. The refractory has a temperature gradient from hot to cold face, which means that the hot face has the maximum temperature and the temperature drops over the thickness of the refractory toward the back or cold side. Slag is seen entering the refractory until it reaches a frozen plane. The freeze plane is a spot in the refractory where the temperature is low enough for the slag to harden and coat on the hot face of the refractory.

This coating is relatively thin (~2–5 mm thickness) in most metallurgical applications. Slag can build in thick areas (>25 mm) when the refractory is not in touch with the metal bath. In certain instances, two different slag layers can be detected outside the refractory after the material removal. The inner layer

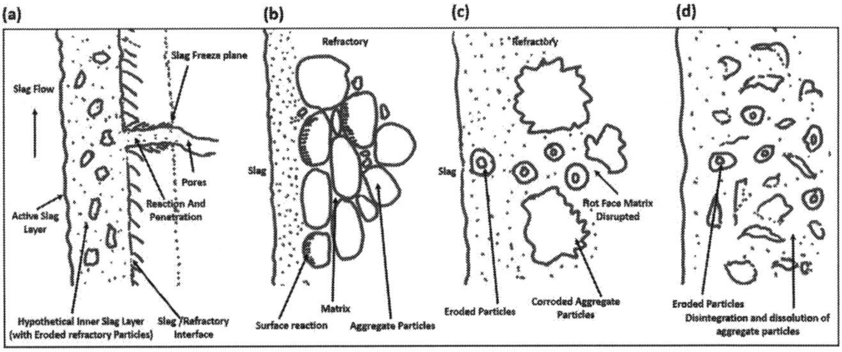

Figure 6.12 (a) Cross-section of the slag/refractory interfacial area, (b) stage I of slag attack, (c) stage II of the corrosion process, (d) stage III of the corrosion process.

often has a refractory solution in the slag that increases the slag's viscosity with the added components. In this circumstance, the self-contained pores can connect them via a channel under the available plane to the hot face. It confirms why isolated pores seem to be filled with slag numerous times. Slag responds in the immediate hot face with the refractory formation of new phases. There is also a reaction behind the hot face in which the slag contacts the refractory on the pores. Metal can also permeate the refractory together with slag in metallurgical applications. Temperature gradient affects slag corrosion; for example, a steep temperature gradient facilitates less slag penetration and local slag–refractory reaction in the interface. Fresh slag movements can also be confined to the slag/refractory interface, and reaction products can be transmitted slowly away from the slag–refractory interface Figure 6.12b shows the schematic representation of Stage I type slag–refractory initial hot face reaction on the interface with restricted penetration, may be reactive with aggregate particles and matrix up to a fraction of millimeters. In this stage, the hot face temperature predominantly impacts the corrosion reaction rate. If the hot face temperature is kept only below the point when corrosion products become liquid (melt), the corrosion is very sluggish or nonexistent. In another way, rapid slag-assisted corrosion is noticed if the refractory interface temperature is over 20 °C than the eutectic point. Usually, the eutectic temperature using thermal analysis is employed to determine the slag melting temperature. Despite hot face temperature as the first variable, the secondary parameter slag impingement rate and chemistry of slag further enhance the corrosion of refractory. Refractory is soluble at high temperature regarding slag chemistry; for example, high basicity protects basic refractory lining. Thus, competitive slag chemistry with refractory composition minimizes slag corrosion and frequently adds supplementary oxides to maintain the slag chemistry.

The gradual hot interface reaction stick slag (Stage I) and thick slag wall initiates slag penetration in refractory because of a wide temperature gradient. Capillary pressure facilitates slag penetration through micropores (may be <10 μm) present on the hot face. In this stage, the refractory becomes extensive corrosive, and the corrosion mechanism shifts to Stage II, as shown in Figure 6.12c. Stage II is usually associated with two phenomena when the temperature gradient is only appreciably wide for the slag penetration; a) complete penetration of molten slag through the hot face region of refractory, b) rupture of refractory aggregate and matrix due to penetration. While molten slag penetrates in bonded refractory, the weak grain boundaries are damaged first, interrupt the linkage between aggregates and matrix, enhance the thermal mismatch between penetrated phase and matrix, and loosen the particles. Weak interface and low density at the local region lead to further slag penetration that again enhances the spallaing because the coefficient of thermal expansion in the area penetrated by slag is different from the coefficient in the unpenetrated portion of the cold side. Continuing thermal excursions (cooling and heating) can lead to spills between penetrated and unpenetrated portions near the dividing line. After splitting, the remnant lining restarts the corrosion process from Stage I to Stage II.

Gradual movement from hot face to interior region results in the last phase of corrosion, usually designated as Stage III as shown in Figure 6.12d. In this circumstance, this incidence assists in reaching the final phase of corrosion toward the end of the refractory lining life with a relatively modest corrosion rate and minimum spalling. It expedites the loosening of particles and minimal binding just 2–4 mm behind the hot face. The matrix is gradually dissolved, and certain leftover aggregates along with adherence of slag.

6.5.1 Refractory Solubility

The most fundamental component affecting slag attack is solubility, which is stated in percent by weight of the solvent slag and is determined using a ternary system. Slag–refractory interaction occurs by the simple solution of the solid phase in refractory or completely changing the chemical composition. Thus, refractory solubility is usually active until the saturation of slag composition and reactivity.

The composition of the two substances in the ternary system, denoted as points R and S in Figure 6.13, is connected through a straight line that unites two points representing these substances when the composition A, B, C are expressed in a ternary diagram. This line does not necessarily show the forming of solution L, but it does not matter as long as the balance is reached and the system is completely liquid. The melting-point isotherms in the system are checked via this straight line from which a melting point corresponding to various possible mixes may be considered.

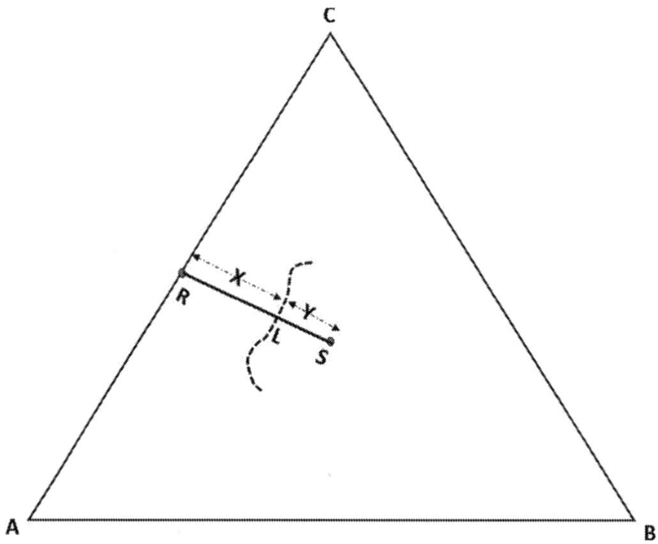

Figure 6.13 Definition of refractory solubility.

In Figure 6.13, x / (x + y) percent of S is the equivalent of y/ (x + y) percent of R at 1200 °C, the relative proportions of the two substances present at the melting point are found by using the part of the line on each side of the intersection. The use of such estimation is discussed in Chapter 2. If R is the refractory composition and S is the slag, then L is R solubility at 1200° in S. This is to say that the solubility (y/x) is 100% pure slag based. Usually, the solubility in kilograms of dissolved refractory per kilos of slag is expressed. The solubility determined in this way is actual solubility, i.e., the quantity dissolved after sufficient time to achieve balance, and it is the most significant quantity of refractory dissolved at the given temperature. It is necessary to comply with the knowledge of three conditions, i.e., (a) the chemical composition of the refractory and slag; (b) the refractory and slag composition within a ternary system; and (c) most significant isotherms of the melting point, to implement this method. The refractory and the slag composition should be designed with the knowledge of refractory slag interactions and modification of the local composition [18].

6.5.2 Slag Composition and Volume Optimization

In ironmaking and steelmaking processes, different qualities of slags are inevitably formed. The process is dependent on numerous essential oxides such as CaO, MgO, SiO_2, Al_2O_3, iron oxides, and other trace oxides. Blast furnace (BF), Base Oxygen (BOF), or Electric Arc Furnace (EAF), and ladle furnace (LF) are the key slag resources for the processing, respectively for the production of iron, primary steel, and secondary steel. SiO_2 and Al_2O_3, which are produced from iron ores as gangue minerals and CaO as a sintering stream, form part of the main component of BF slag. BF slag chemistry is not that involved as mainly iron oxide in the slag and its composition is dependent on the ore and ash gangue material in the coke. Monitoring of impurities in the iron mineral can reduce slag volume; however, economic feasibility is necessary before any additional iron processing costs.

The major component of BOF slag is CaO, introduced as a refining agent to maintain the C/S ratio of more than 2.5, and the process of refining SiO_2 and iron oxides. During the refining and reduction of iron oxide to metallic iron, the EAF slag is divided into two types: oxidation slag and reduction slag. The compositions of components made from gangue components depend on the raw material compositions, whereas the compositions of other components supplied as a refining flux are designed to enhance their refining performance, and so a wide range of slag compositions exist. The slag manufactured in the LF (LFS) comes from steel refining and is generally a heavy metal carrier like chromium, zinc, etc.

The cooling rate essentially determines the slag crystallinity behavior. For example, a slow cooling rate facilitates thick and high stable crystalline slag. However, when cooling is quick, its components are attached to an amorphous structure; thus in the presence of specific substances, the slag is unstable or active.

The BOF process's beginning and resultant slag compositions for steelmaking fluctuate greatly because of the oxygen blowing. This difference can be represented in approximately one of Figure 6.14, $CaO–FeO_n–SiO_2$ phase diagrams. As illustrated in area A of Figure 6.14, it may be seen that the first BOF slag is more acidic and contains more SiO_2 and FeO_n.

The slag basicity of CaO mass percent/SiO_2 mass percent usually needs to be raised to 3–5 in the final slag while the final stage of oxygen blowing to maintain a low level of sulphide and phosphorus contents. FeO_n content in the final layer is frequently as high as 20–30 mass percent to maintain slag fluidity and oxidation, as illustrated in Area C. Despite these consequences, the minimization of slag volume and the optimization of composition are crucial criteria for maintaining the manufacture of quality steel.

BOF slag volume can be reduced through scrap utilization. The BOF slag minimization utilizing pretreatment hot metal with lower silicone concentration is probably not commonly used worldwide because of the scrap utilization. However, the mass ratio of slag to steel can be decreased to a certain degree to lower the amount of silicon in the pretreatment of hot metal adequately for BOF operation. Such a process facilitates the retention of iron throughout the conversion operation so that BOF steelmaking process minimizes the slag volume.

The slag design requires meeting several significant and often conflicting demands on slag properties, in specific chemical composition, viscosity, temperature, and

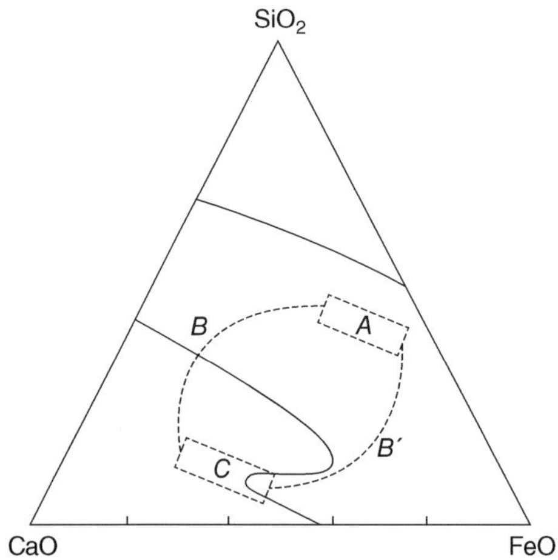

Figure 6.14 The composition ranges of initial and final slag on BOF operation practice: area A: BOF slag in its initial form; area B: BOF slag in the final form of BOF; area C: slag composition at the end of oxygen blowing; Lines B and B′: BOF slag in the boundary lines [19].

surface tension during the continuous steelmaking process. For example, on one hand, the slag should be relatively oxidizing to satisfy decarburization, whereas, on the other hand, slags with high FeO tend to be aggressive toward the refractory. Also, the design relies on slag foaming to shield the electric arc, but this foaming is supposed to be achieved with virtually zero oxygen injection, as otherwise, the FeO is driven up excessively, causing refractory corrosion. The engineering of an appropriate slag is thus vital to the viability of the proposed continuous steelmaking technology. The slag optimization exercise consists of different oxide phases that are discussed in this section. This exercise may be considered an example of a classical approach to employing process metallurgy principles in technology development and designing the slag to protect refractories.

A thermodynamic evaluation is estimated as the minimum FeO required for sufficient decarburization. FeO must be maintained at the lowest possible concentration for a steel carbon level to reduce slag corrosiveness. Product specifications and downstream processing steps primarily decide the carbon level of steel and tap temperature (i.e., not related to slag). Therefore, slag chemistry, particularly the dependence of FeO requirement (for a given C%) on basicity, and MgO content, have significant importance. In this context, a recent article compiled a set of experimental data on the activity coefficient of FeO in slags with a wide range of compositions [20] / with permission of Elsevier. The results, presented in Figure 6.15, indicate that there is a strong correlation between the activity coefficient (with pure liquid FeO as the standard state) and the slag

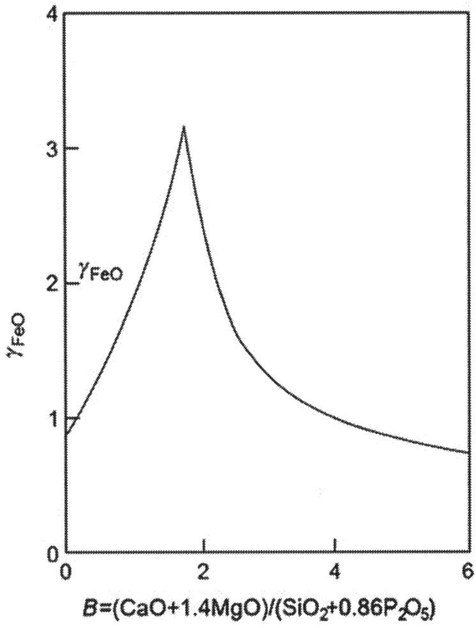

Figure 6.15 Dependence of FeO activity on slag basicity at 1600 °C [20].

basicity, expressed as B = (%CaO + 1.4% MgO) / (%SiO_2 + 0.86% P_2O_5). FeO's activity coefficient peaks at B = 1.8, implying the possibility of FeO concentration minimization at optimum basicity. For example, by increasing the basicity of slag from 1.0 to 1.8, the FeO content of slag may be reduced to half for the same carbon content in steel. However, the practical implication is that the optimum basicity for operating under minimum FeO concentration is between 1.8 and 2.2.

Usually, the carbon content of steel is in equilibrium with various viscosity of slags FeO–CaO–MgO–SiO_2– Al_2O_3, as well as MgO and FeO concentrations. The results are presented as isocarbon curves on a ternary slag phase diagram. As seen in the diagram in Figure 6.16a, the required FeO content for a specific C% depends on the slag composition. It goes through a minimum at basicity (V = [CaO + MgO]/[SiO_2 + Al_2O_3]) of 1.5–2.0, indicating a large activity coefficient of FeO in this range. Such incidence is also consistent with the data presented in Figure 6.16b, noting the small difference is because of the slight difference in basicity content of 12% MgO. Based on this analysis, the basicity of slag is usually maintained at around 1.5–2.0 for minimizing the FeO content. For example, a low *V* ratio compared to conventional EAF slag basicity (2.5–3.5) results in relatively smaller slag volume and minimum lime/dolomite fluxes consumption. Additional benefits of operating in this low basicity (1.5–2.0) regime include a reduced amount of FeO in the slag (due to smaller slag volume) and higher activity coefficient, resulting in higher iron yield and lower energy loss associated with the slag thermal energy. However, it must be noted that a higher

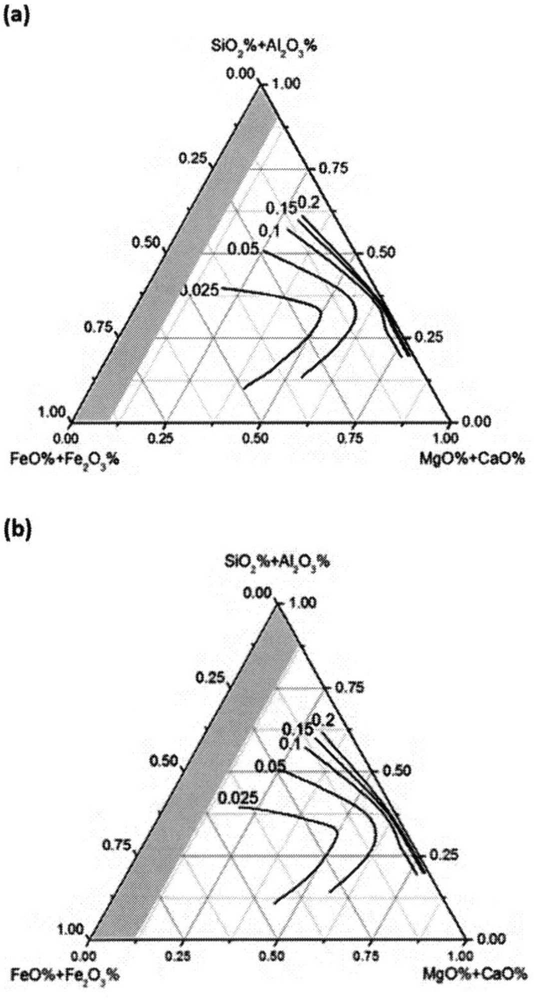

Figure 6.16 Isocarbon curves in $SiO_2-Al_2O_3-CaO-MgO-FeO$ system with (a) MgO = 8%, (b) MgO = 12% Adapted from [21].

MgO concentration is required to saturate low basicity slags with MW (magnesiowustite) and/or C_2S, as depicted in Figure 6.17.

In reality, the slight oversaturation to form solid particles is beneficial for foaming, with MW being the preferred solid. It serves two purposes: improved foaming and reduced slag affinity for the MgO-based refractories. In the basicity range of 1.5–2.0 and for typical low carbon steel with 0.05–0.1% C, the FeO content of slag at equilibrium is 3–8% (Figure 6.17). However, higher FeO concentrations are expected to maintain a driving force for the decarburization reaction in steelmaking. Also, the equilibrium condition is far from being reached due to the

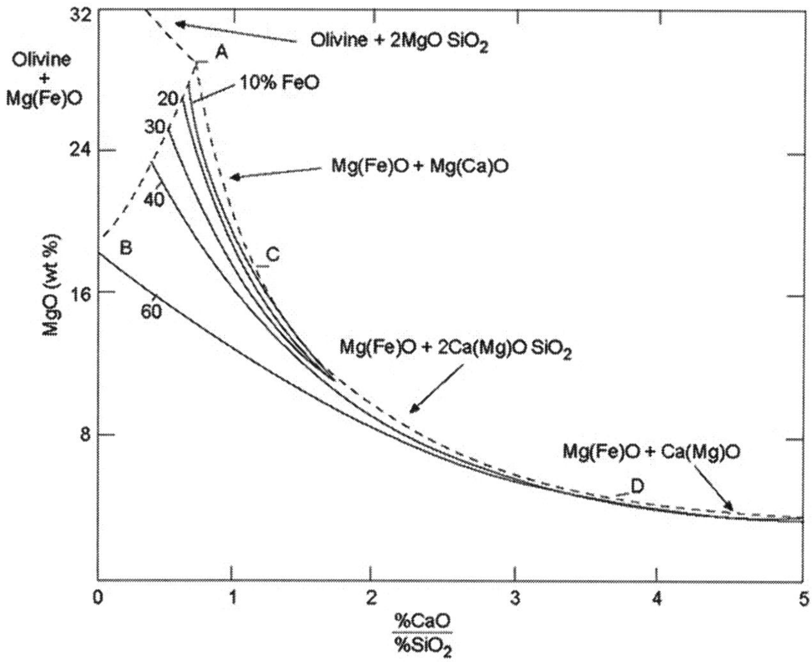

Figure 6.17 Solubility of MgO in FCSM (FeO–CaO– SiO_2–MgO) slags at 1600 °C [22] / with permission of Elsevier.

limited bath mixing in electric furnaces and short residence time. Such slag composition design for any specific vessel is essential to enhance the brick life and process metallurgy.

References

1 Gulbransen, E.A. and Jansson, S.A. (1970). *Heterogeneous Kinetics at Elevated Temperatures* (ed. G.R. Belton and W.R. Warrel). New York: Plenum Press.

2 Bube, R.H. (1960). *Photoconductivity of Solids*. New York: Wiley.

3 White, W.B. (2000). Theory of corrosion of glass and ceramics. *Mater. Sci.* page 22–23.

4 Opila, E.J., Jacobson, N.S., Myers, D.W., and Copland, E.H. (2006). Predicting oxide stability in high-temperature water vapor. *JOM*. 58: 22–28.

5 Gentile, P.S., Sofie, S.W., Key, C.F., and Smith, R.J. (2012). Silicon volatility from alumina and aluminosilicates under solid oxide fuel cell operating conditions. *Int. J. Appl. Ceram. Technol.* 9: 1035–1048.

6 Opila, E.J. and Myers, D.W. (2004). Alumina volatility in water vapor at elevated temperatures. *J. Am. Ceram. Soc.* 87: 1701–1705.

7 Opila, E.J., Fox, D.S., and Jacobson, N.S. (1997). Mass spectrometric identification of Si-O-H(g) species from the reaction of silica with water vapor at atmospheric pressure. *J. Am. Ceram. Soc.* 80: 1009–1012.

8 Layden, G.K. and Brindley, G.W. (1963). Kinetics of vapor-phase hydration of magnesium oxide. *J. Am. Cera. Soc.* 46: 518–522.

9 Thomas, C., Rosales, J., Polanco, J.A., and Agrela, F. (2019). Steel slags. In: *New Trends in Eco-efficient and Recycled Concrete* (ed. Jorge de Brito, Francisco Agrela). Elsevier. 169–190.

10 Sequeira, C.A.C. (2019). *High Temperature Corrosion, Fundamentals and Engineering.* John Wiley & Sons.

11 Nickel, K.G. (1994). *Corrosion of Advanced Ceramics –Measurement and Modelling.* Dordrecht, NL: Kluwer Academic Publisher. 267: 467.

12 Nickel, K.G., Quirmbach, P., and Pö Tschke, J. (2010). *High Temperature Corrosion of Ceramics and Refractory Materials.* Elsevier.

13 Levich, V.G. (1962). *Physicochemical Hydrodynamics.* Englewood Cliffs, NJ: Prentice-Hall. 66–72.

14 Goto, K., Argent, B.B., and Lee, W.E. (1997). Corrosion of MgO—MgAl2O4 Spinel Refractory Bricks by Calcium Aluminosilicate Slag. *J. Am. Ceram. Soc.* 80: 461–471.

15 Goto, K. and Lee, W.E. (1995). The "Direct Bond" in Magnesia Chromite and Magnesia Spinel Refractories. *J. Am. Ceram. Soc.* 78: 1753–1760.

16 Y. Satyoko and W. E. Lee. (1999). Dissolution of dolomite and doloma in silicate slag. *Br. Ceram. Trans.* 98(6): 261–265.

17 Endell, K., Fehling, R., and Kley, R. (1939). Influence of fluidity, hydrodynamic characteristics, and solvent action of slag on the destruction of refractories at high temperature. *J. Am. Ceram. Soc.* 22(1–12): 105–116.

18 Lee, W.E. and Zhang, S. (1999). Melt corrosion of oxide and oxide–carbon refractories. *Int. Mater. Rev.* 44 (3): 77.

19 Zhang, J., Matsuura, H., and Tsukihashi, F. (2014). Chapter 4.4 – Processes for recycling. In: *Treatise on Process Metallurgy Volume 3: Industrial Processes.* 1507–1561.

20 Turkdogan, E.T. (1996). *Fundamentals of Steelmaking, The Institute of Materials.* London, UK.

21 Seetharaman, S., Mclean, A., Guthrie, R., and Sridhar, S. (2014). *Treatise On Process Metallurgy Process Phenomena.* Elsevier.

22 Bennett, J.P., Kwong K-S., Krabbe R., Singh J.P., Skaar E.C., Wiskochil David., Karr C.L., and Wilson E. (2000). An overview of DOE's program to recycle/reuse spent refractories as a slag conditioner and to extend refractory service life in the EAF. No. ANL/ET/CP-103160. Argonne National Lab., IL (US),

7

Operation and Refractories for Primary Steel

7.1 Introduction

Bessemer converter–Open Hearth route is a classic steelmaking process. However, gradually replaced by revolutionized Basic Oxygen Furnace (BOF) process overcomes the multiple handling, poor yield, and environmental pollution. It dramatically reduces the tap-to-tap time from 18 hours to 45 minutes, high throughput many more times, improves quality steel, and is less environmentally hazardous. The basic idea of converter processes for steelmaking is to convert carbon-rich liquid hot metal from blast furnace to low-carbon steel. Various concepts are used like basic oxygen process (BOP), basic oxygen steelmaking (BOS), and basic oxygen furnace (BOF) process.

Steel slags are basic in nature and enriched with basic oxides; thus it is a mandate to work lining by basic refractory for both the BOF and EAF vertical vessels. The prime regions of the converter are the spherical bottom, cylindrical shell, and an upper cone. Refractory choice should maximize the inner volume to achieve an optimum metallurgical process. Some specific dimension may provide insight of the vessel working volume and design.

Usually, around 20% of converter volume is occupied or filled with hot metal and scrap. A couple of relationships between molten bath height (H_b), bath diameter (D_b), working height of the converter (H_w), and throughput capacity in tons (T) is shown in Equations (7.1) and (7.2).

$$\frac{H_b}{D_b} = 0.328(T)^{-0.0148} \tag{7.1}$$

$$D_b = 0.704(T)^{0.386} \tag{7.2}$$

For 150 ton converter throughput, $D_b = 4.87$ m, $H_b = 1.48$ m.

The working height $H_w \approx 3.5 \times$ bath height $= 5.18$ m. Thus, the total working zone height is $(1.48 + 5.18) \approx 6.7$ m. If the bottom refractory (working + backup)

Fundamental Design of Steelmaking Refractories, First Edition. Debasish Sarkar.

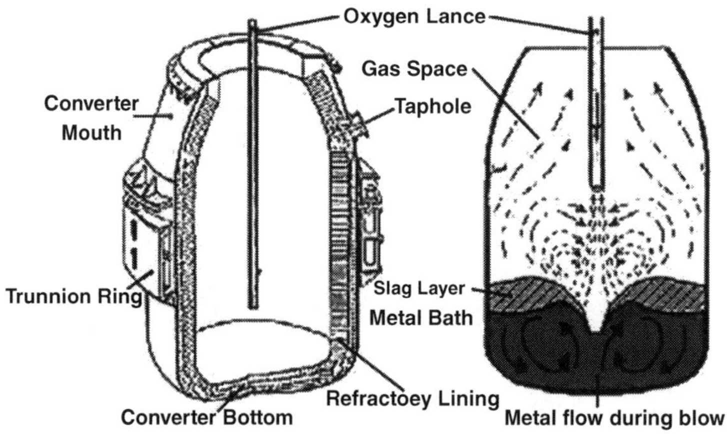

Figure 7.1 Schematic representation of BOF consists of important zones and a view of available space for the metallurgical process.

and shell thickness are around 1–1.5 m, the total height of the converter becomes approximately 8 m. Considering total blank height (6.7 m), around (1.48 m/6.7 m) 20% of converter volume is occupied or filled with hot metal and scrap. In practice, the BOF capacity varies from 60 to 400 tons. Most of the BOF have a capacity of more than 200 tons. It can rotate around its horizontal axis to feed and accommodate the molten metal, raw materials, sampling the molten metal, discharging the steel and slag after completion of the process. A typical schematic representation of BOF and their available space during processing is shown in Figure 7.1.

While considering the refractory lining for effective volume calculation, a brief idea on lining may assist in designing the refractory. Lining comprises both permanent lining and above on it working lining that experiences continuous wear during operation. A permanent lining made of different classes of shaped and unshaped refractories with a variant thickness of 100–225 mm is common practice; however, its selection is based on the capacity and total height of the converter. Wear or working lining is just on the permanent lining and maintains the entire refractory lining length. MgO-C refractory bricks consist of high purity, and high crystalline 15% graphite enhances the corrosion resistance. In multi or duplex blowing (combined or hybrid blowing) MgO-C are preferentially employed for the bottom tuyeres and around that zone to protect the severe wear. Refractory bricks are used to line several parts of BOF, either of MgO or Doloma refractories. Conventional bonding systems like coal tar, resin, or organic binders are most common to make such bricks. Despite the bonding between refractory grains, the residual carbon characteristics and network have a significant role in slag resistance. A common operational parameter is needed to consider before designing the refractory for BOF and these are summarized in Table 7.1.

Table 7.1 Several operational parameters to design and select the BOF refractories.

1. Capacity (MT)	14. Process time (min)—charge to tap
2. Lining detail	15. Slag arrester type (duct or any other)
3. Preheating schedule	16. Sub-lance system (Y/N)
4. Input composition	17. LOT scanner (Y/N)
5. Tapping temperature (°C)	18. Secondary operation if any in BOF
6. Peak temperature (°C)	19. Hot repair practice
7. O_2top blowing	20. De-bricking system
8. Top lance details	21. Life (heats)
9. Bottom blowing	22. Yield (%)
10. Slag conditioning	23. Critical area to take care for higher life
11. Slag chemistry (%)	24. Any other pain area to take care
12. Steel grade	25. Data collection date
13. Steel chemistry (%)	26. Waiting period-end of tap and start of charge

Blast furnace–*converter*–secondary metallurgy–casting is a common domain for making high-quality steel in integrated mills. However, several steelmakers have adopted electric arcing process to make steel as it has high flexibility in operation and availability of raw materials. In comparison the BOF, electric arc furnace (EAF) primary steelmaking process engaged to produce ~30% steel around the globe. A competitive share in the production, energy consumption, and CO_2 emission for both BOF versus EAF the process is shown in Figure 7.2. Analysis is represented while common charging materials are direct reduced iron (DRI) or sponge iron (gas or coal) and scrap.

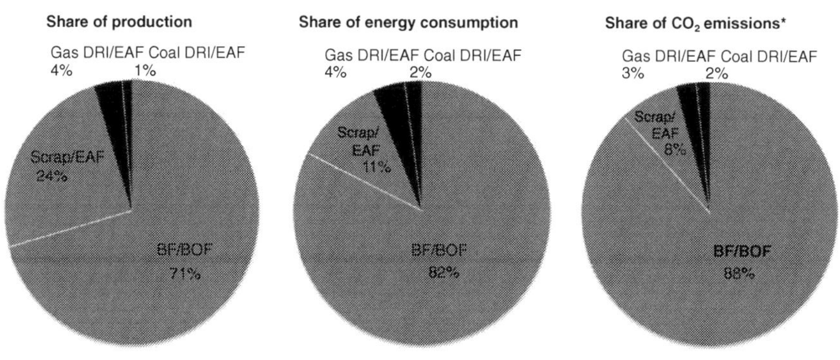

Figure 7.2 A competitive analysis of BOF vs. EAF, (a) production, (b) energy consumption, (c) CO_2 emissions.

BOF process is self-sufficient in energy, whereas EAF, a cylindrical vessel demands electric arcing during steelmaking process. It has dish-shaped refractory lined hearth with having three vertical electrodes that reach down through a dome-shaped removable roof, as shown in Figure 7.3. Usually, the vessel diameter varies with respect to capacity, for example, a 9, 6, and 2.5 m diameter is required for 300-, 100-, and 10-ton steelmaking, respectively. EAFs are versatile, charging everything from all sorts of scrap to hot briquetted iron (HBI), direct reduced iron (DRI), pig iron, and hot metal. EAFs may produce all type of steels: long and flat, carbon and alloyed, for continuous casting, ingot casting, and teeming of castings in molds.

EAF operating conditions decide the division of specific zones. The operational features, slag, and steel chemistry determine the varying nature of refractories for the vessel. Prime zones for EAF are:

a) Hearth: to accommodate initial charge materials and produced liquid steel
b) Slag line: transitional area in between hearth and side wall
c) Upper side wall: above the slag line but below the water cooled panels
d) Tophole: opening zone that provides liquid steel tapping from EAF to ladle
e) Roof: removable portion, also known as delta, provides openings for electrodes to enter the furnace and remove exhaust gas from the furnace.

Usually, MgO or MgO-C refractories are used in EAF for the steel contact zone. Different MgO- and Al_2O_3-based refractories consisting of appreciable thermal shock resistance are used in the roof, taphole, spout, or runner of the EAF.

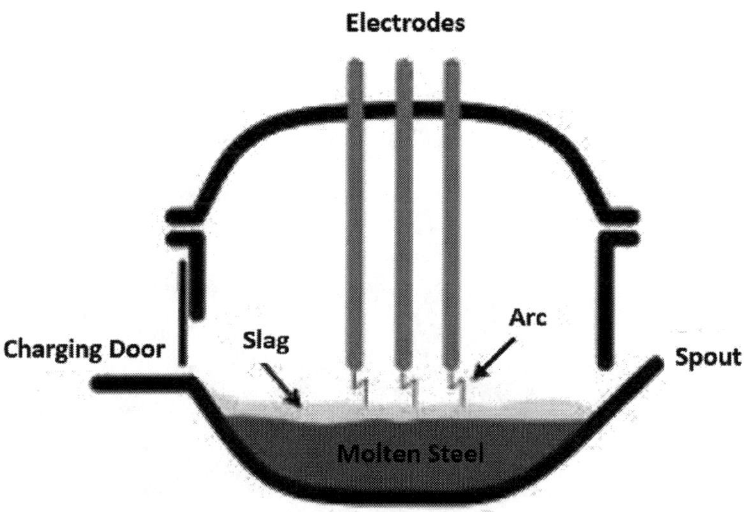

Figure 7.3 A schematic representation of EAF consists of important zones and view of available space for the metallurgical process.

7.2 Operational Features in BOF

The BOF process is autogenous. In other words, it is self-sufficient, and no external energy is required to convert liquid iron (hot metal, HM) while using gaseous oxygen (O_2) to oxidize several unwanted impurities. The oxygen purity should be a minimum 99.9% for low carbon steel to avoid nitrogen absorption by steel. In practice, several BOF processes are adopted for primary steelmaking.

The main processes are top blowing, bottom blowing, and a combination of the two, which is known as combined blowing. Systematic processing implies three predominant operational steps for BOF; charging, blowing, and tapping the metal. A simple schematic shown in Figure 7.4 represents sequential steps to obtain the steel from hot metal through definite charging of hot metal, scrap, and additives, followed by blowing and finally tapping after obtaining the desired composition.

A different mode of blowing is discussed in Section 7.2.2. All operational features may guide how one can select and design a particular refractory for a selected zone.

Charging the furnace

Scrap is charged as a coolant for the metal to steel conversion process. It assists in synchronizing the temperature equilibrium condition during high-temperature rise because of violent exothermic reactions in the furnace. Usually, three to four times hot metal (e.g., 300 ton) is necessary to pour from the ladle compared to scrap weight. Some additives or slag basicity controlling oxides are needed to add during the steelmaking process.

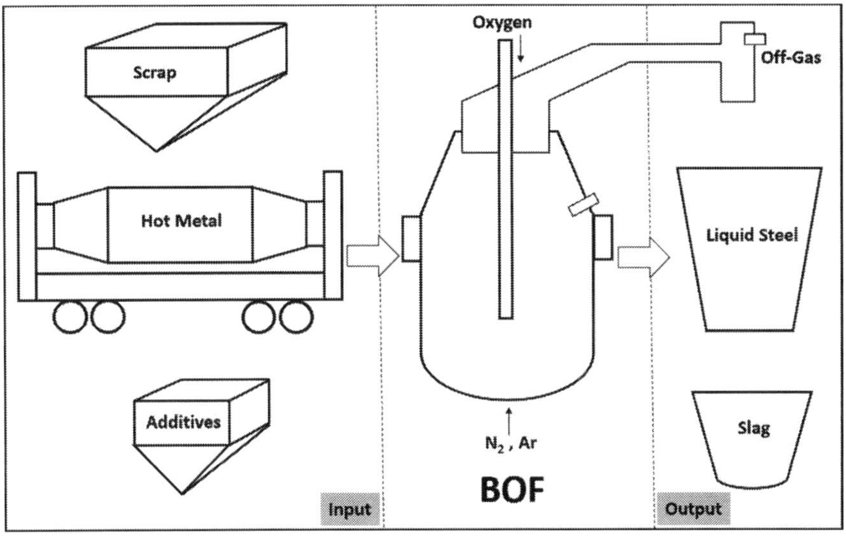

Figure 7.4 Schematic representation of the basic operational features in BOF.

Blowing

While all material is placed in the vessel, the furnace is blown by a precalculated oxygen blast through a lance lowered into the molten metal. Autogenous processes expedite the heat resource through an exothermic reaction when oxygen combines with impurities like carbon (C), silicon (Si), manganese (Mn), and phosphorous (P). Carbon escapes from metal as carbon monoxide gas and other impurities are transformed to their corresponding oxides, and these solid oxides are acidic. Thus, their separation can be done by adding lime (CaO) to the furnace. The resultant mix and subsequent reactions within lime and developed oxides form slag and float on top of the metal layer. The lance position, blowing time, and oxygen flow rate are important criteria to control over the completion of the reaction.

Tapping the furnace

Quality control in specific metal compositions during the process is an essential step. Thus, after the blow has continued for about 20 minutes, the metal sample is essential to collect and analyze to meet the desired chemistry. After completion of the BOF process, the furnace can be tapped. Steel is run out of the tap hole into a ladle, separating it from the lighter slag, which is later collected as waste. Brief process of BOF is

- Input is mainly hot metal and scrap
- Lime is added
  - Cushioning effect for scrap impact area
  - Reaction with Si of hot metal to arrest the same so that adverse effect on MgO-C refractory may be taken care of
  - Reactivity of lime is very important so that its effect can be utilized
- Oxygen blowing for decarburization
- Exothermic reaction by oxygen addition to maintaining the temperature at high level
- Top–bottom oxygen blowing is used to reduce the process time
- Slag decanting and metal tapping
- Some slag is retained after metal tapping
- Slag conditioning by addition of lime and dolomite
- Slag splashing by nitrogen blowing to make coating up to top cone
- Protection of refractory by slag coating

7.2.1 Charging and Blowing

Removal of carbon as well as phosphorous and the increase of the temperature are the basic targets of the BOF steelmaking process. Exothermic reaction follows when oxygen contacts blast-furnace iron, by the ensuing exothermic reactions, especially the oxidation of Si to SiO_2. Thus, added scrap consumes 340 kilocalories per kilogram of scrap to melt in this environment, minimizes the excessive heat due to exothermic reaction, and balances the thermal equilibrium. Therefore, a typical BOF charge consists of about 75% liquid iron and 25% scrap. HM and scrap are charged in one vessel, after which O_2 is blown into the BOF at supersonic velocities to remove impurities in HM.

The oxygen blowing process and subsequent reaction conversion are associated with loss; thus always need more input (molten iron and scrap) than output (steel). For example, a cumulative 1.08 ton feed may yield nearly 1 ton of liquid steel, indicating near 7–8% loss during the conversion process. The computer-controlled automatic feeding and monitoring are adopted in considering the working environment and maintaining temperatures, chemical compositions, and charge weights. During oxygen blowing, the converter is kept vertical and turned on the oxygen and the lance placed downward in a precalculated position. The entire oxygen blowing process, including flow rates, lance height from the molten bath, and lime addition time and quantity, are often synchronized automatically. Some floor data indicates the consumption of oxygen and lime for steel processing. For example, the oxygen flow rates for larger converters usually exceed 800 cubic meters per minute, and each ton of steel consumes 110 cubic meters of oxygen. Steel processing demands lime to control slag basicity and protect the refractory lining. In a typical data, 70 kg of an optimum pebble size of burnt lime is added per ton of steel before blowing. Eventually, this combines with silica and other oxides developed during oxygen lancing and forms near to 150 kg slag per ton of steel. Prior to use of burnt lime, it is essential to regulate the chemical purity content of >92% CaO and low silica. Fluxes like dolomite (CaO. MgO) enhance the MgO content up to 6 wt.%, which results in the minimization of MgO-C working lining corrosion by slag. Usually, beyond 8 wt.% MgO in slag indicates the corrosion of MgO-C bricks. The combustion gaseous product CO exits from the converter and transmits heat back to the bath. Usually, the resultant chemical analysis is done when the BOF steel temperature 1650 °C.

7.2.2 Mode of Blowing

While considering the oxygen steelmaking installations, processes vary from *top blowing only* to *bottom blowing only* with a combined process type, as shown in Figure 7.5. In top-blown BOF converters, a water-cooled top-blowing oxygen lance is provided for blowing oxygen over the surface of the hot metal (molten iron). The oxygen flowrate level is usually 6–10 $m^3\ min^{-1}\ ton^{-1}$ of molten iron. The oxygen pressure varies in the range of 1–1.5 MPa. It can be noted that the service life of a top-blowing oxygen lance normally lies between 400 and 600 heats.

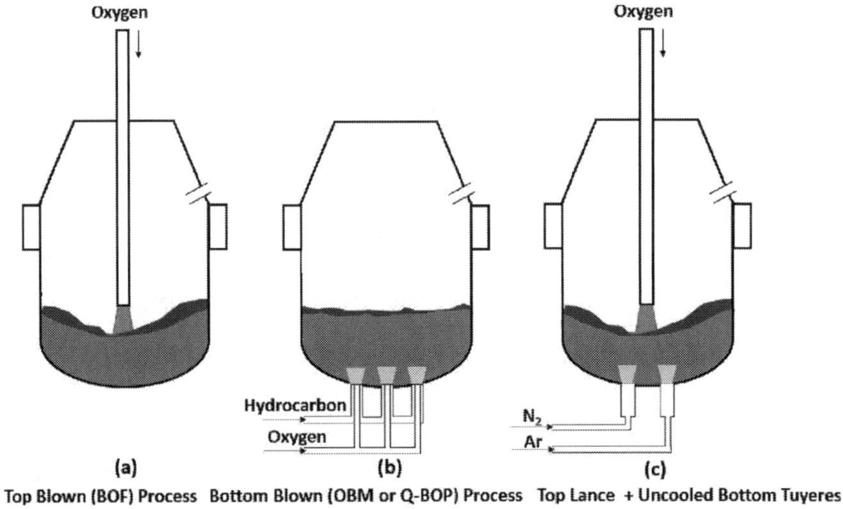

Figure 7.5 The top-blown, bottom-blown, and combined-blown approaches of the oxygen steelmaking process in integrated steel plants.

In the oxygen bottom blowing BOF converters, there is an arrangement of several tuyeres in the bottom zone of the converter for blowing only oxygen or oxygen combined with the lime powder into all parts of the hot metal (molten iron). The bottom tuyeres are cooled using protective hydrocarbon tuyere coolants such as propane, methane, or fuel oil. It is known that BOF converters are steelmaking vessels with basic refractory lining. The refractory wear is predominant in the BOF slag zone, that is, the refractory lining area above the surface portion of the molten iron in contact with the corrosive slag melt.

Compared to the remaining processes, the benefits of top-blown processes mainly include the effective use of slag splashing and the reduced amount of downtime attributed to the more even refractory wear. The common problem associated with the bottom blowing operations is clogging that is caused by the accumulation of slag deposits onto the gas purging channels.

Although the bottom blowing operation has the drawback of clogging problem, it exhibits few extraordinary benefits of charging practices compared to that of the top blowing procedures. One such beneficial feature of the bottom-blown operation is its ability to melt scrap to a size of 2 feet. This process benefit offers the utilization of various types of scrap materials that lower the production cost. The additional benefit of this process is that there is no remaining solid scrap within the vessel after it is converted to the molten liquid. Further advantage of this process is known to be the dolomite lime charging that is either with the use of overhead bins arranged above the converter or as a pulverized mixture combined with burnt lime via bottom tuyeres. It can be noted that the charging practices followed for all remaining raw materials are merely similar to that of the BOF process. Apart from the charging advantages, another benefit of oxygen bottom blowing is the stronger reaction of oxygen with carbon and silicon present in all parts of the iron melt. It leads to the remains of fewer oxide inclusions in the blown metal. The residual content of manganese can also be improved with bottom blowing. Further, the enhanced reaction rates in bottom blowing offer the additional process advantages, including shorter average bath stirring time, slag oxidation control, and no vacuum decarburization requirement. Moreover, all grades of steel articles can be fabricated using bottom blowing similar to the top blowing practices. Based on the advantageous process features of the OBM (*Oxygen Bottom Blowing*) practices, steelmakers certainly point out the major drawback of the oxygen top blowing practice. The process has a weak stirring, and the inadequate mixing effect of the oxygen jet stream makes hot metal bath quite quiescent. It leads to over-oxidation of the slag as well as metal bath. Therefore, to obtain the beneficial features of both techniques, a combined blowing process has been designed with the combination of both top blowing as well as bottom blowing approaches.

Although many of the combined blowing processes are developed based on the LD process, the combined blowing approaches with inert gas bottom blowing are more beneficial. A majority of converters have used combined blowing strategies in recent times. The benefits of the modern combined blowing process approach over the conventional pure top blowing processes is noted as follows:

- accelerated overall blowing efficiency ascribed to the strongly intensified stirring as well as intimate mixing of the metal bath.
- minimum amounts of iron oxide content in the slag. This indicates the over-oxidation of slag is lowered.
- enhanced service life of the refractory lining because these practices avoid overheated iron oxide (FeO) enriched slags.
- better bath homogeneity provides accurate composition and temperature details of the end product.
- decreases slag splashing as well as slag spitting.

For oxygen blowing, these practices use a long water-cooled lance that is approximately 8–10 m in length. This oxygen lance is provided with a convergent–divergent nozzle that blows O_2 into the converter at a high pressure in the range of 11–14 kg/cm^2 and at a supersonic velocity normally of Mach number > 1. The oxygen lance nozzle is an essential part, which has been specially designed to perform the following functions:

- to produce a supersonic jet stream of non-coalescing free O_2 gas for high O_2 ingress in all portions of the molten bath.

- to stimulate agitation in all parts of the molten bath.
- to form a three-phase dispersed mixture with the combination of slag, metal droplets, and gas bubbles.

For the case of converter steelmaking practices with pure as well as various versions of combined blowing approaches, the entrainment of oxygen as a supersonic free gas jet stream into the molten bath is a notable feature. The momentum of the free O_2 gas jet stream is an important flow property that mainly depends on the velocity of the gas stream. This flow property is normally expressed in terms of the momentum flow rate, which indicates the amount of jet stream momentum that gets converted to force after hitting the molten bath. The momentum flow rate ($\dot{m}$) of a free O_2 gas jet stream can be estimated from:

$$\dot{m} = 1.029 \times 10^5 P_o N d_n^2 \left[\left\{ 1 - \left(\frac{p}{p_o} \right)^{0.286} \right\} \right]^{0.5} \tag{7.3}$$

Where ($\dot{m}$) is the momentum flow rate (newton), P_o (bar) is upstream stagnation pressure $\approx 6.755\ V^{0.104}$, P (bar) is surrounding/atmospheric pressure, N is number of nozzles, d_n is the nozzle throat diameter (m) $\approx \frac{7.46}{\sqrt{Nx10^3}} V^{0.446}$, and V is known as capacity (ton). Consider a case of converter vessel with a capacity (V) of 150 ton and four the number of nozzles with a diameter of 0.035 m. The estimated upstream stagnation pressure (p_o) is 11.37 bar. Therefore, the designed four-hole nozzle can produce a total amount of momentum flow rate ($\dot{m}$) of 4057 N. A dimensionless momentum flow rate number (φ) is also often used to describe the quantitative changes occurring in the translation of the momentum of free O_2 gas jet stream into the force after hitting the molten bath, is given as:

$$\varphi = \frac{\dot{m}}{\rho_i g L^3} \tag{7.4}$$

Where, ρ_i is known as the liquid density (kg/m^3), g is referred to as acceleration due to gravity (m/s^2), and L is designated as the lancing distance (m) from oxygen lance tip to the surface of the liquid or molten bath. Decreasing the lance distance (L) increases the value of the dimensionless momentum flow rate number (φ), which provides a quantitative indication of the action of the jet stream on the molten bath at a lancing distance of L, opposing gravity. The two elementary stages of the blowing exercise are known as soft blowing and hard blowing. Soft blow is the shallow penetration of the jet that is blown into the bath from the lance at a longer distance. In contrast, the hard blow refers to the deep penetration of a jet blown into the bath from the lance at a shorter distance. For all kinds of converter-based steelmaking practices, the two basic needs of the lance profile are: (i) the formation of FeO enriched slag by the soft blowing exercise in the initial stage, (ii) hard blowing exercises in the further stages to remove the impurities such as carbon and phosphorous by progressively increasing the oxygen availability in the molten bath. Such deep penetration of oxygen jet into bath melt also avoids the over-oxidation of slag. Although the oxygen lancing with soft and hard blow is essential for hot metal refining in all steelmaking practices, it can be noted that the overall oxygen blowing time does not fluctuate with the converter's capacity. A quantitative measure can characterize the jet penetration into the molten bath as shallow or deeper is known as the depth of penetration (h), which is given as:

$$h = 4.407x\varphi^{0.66} \tag{7.5}$$

For example, oxygen blowing at a lance distance (L) of 3 m produces a shallow jet penetration into 0.23 m depth of the bath. But a deeper jet penetration into 0.46 m depth of the bath can be accomplished by lancing oxygen at a shorter lance distance (L) of 1.5 m. Moreover, it can be noted that prolonged soft jet blowing may lead to slag sloping due to over-oxidation of slag. The impingement of molten bath by a reactive soft and hard oxygen jet results in a series of effects, which are given in Table 7.2 as follows:

Table 7.2 Characteristic features of soft jet and hard jet.

Soft jet	Hard jet
• Oxidation of Fe • Shallow penetration • Slag/metal interaction • Slag formation is promoted • P removal is enhanced	• Deeper O_2 availability in all parts of the bath • Favors oxidation of C but impairs the removal of P • Carbon monoxide (CO) gas generation promotes widespread bath agitation • Droplets are produced and emulsified in the slag

7.2.3 Physicochemical Change in BOF

In order to understand more details on the chemical and thermal phenomena, different reaction zones during combined blows in BOF are illustrated in Figure 7.4 [1]. BOFs are, accordingly, a complex combination of numerous chemical reactions progressing simultaneously with intensive heat production and heat consumption. During the successive stages of blowing, many of the secondary processes are involved with the primary oxidation reaction in the steelmaking practice.

Some of the important secondary processes include: (i) dust/iron ore addition to slag and lime dissolution, (ii) scrap melting and its dissolution in all parts of the hot metal bath, (iii) postcombustion reactions particularly including oxidation of

CO gas by oxide constituents of slag, (iv) heat gain from the exothermic reactions involved in scrap melting, (v) heating of the metal bath/slag, and (vi) heat losses through furnace wall/by furnace gas from the reaction zone to the environment.

During blowing in the converter, the molten iron increases from 1200–1300°C to 1600–1700°C (see Figure 7.6). Eventually, the scrap or DRI (sponge iron) plays a dual role; raw material and coolant. Usually, the final temperature of raw steel may reach to 2000°C in regular oxygen blowing without a scrap. In some literature, the reported oxygen blowing generates as high as 2400–2500°C in impact areas [2, 3]. However, the generated heat spread from the local hot spot zone to the surroundings, thus oxidation reactions are taking place in several zones.

i) Iron melt surfaces experience extensive oxygen impinging at the hot spot, and subsequent distribution of heat and oxygen facilitate the oxidation of impurities and transform them into slag.
ii) Oxygen dissolves in iron melt and acts as oxidizer.
iii) Oxygen enriches dissolved iron droplet mix inside the slag and their continuous movement and circulation expedite the reaction with dissolved oxides and gas components like O_2 and CO_2.

Interestingly, a vigorous emulsion phase forms when mixing both liquid slag and metal droplets that form foaming due to the uprising movement of CO gas. During the primary steelmaking process, a substantial amount of dust in the range of 10–30 kg per ton of steel is forming because of entrainment of charged material,

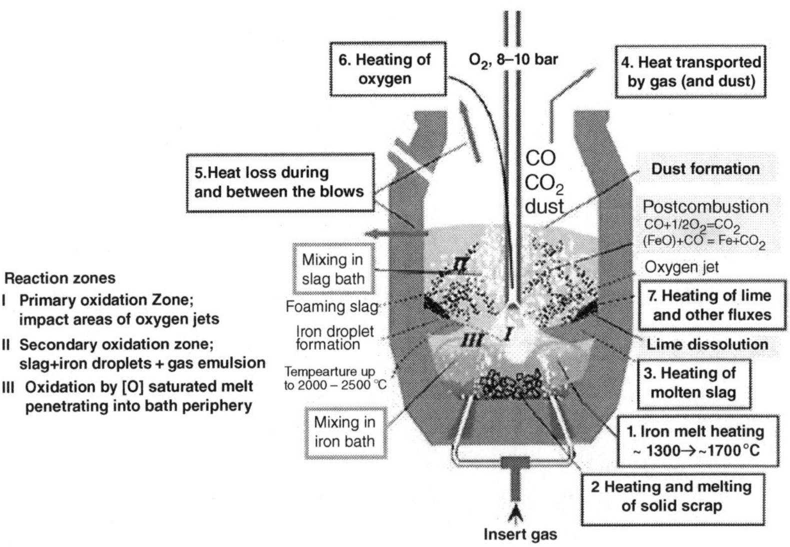

Figure 7.6 Schematic presentation of combined blown LD/BOF converter, chemical and thermal phenomena, and the main reaction zones [1].

vaporization, and mechanical ejection of metal and slag. The following events can be summarized to illustrate the prime phenomena in oxygen converter:

- The initial stage of oxidation of the constituents in the hot metal bath occurs. This involves oxidation of carbon, iron, and other minor elements on the bath surface by blowing a stream of oxygen gas jet onto the bath.
- Melt flow mixing involves intensive oxygen/inert gas blowing to induce fluid flow circulations in all parts of the iron bath to transport the minor iron melt constituents to the bath surface from inside portions of the molten bath.
- Secondary oxidation stage. This point of the steelmaking process is involved with the oxidation of minor element species, especially in all portions of the iron droplets circulating throughout the slag melt/hot metal bath by reacting with oxide constituents of slag/oxygen containing furnace gas.
- Tertiary oxidation stage. In this, greater oxygen available metal melt was transported from the hot spot area to the interior portions of the metal bath to oxidize the minor elements present inside in all portions of the bath.
- Slag melt forms with the dissolution of various slag forming constituents including lime, fluxing agents, and many other oxide components as a result of iron and minor elemental impurities in the iron bath.
- Reaction zone temperature can be raised during the successive stages of blowing into the converter. It can be noted that the reaction zone is a phase assemblage of iron bath, slag, and furnace gas.
- Melting, dissolution, and mixing of scrap with all parts of the molten metal bath.
- Dust generation in converters commonly involve mechanisms including metal/slag ejection, charge entrainment, and metal evaporation.
- Heat losses from the converter that is usually happening during the successive stages of blowing.
- Melt corrosion of furnace refractory lining by slag/metal attack.
- Post combustion reaction of CO gas. This indicates the combustion of CO gas to CO_2 gas by oxide constituents (FeO) in slag as well as oxygen in the converter vessel. It can be noted that the combustion reaction is of an exothermic nature and releases heat that is distributed among steel melt, slag mixture, and gaseous matter.
- Violent oxidation. After 8 minutes of oxygen blowing into the bath, it initially results in the complete removal of silicon. This leads to the highest rate of carbon removal through a violent oxidation reaction and generates large amounts of CO gas that is mixed with the molten slag.
- Slag overflow. It is important to control the overflow of foamy slag from the converter vessel during the various stages of blowing at a higher rate.
- Sub-lance test is a water-cooled sensor lance that is often immersed in the molten steel bath during the final stages of blowing to check and sampling of steel. These test results are primarily used for the following purposes: (i) to

predict the end point composition/temperature; (ii) to stop the oxygen blow after reaching the desired level of steel composition/temperature; and (iii) well-control of charging conditions for quick tapping.

- Burnt lime is often added with a short reblow to raise the temperature or adjust the chemical composition of steel.
- Steel tapping from the converter involves pouring of molten steel through the tap hole into a transfer ladle.
- Steel tapping temperature. For the fabrication of steel articles through ingot pouring/continuous casting techniques, the common practice is to tap the liquid steel at a particular selected temperature that falls within a temperature window. Based on this tapping temperature, steelmakers can predict the temperature losses during the various stages of treating/holding the liquid steel in the ladle. Tapping temperature for the case of low-carbon steels with 0.1% of C is normally around 1596°C, which is 80°C higher than its solidification temperature (theoretically). Whereas for high-carbon steels, tapping temperatures are maintained at certainly much lower temperatures.
- Alloying additions. Deoxidizing agents such as Aluminum/Ferrosilicon are added to the ladle before or at the time of tapping to reduce the dissolved oxygen content in steel. Further, there is a significant oxidative loss of residual manganese (Mn) content of the hot metal during the various stages of blowing. This results in liquid steel with residual Mn content of 0.1%. Therefore, an alloying additive such as ferromanganese is often added to increase the residual content of Mn in steel.

7.2.4 Tapping

At the end of every heat, tapping the liquid steel, draining the slag, and carefully inspecting the converter refractory lining are the common steps of standard operating procedure before starting the next heat. After charging the scrap and hot metal, the converter is tilted vertically for oxygen to blow through the lance. For a given bath level, the lance distance is selected in such a way to fulfill the following operational requirements: (i) the free oxygen gas jet stream should cover a maximum surface area of the metal bath; (ii) simultaneously, it is important to control the concentration of the force at the given iron bath level to avoid sparking (i.e., ejection of tiny iron particles); and (iii) other essentials are, control of slag foaming as well as promoting the progress of oxidation reactions without any interruption during the successive stages of oxygen blowing at shorter lance distances. Thereafter, sampling and analysis of metal/slag are carefully carried out to check the end point. The number of heats in the converter per day are strongly dependent on tap-to-tap time that is profoundly affected by the flow rate of oxygen gas jet, the composition of the hot metal bath, lancing profile, and targeted steel chemistry. For example, almost 30 heats can be drawn per day by maintaining the tapping time for one to another tap in between 30 and 45 minutes. Although the capacity of a converter is an important process parameter, the overall oxygen blowing time and time-lapse between one to another tapping may not significantly vary with the converter capacity. The service life of a converter refractory lining normally lasts for several heats, say a few thousand and more, and it is possible to produce almost 5 million tons of liquid steel per year from a large BOF steel-making shop operating with a minimum number of three converters. In general, there is a weekly maintenance shift for a careful inspection and repair of refractory lining. Relining the converter usually needs less than one week.

7.2.5 Slag Formation

Quality steel depends on the synchronized composition of a good quality slag. Hot iron to molten steel formation demands several processes, including basic slag formation through a controlled reaction of lime, fluxes, and doloma (CaO.MgO) with several oxides like MnO, SiO_2, FeO, P_2O_5 developed during oxygen lancing in hot metal. Figure 7.7 shows the limits and direction of BOF slag composition change principally on the basis of main slag constituents.

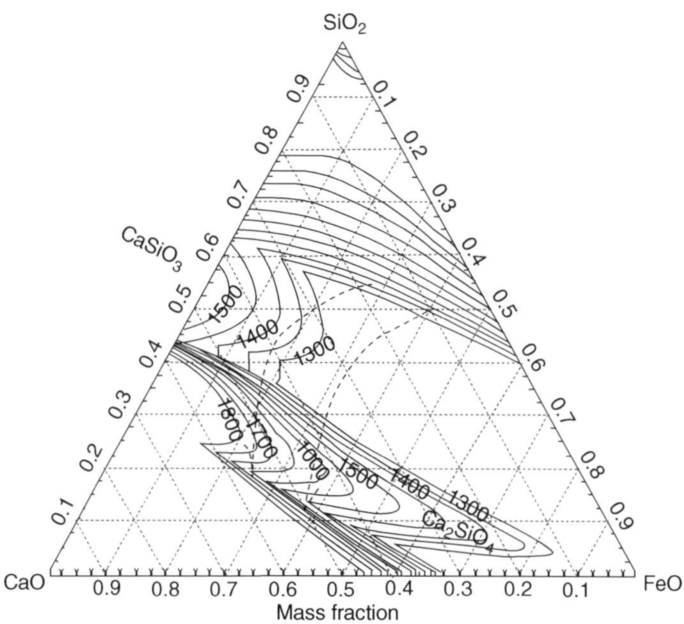

Figure 7.7 Ternary phase diagram shows the typical slag path route with increasing slag temperature between 1300°C and 1700°C. The chief constituents of slag are lime (CaO), silica (SiO_2), and iron oxide (FeO).

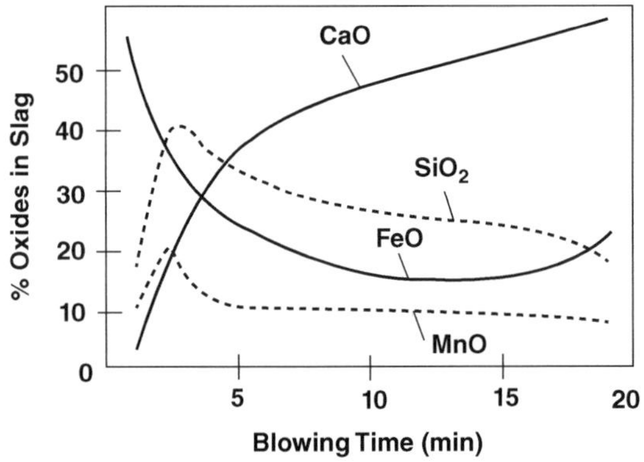

Figure 7.8 Evolution of composition of iron bath while oxygen blowing in 55 ton LD converter.

Classic literature illustrates the gradual change in slag composition with an increment of oxygen blowing time, as shown in Figure 7.8. The concentration of CaO increases in slag, whereas the silica and iron decrease during the prime blowing time zone. While considering the low carbon steels, the steelmaking slags tend to enrich with iron oxide during the end of the blow. Thus, slag is primly enriched with calcium silicate during the beginning of the blow. The oxygen blow enhances the bath temperature with progress and expedites the CaO solubility in slag. Gradual increment of CaO concentration changes the CaO/SiO_2 ratio and approaches to form $CaSiO_3$ and Ca_2SiO_4 that eventually like to precipitate, thus the slag liquidus temperature enhances [4].

Thus, it is quite obvious that the variation of liquidus temperature is attributed to the concentration of different oxides available and their interaction. The addition of doloma (a common practice for dolomite lining) enhances the concentration of MgO that affects the slag liquidus temperature and softens the *calcium silicate noses*.

7.3 Operational Features in EAF

The electric arc furnace (EAF) process exhibits many operational benefits in making steel compared to the basic oxygen furnace (BOF) steelmaking, that are given as:

i) better thermal control with external arc heating over BOF approach that uses heat released from the exothermal oxidation of constituent element species in charge;

ii) greater economic feasibility of EAF process over BOF. This is attributed to the maximum utilization of chemical energy together with electrical energy in EAF;
iii) permits large amounts of alloy additions in EAF process ascribed to its economic feasibility over BOF.

Apart from these advantages, some of the important drawbacks of the EAF steelmaking process are given as:

i) produces steel articles with a carbon content normally well above 0.05% due to less intense mixing of slag/metal in EAF as well as EAF steelmaking is not as oxidizing over BOF approach;
ii) another issue is with the brittleness of EAF steel articles containing higher amounts of nitrogen in the range of 40–120 ppm over BOF processed steels with a residual nitrogen content of 30–50 ppm. In EAF, liquid steel absorbs nitrogen from air in the high-temperature arc zone, which is an inevitable problem. Many other practices have been suggested to control the residual nitrogen content of EAF steel. Some of them include: a) stirring the steel melt with high-speed Ar gas jet or the application of vigorous CO boil to adsorb nitrogen in liquid steel; b) use of short arc for heating, which makes less exposure of liquid steel to air and reduces nitrogen pick up in steel.

Many of the EAF steelmaking shops use electric arc furnaces (EAF) that are operating with DC current attributed to their lower consumption of electrode as well as power over the AC current-based ones. Some of the essential features of the DC current-based EAF's are given as: (i) These furnaces are equipped with two electrodes. The larger one is arranged in such a way as it is extending through center portion of the furnace roof. Whereas the other one is provided in the bottom portion of the furnace, usually known as the counter electrode that is contacting the melt; (ii) Energy-efficient operations with the presence of hot-heel ensures flow of good amount of current through the charge; (iii) DC arc produces a quiet steadier burn that causes less operational disturbances; and (iv) these furnaces are normally provided with 130 tons of operational capacity. Further, some of the shortcomings of the DC current-based EAFs are: (i) limited service life of the bottom electrode; (ii) hearth integrity; (iii) operational difficulties; and (iv) expensive electrical instrumentation.

EAF steelmaking uses scrap steel as the major raw material in the overall charge. For the fabrication of EAF steels with improved ductility, a proper quality of scrap is essential. In such cases, residuals of scrap play an important role. The desirable residual elements include copper (Cu), chromium (Cr), nickel (Ni), molybdenum (Mo), and tin (Sn). Steel scrap with a maximum amount of 0.2% of these desirable residuals is used. To cut down the scrap expenses, some of the EAF shops use recycled scrap as a source of scrap with the combination of directly reduced iron (DR-Iron). It can be noted that the DR-Iron does not contain the desirable residuals, but the recycled scrap is enriched with these desirable entities. Fabrication of low carbon steels with improved ductility characteristics by a graphite electrode containing EAF is considered as a less attractive EAF steelmaking approach. In actuality, the EAF processed steels consist of elemental impurities such as carbon, nitrogen, and residuals in higher proportions.

The graphite electrodes are typically chosen according to the operational capacity of EAF. For example, graphite electrodes of 600 mm diameter are used in EAF vessels of 100-ton capacity. The EAF is provided with a vertically movable mast, which is a structural arrangement that holds the electrodes. The major function of the mast is to regulate the arc length and current flow by monitoring the distance between the scrap/melt and electrode tip. The power supply equipment is installed in a concrete vault near to the furnace. The essential units of this power supply equipment includes a furnace transformer, a step-down type transformer, a tap changer for controlling the voltage of graphite electrode, and vacuum circuit breakers for emergency furnace protection. The furnace transformer is joined with electrodes via heavy water-cooled cables and power carrying arms. After the end of a heat, liquid steel is tapped, and the furnace roof is moved away for a careful inspection and necessary maintenance of the EAF hearth. In EAF steelmaking, the scrap charging procedures use specially designed cylindrical buckets with an open top arrangement for scrap loading and a drop-bottom arrangement for a quick scrap charging into the furnace. The graded distribution of the various sizes of scrap (heavy, medium, light) in the overall charge may have the following potential benefits: (i) lighter scrap in bottom portion of the furnace acts as a cushion that reduces the direct impact of the heavy scrap upon loading; (ii) provides good electrical conductivity in all parts of the overall charge; (iii) minimizes the problem of graphite electrode breakage; and (iv) protects the refractory lining of furnace walls from overheating during the meltdown period. Admixtures such as pig iron slag forming agents are also used with the combination of scrap in the overall charge. It can be noted that the pig iron is a source of carbon that helps in reducing the overoxidation of steel, whereas slag forming agents serve as an aid for the quicker formation of slag.

The various stages of the EAF melt down procedure are outlined as follows:

i) After dumping a full bucket of overall charge as per the suggested charge preparation procedure, the furnace roof is closed, and graphite electrodes are lowered to begin the scrap meltdown process;
ii) At the beginning stage, a low power setting is selected to melt down the lighter scrap that is placed on the top portion of the overall charge. This accelerates the initial bore-in so that the electrodes can penetrate the scrap in a deeper way and promote high power meltdown;
iii) After the completion of high-power meltdown, a second bucket of charge is dumped into the furnace, and the same meltdown procedure is repeated.

At the end of the meltdown step, sampling and analysis of the liquid steel as well as molten slag are carried out to check the end point (composition/temperature). The liquid steel is said to be overoxidized when it contains carbon content of about 0.25%. The basic slag chemistry is normally given based on percentage amounts of oxide constituents of slag. This includes 55% of lime, 15% of silica, and 15–20% of iron oxide. In some cases to meet the end point, carbon percentage in steel can be adjusted by two major approaches. One is oxygen blowing to reduce the carbon content in steel. The other is injecting carbon to enhance the carbon content in steel. After meeting the end point, liquid steel is tapped through a vertical tap hole or spout into a transfer ladle. On the other hand, the molten slag is poured into a slag pot via the furnace rear door. Complete separation of slag from the liquid steel can be accomplished with the use of hot heel practice, which leaves nearly 15% of the liquid steel in EAF.

It is well-known that slag foaming is often created with the use of a lime–carbon mixture or by injecting carbon to protect the furnace walls from radiation damage during the stages of the high-power meltdown period. Slag foaming characteristics strongly depend on the slag chemistry. For example, a good slag foaming requires moderate iron oxide containing slag with a basicity well above 1.5. It is important to note that slag has a vital role in steelmaking in EAF.

The desirable slag chemistry includes CaO, SiO_2, FeO, and MgO as the chief constituents of slag as well as Al_2O_3, MnO, and P_2O_5 as the minor slag constituents. For such basic slag formations, the commonly used slag forming agents are burnt lime, admixture of dolomite/lime, steel scrap containing oxides, directly reduced iron having gangue oxides, carbonaceous ash, and recycled slag. Calculations based on slag mass balance are often important for comparing the experimentally measured slag chemistry that provides a basic understanding of the melt corrosion of refractories. Although the slag chemistry with a basicity of 1.5 has been suggested for a good slag foaming character, sometimes a modified chemistry of slag with a basicity of 2 can decrease the MgO requirement for slag saturation purposes. For a specified slag chemistry with a given FeO content and basicity value, there is the possibility of estimating the amount of MgO needed for saturation of slag from the MgO saturation lines, as shown in the ternary phase diagram in Figure 7.7. Apart from all these aspects, control of slag foaming is another important task, which can be accomplished nowadays by forming slag with a careful injection of carbon and oxygen [5].

A two-slag practice can also be recommended in EAF operations for producing very clean steel that is constituted with much less residual amounts of oxygen and sulfur. The two-slag practice involves draining slag from the initial oxidizing melt down and the subsequent addition of new slag forming agents for the formation of reducing slag. The chief constituents of a reducing slag mixture are given as 65% of lime, 20% of silica, and the rest are proportioned with calcium carbide/alumina. However, this reducing slag forming mixture does not contain any amount of iron oxide, practically. Alloy additives that can be easily oxidized, are also encouraged for better metallurgical process control. The liquid steel is being continuously refined under the action of reducing slag to meet the targeted end point requirement (composition/temperature). For clean steelmaking via a two-slag EAF operation, the total time required for a heat is between 1 and 4 hours. However, the total amount of heat time varies significantly with the refining mode employed in the production of clean steel with a specified end composition. EAF shops that do not follow a two-slag clean steel practice use EAF's for energy-efficient scrap melting operations and depend on ladle refining stations for making clean steel. Further, the EAF's high-temperature working environment often leads to significant oxidative/erosive losses of the electrode material. Depending on the steel grade, the material loss of working electrode may vary at a rate of 3–6 kg per ton of liquid steel, can be noted as a considerable issue in EAF route.

7.4 Refractory Designing and Lining

The basic philosophy behind refractory selection depends on the consistent performance without sacrificing steel composition. In primary steelmaking vessels, several zones may have different interacting atmospheres in the perspective of mechanical and thermal stress as refractory must be accommodated in a different geometry. Prior to refractory selection for a particular zone, it is essential to concentrate on the following parameters:

- Steel chemistry and slag composition
- Thermal and mechanical stress
- Refractory lining and corrosive wear
- Refractory composition and properties

While considering the steel output and refractory consumption, prime focus has been given to exploring the working and permanent lining for BOF and EAF during primary steelmaking and design aspects of common and essential basic refractories. Note that the word *basic* refers to the refractory linings of the steelmaking vessels that are made of alkaline materials, such as magnesite (MgO) and dolomite (CaO–MgO). For a satisfactory service performance in steelmaking environments, these basic refractory linings should exhibit good refractoriness, better abrasion resistance during the stages of charging as well as blowing, and excellent resistance against highly oxidized/basic slags.

7.4.1 Steel Chemistry and Slag Composition

An essential consideration is to monitor both the composition of steel and slag. A wide range of elements manifest steel properties, and it is necessary to know the relevant chemistry of steel during processing. Despite steel chemistry, each quality of steel produces different slags consisting of a wide range of oxides with various percentages. Before discussing the refractory designing parameters, a brief on both provides a better insight.

Iron (Fe) is the chief constituent of steel and exhibits allotropy as a function of temperature. The two allotropic forms of Fe in steel are alpha iron (α–Fe) with body centered cubic (BCC) crystal structure, and gamma iron (γ–Fe) with face centered cubic (FCC) crystal structure, respectively. Designing the steel microstructure with the combination of allotropes of iron and alloying elements (particularly carbon) results in steel articles with novel properties. A brief description of the various alloying elements can emphasize the importance of alloying.

- Carbon is known as the chief hardening element in steel. In general, the addition of carbon has a beneficial effect on improving the hardness and strength properties of steel articles but shows a negative effect on their ductility, weldability, and toughness properties. It has been reported that the enhancement of hardness and strength values of steel articles are almost proportionate with the incremental addition of 0.85% of carbon content. For ultralow carbon (ULC) steels, the usual addition of carbon ranges between 0.002% and 0.007%. In normal carbon steels and high strength low alloy (HSLA) steels, the minimum carbon content is about 0.02%. It can be noted that the steelmakers can go up to a highest level of 0.95% of carbon in plain carbon steels, and 0.13% of carbon in HSLA steels.
- Boron addition improves the hardenability of steel. For the case of low carbon steel products, boron is commonly added to tie up nitrogen, which is beneficial in reducing the yield point elongation for minimizing coil breaks. Also, steel articles with excellent formability characteristics can be produced using appropriate processing practices with approximately 0.009% of boron. It can be noted that the residual amount of boron in steel is normally well below 0.0005%.
- Nitrogen in steel can be entered as an intentional addition or as an impurity with a residual amount normally less than 100 ppm.
- Aluminum is added primarily as a deoxidizer in steel refining. The added aluminum combines with oxygen in steel and forms aluminum oxide inclusions that float in the steel refining slag. This class of refined products is known as aluminum killed steels. Steelmakers recommend that a minimum amount of 0.01% of aluminum is typically required for deoxidation practices to produce aluminum killed steels. Also, aluminum is a grain refining agent in steel in hot-working operations. For example, during the stages of hot-rolling the added aluminum reacts with nitrogen in steel and forms aluminum nitride precipitates that act as grain refiners by significantly inhibiting grain growth in steel. Although the AlN precipitates are beneficial, they have been controlled to reduce their determinantal effects on the properties of steel coil during the downstream processing.

- Phosphorous is added in the production of re-phosphorized high strength steel that finds uses in the fabrication of automotive bodies. The residual amount of phosphorous in steel is well below 0.020%. For further strength enhancement of these steel articles, silicon can be added with the combination of phosphorous. A minimum amount of 0.10% of silicon is normally recommended in automotive bodies, but it is limited to 0.04% in the case of post-galvanizing applications.
- Sulfur is always considered the most determinantal impurity in steel. A residual amount of 0.012% of sulfur is recommended in commercial steels, but it is reduced to an amount of 0.005% for the case of formable HSLA steels.
- Calcium additions are essential for sulphide shape control in steel. This improves the formability characteristics of steel articles. For additional enhancements of strength of HSLA steels, a recommended amount of 0.003% of calcium is added.
- Manganese addition improves strength, hardness, and cold temperature impact toughness of steel articles but reduces their ductility and weldability if it is added in higher proportions. The contribution of Mn toward the enhancement of strength and hardness of steel articles is similar to carbon but to a lesser extent. The Mn content in all commercial steels is normally in the range of 0.2–2.0%.
- Nickel addition predominantly improves corrosion resistance properties of steel articles, mainly including stainless steels. Also, Ni is beneficial for improving the impact strength and hardenability of carbon steels, which is added in smaller amounts.
- Molybdenum additions find uses in increasing the strength properties of boiler as well as pressure vessel steels that typically operates in the working temperature range of 400°C. The combined addition of molybdenum (Mo) with chromium (Cr) is beneficial for the enhancement of high-temperature corrosion resistance as well as creep strength properties of steel articles.
- Copper additions are essential for mainly improving the atmospheric corrosion resistance as well as paint adhesion properties of structural steels, and it contributes for the small enhancement of steel hardenability.
- Chromium additions are advantageous for improving the corrosion resistance as well as oxidation resistance properties of steel articles. The addition of more than 1.1% of chromium develops a protective layer in all parts of the steel surface that help in protecting the steel articles against oxidative wear.
- Niobium is added as a grain refiner that retards grain growth and produces steel articles with finer grain sizes, which improves their mechanical properties, mainly strength, toughness, and ductility.
- Vanadium, niobium, and titanium are the micro-alloying elements commonly added in the range of 0.01–0.10% as singly or in combination with steel articles for significantly improving their strength. Additionally, titanium and niobium have a beneficial role in steel refining operations as stabilizing additives

that combine with carbon and prevent its adverse effects of hardening. This directly results in superior formability and surface quality of ultralow carbon steel articles.

The addition of different alloying elements is obvious and residual material develops a wide range of slags in the presence of existence slag coming from blast furnace as well as plausible corrosion of working lining. To fulfil the steel composition and operational demand, a typical wide range of BOF slag composition ensures the presence of oxides in slag [6]. The distribution of free CaO (f-CaO) is actually heterogeneous. Further, there is a possibility of dissolving nonstoichiometric FeOx and MnOy compounds in f-CaO in BOF slag during the steelmaking process. It can be noted that the resultant phase composition varies with respect to different BOF areas and difficult to predict the consistency. Some typical slag composition from different resources is given in Table 7.3.

Making a prerequisite slag composition is essential to maintain the refractory campaign life. Thus, chemistry and phase analysis of both steel and slag is essential prior to supplying refractory details.

7.4.2 Thermal and Mechanical Stress

The basic understanding of the thermal stress/strain fields generated in all parts of the hot-face refractory lining is often important in order to predict the severity of the *thermally induced refractory cracking* and consequent refractory failure in high-temperature steelmaking environments. In such a scenario of refractory failures, there has been a significant role for wear attack on refractories in service. This class of refractory wear is predominantly associated with two serious issues known as high-temperature

Table 7.3 Chemical analysis of BOF slag in various integrated steel plants in China.

Chemical compositions (wt.%)										
Source	CaO	SiO_2	Al_2O_3	Fe_2O_3 (FeO)	MgO	MnO	P_2O_5	TiO_2	Others	f-CaO
SS	43.29	13.63	3.23	24.34	10.37	1.05	2.78	0.56	0.75	5.68
MS	41.4	9.79	2.69	25.25	13.6	4.67	0.98	0.79	0.83	6.29
CS	37.37	12.21	1.01	27.90	11.5	3.32	4.10	1.02	1.37	4.19
BS	39.42	11.63	0.83	30.43	8.50	2.21	3.23	0.67	3.08	4.81
BOF slag	40~50	10~15	1~5	24~28	10~12	<5	<8	<2	/	<10

Jiangsu Tieben Steel and Iron Co. Ltd. (SS); Baoshan Steel and Iron Co. Ltd. (BS); Ma'anshan Steel and Iron Co. Ltd. (MS); Cangzhou Steel and Iron Co. Ltd. (CS).

mechanical abrasion/erosion and melt corrosion of refractories. Abrasive/erosive damage usually causes material loss that creates discontinuous matrix regions that bond the aggregates and ultimately lead to microstructure heterogeneity in all parts of the worn refractory. Melt corrosion also leads to microstructure heterogeneity by forming various chemical compounds throughout the discontinuous matrix regions bonding the aggregates. Furnace refractory linings are usually subjected to thermal shock (hot shock/cold shock) during various stages of their thermal cycling profile. Steelmaking refractories are known to be multiphase composite materials. When they are subjected to thermal shock, the generation of thermal gradients can lead to a differential thermal expansion behavior of aggregate/matrix parts. This differential expansion behavior causes a significant thermal expansion coefficient mismatch (CTE) in all heterogeneous portions of the discontinuous matrix regions that bond the aggregate parts of the shocked refractory. It induces thermal stress/strain fields from surface to bulk of the refractory and can also lead to refractory cracking at a significant CTE mismatch with the origin of thermal gradients. Almost all steelmaking operations are practically batch processes. There are three common stages in every batch operation: charging, melt processing, and tapping. During these stages, the thermal cycling profile of the steelmaking vessel and working refractory (hot-face) vary considerably for a range of temperatures based on heating/cooling. During these temperature changes, a transient stress field is imposed on the steady state heat refractory condition of refractory hot-face. As stated earlier, such changes in thermal cycling profiles may generate strong thermal gradients between hot/cold faces of the refractory lining and often leads to significant refractory damage.

In a recent article, it is suggested that conducting the simulation experiments may be the best way to study the changes raised in the thermal stress–strain fields of a refractory to predict the thermal shock damage based on its thermal cycling profile in particular steelmaking conditions [7]. The various aspects that generate thermal stresses within a body, including in the refractory material under different process constraints, have been studied by Kingery et al., and detailed as a specific case as follows [8]: (i) freely expanding body, in which thermal stresses are generated due to differential thermal expansion behavior of various portions of the body upon restraining the free expansion of any portion within it; (ii) heterogeneous body without thermal gradients (isothermal condition), in which thermal stresses are caused by a significant CTE mismatch across various heterogeneous entities (aggregates/fines) within the body during the stages of heating\ cooling; and (iii) multiphase composites such as refractories, in which thermal stresses arise due to thermal gradients as well as heterogeneous constituents of various portions (aggregates/fines) of the refractory during the various stages of thermal cycling in service. As stated earlier, the origin of thermal gradients causes differential expansion/contraction of various microstructural components, results in a significant CTE mismatch and generates thermal stresses in almost all regions of the shocked body, as shown in Figure 7.9. He explained in consideration of the origin of thermal stresses in a slab type body during the stages of cooling (see Figure 7.9a) as well as heating (see Figure 7.9b). Cooling generates a significant amount of tensile stresses in all subsurface portions near the surface of the body. In comparison, compressive stresses are significant in all parts of the interior portions of the body. Such a tensile nature of the thermal stress distribution is due to the restraining effect imposed by the subsurface portion on the contracting tendency of the cooling body surface. On the other hand, the direction of thermal stress (tensile/compressive) distribution is reversed upon heating. Similarly, tensile stresses are generated near the surface upon cooling for multiphase composites like refractory components. This is an unconstrained case of refractory materials. However, for a constrained case such as refractory lining, tensile stresses arise upon heating. Such a tensile nature of thermal stress distribution may result in thermal shock failure of refractory lining materials due to their low shear strength characteristics [8].

The rate of cooling/heating is another important factor determining the distribution and severity of the thermal stresses (tensile/compressive) generated in the refractory components. Such studies have been performed by Tsibin et al. [9] and provide the explanation based on the influence of heating rate on the result of thermal stresses within a concrete lining. The test results are presented in Figure 7.10 and can be implicated to the preheating stages of the refractory lining. From Figure 7.10a, you see that there is a significant variation in the pattern of thermal profiles within a concrete lining in convective heating for a range of heating times, τ_1, τ_2, and τ_3. The change in the heating time is given as $\tau_1 < \tau_2 < \tau_3$, indicating a more heating rate with short heating time of τ_1. The formation of steeper thermal gradients from τ_3 to τ_1 is in agreement with the change of heating time as well as heating rate.

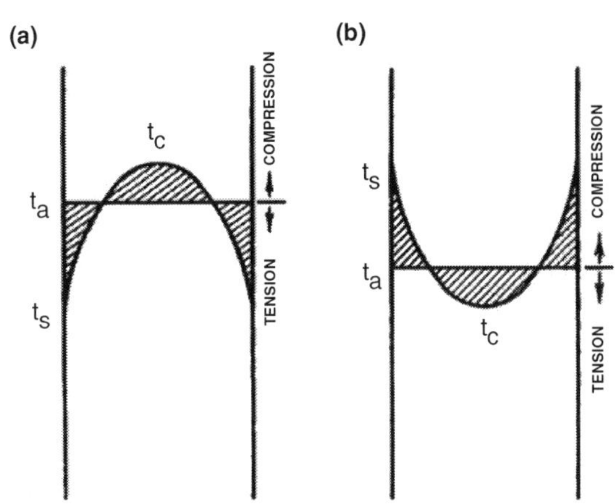

Figure 7.9 Simplified diagram showing the thermal stress distribution profiles upon (a) cooling and (b) heating of a slab. Subscript *s* and *c* represent surface and center, respectively [8].

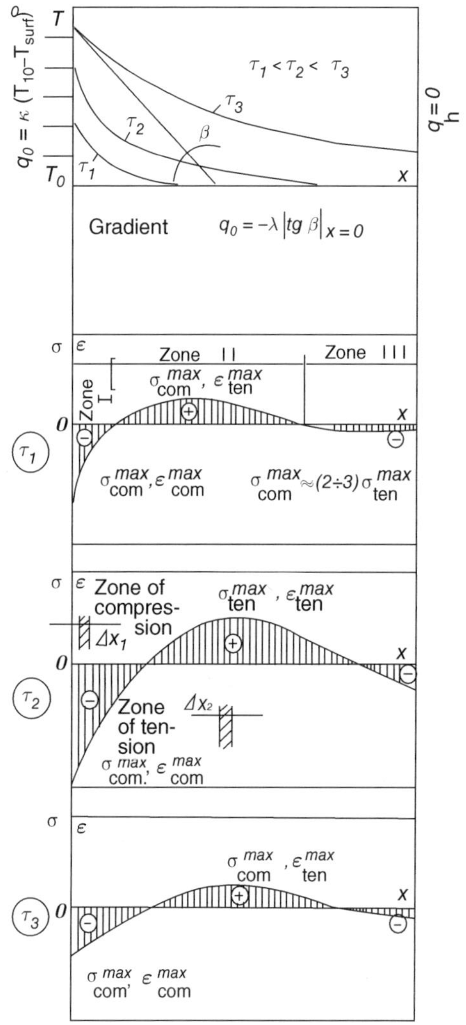

Figure 7.10 Simplified diagram showing the distribution of the thermal stress profile within the concrete lining consisting of various thermal gradients (a) change in the thermal profile pattern across the thickness of the concrete lining with the change of heating time from τ_1 to τ_3, given as $\tau_1 < \tau_2 < \tau_3$; (b)–(d) nature of thermal stress (tensile/compressive) distribution profiles with respect to the change in heating times, τ_1 (b), τ_2 (c), τ_3 (d) [9].

The variation in the distribution of thermal stress profiles (tensile/compressive) with respect to the convective heating times τ_1, τ_2, and τ_3 ($\tau_1 < \tau_2 < \tau_3$), is shown in the Figure (7.10b)–(d). The following observations can be noted: (i) tensile stresses are significant in all regions of the Zone II; (ii) compressive stresses, however, are much more significant in all regions of the Zone I and Zone III; (iii) moving away from the hot-face, it is observed that the severity of the thermal stress (tensile/compressive) has been significantly decreased due to the result of small thermal gradients with increasing convective heating time from τ_1 to τ_3; and (iv) highest thermal stress with maximum severity has been observed for the case of convective heating time τ_2 that results in excessive thermal gradient in lining thickness.

In a similar work class, a finite element method was employed through 2D mathematical model [7]. This is based on the refractory body with biaxial state of thermal stresses (longitudinal tensile/transverse tensile). The test results are provided in Figure 7.11. This analytical data indicates the influence of the temperature distribution profile on the biaxial state of thermal stresses (longitudinal tensile stress/transverse tensile stress) in a refractory body after heating from one end for a heating time 3000 seconds. A wider distribution of thermal stresses of tensile nature in longitudinal and transverse directions can be observed in Figure 7.11(b)–(c). For the case of the biaxial stress state, the cracks may form in the regions containing maximum tensile stresses, but the onset of crack growth is favored in the regions with minimum tensile stresses. Regarding the present situation, it is noted that the generated cracks can easily propagate through the transverse direction rather than via the longitudinal direction due to the lower magnitude of the maximum transverse tensile stress (9.1 MPa) compared to the more magnitude of tensile stresses in the longitudinal direction (11.1 MPa). However, the onset of crack growth is favored in the longitudinal direction with the application of steeper temperature profiles that are often generated in a refractory body with short heating times. Although the experimental findings of Cheng et al. are being implicated in practical applications, such studies assume materials as homogeneous bodies and their properties are independent of temperature. Furthermore, the temperature dependent material properties (density, thermal conductivity, specific heat capacity, thermal diffusivity) are known as thermophysical properties that may significantly influence thermomechanical behavior of materials, including refractories.

Moreover, Ganguly and coworkers [10] certainly pointed out that one of such thermophysical properties, namely thermal conductivity can be significantly varied with the change of temperature in the case of refractory components. Similarly, Joliff and coworkers [11] identified that Youngs modulus is also one of such thermophysical properties that can be varied significantly with the change of temperature. Their findings are based on a series of heating/cooling experiments on a composite body made of Al-inclusion embedded glassy matrix.

7.4.3 Refractory Lining and Corrosive Wear

A long-service BOF refractory lining with minimum maintenance/repair losses is a potential benefit for the continuous BOF operation in the production of liquid steel as per the demand. The modern ways for achieving a BOF refractory lining of long service life with minimum refractory consumption are optimizing the BOF lining design and its maintenance practices and BOF operational discipline. The various parts of the refractory lining in different zones of the steelmaking converter (BOF) can be seen from the BOF cross-sectional view, as shown in Figure 7.12.

Refractory life depends on the several properties of refractories, lining configuration, and operational features. Wear is a common phenomenon in severe

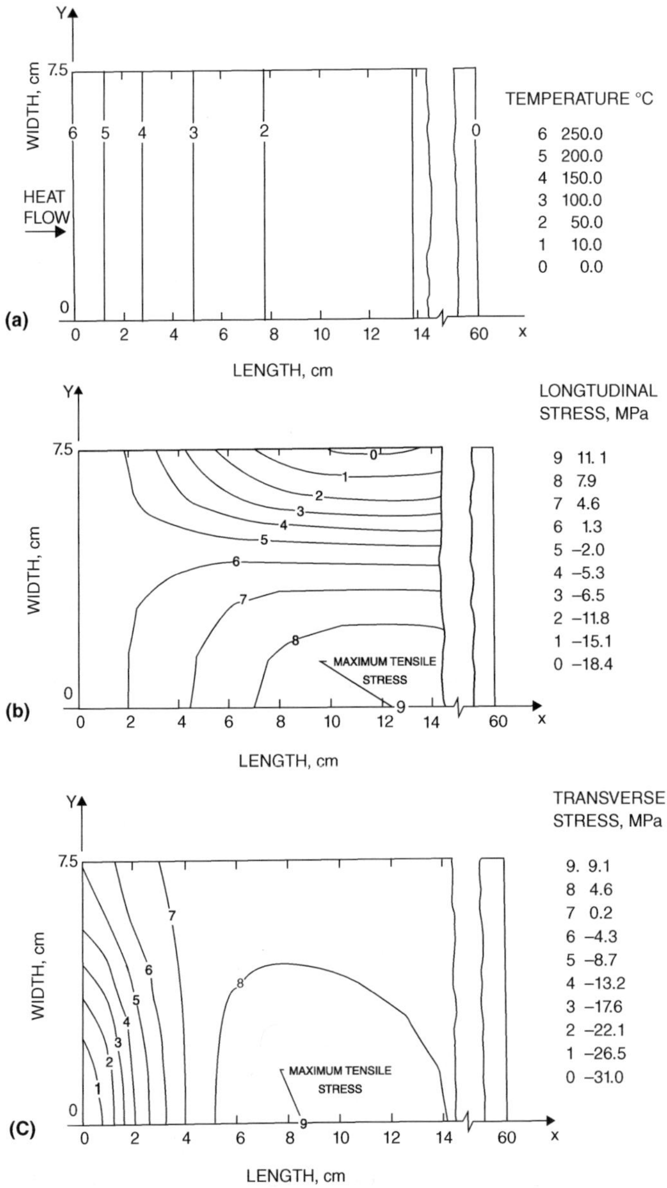

Figure 7.11 Finite element analysis studies for predicting the thermomechanical behavior of a refractory body heated from one end for a heating time of 3000 seconds. Simulation diagrams showing the (a) temperature profile pattern as well as the associated contour counterparts of (b) longitudinal tensile stress, and (c) transverse tensile stress in all parts of the heated body [7].

environments during metal to steel processing, and this wear mechanism depends on either individual or a combination of several effects:

i) The residence time of the metal and slag in the converter
ii) Slag chemistry on the chemical attack of working lining
iii) The temperature of both slag and molten steel
iv) High brick porosity facilitates molten liquid penetration
v) State of oxidation of the melt
vi) Erosion by metal and slag during oxygen blowing and BOF tilting
vii) Impact and abrasion by foreign particles like dusts and gases
viii) Impact of scrap and metal in charging
ix) Oxygen jet impact and penetration
x) Thermal cycling in the gap between two heats
xi) Certain refractory lining damage during de-skulling of the converter
xii) Poor thermomechanical response of refractories

Several wear zones are the result of the above incidences. Liquid steel tapping temperature varies significantly and sometimes needs to superheat for a particular steel quality during secondary steelmaking. Continuous corrosion and wear change the slag composition as well.

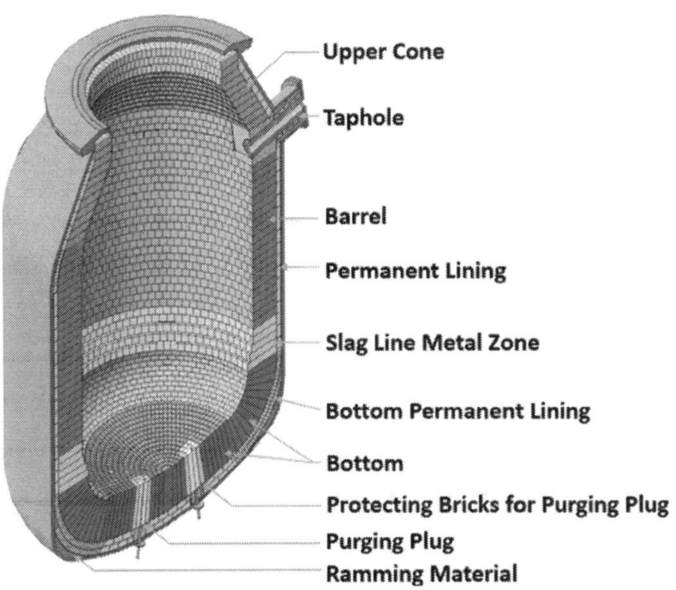

Figure 7.12 Cross-sectional view of BOF to demonstrate the critical zones for refractory lining.

Magnesia Carbon (MgO-C) basic refractory is the prime wear lining; thus basic slag reduces the corrosive wear. MgO solubility limit in slag varies up to 8% depending on temperature and oxidation state. Without the addition of calcined dolomite, the brick lining wears out in contact with slag and reaches to MgO saturation level. Furthermore, in the presence of high content of FeO the MgO content is diluted and temperature increases, resulting in an increment of reaction rate kinetics. Continuous operation expedites the erosion and abrasion effects during steelmaking.

An apposite material and dimension thickness design may reduce the wastage of lining refractory after completion of the converter campaign. However, such a theoretical assumption is violated when uneven wear out happens during operation. Uneven indicates all zones are not wearing out at the same rate. Thus, we need to repair certain zones to get the complete refractory life, and monolithics do this. Typical zones are shown in Figure 7.13.

A balanced lining thickness at an optimum refractory cost to improve lining life is always the target of the steel sector. Different qualities of refractories and lining thickness are the prime criteria for a balanced lining that one can design and finalize after carefully studying the wear pattern of the different zones of the converter. Such lining

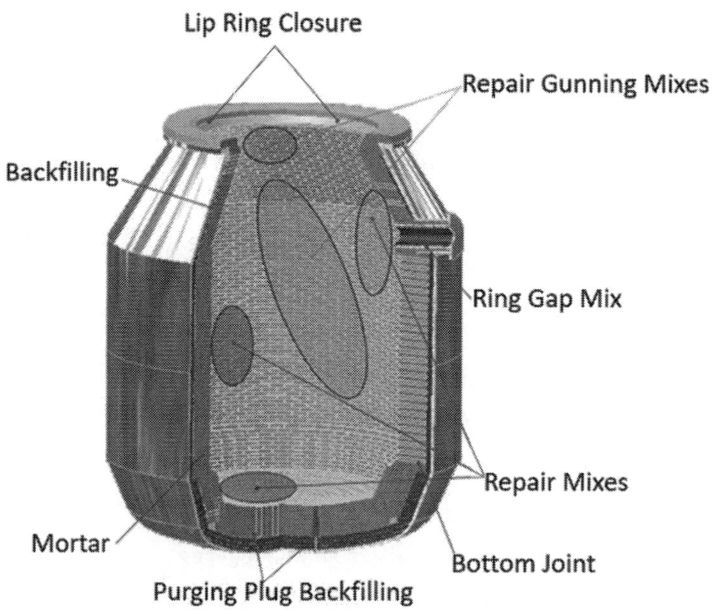

Figure 7.13 All critical zones, including probable damage zones for the repairing by monolithic.

concept is commonly known as *zonal lining*. In this concept, a lesser wear zone is assigned to a relatively low quality or lesser thickness of refractory, whereas high quality, high wear resistance, and high-cost refractory are designed for severe wear segments to balance the entire vessel lining and enhance the quality of the campaign life.

A brief on the wear and lining thickness assessment concerning a definite quality of steel/slag has been discussed in the presence of high silicon content (2.0–5.0 mass%) hot metal produced by the Corex process that eventually will help to understand the relation between performance, lining thickness, and wear during the process [12]. After refining the Si content substantially amount is reduced into semi-steel in the range of 0.5–1.0 mass%.

As shown in Figure 7.14, the desilication process greatly damaged the lining thickness of the converter, in which the slag section lining thickness reduced to 450–550 mm, and also front side-lining thickness decreased substantially. The lining thickness of BOF was detected by a laser contouring system. Such an exercise helps to predict the life of brick and further design and selection for a particular zone of interest. The statistical results of the thickness of the converter are shown in Figure 7.15. In zonal concept design, slag lining experiences more wear out and such variation can be identified with respect to different numbers of heats. A serious slag line section corrosion is noticed after 1997th heat compared to 1966th heat, where the area increases from 2.01 to 3.65 m^2 when lining thickness is less than 400 mm. In continuation of 2103th heat, the front side area lining

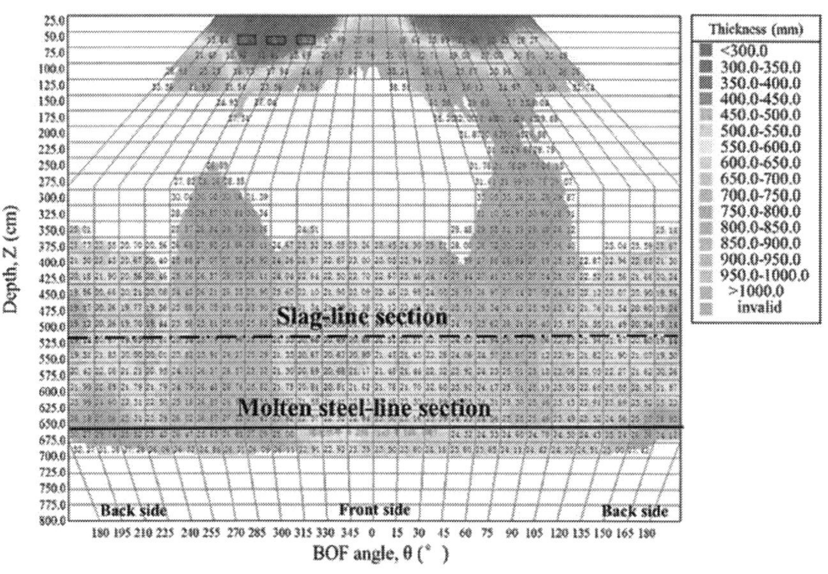

Figure 7.14 Lining thickness after desilication, 2103th heats [12].

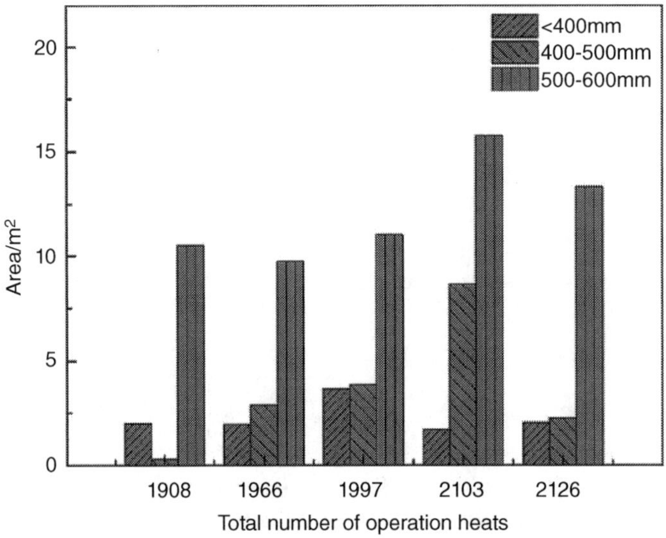

Figure 7.15 Effect of desilication on lining thickness [12].

thickness becomes thinner, moreover, the area for the lining thickness less than 400 mm is 1.72 m^2. During the normal process, the thickness of the slag line section appears thin.

The refractory brick wearing pattern is different for a different segment; thus intermediate repairing is required to complete the vessel campaign. Continuous wear starting from bombardment by scrap charging and the corrosive effect of the molten steel–slag combination. Unexpected worn brick may fall or come out from lining and make a pocket and delay repairing operations. Thus, to prevent discontinuity of vessel operation for repairing brick or pocket, utilization of suitable refractory monolithic gunning or patching on the damaged areas of the vessel in between heats is a common practice. A laser-based temperature profile and topographical of the vessel lining guide the intermediate repair process.

A brief BOF lining corrosion mechanism by slag during desiliconization can be summarized in the following steps, as shown in Figure 7.16:

1) At the beginning of the desiliconization, the current boundary between refractory and liquid slag is solid slag splashed by last heat (Figure 7.16a). Thus, the formation of a solid slag layer restricts direct contact between refractory and liquid slag and protects the refractory lining from immediate corrosion or erosion.
2) The protective layer develops a slag concentration gradient and steady mass transfer between solid and liquid slag at a certain temperature and expedites the solid layer melting and gradual removal from contact with refractory. Moreover, high speed oxygen jets and bath oscillation also induce solid slag

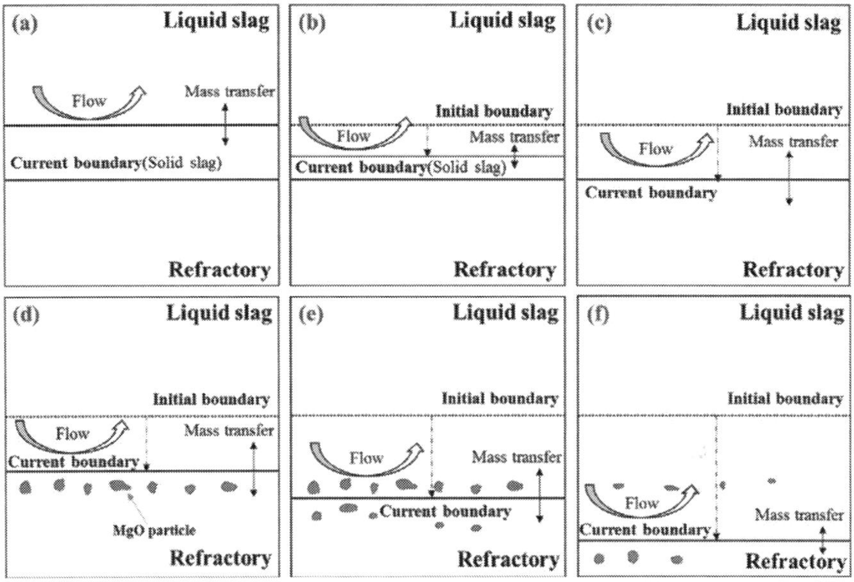

Figure 7.16 MgO-C refractory corrosion in BOF by slag [12].

layer loss from refractory contact (shown in Figure 7.16b). Thus, the solid slag layer gradually disappears, and liquid slag begins to direct contact with the refractory lining (Figure 7.16c).

3) In this stage, a concentration gradient between liquid slag and refractory develops, and direct contact results in a constant mass transfer between them at elevated temperatures. The major phase MgO particles initiate to precipitate at the interface and slowly diffuse into liquid slag (shown in Figure 7.16(d)–(e), which eventually enhances the slag MgO concentration.
4) MgO enriched liquid slag flow and took away the particles from the interface of the refractory brick. MgO melt in the liquid slag and interface area becomes saturated with the composition of liquid slag, and external liquid slag erodes the refractory from outside to inside (Figure 7.16f).
5) The adherence between refractory and slag facilitate the complete disappearance of the solid slag layer and interface, eventually resulting in refractory dissolution into the slag until the campaign life.

7.4.4 Refractory Composition and Properties

A wide range of refractories for different zones including tar-bonded dolomite, pitch-bonded dolomite, pitch-bonded magnesia, and premium quality MgO-C bricks are the most promising and popular refractories for BOF wear or working lining. Usually, high performance working line brick is made of high purity

Figure 7.17 Different shaped MgO-C refractories for BOF. Credit: CIMM GROUP

fused MgO, high crystalline graphite, metallic, and resin bonded to meet the combination of properties to withstand high temperatures and severe converter environments. A balance of properties such as oxidation resistance, hot strength, and slag resistance is expected to continue a campaign life during a rapid change in operation.

A well designated zone such as upper cone, tapping area, barrel, trunnion, scrap impact, lower cone, and bottom is carefully lined by different MgO-C bricks. Different zone along with operational features are discussed in previous sections. Complete zone concept and critical lining parameters for both fixed and moveable bottom BOF have been discussed in our earlier work to select the refractories [13]. Refractory in BOF experiences different chemical, thermal, and mechanical stress environment; thus a wide range of refractory properties may fulfill the performance demand. This property synchronization is possible through the judicial selection of raw materials and their grading and relevant processing protocols. A wide list of BOF refractory, including basic composition, physical properties, and thermomechanical properties, may provide the refractory requirement for BOF.

Figure 7.17 represents different shaped refractories for BOF. It is essential to consider the material selection criteria, and quality of magnesite with respect to zones of the converter, a brief of such information is given in Table 7.4.

Individual or combinations of Large Crystal Dead Burnt Magnesite (LCDBM), and fused magnesite (FM), and Large Crystal Fused Magnesite (LCFM) are prime resources for BOF lining. A judicial choice of raw materials, binder system, graphite flake, and crystallinity may reduce thermal expansion and stress, and enhance life. A clear understanding of different zones is shown in Figure 7.18.

Table 7.4 Essential material selection criteria and materials for different zones of BOF.

Zone	Material selection criteria	MgO quality
Upper cone	Relatively low load of oxidation and erosion, sometimes mechanical problems (de-skulling) higher wear due to oxidation (position of lance) • no reinforcement necessary • steel cladding bricks + burnt MgO-bricks are suitable for the mouth	97% LCDBM, 96% FM
Barrel	Relatively low load of oxidation and erosion • normally no reinforcement is necessary	97% LCDBM, 96% FM
Tapping area	High load of corrosion and oxidation through hot, oxygen-rich slags • high-purity magnesia • fused magnesia • antioxidants	97–98% FM
Slag line	High loads through corrosion and erosion by aggressive and agitated slags • high-purity LCFM or FM	97% FM, 98% LCFM
Bath and bottom	Enhanced load through erosion by agitated steel and thermomechanical problems • fused magnesia • pitch impregnation • antioxidants (around tuyeres) • special constructions (spherical bricklaying)	97% FM, 98% LCFM
Lower cone	Enhanced load through erosion by agitated steel, but less than in the slag-zone • fused magnesia • special double-wedged shapes to tilt the lining	97% FM, 98% LCFM
Trunnion	High load of oxidation due to lack of protecting slag-layer • enhanced C-content • fused magnesia • reinforcing antioxidants	97–98% LCFM
Scrap-impact	Maximum loads through mechanical stress, slagging, and thermal loads • highest grade fused MgO • steel-fiber reinforcement • antioxidants • no back-filling	98% LCFM

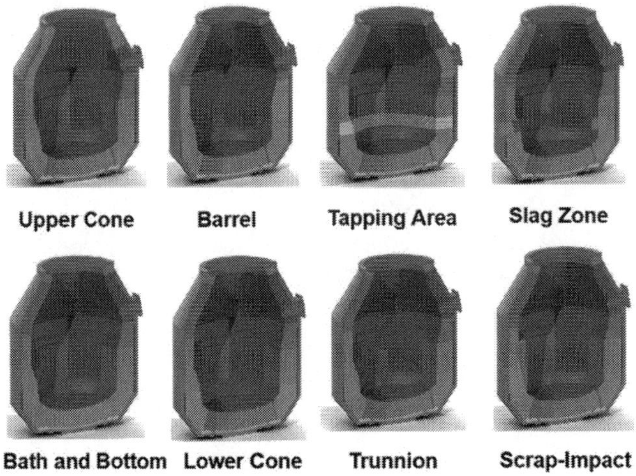

Figure 7.18 Red color represents different zones of BOF.

7.5 Refractory Maintenance Practice

A definite maintenance strategy may enhance the productivity of the basic oxygen furnace (BOF). Maximum automation, enhancing production time by reducing downtimes, and updated vessel availability are the prime targets for most integrated steel mills. Standard protocols are made for BOF refractory maintenance:

- Zero maintenance and almost no maintenance
- Slag splashing
- Slag coating and washing
- Monolithic patching
- Gunning
- Super or intensive refractory gunning

Zero Maintenance: The zero-maintenance converter, with up to 2500 heats campaign lives is usually a common incidence in the steel sector. Under this circumstance, only taphole changing and sometimes mouth area gunning practice can avoid skull formation, and it may adopt if two other vessels are in continuous production and capable of accommodating a maximum number of heats per day.
Slag Splashing: Slag splashing practice was first invented in the USA. The vessel lining life may increase 2–3 times compared to regular practice. While steel tapping is completed, the remaining slag in the vessel is splashed with a high pressure jet of N2 up on different lining areas for 2–5 minutes. The remaining excess slag is poured off before the subsequent charging to avoid the slag build-up in the bottom.

Slag Coating and Slag Washing: With the completion of tapping and slag splashing, the vessel tilted for deslagging; however, a small amount of liquid slag is retained in the vessel bottom. Usually, this slag is enriched with doloma or raw dolomite to cool the slag, enhancing the viscosity and adhesive properties. Thus, several times vibrations or rocking expedites the motion of retaining slag and make a thin slag coat on the converter bottom, bottom joint, tapping pad, and scrap impact zone.

Hot Patching: The wear profile may guide the exact location about the severity of refractory damage and then repairing required to maintain campaign life. Remarkable self-flowing refractory mixes enable solidifying in an optimum time and are excellent for repairing scrap impact zone, bottom joint, and tapping pad. Judicious and competitive patching versus slag-coating to be scheduled, hot patching may persist longer lasting and usually planned and performed in production interruption.

Gunning: Gunning is an excellent method to cover the worn zone throughout the vessel. Robotic gunning leads to accurate and uniform lining thickness and maximizes the exploitation of all refractory linings.

Super Gunning: A high speed and automatic **s**uper gunning can replace slag splashing in available short breaks, eventually reducing production time. High throughputs mix up to 400 kg/minute can be maintained and precise repair for all critical areas is possible.

Prior to maintenance, it is advocated to estimate the wear pattern that reduces the maintenance time and helps to perform the precious operation. An intelligent maintenance concept can help improve refractory efficiency, resulting in higher service life, higher steel production, and more flexibility for plant maintenance (plant logistic) at reasonable refractory costs. See Figure 7.19 for an example. Often, refractory maintenance costs are rated as a minor cost factor when only direct refractory costs are evaluated. In most cases, the influence of the maintenance method contributes far more to the total cost of ownership or TCO [14] costs.

The main factors influencing the lifetime of the lining are; scrap charge, Si-content in hot metal, Oxygen blow rate, lime addition, quality of lime and basicity, lance height, atomization and control of reproduction, Lance type, flux chemistry, hot metal chemistry, the content of FeO and MgO in slag, temperature, re-blows and total time, stops during production, slag splashing or coating and repair practice, charge volume, furnace volume, skull build-up, and lining maintenance.

Different maintenance practices with slag are carried out: slag splashing (which protects all the zones of the BOF), slag foaming (promotes the wall zone protection), and build-up (that protects the bottom zone). Additionally, various refractory products are applied. Compared with other maintenance practices, gunning repair has been established as one of the most effective practices for extending the service life of all kinds of metallurgical process\refining vessels, including BOF. It is attributed to the very accurate application of monolithic

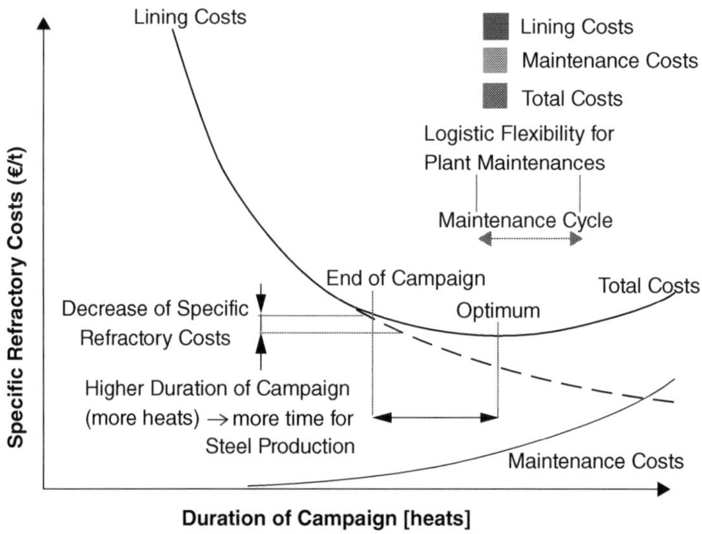

Figure 7.19 RHI reported an intelligent maintenance cycle to reduce the TCO [14].

mixes on specific pre-worn areas. The following factors are essential to monitoring to obtain higher erosion and infiltration resistance in gunning:

- careful formulation of the gunning materials
- minimum rebound loss
- less working stress on masonry personnel
- short preparation time
- gunning the repair material at high flow rates to cut back the repair time
- using two different classes of monolithic mixtures, whereas one for gunning and another for bank repair
- complete practice details of refractory maintenance in BOF, for example, its residual lining thickness data

7.6 Philosophy to Consider Raw Materials

Details MgO-C processing and its pros and cons are discussed in Section 8.6.1. A brief on the selection of MgO, carbon, and antioxidants for different zones of BOF is highlighted. The high purity of MgO with large crystals enhances chemical resistance, but the price is relatively high. Higher purity and quantity of graphite are very important aspects as they provide high chemical and thermal shock resistance. However, it may responsible for lower mechanical resistance and higher

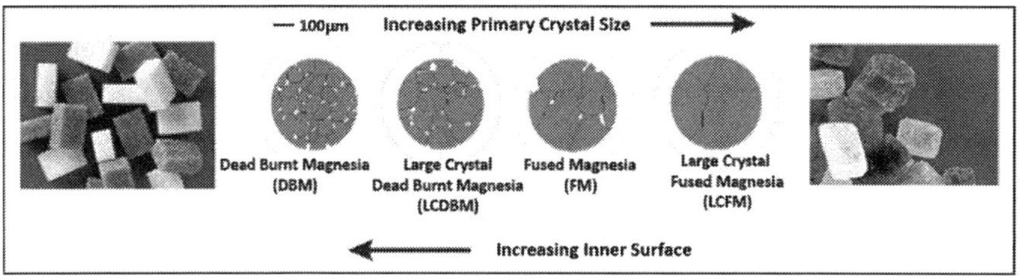

Figure 7.20 Different class of MgO, highlighted in Table 7.4.

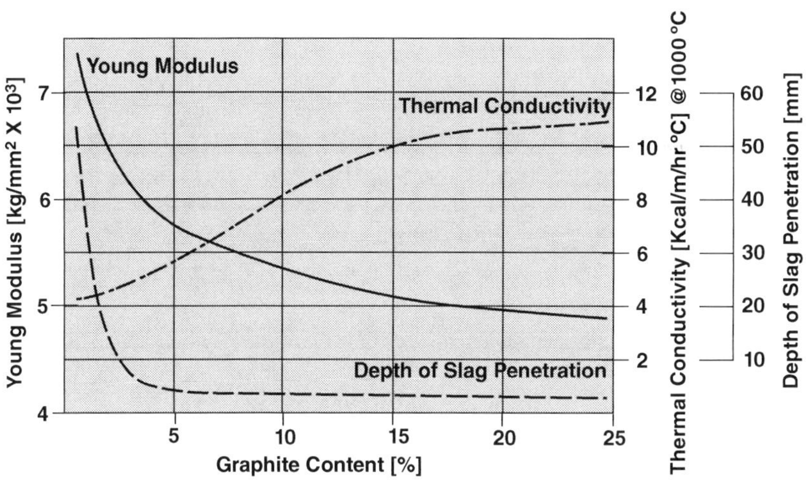

Figure 7.21 Influence of graphite on the brick properties.

price. The selection of antioxidants is required to protect the carbon and provide desired hot modulus of rupture.

MgO quality criteria preferably low amount of low melting silicate phases like Monticellite, CaO/SiO_2 ratio > 1.9 to avoid Monticellite (Ts = 1495°C), low amount on iron, silica, lime, and alumina, and large MgO crystals to reduce the corrosion speed. A brief view of the different class of MgO crystals is shown in Figure 7.20.

The addition of graphite implicates reduction of the slag infiltration, higher thermal conductivity that may provide a risk of induced stress on the steel shell, and higher thermal stress. Graphite content significantly influences mechanical response, slag infiltration performance, and thermal conductivity in Figure 7.21.

For BOF working lining, mainly affected by slag and metal corrosion, MgO-C refractory with about 18% carbon is used. Higher carbon reduces the strength, but corrosion is more important than strength for such applications. However, strength is important for the charging pad of BOF where the liquid metal and all other steelmaking batch materials fall from height. An MgO-C brick with 9%–10% of C is used in this bottom zone. The whole BOF hanging through metallic supports and the effect of stresses due to the hanging condition occurred in the trunnion area (T). The stresses increase due to the vasculation movement of the BOF during steelmaking process. For this reason, refractories in these areas require strength and flexibility, and the bricks used are MgO-C with 12%–15% of carbon content. In the taphole area, the refractories are under the tremendous movement of molten metal during tapping. Corrosion resistance and mechanical strength are important in these areas and strong refractories containing 10%–12% of carbon is used for this application. The wear mechanisms of each zone of the BOF are complex and are caused by a combination of different factors: chemical attack, erosion, abrasion, effects of joint conditions, and friction force caused by thermal stress (mainly in the zones of the barrel and cone of the converter). Some lining zones experience wear faster than others, and therefore, maintenance practice constitutes a relevant factor in prolonging the BOF life.

Antioxidants are added to slow down the detrimental burnout of carbon by reducing the partial O_2-pressure in the C-containing brick. It leads to sintering and densification of the brick structure. Consequently, the brick's strength increased CCS, MOR, and HMOR. Different types of antioxidants are shown in Figure 7.22. Antioxidants leave reaction products, leading to a sintering and densification of the brick structure. New links between magnesia or alumina grains

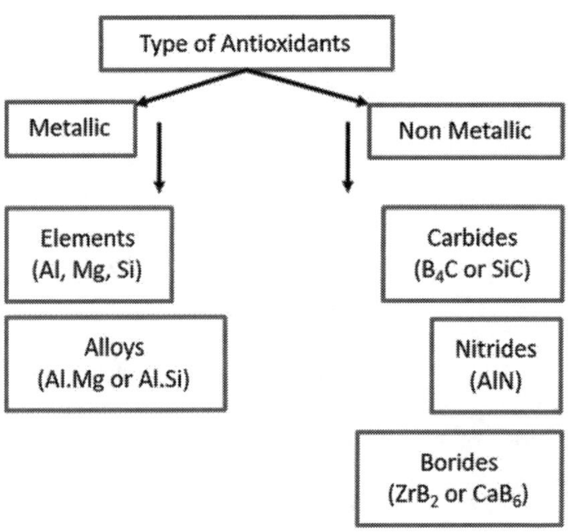

Figure 7.22 Type of additives for MgO-C bricks.

are formed, and the brick's strengths are increased. However, it is worth noting that it has certain negative impacts, such as the sensitivity of the reaction products to water, influence on the chemical wear resistance of the brick, and reduction of the structural flexibility and increased thermal expansion. Thus, an optimum amount and suitable antioxidant selection for well selected areas are important for achieving consistent performance.

7.7 Microstructure-dependent Properties of Refractories

BOF has several zones and experiences multiple wear factors despite slag corrosion. Among several instances:

a) increment of heavy scrap charging expedite abrasion wear of the *impact area*
b) *bottom tuyere* experience excessive thermal spalling although MgO-C have excellent thermal shock resistance
c) promotes the wear of *different areas* when the air tends to mix, although the entire system suppresses the decarburization (oxidation) by sufficient high pressure of CO gas

Concerning the performance of BOF brick, it is well-accepted that MgO-C brick intensely depends on the purity of raw material, graphite fraction, and arrangement of additives such as metallic powders. Adding to these compositional factors, the microstructure of MgO-C bricks, such as the distribution and/or bonding status of grains, also affects material properties. Typical microstructure controlling technologies are listed as follows:

7.7.1 Microstructure Deterioration Inhibition to Improve Slag Corrosion Resistance

MgO-C brick corrosion resistance strongly depends on raw materials impurity and are independent of microstructure. Usually, a dense MgO layer formation in the vicinity of the corroded surface is often recognized and the accompanied boundary layer is composed of impurities and concentrated graphite. This newly formed layer further controls the corrosion rate. Molten FeO facilitates graphite oxidation and loosened the structure and subsequent gap formation that plays an important role in reducing the strength and resultant wear behavior of BOF refractory. An interesting post-heated MgO particle grain observation may provide an insight to the roughening of the MgO grains that results from the Mg gas evaporation and subsequent reaction between MgO grains and graphite. Any impurity content may be adverse to the entire matrix. Alternatively, Mg evaporation promotes microstructure deterioration of MgO-C brick. The diffusion of

Mg(g) and CO(g) through the pore channels of the brick and successive reaction with MgO and C matrix is the rate-determining step of the corrosive phenomenon.

Therefore, densification of the microstructure suppresses the diffusion of the gases, resulting in suppression of the reaction. High densification can be achieved through adequate grain size distribution and arrangement and alteration of manufacturing protocols. It can suppress the microstructure deterioration by thermal fluctuation. A classic slag corrosion study on the influence of the density and their microstructure deterioration reveals that low porosity specimens exhibit superior slag corrosion resistance, whereas less dense structure develops a wide range of porosity after posttreatment (Figure 7.23a).

7.7.2 Slag Coating to Protect the Working Surface

Slag coating on the working surface inhibits the interaction between MgO and graphite and suppresses the microstructure deterioration. A conceptual relation

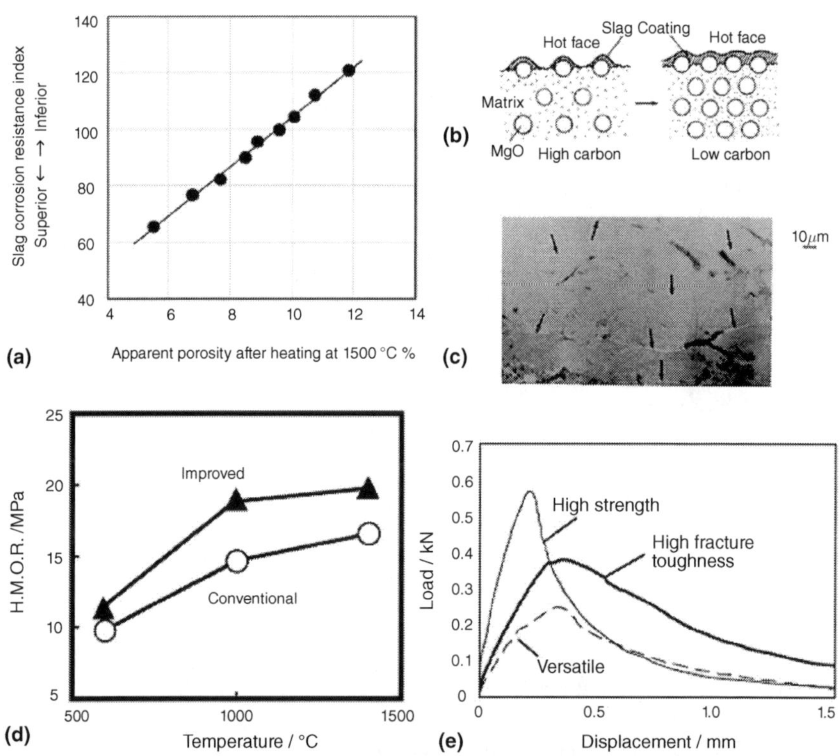

Figure 7.23 (a) Variation of slag corrosion index, (b) conceptual representation of slag coating, (c) pitch filling state in interstices among graphite layers, (d) influence of additive distribution on HMOR, and (e) load-displacement curves of several kinds of MgO-C bricks after firing at 1200°C, the dark line is related to high crack resistance MgO-C refractory [15], SHINAGAWA REFRACTORIES CO.,LTD, 2016.

in between the microstructure and slag coating layer formation is shown in Figure 7.23b. Slag preferentially adheres on the exposed hot surface of MgO grains and enhances the surface area and available sites to promote slag coating. It is worth mentioning that the fine MgO may enhance the surface area, but it can dissolve into the slag easily and not remain on the hot surface; thus an optimum size distribution with a larger grain size is expected for coating on MgO.

7.7.3 Microstructure Reinforcement by Evaporation-Condensation of Pitch

Metallic additives in resin-bonded MgO-C bricks are common for effective strength enhancement at elevated temperatures. However, metallic additives form a substantial difference in elastic modulus of high-temperature working surface and intermediate temperature of the middle area of the brick lining, reducing the thermal spalling resistance. In this circumstance, carbon bond enhancement by pitch addition may be considered as an alternative approach to reinforce the microstructure. Such an approach is beneficial to decreasing the elastic modulus at intermediate temperatures and improving thermal spalling resistance. In actuality, the carbon bond increment is facilitated through the evaporation and condensation effect of pitch. Figure 7.23c shows an example of a laboratory test for the evaporation-condensation effect of pitch. In this experiment, the specimen was prepared by mixing flake graphite and pitch, and heated at 800°C. The obtained microstructure observed in which pitch derived condensed carbon fills the fine interstices of graphite flake and is marked by an arrow.

7.7.4 Whisker Insertion to Reinforce Microstructure

The charging region is severely worn by mechanical abrasion caused by scrap charging. Metal addition enhances hot strength and reduces such abrasive wear, but additive lowers the spalling resistance. Thus, an optimum amount of additive is expected to synchronize hot strength and spalling resistance together. In the presence of various type of additives, MgO-C exhibits a different class of behavior, such as carbidation, nitridation, oxidation, and melting. It depends on temperature, atmosphere, and distribution of additives. Hence, it is important to achieve the microstructure that efficiently utilizes the behavior of metallic additives under the conditions of BOF operation. An optimized particle size distribution and metal additive dispersion improves the hot modulus of rupture for same kind of metal additive (see Figure 7.23d)

7.7.5 Fracture Toughness Enhancement and Crack Propagation Inhibition

High spalling resistance refractory has the capability to restrict the crack initiation and propagation. Despite several reasonings of the flaw formation and their existence, the crack propagation resistance of refractory system is essential for better campaign. Competitive load-displacement curves of typical MgO-C bricks obtained 3-pt bending of 1200°C heat treated specimens is shown in Figure 7.23e. MgO-C is composed of graphite flakes and significantly influences the crack propagation resistance in operation, despite non-wetting behavior. Cumulative attainment of both high strength and high fracture toughness is always a challenging task. The load-displacement plot exhibits the stiff slope related to static elasticity, and load relaxation beyond the maximum load point expresses the crack initiation and propagation. The oxide refractories usually have less fracture toughness compared to graphite refractories. MgO-C exhibits a slow decrease rate of load, whereas the oxide refractories follow abrupt load reduction. While considering the integral of the curve in the latter, i.e., crack propagation stage defines as *work of fracture* usually consumed for the crack propagation. Thus, the larger area indicates material is capable of withstanding high fracture resistance. Usually, the presence of metallic additives increases strength but does not increase the fracture toughness of MgO-C brick. Hence, another microstructure controlling technology is necessary for improving the crack stability coefficient (Rst), which expresses crack propagation resistance.

References

1 Jalkanen, H. and Holappa, L. (2014). Converter steelmaking. In: *Treatise on Process Metallurgy* (ed. Seshadri Seetharaman), 3. Elsevier.

2 Chiba K., Ono A., Saeki M., Ohno T., Yamauchi M., and Kanamoto M. (1993). Development of On-line Manganese Analysis Method for Molten Steel in Converter, *Nippon Steel Tech.* Report, 57: 44.

3 Baptizmanskii, B.I. and Oxotskii, B.B. (1981). *Fiziko-khimicheskie osnovy kislorodno-konverternogo processa*. Kiev: Vyscha Schkola.

4 E. Ristimä ki, Diplomawork, Helsinki (1974). University of Technology, Finland, (in Finnish).

5 Matschullat, T., Rieger, D., Krüger, K., and Döbbeler, A. (2008 May). In: 9th European Steelmaking Conference, Krakow, Poland.

6 Jia, R. and Liu, J. (2016). Simulated experiment study of factors influencing the hydration activity of f-CaO in basic oxygen furnace slag. *Adv. Mater. Sci. Eng.* 2016: 15. Article ID 7529382. doi: 10.1155/2016/7529382.

7 Chang, W.S., Knight, C.E., Hasselman, D.P.H., and Mitchiner, R.G. (1983). analysis of thermal stress failure of segmented thick-walled refractory structures. *J. Am. Ceram. Soc.* 66 (10): 708–713.

8 Kingery, W.D. (1955). Factors affecting thermal stress resistance of ceramic materials. *J. Am. Ceram. Soc.* 38 (1): 3–15.

9 Tsibin, I.P. and Kuznetsov, A.T. (1988). A physical model of thermal failure of refractory concretes. *All-Union Refract. Inst.* 29 (1–2): 43–48.

10 Ganguly, B.K., McKinney, K.R., and Hasselman, D.P.H. (1975). Thermal-stress analysis of flat plate with temperature-dependent thermal conductivity. *J. Am. Ceram. Soc.* 58 (9–10): 455–456.

11 Joliff, Y., Absi, J., Glandus, J.C., Huger, M., and Tessier-Doyen, N. (2007). Experimental and numerical study of the thermomechanical behaviour of refractory model materials. *J. Eur. Ceram. Soc.* 27 (2–3): 1513–1520.

12 Dai, Y., Li*, J., Yan, W., and Shi, C. (2020). Corrosion mechanism and protection of BOF refractory for high silicon hot metal steelmaking process. *J. Mater. Res. Technol.* 9 (3): 4292–4308.

13 Biswas, S. and Sarkar, D. (2020). *Introduction to Refractories for Iron and Steelmaking*. Springer Nature.

14 The Journal of Refractory Innovations Steel & Industrial Edition Autumn 2017, ISSUE 1, RHI Magnesita.

15 Iida, A., Kakihara, M., Suto, M., and Tada, H. (2016). Improvement of MgO-C bricks for BOF based on microstructure investigation. Shinagawa Technical Report, 59.

8

Operation and Refractories for Secondary Steelmaking

8.1 Introduction

The primary steel process, including BOF and EAF, produces a certain limit of steel composition; thus further modification demands composition tuning and minimization of impurities. Several processes are adopted to control the resultant composition, primarily divided into two processes, *without vacuum* and *with vacuum*.

The vacuum process expedites producing ultralow carbon steel and excellent composition tuning of steel. In a broad sense, Ladle Heating Furnace (LHF) and Argon Oxygen Decarburization (AOD) are most common without a vacuum secondary steelmaking process. A ladle heating furnace (LHF) reheats the liquid steel from BOF. It utilizes electric energy through graphite electrodes to enhance and maintain the temperature and ensure steel homogenization temperature and chemistry through inert gas purging. Steel is refined and final composition is achieved. Argon Oxygen Decarburization (AOD) process is adopted for the low carbon content and chromium enriched stainless steel grade and other high grade alloys. In this process, oxygen is blown from the top lance, and a mixture of oxygen with argon is blown from side tuyeres for selective oxidation of carbon over alloy elements. AOD is usually used in conjunction with Induction Furnace (IF)/Electric Arc Furnace (EAF) route.

Most common and adopted vacuum processes are vacuum degassing (VD), vacuum arc degassing (VAD), vacuum oxygen degassing (VOD), and recirculation degassing (RH). Removing dissolved gas such as H_2 and N_2 present in molten steel is essential to improve the steel properties. Despite gas removal, desulfurization is also done to achieve a specific steel quality. Different steelmaking processes experience different degrees of the environment when targeting either reduction of dissolved carbon, phosphorous, sulfur, hydrogen, and nitrogen or the addition of different elements to synchronize the properties. It is prominent that steel sectors are seeking competent refractories to attain hassle free operation and maintenance of refractories; thus versatile refractories are required for secondary steelmaking vessels. Several processes and their corresponding refractories are highlighted in Table 8.1 and discussed in this chapter.

Fundamental Design of Steelmaking Refractories, First Edition. Debasish Sarkar.

Table 8.1 An overview of steelmaking processes to produce different grade steel and their relevant refractories. Secondary steelmaking working lining refractories are Al_2O_3-MgO-C, MgO-C, Dolo-Carbon, Direct bonded Dolomite, Alumina-Spinel, and Magnesia-Chrome.

Primary steelmaking	Secondary steelmaking	Type of vessel	Steel product	Operational parameters	Refractory lining [1–3] Working/hot face	Refractory lining [1–3] Back-up/Safety/Outer
BOF/EAF	Non-vacuum processes	LF	• Carbon steels [4, 5] ○ Si-killed steels ○ Si-Al-killed steels ○ Al-killed steels ○ Rimmed steels	**Deoxidation** [6] • Deoxidizer: Aluminum (Al) wire • Argon gas flow rate:80 N L min^{-1} • Liquid bath height: 2.34 m **Desulfurization** [7] • Flux material: CaO • Height of the ladle: 3.82 m • Molten steel height: 3.26 m • Slag thickness: 0.10 m • Purging rate(Ar): 25–1200 N L min^{-1}	• Slag zone: Magnesia-Carbon • Metal zone#: Alumina-Magnesia-Carbon/ Dolo-Carbon/ Alumina-Spinel #selection of metal zone refractory, based on type of killing. • Bottom: Alumina-Magnesia-Carbon	• Castable based lining with high Alumina-Spinel (or) • Brick based lining with Magnesia-Spinel/ Magnesia/ Alumina-Spinel
		AOD	• Stainless steels [1, 8] ○ Austenitic (16–20% Cr; > 8% Ni) ○ Ferritic (11.5–17% Cr;) ○ Duplex (>22% Cr; 4–6% Ni) ○ Martensitic ○ (11.5–17% Cr; < 2.5% Ni)	**Decarburization/Cr-level** [9] • Steel bath weight—19.4 tons • [C] content (initial)–1.12wt.% • [Cr] content (initial)–16.74wt.% • 1st blow (O_2:N_2)–800:400 N m^3/h • 2nd blow (O_2:Ar)–600:600 N m^3/h • **End of 1st blow** ○ [C] content—0.2038wt.% • **End of 2nd blow** ○ [C] content—0.0405wt.% ○ [Cr] content—16.27wt.%	• Upper cone: Direct bonded Doloma • Slag line: MgO enriched, Direct bonded • Doloma • Bottom: Direct bonded Doloma • Taphole: Magnesia-Chrome (Rebonded)	• Brick based lining with Magnesia

Vacuum processes	VD	• Carbon steels [10, 11] • Alloy steels [12]	**<u>Desulfurization/Dehydrogenation</u>** [13] • [S] content (initial)–0.023wt.% • [H] content (initial)–3.4 ppm. Ar-gas flow rate:130–200 l/min • Process temperature: 1873 K • Vacuum pressure: 2×10^{-3} bar • Vacuum treatment: about 19 min. • **<u>After vacuum treatment</u>** ○ [S] content—0.007wt.% ○ [H] content—0.85 ppm.	• <u>Slag zone:</u> Magnesia-Carbon • <u>Metal zone:</u> Alumina-Magnesia-Carbon/ Dolo-Carbon/ Alumina-Spinel • <u>Bottom:</u> Alumina-Magnesia-Carbon	• Castable based lining with high Alumina-Spinel (or) • Brick based lining with Magnesia-Spinel/Magnesia/ Alumina-Spinel
	VAD	• Carbon steels [14, 15] • Alloy steels [14] • Stainless steels [1]	**<u>Dehydrogenation</u>** [16] • [H] content (initial)–4.7 ppm. Ar-gas flow rate: 0.003 N $m^3 s^{-1}$ • Vacuum pressure: 1.3×10^{-3} bar • Vacuum treatment: about 20 min. • **<u>After vacuum treatment</u>** ○ [H] content—1.57 ppm.	• <u>Slag zone:</u> Magnesia-Carbon • <u>Metal zone:</u> Alumina-Magnesia-Carbon/ Dolo-Carbon/ Alumina-Spinel • <u>Bottom:</u> Alumina-Magnesia-Carbon	• Castable based lining with high Alumina-Spinel (or) • Brick based lining with Magnesia-Spinel/ Magnesia/ Alumina-Spinel

(Continued)

Table 8.1 (Continued)

Primary steelmaking	Secondary steelmaking	Type of vessel	Steel product	Operational parameters	Refractory lining [1–3] Working/hot face	Back-up/Safety/Outer
		VOD	• Ultra-low Carbon steels [1] • Stainless steels [1]	**Decarburization/[N] Desorption** [17] • [C] content (initial)–4000 ppm. [N] content (initial)–120 ppm. • [Cr] content—16wt.% • Ar-gas flow rate: 300–1500 N l/min • O_2-gas flow rate: 500–2000 N m^3/h • Process temperature: 1923–1973 K • Vacuum pressure: 0.1–20 kPa • Vacuum treatment: about 40 min. • **After vacuum treatment** ○ [C] content ~ 100 ppm. ○ [N] content ~ 100 ppm.	• Slag zone: Direct bonded Doloma/ Magnesia—Chrome • Metal zone: Direct bonded Doloma/ Magnesia-Chrome • Bottom: Direct bonded Doloma/ Magnesia-Chrome	• Castable based lining with high Alumina-Spinel (or) • Brick based lining with Magnesia-Spinel/ Magnesia/ Alumina-Spinel
		RH-OB	• IF steels [1] • Ultralow carbon steels [1] • Alloy steels [1] • Heavy plate steels [1] • HIC resistant steels [1]	**Desulfurization** [18] • Flux material: $CaO\text{-}CaF_2$ • Materials size: —100 mesh • Blowing rate: 100–130 kg min^{-1} • Blowing time: 8–12 min • Lance height: 2 m • Ambient pressure: 1–2 Torr **Decarburization** [18] • Flux material: Fe_2O_3 • Materials size: —100 mesh • Blowing rate: 20–60 kg min^{-1} • Blowing time: 8–13 min • Lance height: 3 m • Ambient pressure: 1–2 Torr	• Upper vessel: Magnesia-Chrome • Lower vessel: Magnesia-Chrome • Bottom: Magnesia-Chrome • Throat: Magnesia-Chrome • Snorkel: Magnesia-Chrome	• Corundum/High-Alumina castable reinforced with stainless steel grid

8.2 Steel Diversity, Nomenclature, and Use

Socioeconomic demands different steel quality for diverse horizons, and the American Iron & Steel Institute (AISI) categorized four fundamental catagories; carbon steel, alloy steel, stainless steel, and tool steel, in their chemical compositions. Chemical composition encompasses different physical, chemical, and mechanical properties. The prime constituents of steel are iron and carbon, and properties are controlled by the amount of carbon and the additional alloys. Society of Automotive Engineers (SAE) and American Iron and Steel Institute (AISI) developed four-digit code numbering systems to identify the carbon and alloy system. Some alloys are also designated by five codes. Eventually, a particular number can identify the steel grade and its utility mode. The first digit *1* indicates carbon steel, and different steel grades are designated by different digits, as shown in Table 8.2. AISI and SAE systems henceforth *1xxx*, designate as carbon steels. Furthermore, this is subdivided into four categories in the perspective of underlying properties:

- 10xx series encompasses carbon steel with a maximum 1 wt.% Mn
- 11xx series indicates re-sulfurized carbon steel (to improve machinability)
- 12xx series indicates re-sulfurized and re-phosphorized carbon steel (easy machining)
- 15xx series is associate with non-re-sulfurized and high Mn (up to 1.65%) carbon steel

The second digit indicates the percentile (1 equals 1%) concentration of major elements. The last two digits confirm the carbon concentration. SAE 5130 is a Cr alloy steel consisting of about 1% Cr and approximately 0.30% C. The steel quality and numbering system is represented in Table 8.3.

Table 8.2 Steel nomenclature.

SAE designation	Type
1XXX	Carbon steels
2XXX	Nickel steels
3XXX	Nickel-chromium steels
4XXX	Molybdenum steels
5XXX	Chromium steels
6XXX	Chromium-vanadium steels
7XXX	Tungsten steels
8XXX	Nickel-chromium-vanadium steels
9XXX	Silicon-manganese steels

Table 8.3 Details of classification and numbering of steel.

Type	Designation	
Carbon steels	10XX	Plain carbon, Mn 1.00% max
	11XX	Re-sulfurized free machining
	12XX	Re-sulfurized/re-phosphorized free machining
	15XX	Plain carbon, Mn 1.00–1.65%
Manganese steels	13XX	Mn 1.75%
Nickel steels	23XX	Ni 3.50%
	25XX	Ni 5.00%
Nickel-chromium steels	31XX	Ni 1.25%, Cr 0.65–0.80%
	32XX	Ni 1.75%, Cr 1.07%
	33XX	Ni 3.50%, Cr 1.50–1.57%
	34XX	Ni 3.00%, Cr 0.77%
Molybdenum steels	40XX	Mo 0.20–0.25%
	44XX	Mo 0.40–0.52%
Chromium-molybdenum steels	41XX	Cr 0.50–0.95%, Mo 0.12–0.30%
Nickel-chromium-molybdenum steels	43XX	Ni 1.82%, Cr 0.50–0.80%, Mo 0.25%
	47XX	Ni 1.05%, Cr 0.45%, Mo 0.20–0.35%
Nickel-molybdenum steels	46XX	Ni 0.85–1.82%, Mo 0.20–0.25%
	48XX	Ni 3.50%, Mo 0.25%
Chromium steels	50XX	Cr 0.27–0.65%
	51XX	Cr 0.80–1.05%
	50XXX	Cr 0.50%, C 1.00% min
	51XXX	Cr 1.02%, C 1.00% min
	52XXX	Cr 1.45%, C 1.00% min
Chromium-vanadium steels	61XX	Cr 0.60–0.95%, V 0.10–0.015%
Tungsten-chromium steels	72XX	W 1.75%, Cr 0.75%
Nickel-chromium-molybdenum steels	81XX	Ni 0.30%, Cr 0.40%, Mo 0.12%
	86XX	Ni 0.55%, Cr 0.50%, Mo 0.20%
	87XX	Ni 0.55%, Cr 0.50%, Mo 0.25%
	88XX	Ni 0.55%, Cr 0.50%, Mo 0.35%
Silicon-manganese steels	92XX	Si 1.40–2.00%, Mn 0.65–0.85%, Cr 0–0.65%
Nickel-chromium-molybdenum steels	93XX	Ni 3.25%, Cr 1.20%, Mo 0.12%
	94XX	Ni 0.45%, Cr 0.40%, Mo 0.12%
	97XX	Ni 0.55%, Cr 0.20%, Mo 0.20%
	98XX	Ni 1.00%, Cr 0.80%, Mo 0.25%

Carbon steel is further categorized as low C steel or mild steel, medium C steel, and high C steel. Low C steel typically contains 0.04–0.30% carbon, the largest carbon steel class for different shapes of casted products from flat sheet to structural beam. A wide range of elemental addition or reduction is necessary to meet the desired properties for a particular application. For example, drawing quality steel demands low carbon content and aluminum addition; structural steel has relatively higher carbon and manganese content. Medium carbon steel typically has a C range of 0.31–0.60% with Mn range of 0.06–1.65%. It is more robust, less ductile, and challenges form, weld, and cut. Heat treatment expedites the hardening and tempering of this grade of steel. High carbon steel, usually known as *carbon tool steel*, consists of carbon in 0.61–1.50%. It is difficult to bend with carbon increment, and heat-treated high carbon steel exhibits extremely hard and brittle.

Alloy steel is made by adding one or more alloying elements (other than C) lie, Si, Al, Ni, Mn, Cr, Ti, Cu, and Cr. The purpose and properties are different from regular carbon steel and have economic viability. Stainless steel is more prone to corrosion resistance than carbon or alloy steel. Chromium, nickel, and molybdenum are common alloying ingredients, and cumulative alloy content often varies 15–30%. It is easy to machine, and several common applications with a wide variety of shapes are available for hardware appliances, structural and architectural, food handling and processing, medical instruments, etc. In more detail, stainless steel can be classified as austenitic, ferritic, martensitic, precipitation-hardening, and duplex stainless steel.

Austenitic stainless steel comprising chromium and nickel is identified as a 300 series type, whereas chromium, nickel, and manganese are identified as 200 series. Usually, hardening facilitates by cold working rather than heat treatment. Annealed austenitic phase is essentially nonmagnetic; however, some may have slight magnetic behavior after cold working. It exhibits excellent corrosion resistance, excellent formability, and increased strength because of cold working. The most widely used austenitic steel is type 304 (18–8 stainless) with a nominal composition of 18% chromium and 8% nickel. Ferritic stainless steels are straight-chromium 400 series type and cannot be hardened by heat treatment; however, moderate hardening can only be done by cold working. Usually, they are magnetic, have good ductility, and are resistant to corrosion and oxidation. Martensitic stainless steel can be hardened by heat treatment. It is magnetic and resistant to corrosion in mild environments. It has reasonable ductility, and adequate heat treatment may expedite the tensile strength exceeding 1380 MPa.

Precipitation-hardening stainless steels are commonly chromium-nickel types with coexisting other alloying elements like aluminum or copper. It can be hardened by solution treating and aging to high strength. An equal part of coexisting austenite and ferrite phase persists after annealing, known as duplex stainless steel. It has excellent corrosive resistance to chloride stress corrosion cracking, pitting and crevice corrosion resistance, and high yield strength compared to conventional grades. Tool steel should have high hardness and abrasion resistance quality. It is available in different shapes, including flat bar, round and square bar, and is extremely hard and can be used for specific tool applications, including dies (stamping or extrusion), mold making, cutting tools, and impact devices like hammers, knives, etc.

8.3 Vessels for Different Grades of Steel

A generic question enlightens the specific process that can produce the desired grade of steel; thus variety refractory is essential to control the slag corrosion resistance and steel composition. Table 8.1 highlights the required vessel type for different types of steel. In brief, Ladle Furnace (LF)—the quality link between melting and casting; Argon oxygen decarburization (AOD)—excellent solution for stainless steel; VD/VOD plants—Vacuum Degassing for maximum metallurgical flexibility; RH system—the vacuum recirculation process. A quick look at three fundamental processes is given in Figure 8.1.

Figure 8.1 (a) Ladle furnace, (b) VD/VOD furnace, (c) RH system.

Steel manufacturers experience continuous competition and pressure from global market demand to produce narrower composition ranges, precise inclusion size, and their distribution for reproducible service properties. It can only be achieved by adequate process control and vessel lining [19]. Despite processing parameters synchronization, the steel cleanliness and purity also depends on the selection of refractory products. Certain metallic residuals or nonmetallic impurities have a marked influence on the physical and mechanical properties of steels. Figure 8.2 summarizes the role that nonmetallic elements could have on various properties of the metal. Consequently, the steelmaker must conceive more and more complex elaboration modes to eliminate these elements and limit pollution risks.

Continuous process improvement and optimization make it possible to produce a precise ppm level composition control of elements of C, H, N, O, P, and S of sensitive steel grades. For example, vacuum treatment enables liquid steel production with less than 20 ppm carbon content. Similarly, the possibility of producing a few ppm content sulfur level for hydrogen-induced cracking (HIC) steel or oxygen content for bearing grade steel enhances the working life.

Table 8.4 represents a typical limit of several elements for current steelmaking processes. Refractory selection is an important issue to control over such residual elemental content, and the impact of refractory products on the steel may be assessed at three levels:

- the possibility to keep the chemical composition of the liquid steel within the specified range for a given process;

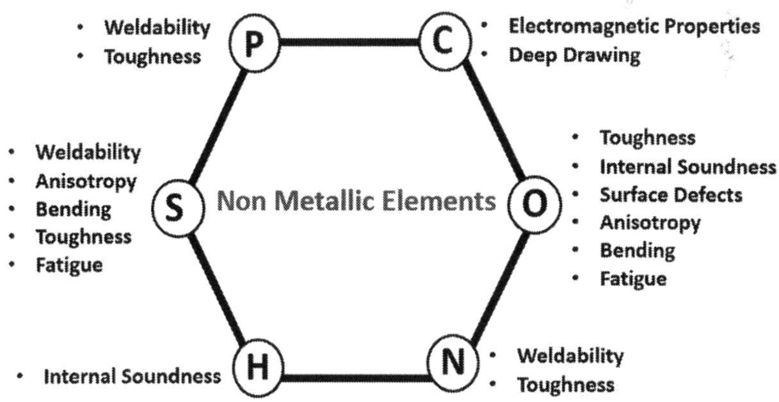

Figure 8.2 Influence of nonmetallic elements on steel properties.

Table 8.4 Lower limits of different residual elements in steel.

Elements	P	C	S	N	H	O
ppm	10	5	5	10	<1	5

- the achievement of the required metal cleanliness, i.e., the amount and nature of nonmetallic inclusions;
- the prevention of defects concerning the steel surface.

Thus, selecting every operational feature and its refractory greatly influences the control over the steel composition. However, control over in-situ slag composition monitors the refractory life and steel composition.

8.4 Operational Features of Vessels

Modern steelmaking of today is therefore characterized by a primary process designed primarily to melt and rough refine hot metal, scrap, or sponge iron, by a product-oriented secondary refining stage to finally adjust the melt and by the continuous casting of the steel. Now, primary steelmaking is used for rapid scrap melting and gross refining only and leaving further refining and control of gases by secondary steelmaking. Secondary steelmaking processes are used to achieve one or more of the following requirements:

- Improvement in quality,
- Improvement in production rate,
- Decrease in energy consumption,
- Improve economy by better yield and higher availability of the various processes,
- Use of restively cheaper raw materials,
- Use of alternate sources of energy, and
- Improvement of recovery of alloying elements.

The objectives of secondary steel refining are as follows:

- Homogenization of the steel bath temperature and chemical composition,
- Improved alloy yield and composition control within narrow limits,
- Decarburization,
- Dephosphorization, deoxidation, desulfurization, and improved steel cleanness, i.e., inclusion control,
- Degassing, i.e., removal of nitrogen and hydrogen gases,
- Adjustment of teeming temperature, and
- Providing sufficient holding time during treatment and teeming.

The metal-to-steel refining process in BOF is done through oxygen blowing, and oxygen dissolves in steel. There are several sources of oxygen to dissolve in steel, including oxygen blowing, steelmaking slag, rust on steel, atmospheric O_2 dissolved in steel during teeming, and oxidizing refractories. The oxygen solubility in liquid steel at 1600 °C is 0.23%, which further decreases to 0.003% in solid during solidification. This excess oxygen produces defects like blow holes and nonmetallic oxide inclusions in solidified casting and reduces the mechanical response. Thus, removing this excess (0.227%) oxygen from steel is mandatory, known as deoxidation. Carbon steels may be subdivided into three groups according to the degree of deoxidation:

- Killed steel: Complete O_2 removal that does not affect gas porosity (blowholes) in solidification.
- Semi-killed steel: Incomplete deoxidized steel consists of a certain amount of O_2 that forms CO during solidification.
- Rimmed steel: Partial deoxidized or non-deoxidized low C steel evolve sufficient CO in solidification and exhibit good surface finish.

The deoxidation process can be done by a single element such as Si, Al, Mn, etc., or by a mixture of elements such as Si/Mn, Ca/Si/Al, and many more; and these are known as simple deoxidation and complex deoxidation, respectively. The deoxidation process is also known as precipitation-deoxidation, as there is form oxide during the process. The deoxidation process under vacuum is called vacuum deoxidation. Bath stirring is an important aspect and direct deoxidation in ladle is known as ladle deoxidation.

Secondary steel refining is carried out even in a transfer ladle fitted with porous plug for inert gas (i.e., argon) purging. Argon supply line is connected to this plug, just before tapping. Similarly, injection treatment can be carried out in the transfer ladle itself. However, the bath tends to lose temperature in all these treatments, and hence, these cannot be carried out for too long a duration, even if required. There are three types of processes: (1) ladle furnace, (2) injection ladle metallurgy, (3) degassing processes.

8.4.1 Ladle Furnace (LF)

Ladle secondary steel refining is an important process for ensuring steel quality. A ladle furnace is a vessel where primary steel transforms to controlled and desired precise composition through secondary refining operations, and its primary functions are:

- Liquid steel reheating through electric power conducted graphite electrodes
- To control chemical composition and temperature on a very narrow level
- Gas stirring to make homogenized steel temperature and chemistry throughout the vessel
- Control over slag formation that protects refractory from arc damage; concentrates and transfers heat to the liquid steel; traps inclusions; and arranges for desulfurization.

The major functions during secondary steel composition are:

- Alloy additions to provide bulk chemical control,
- Cored wire addition for changing the morphology of inclusions,

- Separation of inclusions for high cleanliness,
- Removal of gases like hydrogen and nitrogen,
- Provide a means for dephosphorization,
- Act as a buffer for downstream steelmaking,
- To make very low level of sulfur (below 0.005%) by efficient desulfurization,
- Addition of micro-alloying elements in proper sequence to have maximum recovery.

A typical schematic of ladle-like furnace (as shown in Figure 8.3) is provided with (i) bottom porous plug for argon purging and (ii) electrodes inserted lid facilitates arcing for bath heating. The other lid may be required to connect the vacuum line if required. The porous plug essentially assists in purging gas for the stirring of the molten metal to promote bath temperature and composition homogenization.

Argon gas performs normal stirring operations through a purge plug arrangement in the bottom of the ladle to help flotation of nonmetallic inclusions and faster rate of desulfurization. Sulfur level in the bath comes down to 50 ppm after ladle furnace treatment. Argon gas is passed through a porous plug at the rate of 0.18 Nm^3/min from beginning to end. The total time required in a ladle furnace is 40–50 minutes, and power consumption is about 50 kWh/t. It can carry out

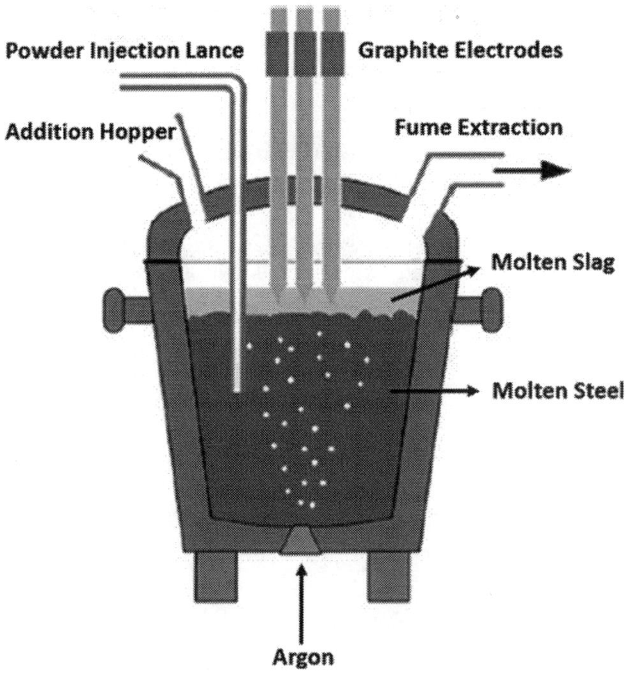

Figure 8.3 Schematic view of ladle furnace and important operational arrangement.

stirring, vacuum treatment, synthetic slag refining, injection, purging gas, etc. Such a process is capable of heating independently during the process without temperature loss. Fumes and dust generated during heating and alloying operations at the LF will exit the water-cooled ladle roof. Eventually, a ladle furnace is equipped with all these arrangements for successful operation [20].

8.4.1.1 Eye-Opening

Inert gas injection is an effective method for homogenizing chemistry, enhancing chemical reaction rates, eliminating thermal stratification, and removing nonmetallic inclusions. Argon gas injection from bottom plugs creates the recirculatory flow that increases the mixing of steel and leads to homogenization of chemical constituents and temperature. The argon flow also helps in the agglomeration of inclusions and enhances the formation of bigger inclusion particles. These particles then float toward the slag layer at the top of the ladle and stick with it, thereby reducing the inclusion level. Further, rising gas bubbles due to argon flow may cause other phenomena such as breaking the slag layer and forming eye-opening at the top. Due to turbulence, the slag layer distorts and enhances the intermixing of slag and steel, thereby increasing the interfacial reactions. The most important reactions are desulfurization at the interface and oxidation of the melt at the eye-opening.

8.4.1.2 Injection of Active Agents

The objective of this process is to achieve better quality steels with controlled chemistry and superior and homogeneous mechanical properties, are as follows:

- To bring down the level of basic impurities such as carbon, sulfur, phosphorus, nitrogen, hydrogen, oxygen, and tramp elements (like copper, tin, antimony, bismuth, etc.); and
- To reduce the nonmetallic inclusions or to modify them so that their harmful effects are counterbalanced.

In this context, injection metallurgy, i.e., treatment of molten steel with powdered metallic materials and fluxes coupled with gas stirring. Gas stirring is gaining acceptance in the steel industry since it involves less investment, is easy to operate, and facilitates quicker processing at reasonable reaction efficiency for a wide span of heat sizes. Adoption of injection metallurgy; thus leads to significant time saving as it relieves the main steelmaking units (LD/BOF or EAF) partly or completely from performing the time-consuming refining and finishing of steel heats.

There is a special interest in the control of nonmetallic inclusions because of their harmful effects on the subsequent stages and their influence on the properties of the final product. By controlling the amount, size, and chemical composition of the inclusions, it is possible to obtain a final product of good quality without inclusions.

The need to produce low sulfur steels with control of sulphide inclusions and control the tundish nozzle clogging during the concast, the ladle injection process was developed. Treatment of liquid steel with calcium has become an important means of deoxidation and desulfurization, to very low levels and control of the shape, size, and distribution of oxide and sulphide inclusions. Benefits directly attributable to calcium treatment include greater fluidity, smoother continuous casting by reducing nozzle blockage, and improved cleanliness, machinability, ductility, and impact strength in the final product.

Calcium treatment of liquid steel is started to prevent nozzle clogging in aluminum-killed plain carbon steels, and calcium helps to liquefy the alumina. In this process, nozzle clogging is also reduced; particularly in billet and thin slab casting, where the liquid steel must pass through narrow refractory spaces. Calcium treatment is also done extensively for inclusion modification to improve the properties of the product. Calcium is usually added in the alloyed form like Ca–Si, Ca–Fe, Ca–Si–Ba, and Ca–Si–Ba–Al to restrict the vapour pressure of calcium in the molten bath. Most commonly used injection powder is Ca–Si (30% Ca), alloying with Si helps to reduce vapour pressure of Ca at steelmaking temperatures.

It is necessary to inject Ca–Si deep into the bath to ensure optimum recovery of Ca. Other reagents are: CaO, CaO–CaF_2–Al, CeO–SiO2, Mg–CaO, CaC_2, and combinations of them. The main purpose of particle injection is to feed a fine-grained material deep into the molten steel. Major factors influencing the injection method:

i) Nature of the powder: The nature of the powders used is governed primarily by the purpose of the injection.
ii) Powder size distribution: Optimum powder size must be determined. Fine size powder provides a large reaction surface, and coarse powders create a small reaction surface.
iii) Powder/gas ratio: Solid particles are mostly injected by a stream of argon gas. The gas bubbles rise through the liquid. The particles trapped in the bubble hit the gas–liquid interface where they will either pass through the interface or remain trapped inside the bubble. The rising solids might vaporize (e.g., Mg), melt and dissolve (Ca, CaC_2, Ca–Si) and then vaporize.
iv) Injection depth: Reaction zones are:
    a) Jet zone in front of nozzle where three-phase dispersion liquid metal-solid-gas is formed;
    b) Bubble plume where bubbles, solid, and/or partially molten or reacted particles and liquid steel rise and react;
    c) Break through zone at the melt surface where there is reaction with air. Reaction products are absorbed into the top slag.

There are two popular methods of injection: (1) cored wire injection, and (2) submerge injection through lance. A typical schematic representation of desulfurizing by both cored wire injection desulfurization unit and powder injection desulfurization unit is represented in Figure 8.4. Despite wire injection, injection of desulfurizing element, or compound (CaC_2, CaSi, CaF_2 + CaO, Ca, Mg) powders combined supply of argon blowing in molten steel is the most effective method of sulfur removal.

In the secondary steel refining process, the processing of killed steel is by deoxidizing the metal element (Si, Al) addition. It ensures no existence of free oxygen that creates porosity during welding and reduces joints' strength. The deoxidized (killed) molten steel ladle is placed on the injection stand, and the lance may lower or immerse into the melt to inject desulfurization agent in the argon stream that eventually expedites stirring and promotes desulfurization. The covered desulfurizing slag has high sulfur solubility and can accommodate sulfides during the desulfurizing process. Stirring makes bath thermal and chemical homogenization. The core wire method demands stirring argon by purging the plug through the bottom of the ladle furnace. The developed fumes extract through the cover during operation. Ultralow sulfur concentration (0.0002%) can be achieved through injection desulfurizing agents.

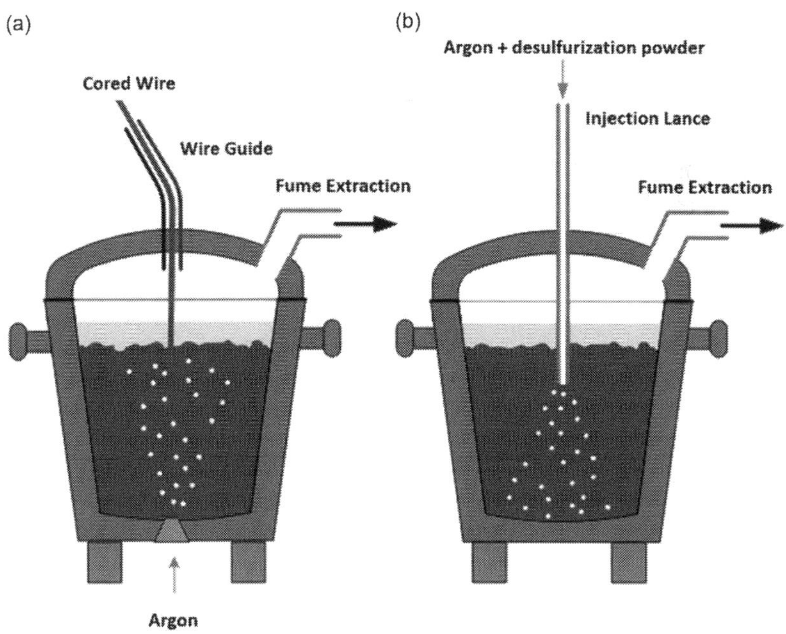

Figure 8.4 Ladle desulfurizing by (a) cored wire injection desulfurization unit with Ar stirring through porous plug, and (b) powder injection desulfurization.

8.4.2 Argon Oxygen Decarburization (AOD)

The Argon oxygen decarburization (AOD) produces stainless steel and other high-grade alloys by oxidizable elements like chromium and aluminum. It comprises high metallic yield, cleaner steel with high productivity, accurate chemistry control down to 0.01% carbon and lower, rapid desulfurization to < 0.001%, and lead removal to < 0.001%. A schematic representation of AOD vessel consisting of top-blowing arrangement is shown in Figure 8.5. The transferred molten metal in AOD is usually processed in three refining steps: a) decarburization, b) reduction, and c) desulfurization.

The decarburization and refining may facilitate reaching the carbon level of less than 0.05%. Herein, the oxygen is mixed with Ar or N_2 inert gases and injected through submerged tuyeres. The mixed Ar minimizes the unwanted oxidation of other precious elements like Cr. AOD vessel is attached with several sidewall nozzles where Ar gas flows through an intermediate slit between two pipes and O_2 gas is blown through an inner pipe. However, Ar gas is also injected through an inner pipe stream toward the end of the blow to dilute oxygen gas. It is important to mention that the partial pressure of reaction product gas CO preferentially is

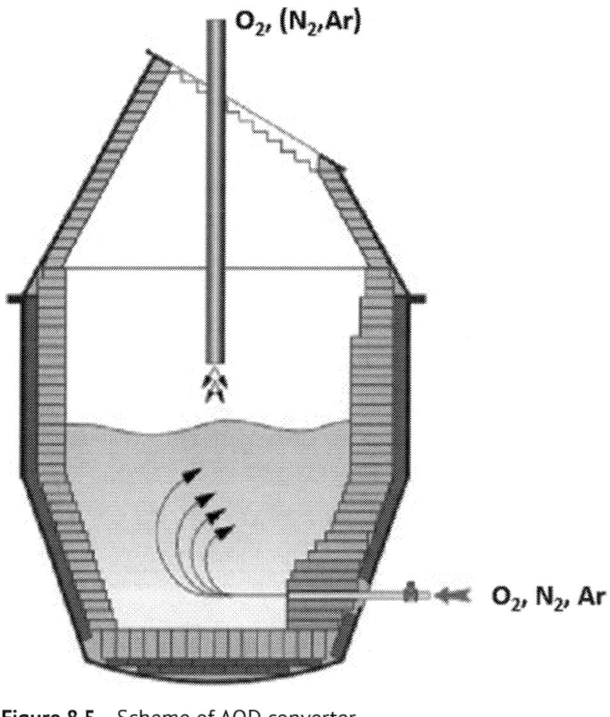

Figure 8.5 Scheme of AOD converter.

minimal to enhance decarburization progress compared to chromium oxidation. It can be done by oxygen content minimization in the blown mixture toward the end of the blow to minimize chromium oxidation and end up in the slag. The following reactions are considered the favorable thermodynamic considerations during dilution or the presence of vacuum:

$$C + O = CO_g \tag{8.1}$$

$$2Cr + 3O = Cr_2O_3 \tag{8.2}$$

by combining

$$3C + Cr_2O_3 = 2Cr + 3CO_g \tag{8.3}$$

The equilibrium constant for the reaction is:

$$K = \frac{[Cr]^2 [CO]^3}{[C]^3 . [Cr_2O_3]} \tag{8.4}$$

It is mandatory to reduce the partial pressure of [CO] gas to keep constant Cr content and decrease C content in steel, as mentioned in the equilibrium Equation (8.3). It can be done by diluting gas like in AOD-type processes or by reducing the total pressure by vacuum treatment.

8.4.3 Vacuum Ladle Degassing Process

Vacuum ladle degassing utilizes a high degree of vacuum assembly to remove CO gas and reduce the partial pressure of CO gas and reaction equilibrium shift toward the formation of more CO. It expedites the deoxidation reaction by dissolved carbon in steel according to the reaction [C] + [O] = {CO}, where [C] and [O] are dissolved elements in steel and CO is the gaseous species. While CO gas bubbles in the liquid steel move upward and float, other advantages are also noticed. The dissolved hydrogen diffuses into the CO bubbles and is removed by a vacuum pump during evacuation. The intensive movement of CO gas also expedites the steel refining from nonmetallic inclusions that agglomerate, float up, and are adsorbed in the slag. It also favors the removal of gaseous nitrogen and nitride inclusions. Homogenous microstructure, low inclusion content, and low gas porosity can be obtained through this process to manufacture ball bearings, large steel ingots, rails, and other high-quality steels.

8.4.3.1 Ladle Degassing (VD, Tank Degassing)

A ladle with molten steel is placed in a vacuum chamber during the tank degassing process. Usually, the ladle is equipped with a bottom mounted refractory porous plug and argon gas purging for homogenous stirring during vacuum treatment. In this arrangement, an additional hopper with a vacuum lock on the chamber cover is attached to add alloying elements and/or slag components, as shown in Figure 8.6. Intensive stirring by bottom argon purging expedite the [C] + [O] = {CO} reaction and desulfurization under vacuum condition. Slag possesses high sulfur solubility. The entire bubbling of both Ar and CO favors floating and removal of nitride inclusions and entrapped nitrogen gas.

The ladle degassing process has several benefits: degassing, decarburization, desulfurization, deoxidation, precise alloying, and removal of nonmetallic inclusions preferentially oxides and nitrides, and thermal and chemical homogeneity.

8.4.3.2 Vacuum Oxygen Decarburization (VOD)

Additional vertical water-cooled lance for oxygen blowing on molten steel surface along with ladle degassing arrangement to produce stainless steel is usually known as Vacuum Oxygen Decarburization (VOD), and a typical schematic is shown in Figure 8.7. With consideration of vacuum environment and oxygen partial pressure, the oxygen reacts with the dissolved carbon [C] + [O] = {CO} than by chromium, a prime constituent of stainless steel. It allows effective decarburization with minor chromium loss. The exothermic oxidation reaction enhances temperature without any external energy resource and the effective

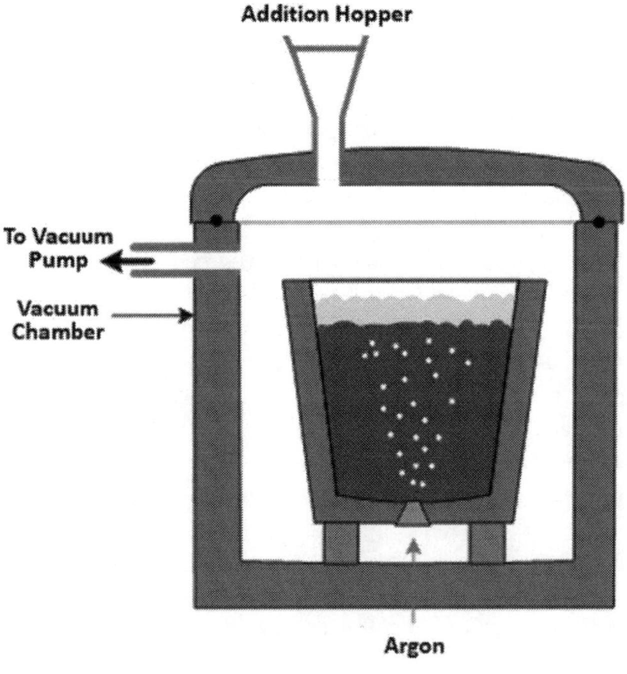

Figure 8.6 Ladle degassing unit (VD).

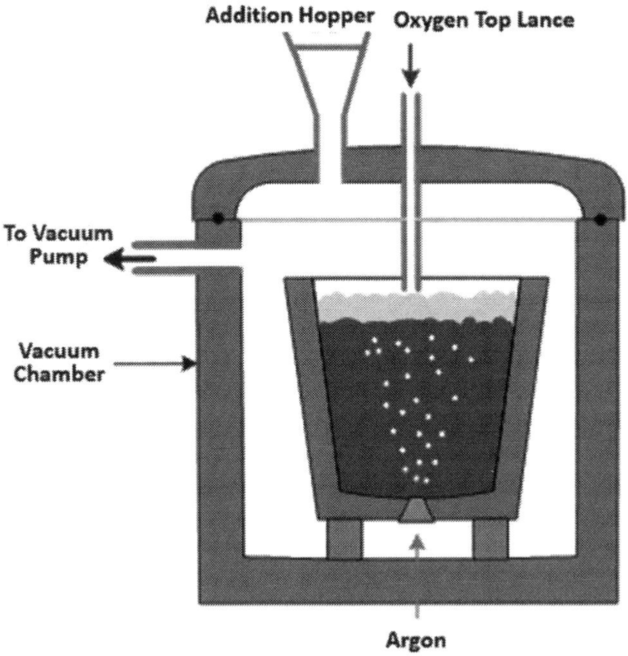

Figure 8.7 Vacuum oxygen decarburization (VOD).

liquid steel temperature. Despite oxygen assisted decarburization (oxidation), deoxidizers are also added to remove excessive oxygen in steel, followed by desulfurization slag addition and continuous stirring by an inert gas through the bottom purging plug to expedite the sulfur removal from the steel.

Vacuum Oxygen Decarburization (VOD) process has several benefits including deep carbon removal (decarburization), Precise alloying, low losses of Cr in steel treatment, reheating, hydrogen removal (degassing), desulfurization, removal of nonmetallic inclusions preferentially oxides and nitrides, and thermal and chemical homogeneity.

8.4.3.3 Vacuum Arc Degassing (VAD) Process

Vacuum arc degassing (VAD) process (Figure 8.8) is a chamber degassing process with electrodes added for liquid steel reheating. Reheating by electrodes, vacuum chamber assembly, stirring through Ar gas purging by bottom porous plug and alloy addition can be done in a single unit.

In the beginning, lime is added to the molten steel ladle, electric arcing is employed at a pressure of 250–300 mm Hg to raise the temperature and fuse the lime followed by short duration degassing, additions for chemistry adjustment and

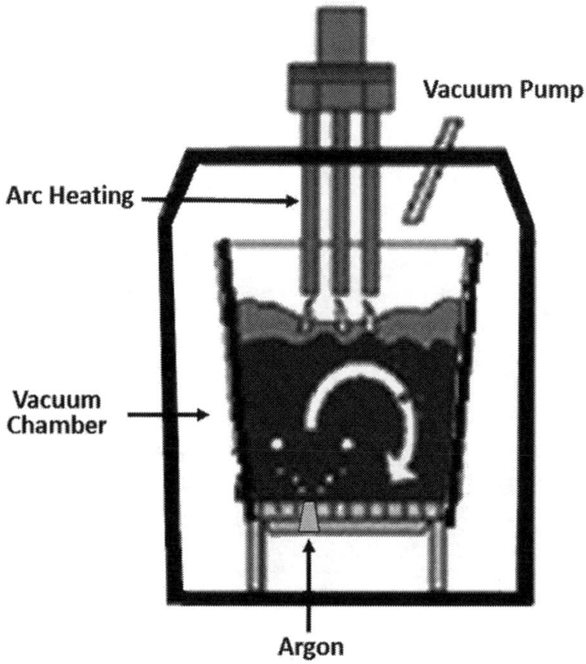

Figure 8.8 Vacuum arc degassing (VAD) process.

deep degassing to pressures as low as 1 mm Hg. A continuous Ar stirring is employed throughout the operation with adequate adjustment of flow rate during different stages of operation in the VAD process. The heating rate is around 3–4 °C/min, and during heating, the rate of argon gas flow is kept on the lower side. In this system, under vacuum, carbon–oxygen reaction under the high-temperature arc is of great help in achieving low [O] content without any solid reaction product. Removal of [H] depends on (i) argon flow rate, (ii) pressure of vacuum chamber, and (iii) initial [H] content. Intense mass transfer by argon and low partial pressure of hydrogen gas because of dilution of liberated carbon monoxide gas promotes the lowering H_2 levels as low as 1.5 ppm. High degree of desulfurization, as high as 80% for steel production with a very low level up to 0.005% is possible by this process.

These days, the VAD process is widely used to produce clean steel. Liquid steel is held in a ladle, which is put inside a vacuum chamber. Ladle is not filled completely, up to 30% of the total volume (i.e., ¼ height) should be provided free board; otherwise, liquid metal may spill over the edges. Steel may be stirred by bubbling an inert gas or an electromagnetic stirrer while being exposed in a vacuum. The teeming ladle, after tapping, is kept in a vacuum chamber, which is then evacuated. Metallurgical reactions such as degassing, deoxidation, decarburization, desulfurization, and alloying take place under vacuum conditions. The ladle is removed after a predetermined time when the correct teeming temperature of steel is reached. Teeming is carried out in an open atmosphere. While pressure drops (in a vacuum), turbulence and vigorousity increase within the ladle, and the bath appears to be boiling.

8.4.3.4 Recirculation Degassing (RH)

Recirculation degassing is the prime process mechanism of the Ruhrstahl-Heraeus (RH) process. It has a different arrangement than the conventional degassing system, in which a vacuum chamber consisting of two snorkels is attached to the bottom part of the chamber, as shown in Figure 8.9a.

Among two snorkels, one is equipped with a pipe through the refractory lining in an attempt to supply argon during operation. Under atmospheric pressure, the snorkels immerse into the molten steel and liquid metal move upward and fill the chamber up to 1.3 m/4.2 ft. In this circumstance, the injected Ar promotes forming a bubble and floating up in one of the snorkels (up-leg) and upward force expedites the melt to rise in the snorkel. In the continuous operation, the molten steel flows back down through the second snorkel (down-leg) to the main ladle unit and continues with a circulation rate near 150–200 t/min. The RH vacuum chambers are usually equipped with addition hoppers to add alloying elements and/or desulfurization slag.

8.4.3.5 Recirculation Degassing with Oxygen Top Lance (RH-OB)

Analogous to VOD, vertical water cooled lance is equipped with conventional RH for oxygen blowing on the molten surface of the steel (Figure 8.9b). Oxygen accelerates decarburization (C + O = CO) and dephophorization, and these exothermic reactions enhance the molten steel temperature and maintain the required temperature without any additional energy source.

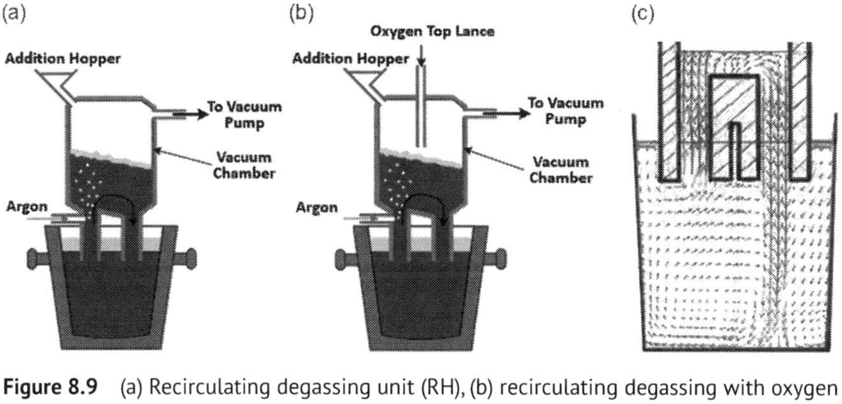

Figure 8.9 (a) Recirculating degassing unit (RH), (b) recirculating degassing with oxygen top lance (RH-OB), (c) schematic view of steel circulation in ladle and snorkel.

A competitive RH and RH-OB process emphasizes the basic differences in the ouput. Despite common advantages like hydrogen removal, sulfur removal, precise alloying, nonmetallic inclusion removal, thermal and chemical homogeneity, the RH assist oxygen removal whereas RH-OB expedites fast carbon removal, phosphorous removal, and reheating.

A deep impulse of steel flow back into the ladle is a precondition for a good homogenization as shown in Figure 8.9c, and it is essential to concentrate on the following operational features:

- To be as deep as immersed with snorkels
- The number of working lift gas tubes are important
- Monitoring Ar-lift gas flow and pressure for each line
- Operate with a single lift gas regulation system

8.4.3.6 Ladle to Mold Degassing

Direct mold is placed in a vacuum chamber, as shown in Figure 8.10. While the molten steel is poured from a tundish to a mold placed in a vacuum chamber, at the same time, the tundish is also filled from a steel ladle. Under vacuum

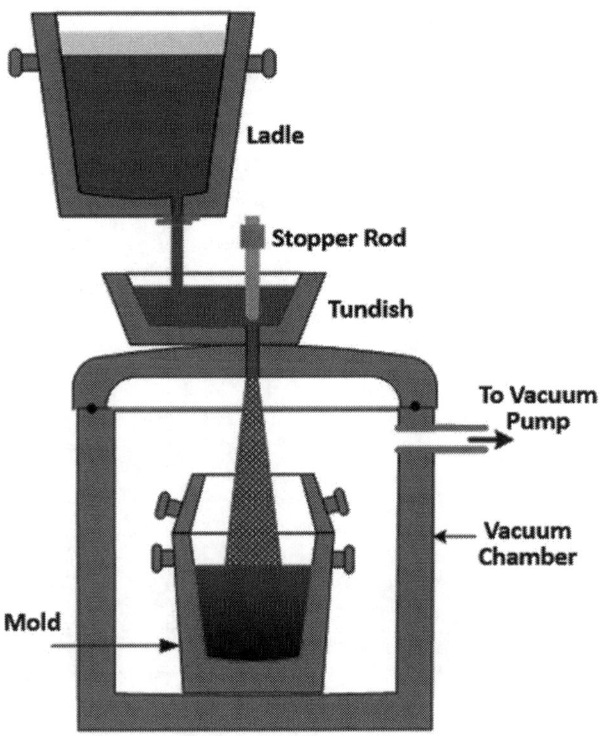

Figure 8.10 Ladle to mold degassing.

conditions, the steel stream appears to be boiling attributed to the deoxidation reaction (C + O = CO), and dissolved hydrogen diffuses into CO bubbles, and continuous evacuation by a vacuum pump.

8.4.4 Stirring and Refining Process in Degassing

Continuous stirring during steel refining is a critical operational parameter for homogenizing of the thermal history and composition. The top lancing method is used for steel refining in the ladle with powder injection and stirring the melt with high argon flow rates. A porous refractory plug is used for gas bubbling at moderate rates, e.g., $< 0.6\ Nm^3/min$. It is fitted usually on the bottom of the ladle, as shown in Figure 8.11. There are various designs of porous plugs as shown in Figure 8.12 [21].

Stirring is mandatory to maintain chemical and thermal homogeneity. It can be done either through a porous plug fitted at the bottom of the ladle or from the top by a submerged refractory tube. Floor practice emphasized that the bottom purging facilitates enhanced gas removal rate and effective degassing during manufacturing of fully killed steel. It is worth noting that a vigorous gas removal environment also causes splashing of liquid steel; and thus needs to maintain an optimum filling of molten steel. Usually, around 25% of ladle height is kept as a free board to accommodate the splashed droplets of liquid steel. Pressure is maintained in between 1 and 10 mm Hg for effective degassing. During degassing, additions are made for alloying and deoxidation. Sometimes, the additional ladle heat treatment is required to compensate for the heat loss due to degassing. Carryover slag from the primary steelmaking vessel is to be as minimal as possible

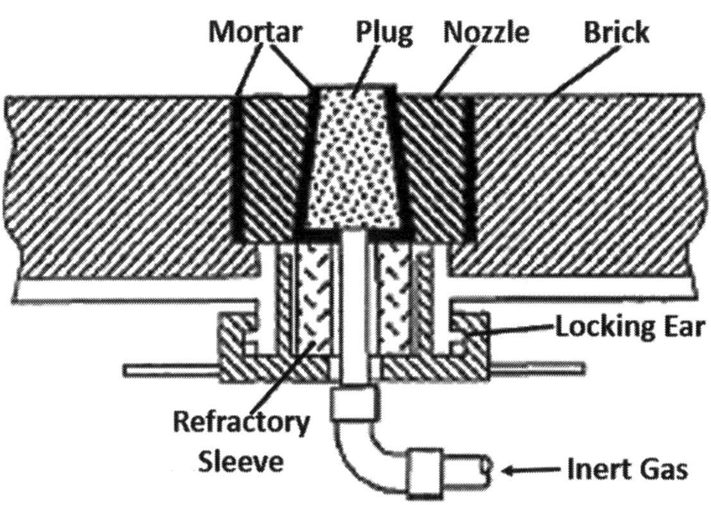

Figure 8.11 Porous plug assembly at the bottom of the ladle.

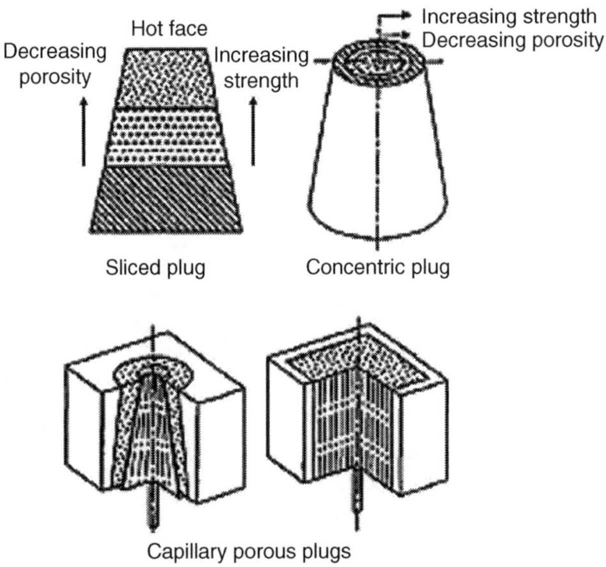

Figure 8.12 Typical examples of porous plugs.

for effective degassing. The FeO content in carrying over slag is enriched with oxygen, and in equilibrium conditions, it transfers the oxygen to steel, resulting in increments of dissolved O_2 in molten steel.

The argon gas connection to the ladle is established when the ladle is set in place inside the vacuum chamber. A vacuum pumping system evacuates the vacuum chamber to the required operating pressures. The fluid motion and turbulence intensity during vacuum treatment of molten steel has a direct impact on homogenization, mass transfer, removal of oxides and nitrides inclusions, the performance of refractory lining wear, slag entrapment, and reaction with atmosphere. The evolved and rising gas bubbles due to degassing (N_2, H_2, and CO) and injected argon gas are the principal source of stirring. In vacuum degassing, the chamber pressure is very small as compared to the ferrostatic head of the liquid in vessel. Consequently, the bubbles get a scope to expand enormously when they rise to the exposed surface, and this is known as bubble bursting, which expedites liquid metal droplets to get ejected into the vacuum chamber in large numbers.

The mixing of liquid steel with gas injection to achieve homogenization of melt temperature and composition is due primarily to the dissipation of the buoyancy energy of the injected gas. The effective stirring power of gas is given by the following thermodynamic relation derived by Pluschkell [22].

$$\varepsilon' = \left(\frac{n'RT}{M}\right)\ln\left(\frac{P_t}{P_\circ}\right) \tag{8.5}$$

Where ε' = stirring power, Wt^{-1}; n' = molar gas flow rate; R = gas constant, 8.314 $Jmol^{-1}K^{-1}P$; T = temperature, K; M = mass of steel, tonne; P_o = gas pressure at the melt surface, atm; P_t = total pressure at injection depth; $H(m) = P_o + \rho gH$, ρ = steel density, 6940 $kg.m^{-3}$ at 1600°C; g = 9.81 ms^{-2}. In consideration of the above equation and data, the modified equation can be represented as:

$$\varepsilon' = 14.23\left(\frac{V'T}{M}\right)\log\left(1+\frac{H}{1.48P_\circ}\right) \tag{8.6}$$

Where V' = gas flow rate, $Nm^3.min^{-1}$.

The gas stirring time to achieve 95% homogenization is called the mixing time, t. Several theoretical and experimental conditions have been studied to formulate the mixing time in terms of the gas stirring power, ladle diameter, and depth of gas injection in the melt. The following relation is from the work of Mazumdar and Guthrie [23].

$$t(s) = 116(\varepsilon')^{-1/3}(D^{5/3}H^{-1}) \tag{8.7}$$

Where D = average ladle diameter, H depth of gas injection.

Calculated mixing times are given in Figure 8.13 for the simplified case of $D = \sim H$. In a 200 t heat, the melt homogenization will be achieved with argon gas bubbling for 2.0–2.5 min at the rate of 0.2 Nm^3/min.

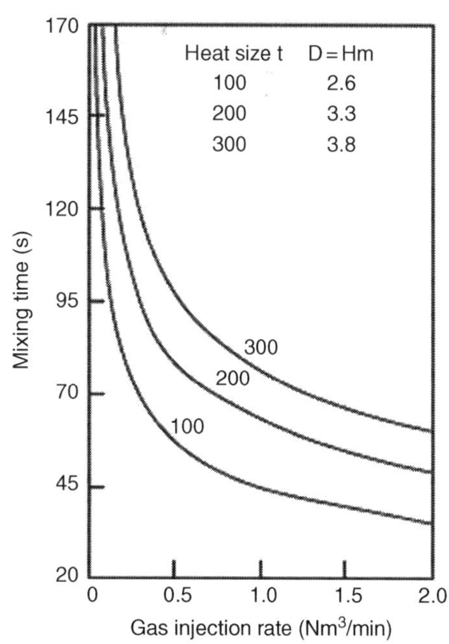

Figure 8.13 Calculated mixing times for 100, 200, and 300 ton heats [20].

8.4.5 Composition Adjustment by Sealed Ar Bubbling with Oxygen Blowing (CAS–OB)

Composition Adjustment by Sealed Ar Bubbling with Oxygen Blowing (CAS–OB) is designed for heating and alloying of steel. It has several advantages: (i) alloying material yields high and can be easily predicted, (ii) aluminum consumption is low, (iii) the target temperature required for casting can be achieved consistently, and (iv) the amount of oxygen content present after treatment is low. Some of the disadvantages of the CAS-OB process are (i) prior transportation of ladle to the station needs to add slag formers, and (ii) the removal of sulfur is not possible in the process.

Compared to other heating processes (e.g., REHeating and IR-UT), the total investment costs of setting up the CAS-OB station are high. However, the heating rate in the CAS-OB process is higher than in other processes. Also, there is a slag sticking problem to the bell structure, due to the weight and volume of bell enhance. So, there may be the chance of some undesirable effects during the operation of CAS-OB. This process is used for alloy additions in the inert atmosphere, which contains argon gas. It also allows the addition of Al and O_2 gas blown through a top lance. After reaction, these form Al_2O_3 and due to the exothermic nature of reaction a considerable amount of heat is also generated. Thus, the CAS-OB process helps in chemical heating of steel in liquid phase. The schematic view of working principle of the CAS-OB process is in Figure 8.14.

The chemistry of liquid steel is adjusted by adding carburizer and ferroalloys along with the heating process of liquid steel. During heating, granular aluminum, carburizer, and ferroalloys are fed continuously, and oxygen is injected into the bell. Due to the time gap between the additions of different materials and the time of measuring temperature, a loss of heat occurs and the actual rate of heating is slightly lower for metal than the estimated value.

Some of the special features of CAS-OB process are:

- Bell diameter is very important for securing slag free region in the slag layer. During bottom bubbling, if the diameter of the open eye is known, we can easily determine the bell diameter. Open eye diameter can be calculated from the geometry of the bubbling plume (plume cone), which usually depends on the gas flow rate.
- The important step in the CAS-OB process is opening the bottom bubbling plug to secure an open eye. A bell position changing system is used these days to avoid the risk of bubbling plug opening failure. With the help of the recent adopted system, bell position can be switched easily to a reliable bubbling area.
- The oxygen content in the CAS-OB process is approximately equal to the bubbling process and ladle furnace, but it is slightly inferior with respect to the RH process.

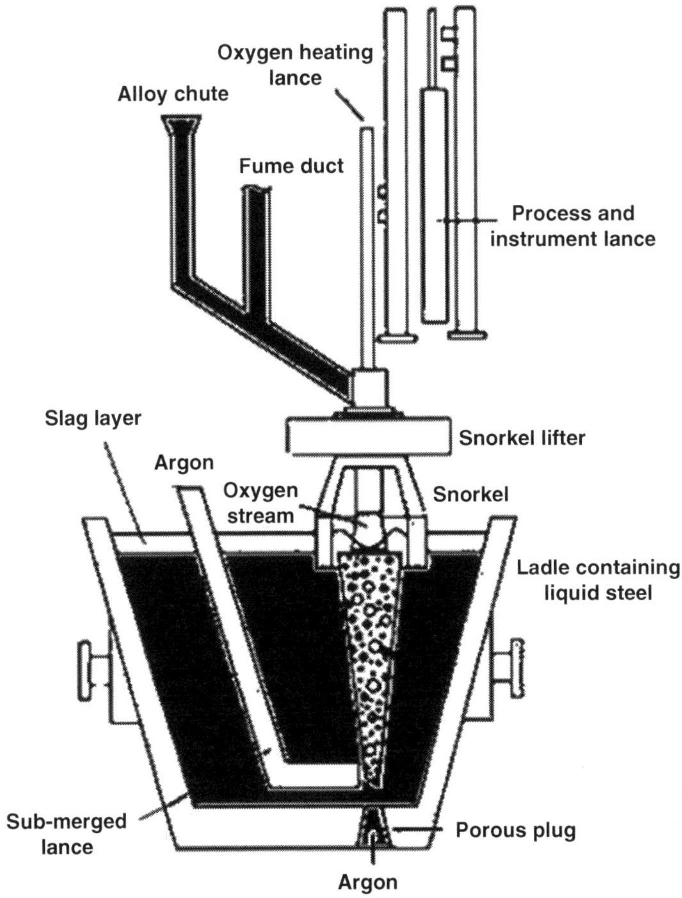

Figure 8.14 Schematic representation of the working principle of CAS-OB process.

- During the CAS-OB process, a few oxide materials are attached to the lower part of the bell, and the skull is attached to the upper part. Oxide growth in the lower part of the bell as oxide materials are attached to the lower part may lead to a clash with the rim of the teeming ladle, and ultimately the breakdown of the bell may take place.

8.4.6 RH Snorkel

Generally, the RH snorkel is cylindrical in shape and consists of a monolithic refractory on the outer side and lower edge, bricks on the inner side, and gas inlet pipes for lifting the molten steel. There is also a water-cooled flange between the lower vessel and the snorkel, and under the flange, there is a cylindrical iron shell, which is used to fix the bricks and prevent air suction. The steel shell is positioned between the bricks and the outer monolithic refractory. Generally, two snorkels are used together. One side is the up leg snorkel, which is used to lift the molten steel in a ladle into the lower vessel by gas injection, and the other side is the down leg snorkel, which returns the molten steel in the lower vessel to a ladle. After using an up leg snorkel, it is changed to a down leg snorkel by stopping the valve of the lift gas pipe. In case of refractory damage, snorkels are usually repaired by gunning refractories, but when the damaged area is too extensive, or the gas pipes of the up leg snorkel are damaged, the snorkel must be replaced as it is beyond repair. In particular, sudden damage to an up leg snorkel requires unscheduled replacement, which harms stable steel production.

A typical schematic representation of a snorkel is given in Figure 8.15. Snorkel is a complicated refractory assembly that is inner lined with bricks and outer lined with castables over the cylindrical steel core shell. One set of snorkels consists of one inlet snorkel and one outlet snorkel. Inlet snorkels are equipped with an Argon pipe assembly to facilitate argon gas purging at the time of application in RH degasser. RH degassing technology is mainly used in the secondary steelmaking stage to make the Auto grade and higher grade steels.

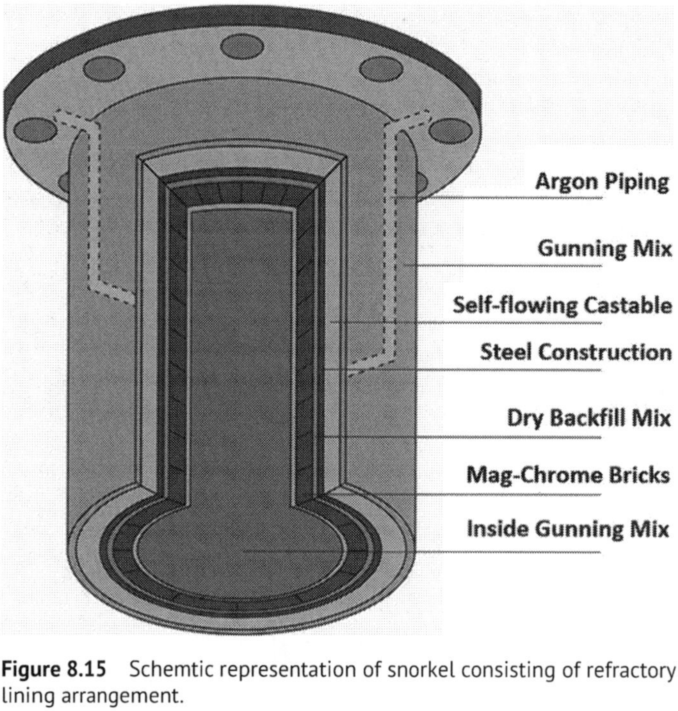

Figure 8.15 Schemtic representation of snorkel consisting of refractory and Ar gas lining arrangement.

8.5 Designing Aspects of Refractories

Several critical factors mandate selecting and designing the refractory for secondary steelmaking that eventually controls the lining life. Refractory designing factors can be further classified into *half dozen factors*, such as Operational parameters, Thermal and Mechanical Stresses, Thermal distribution, Lining Pattern and zoning effect, Steel and Slag chemistry, and refractory quality and composition. A typical refractory lining pattern is shown in Figure 8.16.

Optimum operational parameters facilitate achieving both targeted steel composition and refractory life. A brief operational history and targeted steel

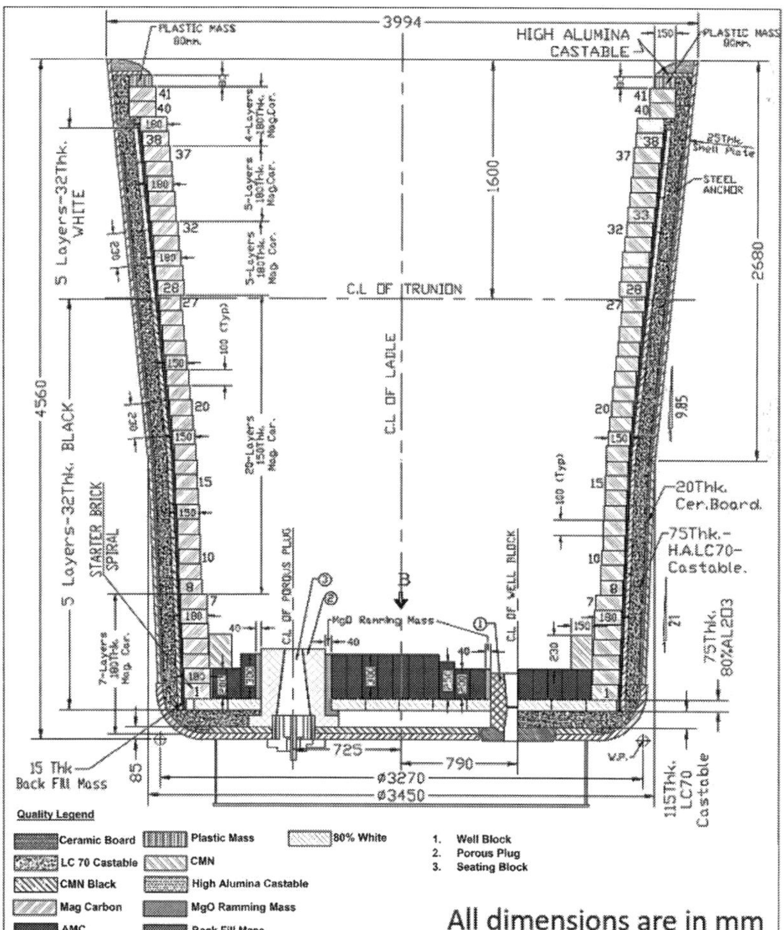

Figure 8.16 A typical ladle furnace design and details the type of refractory lining and their dimensions.

composition are essential benchmarks before refractory design and manufacture for a particular steel-manufacturing organization. The following in-depth *Dozen Factors* may assist in developing suitable refractory for LF, VD, RH etc.:

a) Process route and final ladle temperature
b) Vessel capacity
c) Lining details and thickness
d) Treatment time—bath temperature, arcing, ladle preheating, ladle circulation time
e) Purging details—stirring, purging time, mode of purging
f) Steel grade/chemistry
g) Slag chemistry/basicity
h) Slag volume and contact time with metal
i) Tap temperature and tap area
j) Casting time
k) Life requirement
l) Repair practice

Optimization of ladle life by considering operational parameters, shop logistics, and cost of refractory is a prime target to produce clean steel with minimum consumption of refractories. Identification of stresses at different areas in the working lining of the ladle concerning operational parameters and shop logistics followed by the design of refractory lining can optimize the lifetime of the refractory. Before refractory composition design, one must concentrate on physical, mechanical, and thermal properties as bricks are experiencing different degrees of slag penetration, solubility, mechanical impact, thermal stresses, and corrosion. A schematic representation of the ladle shows the probable stresses and reasons for failure during operation (Figure 8.17).

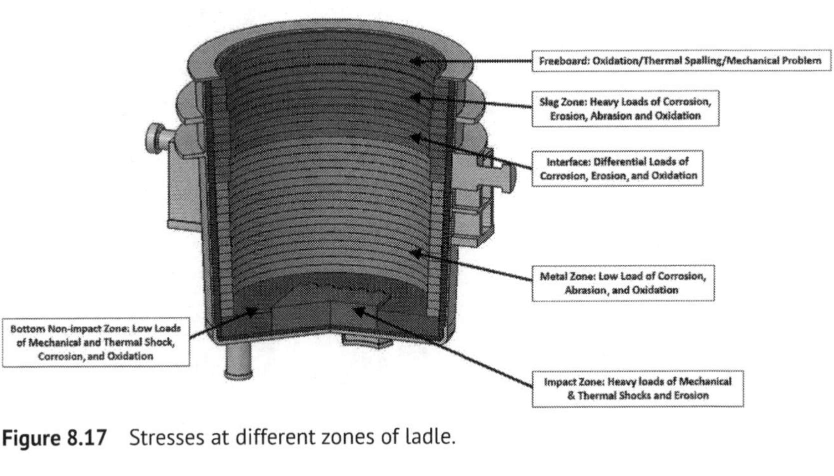

Figure 8.17 Stresses at different zones of ladle.

Despite the mechanical impact, thermomechanical spalling facilitates the stresses and loosening of the lining, and results in liquid penetration and wear. Mechanical destruction by stresses, generated by thermal expansion or mechanical shocks; filling of open pores of the refractories with liquids or sublimating gases. Deterioration of inner structure because of different thermal expansion properties and/or reactions accompanied by a change in volume. This phenomenon is influenced by:

- elastic properties and strength of the refractory
- thermal conductivity of the refractory
- thermal expansion of the refractory
- temperature and temperature distribution
- geometrical conditions (size, shape, etc.)

A simple schematic represents the cracking mode during expansion that experiences tensile stresses (see Figure 8.18). After repeated cooling and reheating, the ladle lining develops *cobble stone* wear because of thermal stresses acting on the edges and corners of bricks. Lining pattern and cobble stone wear of MgO-C is shown in Figure 8.19.

Effect of predominant operational parameters such as tapping weight and temperature, secondary operation and time, additional elements, steel, and slag chemistry significantly influence working lining.

Tapping weight:

- Higher tapped weight imparts high mechanical load at the impact zone
- Higher erosion for higher heat size
- Variation in tapped weight results in bath level variation, which affects *zonal lining* concept

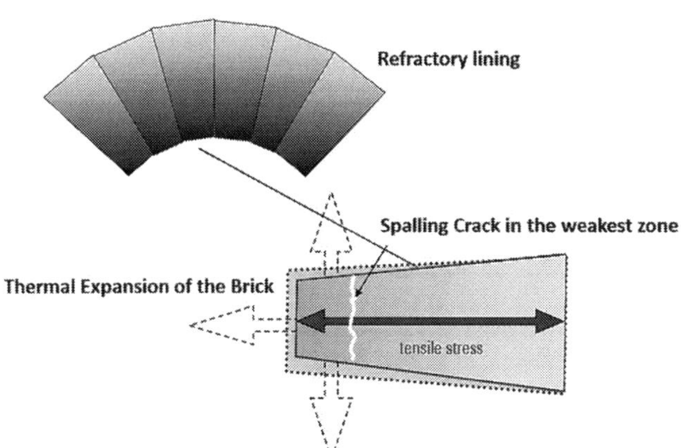

Figure 8.18 Crack in lining due to thermal expansion and formation of tensile stress.

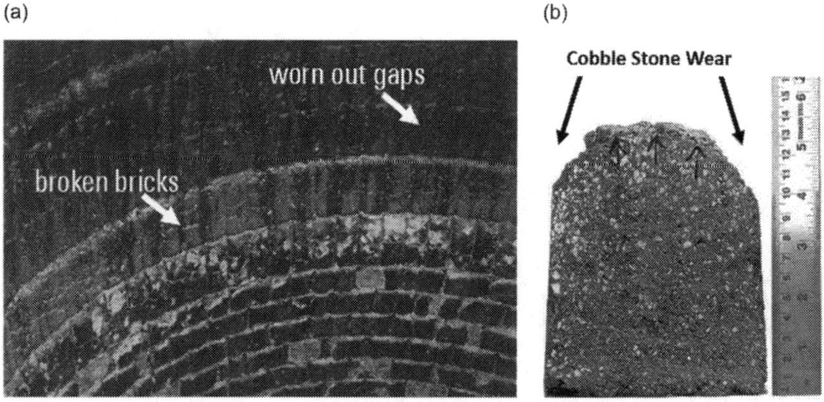

Figure 8.19 (a) MgO-C lining with having broken bricks and worn out gaps, (b) formation of Cobble stone wear after dismantling and cross-section.

Tapping temperature: High tap temperature results

a) higher erosion
b) higher thermal shock
c) higher oxidation

Secondary operation and treatment time:

- Higher erosion for higher treatment time
- Extensive purging results in higher abrasion
- High arcing time in LF results in localized erosion at the proximity of the electrode
- Higher vacuum level at VD/RH weakens the C bond in C bonded refractory

Treatment by addition:

- Higher erosion by more fluorspar addition
- High Ca ppm for Ca-Si treatment results high erosion of high alumina refractory

Steel chemistry:

- Low carbon steel results in more erosion of carbon bonded refractory by carbon pick-up from lining
- High oxygen ppm in steel results in high erosion of carbon bonded refractory by oxidation

Slag chemistry:

- Low MgO slag results in high erosion in MgO-based refractory by MgO pick-up from lining
- Low slag basicity results in high erosion in basic refractory

MgO is one of the prime phases for working lining refractory for secondary steel-making (see Table 8.3). Thus, the solubility of MgO in slag reduces the low brick life that eventually affects the continuous operation. This solubility depends on two important parameters such as temperature and slag basicity. A study reveals that the MgO solubility from brick increases with increasing slag temperature and decreasing slag basicity, as shown in Figure 8.20. The MgO solubility in slag follows; %MgO at temperature T°C = %MgO at 1600°C + [0.0175 × (T – 1600)]. Thus, %MgO solubility becomes 5.25% at 1500 °C when 7% at 1600 °C. Usually, the MgO solubility increases by 1.75% per 100 °C. Steel manufacturer predominately maintains control over slag basicity through lime enrichment, and one follows the concepts of *basicity* through this section.

Basicity can be understood either from ionic or from the molecular nature of slag. The ionic nature of slag assumes slag to consist of ions. In slags, acidic oxides can accept one or several O^{2-} ions, whereas a basic slag is a donor of O^{2-} ions. For example, 1 mole of SiO_2 can accept 2 moles of O^{2-} ions so that each tetrahedron in hexagonal structure becomes independent of each other. Similarly, each mole of P_2O_5 can accept 3 moles of O^{2-} ions. Thus:

$$SiO_2 + 2O_2^- = SiO_4^{4-} \tag{8.8}$$

$$P_2O_5 + 3O_2^- = 2(PO^4)^{3-} \tag{8.9}$$

Amphoteric oxides behave as bases in the presence of acid or as acids in the presence of a base:

$$Al_2O_3 + O_2^- = Al_2O_4^{2-} \tag{8.10}$$

$$Al_2O_3 = 2Al^{3+} + 3O_2 \tag{8.11}$$

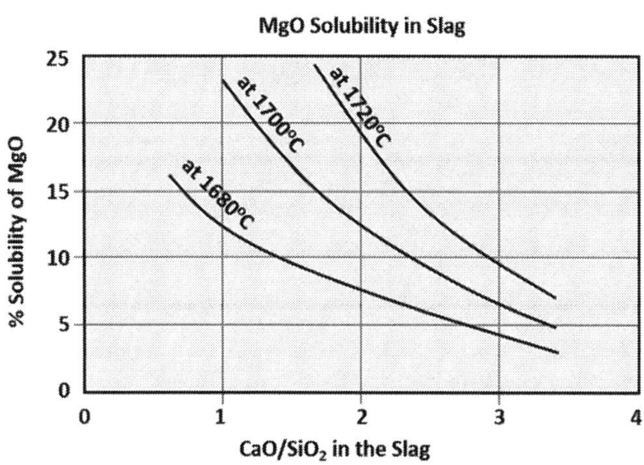

Figure 8.20 The effect of slag temperature and basicity.

Bases can supply O_2^- ions

$$CaO = Ca^{2+} + O_2^- \tag{8.12}$$

In a neutral slag, enough oxygen ions will be present to ensure that each tetrahedron remains independent of each other. In binary CaO–SiO_2, slag will become neutral when CaO is 66.7%, corresponding to the formation of 2CaO.SiO_2. Slag will be basic only when CaO content is more than 66.7%. Basicity can be expressed in terms O_2^- ions that are in excess of that required; thus satisfying the requirements of acidic oxides. In slag:

$$n_{O^{2-}} = (n_{CaO} + n_{MgO} + n_{MnO} + n_{FeO}) + \ldots - (2n_{SiO_2} + 3n_{P_2O_5} + n_{Al_2O_3}) \tag{8.13}$$

In industrial practice ionic definition of basicity is not useful; the molecular approach is more useful. The molecular approach assumes slag to consist of chemical compounds. The basicity of slag is:

$$B = \text{weight\% CaO / weight\% } SiO_2$$

In the presence of different basic oxides, the different strengths of the basic oxides should be considered. In a slag that contains CaO, MgO, SiO_2, and P_2O_5, the basicity is:

$$B = (\text{weight\% CaO} + \text{weight\% MgO})/(\text{weight\% } SiO_2 + \text{weight\% } P_2O_5)$$

In slag/metal reactions that involve desulfurization and dephosphorization, the concept of free lime in slag is useful. Free lime is available after the formation of a neutral compound like 2CaO.SiO_2, and 3CaO.P_2O_5.

For 100-ton hot metal with 1% Si and 0.2% P, the calculation shows that free CaO in slag would be available when CaO content exceeds 4540 kg.

Thus, 1000 kg Si can produce ($Si + O_2 = SiO_2$); $(60/28) \times 1000$ kg $SiO_2 = 2142$ kg SiO_2.

Required CaO to produce 2CaO.SiO_2 = $(112/60) \times 2142$ kg = 3998.4 kg.

Similarly, 200 kg P can produce ($4P + 5O_2 = 2P_2O_5$); $(284/124) \times 200$ kg $P_2O_5 = 458$ kg P_2O_5.

Required CaO to produce 3CaO.P_2O_5 = $(168/142) \times 458$ kg = 541.86 kg.

Total CaO required = 3998.4 kg + 541.86 kg = 4540.26 kg.

Working linings in steel ladles are subjected to highly fluctuating temperatures, and each thermal cycle is further divided into two stages; *vessel filling and steel composition refinement*, and *vessel empty and waiting for next refining process*. During the first stage, the inner face (*IF*) of the bricks is in contact with molten steel and its temperature reaches about 1650°C. The opposite face (i.e., the outer face *OF*) of the initial brick (150 mm in width) is in contact with the monolithic lining. The maximum temperature depends on insulating degree of the lining and the current thickness of the brick, which reaches about 800°C during the first cycles. When the vessel is emptied, slag covers the inner face of the bricks and impregnates the refractory through the porous network. During the second stage, the bricks cool down. The minimum temperature at the inner and outer faces depends on the waiting time. It can be estimated at approximately 1200°C at the inner face and 650°C at the outer face. In consideration of temperature distribution from working lining to back-up lining, a model coined as ladle furnace on-line reckoner (LFOR) has been proposed based on material and heat balance from statistical analysis of 100 heats of ladle, as shown in Figure 8.21 [24]. The analysis was carried for three conditions, after preheat, at on-line purging (OLP) out and at LF-out. The model is well-fitted with experimental data.

There are several important parameters to describe the thermal model, and they can be classified as: a) Heat loss through a refractory wall, inner zone to outer shell temperature, b) Heat loss due to operational features like holding and purging, c) Unusal heat again during arcing, and d) Heat effects due to additions.

Heat loss and temperature drop through refractory lining can be estimated as the refractory temperature profile change with time. A one-dimensional heat transfer Equation (8.14) for the sidewall and bottom wall can solve with the following assumptions. Initial condition → $t = 0$, $T_{wall} = T_{preheat}$, and Boundary condition → $t = t$, $T_{wall} = T_{Liquid\ steel}$:

$$\rho C_p \frac{dT}{dt} = h(T_{surf} - T_{\infty}) \tag{8.14}$$

While the temperature difference is substantial, the heat loss rate is very high in the initial stage and gradually slows down and follows a steady state. Due to refractory loss during the steelmaking process, the sidewall thickness gradually decreases with ladle life and a critical consideration during analysis. In recent literature, the temperature stratification in a steelmaking ladle was analyzed considering three distinct zones in the flow [25]. Three zones are classified as the wall boundary layer, the bottom stagnation zone, and the central plug flow, and their analysis predicts that the flow in ladle can be divided into three regions.

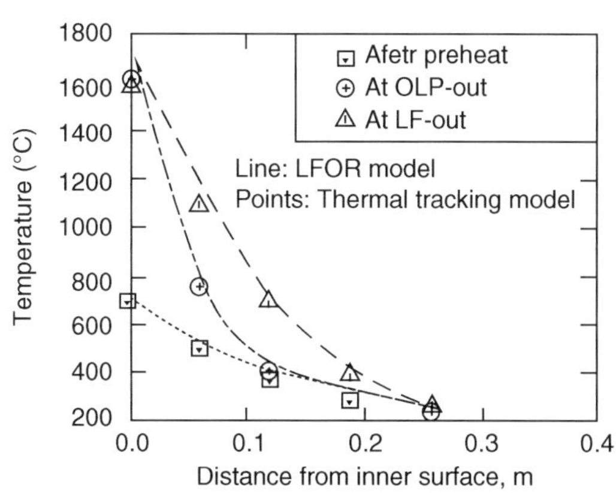

Figure 8.21 A typical temperature–distance profile in ladle refractory lining in operation [24].

Firstly, a downward flowing boundary layer forms near the ladle wall. Secondly, a large upward flow zone forms over the rest of the ladle. Thirdly, a relatively quiescent zone forms in the base of the ladle. This zonal system is shown in Figure 8.22. Heat transfer in this zone is dominated by convection. Further, the zone 3 is dominated by conductive heat loss to the ladle base. This zone is quiescent and stable due to the positive thermal expansion coefficient of molten steel. Heat transfer is essentially one-dimensional.

Before starting production, a lining pattern is another consideration in selecting the refractory shape. For example, Spiral lining is being used for better perfection with no split bricks, and ring lining is generally preferred as it is user-friendly. However, in any case it is necessary to maintain:

- Structural strength of lining is developed by proper choice and application of backfilling mass and castable for fixing of the top retainer plate. A suitable backup mass also accommodates expansion of working lining; thus packing density of filling mass is also required to consider.
- Well block/housing block fixing should be carried out properly so that metal can be fully decanted during teeming.

Some critical shapes of refractory bricks are shown in Figure 8.23, and their advantages and disadvantages are mentioned in Table 8.5.

Usually, zoning represents a different area in a vessel per the application stress. Different grade and quality refractories are selected to take care of zonal stress that eventually reduces the wear. At the same time, the application of the vessel is essential to monitor to optimize performance and get the maximum utilization of refractories.

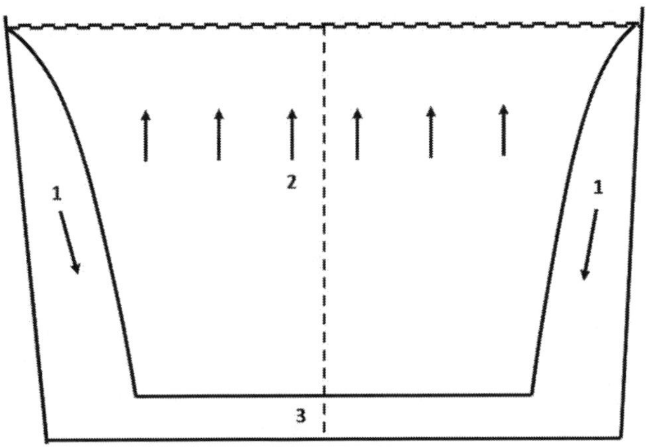

Figure 8.22 Zones of flow predicted to form in a ladle under natural convection.

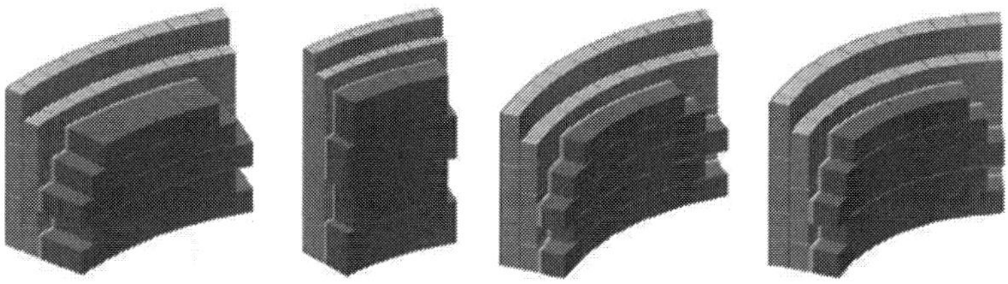

Figure 8.23 Conventional refractory shapes for working lining.

Table 8.5 Represents the advantages and disadvantages of different shapes of working lining for secondary steelmaking.

Shape	Advantages	Disadvantages
P-Shape	a) Quick installation b) Dimensional tolerance c) Low number of joints	a) Low thermal shock resistance b) High brick weight
Key for ladle	a) Good thermal shock resistance	a) More labor cost for lining b) Dimensional tolerance
Mini key	a) Good thermal shock resistance b) Dimensional tolerance	a) More labor cost for lining b) More number of joints
SU-Shape	a) Good thermal shock resistance b) Less shape variation	a) Dimensional tolerances b) Step formation in lining

Each zone may be subdivided into smaller zones depending on the operational practice and shop logistics is termed a micro-zone. Each micro-zone can be treated in the same way as zone by suitable quality selection and mode of application. Such micro-zones differ from plant to plant for obvious reasons on the operational ground and are 100% tailor made. Usually, *micro-zoning* is an important design concept to produce clean steel with minimum consumption of refractories. This process assists in obtaining better alloy recovery, accurate temperature control, excellent desulfurization, low NMI, and reduction of cost pressure on the steel producer; and thus on the refractory supplier.

Life Optimization and Improvement of Cost-Performance Ratio for micro-zoning in working lining of steel ladle are monitored to improve the cost-performance ratio. During this analysis the following points are essential for monitoring:

- Study of life potential of all micro-zones for shop logistics
- Based on shop logistics and root cause analysis of lowest life, optimization of life in all micro-zones can be done by quality adjustment, modification in operation and/or application, and any change in cold or hot repair practice, etc.

- Identification of the most critical micro-zone where life potential is low and other zones where life potential is very high
- Target the highest possible life by shop logistics analysis
- Quality optimization in all micro-zones to get the best possible life at the lowest possible cost of refractory

A classic field trial in India has been conducted to understand the influence of the micro-zoning concept. Life of ladle was restricted to around 9000 minutes holding period (tap to lift). Major operational parameters for steel ladle is listed in Table 8.6. In order to understand the micro-zoning effect MgO-C Grade B has been replaced by two different quality MgO-C bricks, D and E, and their details listed in Table 8.7.

Left over thickness (LOT) was estimated before and after the micro-zoning concept and analyzed the influence of the micro-zoning effect in a 70-ton ladle; different analysis modules are shown in Figure 8.24. The critical and noncritical zone have been identified among a total number of 34 layers, where the life potential of the noncritical area of the slag zone is quite high and remains unutilized. The most critical zone identified is layer number 20–26 at the slag zone purging

Table 8.6 The operational parameters for steel ladle.

Ladle capacity [mt]	70
Tapping temperature [°C]	1640
Secondary operation	LF and VD
Share of VD operation [%]	100
Holding period (tap to lift time per heat) [min]	160
Steel grade	LC, MC, HC
Slag basicity	> 2.5
Yield from tap to cast [%]	98

Table 8.7 The variation of slag zone brick composition.

Test	MgO-C grade B	MgO-C grade D	MgO-C grade E
C [%]	> 14.0	> 14.0	> 14.0
Chemical analysis of magnesia used			
MgO [%]	96–98	97–99	95–97
Al_2O_3 [%]	< 0.3	< 0.2	< 0.3
Fe_2O_3 [%]	< 0.8	< 0.6	< 0.8
CaO/SiO_2	> 1.9	> 1.9	> 1.9

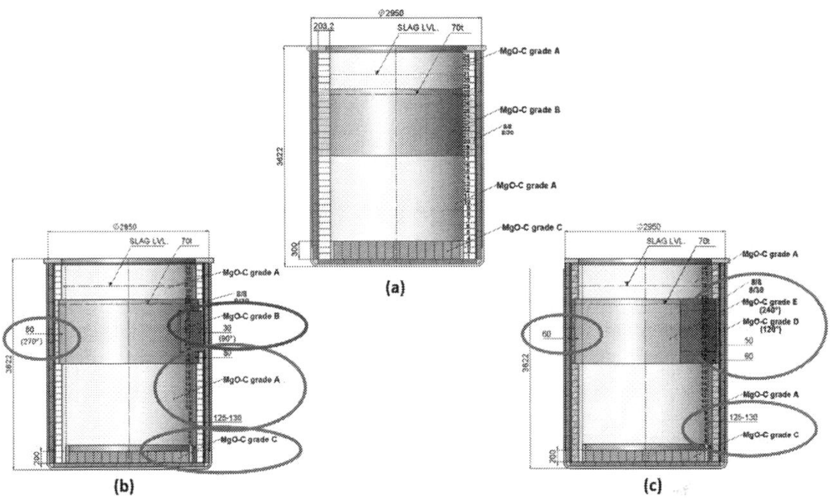

Figure 8.24 Refractory lining (a) before micro-zoning, (b) left over thickness (LOT) analysis after end of slag zone life (before micro-zoning), and (c) LOT analysis after end of slag zone life (after micro-zoning).

side in a 90 degree area. Higher turbulence of slag is the root cause of high erosion of refractories.

Micro-zoning concept was employed and region B has been replaced by grade D and E. MgO-C grade D of higher resistance to slag corrosion was used in 120 degree area at S/Z PP side. Slag zone—rest area (240 degree) MgO-C Grade E with normal resistance to slag corrosion. Switch over to the micro-zoning concept with two different grade bricks in a metal-slag interface that increases lifetime from 57 to 77 heats, and tap-to-lift time rises from 9120 minutes to 12,320 minutes. In a more focused way, considering the Marangoni effect, suitable refractory eventually results in an excellent lining pattern to avoid premature failure of metal-slag zone interface bricks.

In this regard, the lining of number of slag zone bricks can be calculated through simple concept. Let, ladle capacity 70 ton, the slag generation 4% of ladle capacity = 2.8 ton, foamy slag density = 0.8 g/cm^3 or 0.8 ton/m^3; thus slag volume becomes 3.5 m^3 (say X), ladle volume/course = ¼ (π d^2) × h = Y (say). Slag zone thickness = X/Y layers. Let, d = 2.95 m, h = 0.1 m, then Y = 0.68 m^3. Number of course in slag zone = X/Y = 3.5/0.68 = ~5 numbers.

Selection of quality and composition depends on zone selection and the need to follow the road map:

- Proper selection of zones by study of operational parameters and shop logistics
- Determination of lining thickness at different zones by study of wear profile and life demand
- Selection of quality at different zones to optimize life

Selection of zones and refractory quality reduce the counter-act stresses in respective area. Table 8.8 shows the different quality of MgO-C refractory to enhance the ladle life;

In conclusion, the refractory performance demands important features, including:

- Impact area: High mechanical strength
- Purging side metal zone: High resistance to abrasion
- Purging side slag zone: High resistance to erosion by slag in turbulence
- Slag zone: High resistance to erosion by slag
- Good poking of back-filling mass during lining
- Proper locking of lining with effective poking of castable with retainer plate at top
- Key bricks to put at opposite side of purging
- Brick to brick flash joint by proper lining

Table 8.8 Selection of MgO-C refractory for different zones.

Zone	Required quality	Recommended quality
Free board	• High resistance to oxidation and thermal spalling • High mechanical strength	MgO-C brick with 95–96% MgO (sintered) and 8–10% carbon
Slag zone	• High resistance to corrosion by slag and metal • High resistance to abrasion by turbulent slag/metal especially at purging side	• MgO-C brick with 97–98% MgO (Fused) and 12–15% carbon • Periclase of high crystal size is to be judicially utilized
Metal zone	High resistance to corrosion and abrasion by liquid metal for a long period	• MgO-C brick with 96–97% MgO (Fused and sintered) and 8–10% carbon • Alumina-Mag-Carbon (AMC) by suitable combination of Al_2O_3, MgO and graphite to have high strength and progressive + PLC character for joint-free structure in lining during use
Interface	High resistance to differential erosion	MgO-C brick by enrichment of periclase crystal size, purity, and suitable choice and content of graphite
Bottom impact	• High coked strength • High resistance to thermal spalling	MgO-C or AMC or MAC brick with 5–10% carbon
Bottom non-impact	High resistance to thermal spalling	MgO-C brick with 96–97% MgO by suitable combination of fused and sintered magnesite and 8–10% carbon

- Zonal lining to maintain properly
- Consistent tap weight to have same bath level
- MgO in slag to maintain 9–10%
- Slag basicity to maintain > 2.5
- Proper planning to have less holding period

8.6 Refractories for Working Lining

8.6.1 Magnesia–Carbon Refractories

A range of high-temperature steelmaking process vessels as well as high-duty functional components, including electric arc furnaces (EAF), basic oxygen furnaces (BOF), tap hole sleeves, transfer/teeming ladles, slide gate plates, monoblock stoppers, and many other continuous casting articles use MgO-C refractories. Such lucrative properties are the result of high refractoriness, excellent corrosion resistance against slag/metal, and good thermal shock resistance with satisfactory hot strength characteristics. The manufacturing procedure for the fabrication of MgO-C refractories and their postmortem analysis are outlined as follows. Commercially available MgO-C refractories consist of 8–30 wt.% carbon in their brick structure. MgO-C refractory fabrication process flow diagram is shown in Figure 8.25.

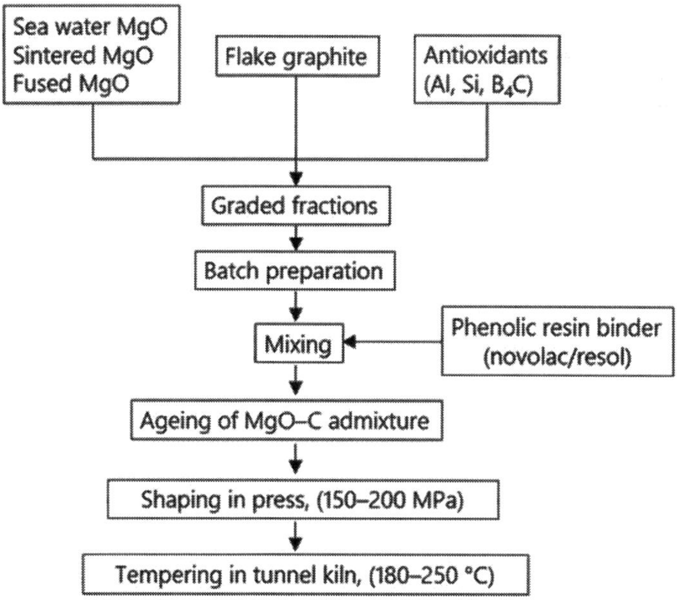

Figure 8.25 Process flow diagram for the fabrication of resin bonded MgO-C refractories.

The basic formulation of MgO-C refractory composition, is composed of magnesia, graphite, organic binder, and antioxidant additives as ingredient materials. High-purity sintered MgO, sea water MgO, and fused MgO with physio-chemical characteristics, including CaO/SiO_2 ratio > 2, low B_2O_3 content, and large periclase crystallite size, are the globally considered raw material sources for MgO. Flake graphite with 94–97 wt.% fixed carbon content, is used as a chief carbon source. Novolac or Resol type phenolic resins are used as organic binders. For improving the refractory oxidation resistance/hot strength character, various forms of antioxidants can be used as additives. They commonly include metal/alloy powders (Al, Si, Mg, Al-Mg) and carbide powders (B_4C, SiC). In actuality the particle size distribution for a given refractory composition consists of large (4–6 mm), coarse (1–3 mm), intermediate (0.5–1 mm), and finer fraction (< 500 μm) of MgO, as well as fine flake graphite and antioxidant additive powders.

These graded fractions are mixed with an organic resin binder (novolac/resol) in a pan mixer at room-temperature as per the given mixing scheme (Table 8.9) to prepare a macro-homogeneous refractory mixture. This homogeneously formulated mix was aged in ambient conditions for a few hours and uniaxially pressed to refractory articles of various shapes at a pressure of 150–200 MPa using hydraulic or friction–screw press. The shaped refractory articles are tempered in a tunnel kiln in the temperature range of 180–250°C for 24 hours. The burnout of volatiles and polymerization of resin binder during tempering stages results in a three-dimensional interlocking carbon–carbon chain network throughout the tempered refractory matrix. This leads to the production of resin bonded MgO-C refractories with significantly improved properties.

A case study on postmortem analysis of MgO-C refractories working at ladle slag zone area is presented. The chemical analysis of the ladle slag is given as (CaO: 21.3%, Al_2O_3: 21.16%, SiO_2: 27.50%, MgO: 6.18%, FeO: 3.38%, MnO: 4.20%, basicity ~ 1.0). It can be noted that the matrix portion of this refractory contains fine MgO, graphite, and in-situ formed magnesium aluminate spinel. In comparison, the aggregate portion of this refractory is composed totally of MgO

Table 8.9 Mixing practice of MgO-C refractory batches as per the industrial practice.

Step	Mixing sequence	Mixing time
1	Large (4–6 mm) + coarse (1–3 mm) + intermediate (0.5–1 mm) MgO aggregates	5 min
2	Fine flake graphite powder + fine antioxidant additives (Al, Si, B_4C)	5 min
3	Phenolic resin binder (Novolac/Resol)	10 min
4	Fine MgO powder	20 min

aggregates. The corrosive damage of these refractories can be divided into the following four stages:

i) Air oxidation of the MgO-C refractory

This involves the direct oxidation of carbonaceous matter with oxygen containing air at temperatures lower than 1400°C, during the prolonged schedule of preheating, as per the following chemical reactions:

$$2C_{(graphite)} + O_{2(g)} = 2CO_{(g)} \tag{8.15}$$

$$2CO_{(g)} = CO_{2(g)} + C_{(soot)} \tag{8.16}$$

$$C_{(graphite)} + CO_{2(g)} = 2CO_{(g)} \tag{8.17}$$

The loss of carbon results in the formation of a porous decarburized layer (also known as oxidized layer) on the refractory hot-face, which leads to promoting slag penetration through hot-face of the slag zone MgO-C refractory.

ii) Slag penetration through the MgO-C refractory

This accounts for further loss of carbonaceous matter due to carbothermic reduction of MgO in the refractory structure as well as FeO in the penetrating slag at temperatures well above 1400°C, per the following chemical reactions:

$$C_{(graphite)} + MgO_{(s)} = Mg_{(g)} + CO_{(g)} \tag{8.18}$$

$$C_{(graphite)} + FeO_{(l)} = Fe_{(l)} + CO_{(g)} \tag{8.19}$$

Such carbon loss reactions considerably enhance the open porosity of the decarburized layer, leading to aggressive slag penetration into the refractory structure, indicating the beginning of slag attack.

iii) Dissolution of the MgO-C refractory matrix into the slag

At this stage of corrosion, the penetrated slag mainly reacts with various constituents of the matrix phase (MgO, $MgAl_2O_4$), forming new corrosive compounds that include: forsterite (Mg_2SiO_4), magnesio-wustite ($MgFe_2O_4$), monticellite ($CaMgSiO_4$, abbreviated as CMS phase), merwinite ($Ca_3MgSi_2O_8$, abbreviated as C_3MS_2 phase), calcium aluminates (CA: $CaO.Al_2O_3$, CA2: $CaO.2Al_2O_3$), calcium alumino silicates (CAS_2: $CaO.Al_2O_3.2SiO_2$), and complex spinels (formula: $(Mg,Mn,Fe)(Fe,Al)_2O_4$)) across the refractory-slag interface, per the following chemical reactions:

$$2MgO_{(refractory)} + SiO_{2(Slag)} = Mg_2SiO_{4(Corrosioncompound)} \tag{8.20}$$

$$MgO_{(Refractory)} + Fe_2O_{3(Slag)} = MgFe_2O_{4(Corrosion\ compound)} \quad (8.21)$$

$$MgO_{(Refractory)} + CaO_{(Slag)} + SiO_{2(Slag)} = CaMgSiO_{4(Corrosion\ compound)} \quad (8.22)$$

$$MgO_{(Refractory)} + 3CaO_{(Slag)} + 2SiO_{2(Slag)} = Ca_3MgSi_2O_{8(Corrosion\ compound)} \quad (8.23)$$

$$CaO_{(Slag)} + MgAl_2O_{4(Refractory)} = CaO.Al_2O_{3(Corrosion\ compound)} + MgO \quad (8.24)$$

$$CaO_{(Slag)} + 2MgAl_2O_{4(Refractory)} = CaO.2Al_2O_{3(Corrosion\ compound)} + 2MgO \quad (8.25)$$

$$CaO.Al_2O_{3(Corrosion\ compound)} + 2SiO_{2(Slag)} = CaO.Al_2O_3.2SiO_{2(Corrosion\ compound)} \quad (8.26)$$

$$Fe_xO_{(Slag)} + MnO_{(Slag)} + MgAl_2O_{4(Refractory)} = (Mg,Mn,Fe)(Fe,Al)_2\ O_{4(Corrosion\ compounds)} \quad (8.27)$$

The progressive formation of these corrosive compounds shifts the local slag-refractory interfaces to unattacked portions of the refractory, which directly results in dissolution of the matrix phase constituents into the slag.

iv) Dispersion/Washing of the MgO aggregates into the slag

The gradual dissolution of the matrix phase constituents into the infiltrated slag during the cyclic operation of ladles, leading to deterioration of the MgO aggregate packing structure and generation of new interface sites between corrosive slag and MgO refractory aggregates. The continuous action of this process, leading to dispersion/washing of MgO aggregates into the slag. This stage of corrosive damage is significant and results in MgO–C refractory erosion to a greater depth.

8.6.2 Alumina–Magnesia–Carbon Refractories

Graphite containing alumina-magnesia-carbon (Al_2O_3-MgO-C) refractories are an important class of steelmaking refractories in alumina–carbon system (Al_2O_3-C) system with end use in bottom-impact zone and barrel-metal line portion of secondary refining ladles, which are commonly working as ladle furnaces (LF) in non-vacuum conditions as well as vacuum degassing (VD/VAD) units. This is ascribed to the advantageous features of these refractories, including high refractoriness, good slag/metal corrosion resistance, better high-temperature strength, and excellent thermal shock resistance with good expansion characteristics (that counteracts brick joint-wear problem). An outline of manufacturing procedure [8, 9] as well as postmortem analysis of these refractories is given as follows. Commercially available resin bonded Al_2O_3-MgO-C refractories are generally fabricated with a total carbon content of 4–10 wt.%. The manufacturing process of Al_2O_3-MgO-C refractories is almost similar to that of the earlier discussed MgO-C refractories, as shown in the process flow diagram (Figure 8.26).

The composition formulation for the fabrication of these refractories, generally composed of alumina, magnesia, flake graphite as the chief carbon source, novolac/resol phenolic resins as organic binders, and antioxidant additives, including metal/carbide powders (e.g., Al, Si, B_4C). High-purity tabular/sintered Al_2O_3 and fused Al_2O_3 are the commonly selected Al_2O_3 sources. On the other hand, high-purity sintered MgO and fused MgO with desirable physiochemical characteristics (low B_2O_3 content, $CaO/SiO_2 > 2$, large periclase crystallite size) are the commonly selected MgO varieties. For the fabrication of Al_2O_3-MgO-C refractories with enhanced properties, a careful selection of particle size distribution is often important, which profoundly affects the extent of magnesium aluminate spinel ($MgAl_2O_4$) formation and associated brick expansion characteristics. The particle size distribution for these refractories formulation can be selected as large (3–5 mm), coarse (1–3 mm), intermediate (0.5–1 mm), and fine (< 500 μm) fractions of Al_2O_3 and MgO, combined with fine graphite powder as well as antioxidant additives. These graded ingredient materials were thoroughly mixed with a novolac/resol type organic binder in a pan mixer at ambient conditions for the preparation of a macro-homogeneous mix, as per the mixing sequence given in Table 8.10.

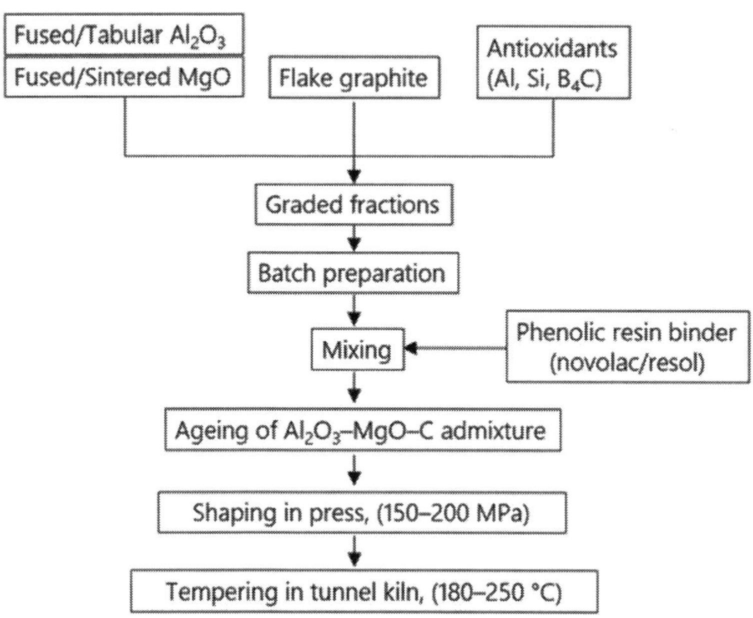

Figure 8.26 Process flow diagram for the fabrication of resin bonded Al_2O_3-MgO-C refractories.

Table 8.10 Mixing practice of Al_2O_3-MgO-C refractory batches as per the industrial practice.

Step	Mixing sequence	Mixing time
1	Large (3–5 mm) + coarse (1–3 mm) + intermediate (0.5–1 mm) Al_2O_3 / MgO aggregates	5 min
2	Fine flake graphite powder + fine antioxidant additives (Al, Si, B_4C)	5 min
3	Phenolic resin binder (Novolac/Resol)	10 min
4	Fine Al_2O_3 powder + fine MgO powder	20 min

The well-prepared homogeneous mix was aged at an ambient environment for a few hours and pressed to various shaped refractory brick structures at a uniaxial pressure of 150–200 MPa with the use of hydraulic or friction-screw press. The release of volatiles during binder burn-out and polymerization of resin, leading to the production of homogeneous formation of 3D-interlocking carbon-carbon chain network in all parts of the tempered refractory matrix. The shaped refractory bricks are tempered in a tunnel kiln at a temperature between 180°C and 250°C for a period of 24 hours, which results in resin bonded Al_2O_3-MgO-C refractories with enhanced properties.

A case study on postmortem analysis of Al_2O_3-MgO-C refractories working at ladle bottom/impact zone area, is discussed. The chemical analysis of ladle slag is given as (CaO: 55.7%, Al_2O_3: 27.2%, MgO: 8.3%, SiO_2: 7.7%, FeO: 0.53%, Cr_2O_3: 0.03%, MnO: 0.07%, basicity ~ 1.6) [26]. It can be noted that the matrix portion of this refractory consists of fine Al_2O_3, MgO, graphite, and in-situ formed reaction product $MgAl_2O_4$. On the other hand, the aggregate portion of this refractory contains large, coarse, intermediate grades of Al_2O_3 as well as MgO aggregates. The corrosive wear process of these refractories can be categorized into the following four stages:

i) Air oxidation of the Al_2O_3-MgO-C refractory

The air oxidation behavior of the bottom lined Al_2O_3-MgO-C refractories is similar to that of the MgO-C refractories due to common carbon phase constituents in their refractory matrix, during ladle pre-heating. In this case also, the formation of porous decarburized layer as per chemical reactions shown in Equations (8.1)–(8.3), promotes slag intrusion through hot-face of the refractory.

ii) Slag penetration through the Al_2O_3-MgO-C refractory

For the case of Al_2O_3-MgO-C refractories, they also exhibit a similar way of carbon loss as indicated by MgO–C system, which comes out from the chemical reactions associated with the reduction of MgO from refractory and FeO from slag, as shown in Equations (8.4)–(8.5). This accounts for common matrix phase constituents in both systems. Additionally, the carbothermic reduction of Al_2O_3 can occur at this time of slag attack. However, thermodynamic feasibility of such reaction is a function of many process variants, as discussed elsewhere [8]. This loss of carbon greatly increases the open porosity of the decarburized layer and promotes the slag penetration through hot-face of the Al_2O_3-MgO-C brick.

iii) Dissolution of the Al_2O_3-MgO-C refractory matrix into the slag

At this stage of slag attack, various constituents of the refractory matrix phase (Al_2O_3, MgO, $MgAl_2O_4$) react with the infiltrated slag, and results in the formation of new corrosive compounds, such as: calcium aluminates (CA: $CaO.Al_2O_3$, CA2: $CaO.2Al_2O_3$, CA6: $CaO.6Al_2O_3$), forsterite (Mg_2SiO_4), magnesio-wustite ($MgFe_2O_4$), monticellite ($CaMgSiO_4$, abbreviated as CMS phase), calcium alumino silicates (CAS_2: $CaO.Al_2O_3.2SiO_2$), and complex spinels (formula: $(Mg,Mn,Fe)(Al,Cr)_2O_4$)), as per the given chemical reactions:

$$CaO_{(Slag)} + Al_2O_{3(Refractory)} = CaO.Al_2O_{3(Corrosion\ compound)} \tag{8.28}$$

$$CaO_{(Slag)} + 2Al_2O_{3(Refractory)} = CaO.2Al_2O_{3(Corrosion\ compound)} \tag{8.29}$$

$$CaO_{(Slag)} + 6Al_2O_{3(Refractory)} = CaO.6Al_2O_{3(Corrosion\ compound)} \tag{8.30}$$

$$2MgO_{(Refractory)} + SiO_{2(Slag)} = Mg_2SiO_{4(Corrosion\ compound)} \tag{8.31}$$

$$MgO_{(Refractory)} + Fe_2O_{3(Slag)} = MgFe_2O_{4(Corrosion\ compound)} \tag{8.32}$$

$$MgO_{(Refractory)} + CaO_{(Slag)} + SiO_{2(Slag)} = CaMgSiO_{4(Corrosion\ compound)} \tag{8.33}$$

$$CaO_{(Slag)} + MgAl_2O_{4(Refractory)} = CaO.Al_2O_{3(Corrosion\ compound)} + MgO \tag{8.34}$$

$$CaO_{(Slag)} + 2MgAl_2O_{4(Refractory)} = CaO.2Al_2O_{3(Corrosion\ compound)} + 2MgO \tag{8.35}$$

$$CaO.Al_2O_{3(Corrosion\ compound)} + 2SiO_{2(Slag)} = CaO.Al_2O_3.2SiO_{2(Corrosion\ compound)} \tag{8.36}$$

$$\begin{aligned} FeO_{(Slag)} + MnO_{(Slag)} + Cr_2O_{3(Slag)} + MgAl_2O_{4(Refractory)} \\ = (Mg,Mn,Fe)(Al,Cr)_2O_{4(Corrosion\ compounds)} \end{aligned} \tag{8.37}$$

As the slag corrosion process progresses, the extent of formation of these corrosive compounds is significant. This shifts the local slag–refractory interface sites from the attacked portions to the unattacked portions of the refractory matrix, leading to the dissolution of the refractory matrix phase constituents into the slag.

iv) Dispersion/Washing of the Al_2O_3/MgO aggregates into the slag

During the campaign of the steel ladle, dissolution of the matrix phase constituents into the slag, is a continuous process and occurs to a certain extent. The progressive action of this matrix phase dissolution significantly weakens the packing structure formed between Al_2O_3-Al_2O_3, Al_2O_3-MgO, and MgO-MgO aggregates and generates the new local interfaces between the aggregate part of the refractory and corrosive slag. This kind of corrosive damage is determinantal, which results in dispersion/washing of Al_2O_3 as well as MgO aggregates into the slag. Such significant refractory losses can lead to greater refractory erosion.

8.6.3 Dolo–Carbon Refractories

Refractories in CaO-MgO system, for example doloma (CaO: MgO = 60:40, percentage by weight) find important applications in various secondary steelmaking vessels as well as many continuous casting equipments, including ladle furnace (LF), vacuum degassers (VD/VAD), decarburization converters (AOD/VOD), nozzles, shroud, and slide-plates. These refractories exhibit high refractoriness, excellent corrosion resistance against silicate slags, good thermal shock resistance, and high-temperature strength properties. Addition of graphite/carbon black to doloma, results in a new tempered class of unfired refractories as like MgO=C and Al_2O_3-MgO-C, namely carbon bonded doloma (Dolo-Carbon) refractories with significantly improved thermophysical, thermochemical, and thermomechanical properties. The classification, manufacturing procedure, and postmortem analysis of Dolo-Carbon refractories is outlined as follows [27, 28]. Two different classes of tempered Dolo-Carbon refractories are manufactured. The earlier class of tempered refractories are known as pitch bonded Dolo-Carbon refractories, which use coal tar/petroleum pitch as an organic binder to bond the doloma aggregates. On the other hand, the new class of these tempered refractories are known as resin bonded Dolo-Carbon refractories, which use organic binder like novolac/resol type phenolic resins for bonding the granular doloma aggregates. The recent steelmaking practices widely use resin bonded Dolo-Carbon products, which are known for their ease of processing, enhanced properties, low carcinogenic emissions, industrial hygiene, and eco-friendly nature over the pitch bonded materials. Commercially available resin bonded Dolo-Carbon refractories contains a total carbon content of 3–6 wt.%. The ingredient materials used for formulating the composition of these refractories are sintered dolo (CaO:MgO = 60:40, percentage by weight) as the source of doloma, graphite/carbon black as the chief sources of carbon, novolac/resol phenolic resins as organic binders, as shown in the process flow diagram (Figure 8.27).

The particle size distribution of these refractories generally consists of coarse (1–3 mm), intermediate (0.5–1 mm), and fine (< 500 μm) fractions of sintered

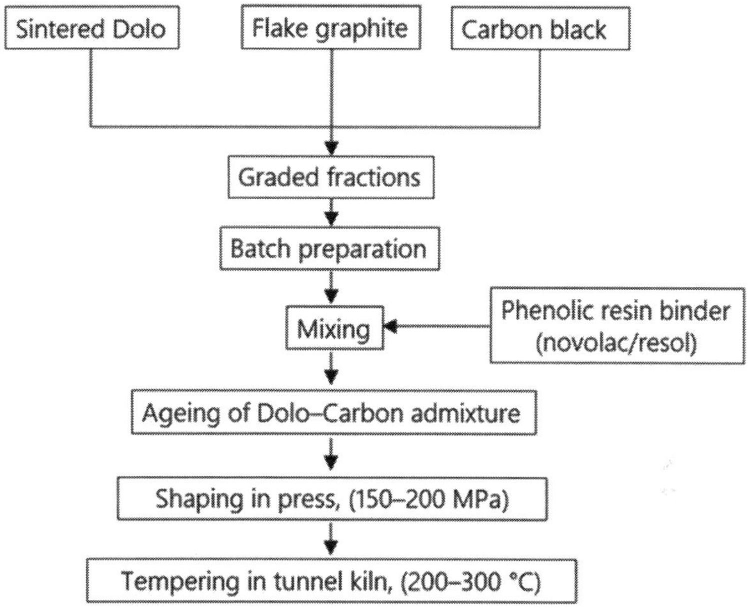

Figure 8.27 Process flow diagram for the fabrication of resin bonded Dolo-Carbon refractories.

dolo, and finer scale graphite as well as carbon black. The graded ingredient fractions are thoroughly mixed with the phenolic resin binder in a pan mixer at room-temperature to prepare a macro-homogeneous Dolo-Carbon mixture. The homogeneously blended mixture was aged at ambient conditions for few hours and uniaxially pressed to various refractory shapes at a pressure of 150–200 MPa in a hydraulic or friction–screw press. The pressed Dolo-Carbon refractories are tempered in the temperature range of 200–300°C for 24 hours in a tunnel kiln. The release of volatiles from binder burn-out as well as homogeneous formation of interlocking 3D carbon-carbon chain framework throughout the tempered matrix due to polymerization of phenolic resin binder during the stages of tempering, leading to produce the resin bonded Dolo-Carbon refractories with significantly enhanced properties. A case study on postmortem analysis of carbon bonded doloma (Dolo-Carbon) refractories working at ladle metal zone area, is detailed. The chemical analysis of the calcium aluminosilicate slag is given as (CaO: 56%, Al_2O_3: 33%, SiO_2: 11%, basicity ~ 1.3) [14]. It can be noted that the matrix portion of this refractory, majorly contains fine sintered dolo, graphite/carbon black. Whereas the aggregate portion of this refractory is made up totally of coarse and intermediate dolo aggregates. The corrosive damage of these refractories in steelmaking slags, can be grouped into the following four important stages:

i) Air oxidation of the Dolo-Carbon refractory

For the case of metal zone lined Dolo-Carbon refractories, their air oxidation behavior at the time of ladle preheating, also similar to that of the earlier discussed MgO-C and Al_2O_3-MgO-C refractories. The oxidative loss of carbon as per Equations (8.1)–(8.3), generates a porous decarburized layer on the hot-face of the refractory that promotes the penetration of slag into the Dolo-Carbon refractory structure.

ii) Slag penetration through the Dolo-Carbon refractory

The metal zone lined Dolo-Carbon refractories, also show similar carbothermic reduction reactions that include reduction of MgO from refractory (cf. Equation (8.4)) and reduction of FeO from slag (cf. Equation (8.5)), as exhibited by the earlier studied MgO-C as well as Al_2O_3-MgO-C refractories. This significant carbon losses, considerably opens up the air oxidized porous structure of decarburized layer, generated on the refractory hot-face. Such a porous hot-face allows an aggressive slag intrusion/penetration into the Dolo-Carbon refractory.

iii) Dissolution of the Dolo-Carbon refractory matrix into the slag

At this period of corrosive damage, the infiltrated slag reacts with all constituents of the matrix phase (CaO, MgO) and produces various forms of new corrosive compounds, including: calcium aluminates (CA: $CaO.Al_2O_3$, CA2: $CaO.2Al_2O_3$, CA6: $CaO.6Al_2O_3$, C12A7: $12CaO.7Al_2O_3$), calcium silicates (C2S: Ca_2SiO_4), calcium alumino silicates (CAS_2: $CaO.Al_2O_3.2SiO_2$), calcium magnesium silicates ($CaMgSiO_4$, abbreviated as CMS phase), forsterite (Mg_2SiO_4) and magnesium aluminate spinel ($MgAl_2O_4$), across the interface regions located between slag and refractory, as per the following chemical reactions, are given as:

$$CaO_{(Refractory)} + Al_2O_{3(Slag)} = CaO.Al_2O_{3(Corrosion\ compound)} \tag{8.38}$$

$$CaO_{(Refractory)} + 2Al_2O_{3(Slag)} = CaO.2Al_2O_{3(Corrosion\ compound)} \tag{8.39}$$

$$CaO_{(Refractory)} + 6Al_2O_{3(Slag)} = CaO.6Al_2O_{3(Corrosion\ compound)} \tag{8.40}$$

$$12CaO_{(Refractory)} + 7Al_2O_{3(Slag)} = 12CaO.7Al_2O_{3(Corrosion\ compound)} \tag{8.41}$$

$$CaO.Al_2O_{3(Corrosion\ compound)} + 2SiO_{2(Slag)} = CaO.Al_2O_3.2SiO_{2(Corrosion\ compound)} \tag{8.42}$$

$$2CaO_{(Refractory)} + SiO_{2(Slag)} = Ca_2SiO_{4(Corrosion\ compound)} \tag{8.43}$$

$$MgO_{(Refractory)} + CaO_{(Slag)} + SiO_{2(Slag)} = CaMgSiO_{4(Corrosion\ compound)} \tag{8.44}$$

$$2MgO_{(Refractory)} + SiO_{2(Slag)} = Mg_2SiO_{4(Corrosion\ compound)} \tag{8.45}$$

$$MgO_{(Refractory)} + Al_2O_{3(Slag)} = MgAl_2O_{4(Corrosion\ compound)} \tag{8.46}$$

There is an extended formation of these corrosive compounds across these interface reaction sites, with the progress of slag corrosive damage. This leads to shifting of reaction interface between slag and refractory from the slag attacked zone to unattacked zone of refractory, which directly results in the dissolution of the matrix phase constituents into the infiltrated slag.

iv) Dispersion/Washing of the MgO aggregates into the Slag

During the cyclic service of the steel ladle campaign, the dissolution of various constituents of the matrix phase into the slag, is a continuously occurring process. The progressive dissolution of matrix phase constituents into the slag, relatively deteriorates the packing structure formed between the MgO-MgO aggregates. This leads to the formation of relatively new weaker interfaces between the local MgO aggregate portions of the refractory and slag. This dissolution process is detrimental, leading to dispersion/washing of MgO aggregates into the slag. For this case, the loss of refractory is significant, resulting in substantial refractory erosion.

8.6.4 Magnesia–chrome (MgO-Cr_2O_3)

Four kinds of refractory magnesium chromite are recognized in Table 8.11. Clinker form either sintering or fusion of MgO and chromite.

The raw material chromite (commonly termed chromium ore) is naturally crushed and washed minimally. It is (Mg,Fe) $(Cr,Al)_2O_4$ complex solid solution spinel, but silicate impurities often exist. The directly bonded bricks are silicate-bonded bricks having 3–10 wt.% SiO_2. A directly bound brick is one in which MgO is directly attached to chromite without a silicate coating. More direct MgO-chromic spinel bands arise when the firing temperature increases and Fe_2O_3 and Cr_2O_3 diffuse into secondary spinel MgO grains, which are primary (rough material) and secondary (because of the processing reaction) spinel are present in the microstructure. Semi rebond bricks are manufactured similarly with sintered MgO and chromium ore. Rebonded bricks are made of fused material and are replenished at high (~1700°C) temperature. A new development is MgO-Cr_2O_3, a sintered brick with similar characteristics to fused brick with attractive properties. Current applications of MgO-Cr_2O_3 bricks in the hot areas of cement ovens,

Table 8.11 Classification of commercial magnesia-chrome bricks.

Type	Raw materials		
	MgO	Chromite	Fused magnesia-chromite
Silicate or direct bonded	✓	✓	✗
Semi-rebonded	✓	✓	✓
Rebonded	✗	✗	✓
Fused cast	✓	✓	✗

glass furnacing checkers, secondary stainless steel vacuum degassers, and nonferrous metal manufacturing such as copper converters. However, because of water solubility and the carcinogenic nature of Cr^{6+}, which may be formed in service, many of these applications are often MgO-Al_2O_3 spinel bricks considered as alternative refractories.

A different slag resistance can be expected while the direct bonded MgO-Cr_2O_3 refractories interact with two slags, such as C/S = 0·5 and C/S = 2·0. Usually, chromite is more resistant to attack in low slag C/S ratio than periclase. The periclase crystals erode and progressively disintegrate when the low C/S slag penetrates the refractory, floating away from the heated surface. In a recent study, Kyoden et al established a link between MgO-Cr_2O_3 and slag resistance [29]. Wiederhorn et al. hypothesized that adding a former network like Al_2O_3 to the brick leads to more viscous and, therefore less penetrating slag locally [30]. The addition of Fe-Cr to the mixture is another technique to enhance MgO-Cr_2O_3 slag resistance. The increase in the density and corrosion resistance of bricks may be enhanced by the oxidation of Fe–Cr powder during in-situ sintering. Thus, greater direct bonding and the usage of melting grain have enhanced MgO-Cr_2O_3 bricks' drain resistance. The chromite and MgO phases are more resistant to fundamental slag, and MgO-Cr_2O_3 refractories generally are more resistant to basic slags. While composition and additional components such as Fe_2O_3, Al_2O_3, and CaO varies, the primary and secondary chromites can be corroded at different speeds. For example, Fe may have more soluble chromium spinels on the slag even if the dissolution of Al_2O_3 may lead to viscous, local slag at the interface of refractory/slag to reduce penetration.

8.6.5 Spinel Bricks

$MgAl_2O_4$ spinel refractories have good thermal resistance and refractoriness and are increasingly employed in steel and cement sectors, which is partly driven by the need to replace linings containing chromium in MgO-Cr_2O_3 refractory. Al_2O_3–spinel systems are usually utilized in steel production, while MgO–spinel systems are now also used in secondary steel-manufacturing ladles. Spinel particles are commercially produced by fusing or sintering. Melting spinel frequently contains Na_2O, which might result in a discrete β-alumina phase. The corrosion resistance is improved by the fused MgO grain and fine Fe_2O_3 and TiO_2 added to the MgO–spinel bricks. Fine powder supplements achieve better texture due to the extra binding between MgO and spinels. The alumina component of the spinel has more easily been dissolved in slag than the chromic components in magnesia-chrome bricks. However, field tests reveal that CaO and Al_2O_3 rich slag interaction with MgO-spinel bricks have better corrosion resistance than MgO–Cr_2O_3. Spinel stoichiometry affects corrosion behavior. The high Al_2O_3 content spinel refractories have high steel slag penetration resistance. Such a phenomenon suggests a nonstoichiometric spinel defect structure with a charge overload that compensates for cation vacancies at octahedral locations.

A static corrosion study of MgO–spinel bricks was recently examined in the presence of CAS slag in the temperature range of 1400–1450 °C [31]. The corrosion mechanism was established through thermodynamic computations and postmortem analysis of corroded bricks. The fine MgO matrix was initially dissolved in local slag, reacting to secondary spinel with alumina already present in the slag. Further MgO in the slag solution and Al_2O_3 elimination through subsequent spinel precipitation leads to CMS and C_3MS_2 crystallization. The microstructure characterization disclosed the secondary spinel, CMS, C_3MS_2, and MgO formed in this sequence because of the change in the composition after slag penetration in brick. Such investigations underline the complicated interaction between taking cations from slag into solid phases (such as Fe and Mn into spinels) and the dissolving of solids in slag that alters the slag's local composition (initially fine reactive matrix phases). Eventually, this interaction facilitates the development of a liquid product that is in touch with solid refractory phases and causes penetration and corrosion.

References

1 Biswas, S. and Sarkar, D. (2020). *Introduction to Refractories for Iron-and Steelmaking*. Springer Nature.

2 Vert, T. (2016). *Refractory Material Selection for Steelmaking*. John Wiley & Sons.

3 https://www.rhimagnesita.com/wp-content/uploads/2019/11/steel-stainless-1909-en-190917-eb-mon.pdf

4 Eiji, M. and Takayasu, O. (1967). Low-carbon aluminum killed steel for high temperature applications. Sumitomo Metal Industries Ltd. U.S. Patent 3,357,822.

5 Yang, W., Cao, J., Wang, X.H., Xu, Z.R., and Yang, J. (2011). Investigation on non-metallic inclusions in LCAK steel produced by BOF-LF-FTSC production route. *J. Iron Steel Res. Int.* 18 (9): 6–12.

6 Beskow, K., Jonsson, L., Sichen, D., and Viswanathan, N.N. (2001). Study of the deoxidation of steel with aluminum wire injection in a gas-stirred ladle. *Metall. Mater. Trans. B* 32 (2): 319–328.

7 Singh, U., Anapagaddi, R., Mangal, S., Padmanabhan, K.A., and Singh, A.K. (2016). Multiphase modeling of bottom-stirred ladle for prediction of slag–steel interface and estimation of desulfurization behavior. *Metall. Mater. Trans. B* 47 (3): 1804–1816.

8 Heikkinen, E.P. and Fabritius, T. (2012). "Modelling of the refining processes in the production of ferrochrome and stainless steel, *Recent Researches in Metallurgical Engineering-From Extraction to Forming.*" *InTech* 186: 65–88. https://

cdn.intechopen.com/pdfs/32751/InTech-Modelling_of_the_refining_processes_in_the_production_of_ferrochrome_and_stainless_steel.pdf

9 Wei, J.H. and Zhu, D.P. (2002). Mathematical modeling of the argon-oxygen decarburization refining process of stainless steel: part II. Application of the model to industrial practice. *Metall. Mater. Trans. B* 33 (1): 121–127.

10 Li, Z.D., Zhou, S.T., Yang, C.F., and Yong, Q.L. (2019). High/very high cycle fatigue behaviors of medium carbon pearlitic wheel steels and the effects of microstructure and non-metallic inclusions. *Mater. Sci. Eng.: A* 764: 138208.

11 Korbáš, M. and Čamek, L. Evaluation of the results of steel purity achieved in steel production using sse method.

12 Capurro, C., Cerrutti, G., and Cicutti, C. (2015). Influence of vacuum degassing on steel cleanliness. In *Conference: 9th International Conference on Clean Steel* Budapest, Hungary.

13 Hallberg, M., TLI, J., and Pg, J. (2004). A new approach to using modelling for on-line prediction of sulphur and hydrogen removal during ladle refining. *ISIJ Int.* 44 (8): 1318–1327.

14 Moriya, T. and Tawara, M. (1977). On the operations of LD-VAD process and product qualities. *Tetsu-to-Hagané* 63 (13): 2070–2076.

15 Jha, K.N., Sardar, M.K., Jha, N.N., and Chakraborty, S. (2003). Hydrogen control during steel making for medium carbon wheels. *Scand. J. Metall.* 32 (6): 296–300.

16 Karouni, F., Wynne, B.P., Talamantes-Silva, J., and Phillips, S. (2018). Hydrogen degassing in a vacuum arc degasser using a three-phase eulerian method and discrete population balance model. *Steel Res. Int.* 89 (5): 1700550.

17 Kitamura, T., Miyamoto, K., and Tsujino, R. (1996). Mathematical reaction model for nitrogen in vacuum degasser desorption and decarburization. *ISIJ Int.* 36: 395–401.

18 Emi, T. (2014). Optimizing steelmaking system for quality steel mass production for sustainable future of steel industry. *Steel Res. Int.* 85 (8): 1274–1282.

19 Poirier, J. (2015). A review: influence of refractories on steel quality. *Metall. Res. Technol.* 112: 410.

20 Dutta, S.K. and Chokshi, Y.B. (2020). *Basic Concepts of Iron and Steel Making*. Springer Nature Singapore Pte Ltd.

21 Turkdogan, E.T. Fundamentals of Steelmaking (The Institute of Materials, London, 1996, Paperback edition first published in 2010 by Maney Publishing, Leeds, UK), 250.

22 Pluschkell, W. (1981). Grundoperationen pf annenmetallurgischer Prozesse. *Stahl Und Eisen* 101 (97–103): in German.

23 Mazumdar, D., Guthrie, R.I.L.(1985). Hydrodynamic modeling of some gas injection procedures in ladle metallurgy operations. *Metall Mater Trans B* 16, 83–90.

24 Nath, N.K., Mandal, K., Singh, A.K., Basu, B., Bhanu, C., Kumar, S., and Ghosh, A. (2006). —Ladle furnace on-line reckoner for prediction and control of steel temperature and composition. *Ironmak. Steelmak.* 33 (2): 140–150.

25 (2016 July). 2016 IJLERA – All rights reserved 5 | page ladle furnace refractory lining: a review Dashrath Singh Kathait. *Int. J. Latest Eng. Res. Appl (IJLERA)* 01 (04): 05–12. www.ijlera.com. *ISSN: 2455-7137*. Dashrath Singh Kathait

26 Calvo, W.A., Pena, P., and Martinez, A.G.T. (2019). Post-mortem analysis of alumina-magnesia-carbon refractory bricks used in steelmaking ladles. *Ceram. Int.* 45 (1): 185–196.

27 Schacht, C. (ed.) (2004). *Refractories Handbook*, 178. CRC Press.

28 Le Coq, X., Dupré, B., Gleitzer, C., Adam, R., Scheidt, F., and Tassot, P. (1990). Slag corrosion of dolomite-carbon refractories. *Steel Res.* 61 (12): 593–597.

29 Kyoden, H., Ichikawa, K., and Iwado, H. (1985). Taikabutsu (Refractories). 37: 284–290.

30 Wiederhorn, M. and Krause, F. (1988). Effect of coal slag on the microstructure and creep behavior of a magnesium-chromite refractory, *Am. Ceram. Soc. Bull.* 67: 1201–1210.

31 Goto, K., Argent, B.B., and Lee, W.E. (1997). Corrosion of MgO-MgAl2O4 Spinel Refractory Bricks by Calcium Aluminosilicate Slag, *J. Am. Ceram. Soc.* 80: 461–471.

9

Precast and Purging System

9.1 Introduction

Shaped (bricks) and unshaped (monolithics) are well-known refractory materials classifications. Bricks and monolithics have their pros and cons in refractory applications. Usually, shaped refractories can be further classified into two shapes, standard shapes and special shapes. Refractory manufacturers and consumers have recognized several standard shapes for different kilns and vessels. However, special shapes have specific designs and particularly suitable zones for kilns and vessels. Herein, a successful design depends on both the combination of shape and composition. The same shape may not be suitable for different vessels. Refractory bricks are almost always made by pressing technique and maintain uniform properties. High thermal strength, resistance to mechanical impact, and chemical attack impart excellent performance for critical regions like archways, belly-band areas, and many more. Predefined properties can be achieved by adjusting design, composition, and processing conditions. Special shapes have relatively critical shapes, and demand is relatively low compared to standard shaped refractories. It is most often made by hand molding and exhibits a variety of properties.

Unshaped refractories are commonly known as monolithic, and they form jointless lining with a tailormade shape in an application or predefined shape (precast). While considering the precast shape, castable is the prime concern to discuss. Castables made of aggregate and binders capable of flowing are favorably used to manufacture precast shapes. Despite castables, several monolithics such as mortars, plastic refractories, gunning mixes, ramming mixes, fettling mixes, and spray mass are common monolithics for steelmaking refractories. Castable provides better volume stability, low permeability, thermal spalling, mechanical impact, and wear resistance once proper casting and installation are done. However, compared to bricks, mismatched thermal expansion behavior results in

Fundamental Design of Steelmaking Refractories, First Edition. Debasish Sarkar.

different thermomechanical responses. Rapid repair and maintenance, even during operation, is one of the advantages of castable. Instead of on-site castable applications for different zones, several precast shapes are well recognized for different applications, and their design parameters are essential for better performance and premature failure.

While considering different precast shapes, a purging plug is a tricky and important precast-prefired refractory for the steelmaking process. All together, design and operational features are discussed from the perspective of purging plugs designing.

9.2 Composition Design of Castables

Alumina is the prime raw material for steelmaking precast shapes. A wide range of alumina castables containing different alumina from 50 wt.% to 95 wt.% is available for conventional and low cement to cast either in-situ or make predefined precast shapes for different zones. Usually, the prime ingredients are the alumina aggregates consisting of different grading to maintain packing and flowability, a predefined quantity of fines to adjust the targeted composition and enhance the mobility, cement, or other liquid as binder, deflocculant, and water. Casted product is hardened through hydration of cement, and intermediate temperature reduces the strength, and further high-temperature heating enhances the strength of the entire shape.

A rapid increase in the demand for unmolded refractories should be separated among fundamental achievements in the field of refractories for the last decade. This progress is significantly connected with the creation and use in this period of refractory concretes of the low-cement type (LCRC), ultralow cement concretes (ULCRC), and also cement-free refractory concretes (CFRC) prepared using nontraditional refractory binders. Advantages of low cement castable over high cement castables:

- High alumina castable is made of more alumina enriched aggregates and cement and uses less cement that eventually reduces the calcium content compared to conventional castable. It minimizes the low melting phase and improves the high-temperature performance.
- Low cement consumes less water up to 6–7 wt.% with improved flowability of the entire mix, resulting in low porosity and high dense casting product.
- Low cement and high alumina castable exhibit high room temperature curing strength and continue at medium temperature. The resultant strength of high alumina low cement castable becomes 2–3 times higher than conventional castable.
- It exhibits minimum shrinkage after drying as well as volume stability at application temperature.

In brief, the high alumina castable refractory production process follows as:

Material selection - > Ingredients - > Mixing - > Packaging - > Storage.

It can be summarized as:

- Selection of raw materials to fulfill the specification and performance.
- Precious calculation of aggregates grading, powders, binders, and additives in consideration of raw material formula and control the quality of every component.
- Weighing and homogenous dry mixing of aggregates and powders through continuous stirring and subsequent addition of binders and additives and moisture-proof packing.

However, certain precautions for using high alumina castable refractory for on-site casting or making precast shapes are essential:

- Before water addition to make a high viscous flowable mass of the castable, the dry refractory mix should stir by an automatic mixer for a while to ensure homogenization and removal of air entrapment. Optimum water content is required to avoid the properties' deterioration after excess water addition.
- The binder tends to harden with time once it contacts moisture; thus it is essential to use the water mixed castable within 30 minutes to achieve the best performance.
- Proper anchoring and careful casting of a predefined design thickness are essential for on-site casting.
- After casting, an optimum preheating rate and time is needed to maintain. Rapid heating may expedite the crack formation.
- A careful inspection can ensure the surface characteristics after proper demolding. Any defect formation may demand a rework to ensure the performance of the casting body.

High performance index of castable depends on the density, purity, and morphology of starting raw materials, grading of materials to achieve optimum packing and excellent rheology during casting, and binder quality and quantity for initial strength and resultant performance. However, binder free compositions have resulted in remarkable gains in refractoriness and mechanical properties at high temperatures by decreasing the amount of low-melting-point phases in the alumina–silica–calcia ternary system.

Conventional high Al_2O_3 refractory castables containing 10–25 wt.% calcium aluminate or high alumina cement (HAC) have, for the time being, many service restrictions because of low strength caused by a breakdown of the hydraulic bond at temperatures as low as 200–800 °C. This was one of the reasons for developing low, ultralow, and no cement castables (LCC, ULCC, NCC) with HAC contents of

5–8%, 2–4%, and zero, respectively. These castables have many advantages over conventional types: they have lower CaO content, low total porosity, and better slag corrosion resistance, as well as higher cold and hot strengths and improved abrasion and thermal shock resistance. Substituting HAC with ultrafine silica and/or alumina powders as binders in these castables generally improves their thermomechanical properties with low additional production costs. It has led to their use in many refractory applications, including steel and nonferrous melting furnaces and rotary cement kilns. In LCCs containing fine silica powder in their matrices, a drop in the hot modulus of a rupture usually occurs above 1200 °C owing to liquid phase formation within the CaO-Al_2O_3-SiO_2 system. This drop-in strength can be improved by adding more fine alumina powder at the expense of fine silica. Herein, the system leads to the formation of mullite ($3Al_2O_3.2SiO_2$) as a network of needles at 1300 °C, strengthening the microstructure of the binding matrix at high temperature [1].

9.2.1 Choice of Raw Materials and Properties

The properties of refractory concrete or castable are primarily dependent upon the choice of refractory aggregate and hydraulic cement used. Thus, a wide variety of aggregates are used in refractory castables. A typical list of starting raw materials is given in Table 9.1.

In principle, several refractory grains can be used in refractory castable as aggregates, but in practice, most aggregates contain mainly alumina and silica in various forms. Starting from natural to synthetic materials has different chemical compositions and properties that eventually control the resultant properties of castable. A broad spectrum of synthetic alumina raw materials has been developed, in which tabular alumina and other corundum aggregates, especially white fused alumina, calcined aluminas, and reactive aluminas calcium-aluminate cement. A new family of multi-modal reactive aluminas, dispersing alumina, and hydratable alumina binders are in the forefront to develop wide range of monolitihics. Furthermore, high-purity magnesium-aluminate spinels have facilities to achieve a higher level of hot strength and slag corrosion resistance of corundum containing refractory.

Choice of the ingredients is the significant root cause to achieving the mobility of the entire refractory mix with water for adequate placement and filling up each and every portion in mold or cavity. The addition of more water enhances both the flow and porosity, resulting in a reduction of strength. The desired flowability may be obtained by maintaining the separation of coarser particles from one another by using the suspension of optimum content of fines and microfine additives in combination with deflocculant. This combination helps the flow and high-temperature properties of monolithic refractories.

Table 9.1 Different quality of fundamental raw materials and their chemical compositions and density.

Raw materials	Raw material processing/synthesis	Chemical analysis	Density (gm/cc)
Calcined Bauxite	Raw bauxite is preferentially calcined in a rotary kiln at a temperature of 1700 ± 50 °C to remove moisture and other organic substances. The obtained material is crushed and ground to fulfill the grading for castables.	Chinese bauxite Al_2O_3-86.5%; Fe_2O_3-1.6%; SiO_2-7.01%; TiO_2-3.59%; MgO + CaO-0.45%; K_2O + Na_2O-0.15%	3.01
WTA	Tabular alumina is produced by sintering ball-formed, intermediately burned calcined alumina at a temperature just under the 2040 °C melting point of aluminum oxide.	Al_2O_3-99.34%; Fe_2O_3-0.035% SiO_2-0.03%; Na_2O + K_2O-0.15%	3.61
WFA	White Fused Alumina (WFA) is obtained from the fusion of high purity calcined alumina in electric arc furnaces. Particle morphology found to be an angular type for WFA.	Al_2O_3-99.78%; Fe_2O_3-0.04%; SiO_2-0.02%; Na_2O-0.16%	3.5–3.95
Reactive alumina	It is a fully ground calcined alumina with a substantial portion made of primary crystals less than 1 μm. In their making, reactive alumina requires intensive grinding, such as batch dry grinding similar ceramic lining and media for more than 24 h.	Al_2O_3-99.8%; Fe_2O_3-0.02%; SiO_2-0.02%; B_2O_3-0.01%; CaO-0.03%; Na_2O-0.07%; MgO-0.04%	
Microfine silica	Fumed silica is synthesized by the pyrolysis method in which silicon tetrachloride reacts with oxygen in a flame, and the SiO_2 seed grows in size or aggregates. Spherical shaped silica particles of varying sizes.	SiO_2-98%; C_{free}-0.5–0.7%; Ash-0.5–1%	0.25–0.35
High alumina cement	The method of manufacture of HAC is by sinter clinker process. Conventionally, high alumina cement is obtained by fusing or sintering a mixture of suitable proportions of argillaceous and calcareous materials such as CaO or $CaCO_3$ and alumina (Al_2O_3) at temperatures above 1500 °C and subsequent grinding.	Al_2O_3-79.6%; Fe_2O_3-0%; SiO_2-0.4%; CaO-19.8%; TiO_2-0.1%	

9.2.1.1 Alumina Aggregates

Tabular alumina is a high-purity synthetic-corundum densified without additives through sintering and recrystallization of calcined alumina at a maximum temperature above 2073 K. The name tabular is descriptive of the well-developed (40–400 μm), tablet-shaped coarse α-Al_2O_3 (corundum) crystals usually exhibit stepped growth habits. High purity $Al_2O_3 > 99.5$ wt.%, tabular alumina exhibits:

- Large 40–400 μm interwoven corundum crystals
- Porosity mostly closed
- Very little open porosity (only 2–3%) and consequent low water absorption (<1.5%)
- Absence of microcracks
- Extreme hardness (9 in Mohs scale) of the ultimate corundum crystal
- High melting point (2323 K)
- High grain strength and thermal shock resistance are attributed to the above characteristics of the tabular alumina microstructure
- No hydration, resistance to most alkalis and mineral acids except hydrofluoric acid and phosphoric acids, the latter used for phosphate-bonded corundum refractories
- High thermal conductivity and electrical resistivity

Competitive data in Table 9.2 shows excellent crushing strength, and strength retention (after thermal shock cycles) of tabular alumina grains relative to white fused and brown fused alumina grains. Tabular alumina, sintered alumina, and white fused alumina are made from alumina, and brown fused alumina (BFA) is

Table 9.2 Thermal shock resistance and grain strength of corundum aggregates (5 mm grain).

Aggregate	Thermal spalling resistance (%) undamaged cycles 20-1300-20 °C			Grain crushing strength (kg) after thermal shocking cycles 20-1300-20 °C			Grain strength breaking load [kg]	Apparent specific gravity [g/cm³]
	10 Cycles	20 Cycles	30 Cycles	10 Cycles	20 Cycles	30 Cycles		
Tabular alumina	95	87	73	170	118	80	296	3.66
White fused	68	19	0	21	4	0	105	3.89
Brown fused	62	10	0	38	5	0	195	3.97
Spinel	82	53	20	43	30	12	242	3.26

made from pre-calcined bauxite. Sintered alumina has relatively finer and more uniform (~10–20 μm) crystals, usually in the presence of grain growth inhibitor MgO. It has relatively low thermal shock resistance compared to tabular alumina.

White fused alumina (WFA) is made by fusing alumina feedstock through carbon electrodes. During cooling of fused clinker, soda is segregated in the upper central portion of the clinker as β-alumina and separated carefully to avoid soda percentage increase in the finer size of white fused alumina. Being made of large single crystals, grains of white fused alumina usually contain microcracks induced during crushing and reducing grain strength. WFA has nearly the same total porosity as tabular alumina but persists in more open porosity than TA.

Brown fused alumina is made from fused bauxite, whose nominal Al_2O_3 content is 95 wt.%, and is usually prepared through electrofusion of pre-calcined bauxite. This process facilitates the reduction of oxides of Si and Fe to metal and forms as ferrosilicon. Iron scrap is added to enhance the ferrosilicon separation process. TiO_2 is difficult to remove, as alumina is being reduced as well. There is the possibility of retaining residual aluminum carbide that reacts with water in castable and generates methane.

Reactive alumina results from fully ground calcined alumina containing 20 to >90% primary crystals less than 1 micron. It is more reactive in the sense that they readily sinter to the highest achievable density at a temperature lower by 100–200 K than is required for sintering powders of large crystal size and less reactivity. In castable, it facilitates mullite formation in the presence of microfine silica and CA6 formation.

9.2.1.2 Microfine Silica and Deflocculants

Although the overall composition of microsilica is SiO_2, the surface of a microsilica particle is not only siloxane; rather, it is partially hydroxylated and hydrated. Usually, microsilica surface is covered by silanol groups, making the microsilica easy to disperse in aqueous systems. Due to the dissociation of the silanol groups, the negative surface charge increases with pH up to approximately 7. At higher pH, more than 7, the zeta potential flattens out, and the microsilica starts to dissolve at higher pH.

Refractory castable can be visualized as an aggregate structure with voids filled by an aqueous slurry of microsilica, cement, and other fines. The negative charge of the microsilica makes it vulnerable to attack by cations, which cause gelling or solidification of the system. Different cations can be used for gelling microsilica slurry. The characteristics of the resultant slurry are dependent on the charge and size of the cation that attaches. Monovalent cation forms soft gelation and is easy to liquefy by agitation, while polyvalent cation develops a three-dimensional network that appears harder and brittle. Cations are generally available from cement and aggregate. The initial setting of most castables is closely related to gelling of the microsilica containing the slurry phase.

In the case of ultralow and no-cement systems with microsilica, most of the setting depends on this solidification, although it is often triggered by small amounts of additives that promote the gelling. Such additives can be small quantities of cement. The absorption of polyvalent cations on the silanol sites is most likely a dynamic process and requires a certain concentration in the water phase before absorption gets significant.

The tendency to react with cations to produce a more or less rigid gel creates a problem if used in densely packed castable systems where cement is present. The calcium and aluminate cations from the cement dissolve and attack the otherwise fluid bond phase. The calcium ions must be prevented from absorbing the microsilica to prevent this coagulation. Such behavior may be done at pH below 5, but as this may affect the setting, adding a proper amount of a polyelectrolyte or a surface-active agent is common. The surface active agent, normally termed deflocculant is believed to absorb to the surfaces; thus preventing the absorption of calcium, but it also creates an equal, often negative charge of the particles of the bond system. As the particles then repel each other, casting can occur at reduced water additions. Deflocculants commonly used are polyphosphates, e.g., sodium hexameta phosphate (Calgon) at approximately 0.2 wt.% addition level, and polyacrylates (Darvan 811D, 0.05 wt.%). The deflocculating mechanisms involved normally fall into two main groups or combinations thereof:

1) Steric stabilization, the deflocculant attaches to the fines, and its size prevents agglomeration of particles.
2) Electrostatic stabilization, the deflocculant creates an equal electrical charge on the surface of the fines. Electrostatic repulsion prevents particles from agglomeration.
3) Combination of 1 and 2.

Additional to the above mechanisms, deflocculants make it possible to maintain the flow of refractory castables for some time, sufficient to place the castable before setting (gelling) commences. Sometimes setting may get adversely affected, though, giving unacceptably long or short set times. Several mechanisms have explained this. For a long set time with polycarboxylate ethers, one explanation is that the surface of the cement gets shielded by the side chains and that longer side chains consequently retard more than shorter ones.

9.2.1.3 Deflocculants and Flow of Castable

LCC based on white fused alumina with 6% CAC and 8% microsilica is considered a representative castable composition to check the quality of microsilica or cement or other components of low cement castables. The castable is composed to follow an Andreasen type PSD with a q-value close to 0.25, giving flow properties close to self-flowing at 4.15 wt.% water (=13 volume %). The standard LCC mix (Table 9.3) was used with the variable deflocculant (FS20) to understand the castable flow.

Table 9.3 Standard LCC castable composition.

Constituents	[wt.%]
Microsilica (18 or 24 m^2/g)	8
CA—14[wt.%]	6
White fused alumina. 3–5 mm	10
White fused alumina. 0.5–3 mm	32
White fused alumina. 0–0.5 mm	16
White fused alumina. −74 micron	12
Calcined alumina: CT 9FG	16
Deflocculant: Castament FS20	0.01–0.1*
Water [wt.%]	4.15

*Standard dosage, 0.05%

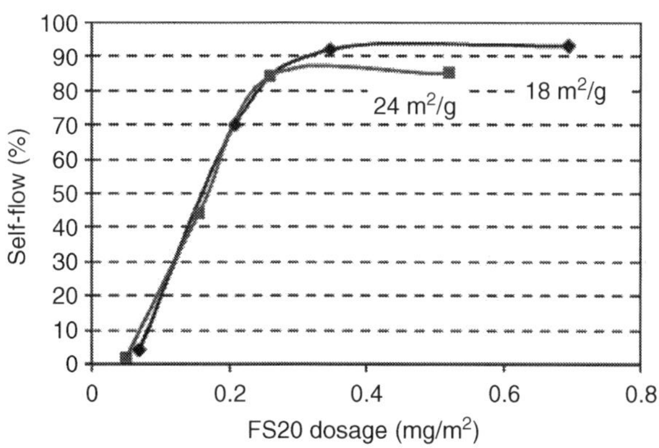

Figure 9.1 Self-flow of WFA-based LCC as a function of deflocculant addition [2]. Credit: Metamin Mümessillik Sanayi ve Ticaret A.Ş

Two different microsilica qualities with surface area of 18 and 24 m^2/g were used in this study. In Table 9.3, the composition of this mix is given. In Figure 9.1, the resultant self-flow values are shown as a function of deflocculant addition per microsilica surface area (mg/m^2) [2].

The two microsilica qualities give very similar results when expressed as a function of surface area, and only the maximum self-flow differs somewhat. Although lower self-flow was obtained with the microsilica with the higher surface area, and direction of lower flow at a higher surface area beyond optimum dosage of deflocculant is attributed to the coagulation and bit of lowering the flow behavior of castable.

9.2.1.4 Deflocculant Type and Dosage

Above, a good correlation between microsilica surface area, flow, and amount of additive was shown. Those findings were based on flow measurements of LCC based on fused alumina, but the same correlations exist for bauxite based systems.

In Figure 9.2a, the self-flow of castables 1 and 2 (Table 9.4) is presented as a function of the addition of dispersant in mg per square meter of microsilica surface area. In practical terms, an addition of 1 mg/m^2 equals an addition level of approximately 0.2%. Earlier results on white fused alumina systems (Table 9.3) have shown similar dependencies for FS20 and Calgon, as shown in Figure 9.2b.

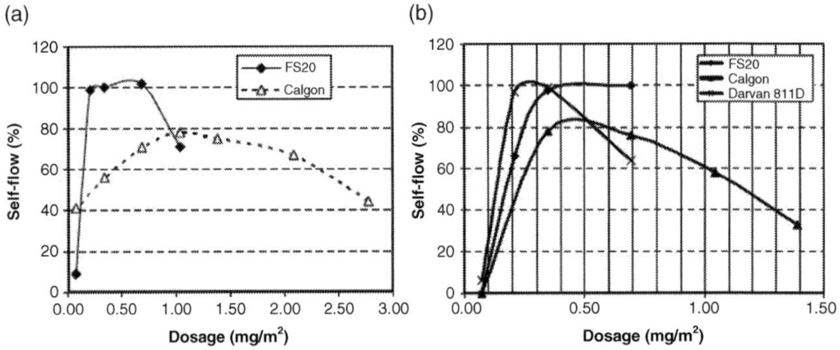

Figure 9.2 (a) Effect of dosage of dispersants on self-flow of bauxite-based LCC (number 1 and 2 in Table 9.4), (b) Effect of dispersant addition on self-flow of white fused alumina-based castables with 6% cement and 8% microsilica, 4.15% water [2].

Table 9.4 Bauxite-based LCC compositions.

(%)	1	2
Chinese bauxite 1–4 mm	30	30
Chinese bauxite 0–1 mm	30	30
Fused Al. −74 mic	18	18
Secar 71	6	6
Calcined alumina	8	8
Microsilica (971U)	8	8
FS20	0.01–0.015	-
Calgon	-	0.01–0.40
Water	5.50	5.50

While comparing Figures 9.2a and 9.2b, the characteristic self-flow versus dosage behavior appearance for both FS20 and Calgon is quite similar despite being skewed. The slight distortion in the curve during Calgon addition because of the probable reaction of phosphate with impurities of the bauxite implies higher dosages compared to the white fused alumina system. In addition, Darvan 811D is a polyacrylate, whereas the FS20 is a carboxylate-ether and molecular interaction differs at an optimum level and a flattened plateau results in FS20. Such incidence is attributed to the mode of stabilization of colloidal slurries of the two different additives. Darvan is by electrostatic repulsion, FS20 by a combined steric and electrostatic mechanism.

9.2.2 Choice of Binders

Calcium aluminate cement (CAC), hydratable alumina (HA) and colloidal silica (CS) are the prime binders for low cement to no-cement castables. A particular binder-based castable is being used in consideration of the application zone.

Portland cement is used at a maximum of 920 K but starts setting more rapidly than calcium aluminate cement. However, high alumina cement requires only 24 h to develop 70–80% full strength when properly cured. This characteristic of CAC has been utilized in non-refractory applications such as runway repair in the airfield. The calcium aluminate cements are mixtures of different mineral phases. Al_2O_3 content varies low (39–50 wt.% Al_2O_3), intermediate (55–66 wt.% Al_2O_3) and high purity (70–90 wt.% Al_2O_3) in different grades of CAC, in which iron and silica adversely affect the cement properties. The common impurities iron and silica are responsible for controlling the service life in the range of 1700–1820 K. The most important hydraulic phase is mono-calciumaluminate-CA. It directly influences the setting and hardening properties of cement or castable. Traditional refractory grade CAC contains phases like CA_2, C_2AS, $C_{12}A_7$, C_2S, C_2AS, and C_4AF. These different CA-phases exhibit different rates of hydration or setting characteristics. CA-phases hydrate sufficiently fast, CA_2 has moderate hydration rate, and $C_{12}A_7$ hydrates very rapidly and can be used to control the setting rate of CA-cements when used in small quantities. Excess free lime results in undesirable effects such as volume expansion, increased setting time or reduced strength; thus controlling free lime (CaO) less than 0.3 wt.% is desirable. C_2S, C_2AS, and C_4AF are hydrated during curing and dehydrated during drying and the initial firing stage. They make an insignificant contribution to strength development during curing and drying. However, these react with aggregate at elevated temperature and form low melting phases that adversely affect the refractoriness and hot strength of castable. Low and ultralow cement castable are usually formulated with a 70 wt.% Al_2O_3 cement that is properly dense clinker with no admixture. However, additives like, set control and property modifiers may be blended according to the demand of customers. When reacting with water, CA, CA_2, and $C_{12}A_7$ form the same CA hydrated phases and contribute to the strength development of refractory castables. The ambient temperature significantly impacts working time and the amount of water required for obtaining workable consistency and desired flow properties. With increasing temperature, all three phases transform into metastable (CAH_{10} and C_2AH_8) and stable (C_3AH_6) phases. At the same time, alumina gel gradually crystallizes between 27 °C and 32 °C. The system follows different stages of the reaction sequence, as shown in cumulative Equation (9.1).

$$\begin{aligned}&C_{12}A_7/CA/CA_2 + H_2O, < 294\ K\\&CAH_{10}(\text{metastable hexagonal prisms}) + AH_x(\text{gel}), (< 21\ °C)\ 294 - 308\ K\\&C_2AH_8(\text{metastable hexagonal plates}) + AH_x(\text{gel}) + AH_3(\text{crystalline}), (21 - 35\ °C) > 308\ K\\&C_3AH_6(\text{stable cubic trapezohedras}) + AH_3(\text{crystalline}), (> 35\ °C)\end{aligned} \tag{9.1}$$

Hydration reactions are exothermic. In the time–temperature profile Figure 9.3, a remarkable temperature increment is observed beyond 150 minutes, and maximum temperature upto 45 °C is observed and developed enough green strength for demolding after 260 minutes.

The casted concrete is dried at 104–110 °C to remove all free water and subsequent metastable phases form at different temperatures (°C): 200–350 ($C1_2A_7$), 600–1000 (CA + A), 1000–1300 (CA_2 + A), and 1400–1650 (CA_6 + A). Herein, the phases transform to cubic hydrate C_3AH_6 and crystalline AH_3

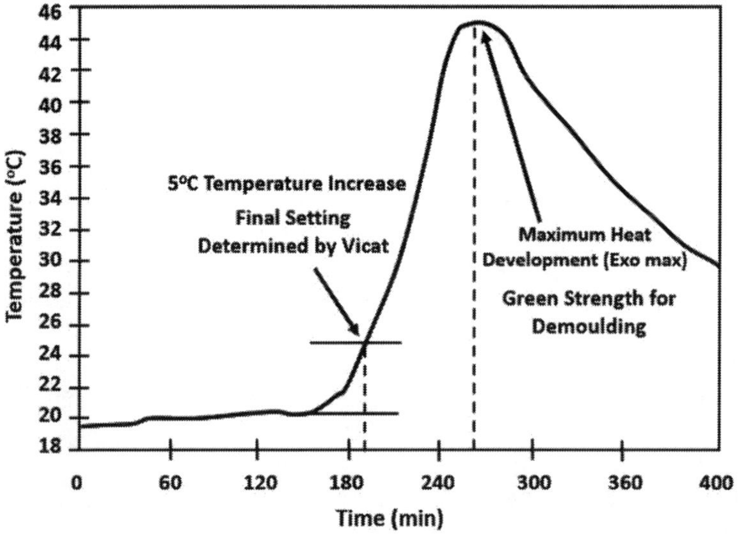

Figure 9.3 Exothermic reaction in 80% tabular alumina 20% cement (70% alumina) [3].

(gibbsite), resulting in compressive strength increment. Upon heating, hydrated cement compounds and alumina hydrates in the cured and dried concrete lose their water of crystallization, and lime and alumina reappear. As the temperature increases, the lime and alumina react according to the following reaction sequence, similar to the sintering of cement clinker. When a sufficient amount of alumina is available in the matrix or aggregates, CA_2 reacts with alumina to form non-hydrating, refractory CA_6 ($CA_6 + Al_2O_3$, mp 2103K). Through the sequence of CA → CA_2 → CA_6, where a ceramic bond is formed between the matrix and the aggregates.

Compared to high cement castable containing 2.5–6.0% CaO (10–15% HAC), an advanced high alumina cement hydraulically solidifying concretes with a minimum content of CaO, and advantageous factor at high-temperature application. With employing such low CaO containing cement castable (70–80 wt.% Al_2O_3 and 18–28 wt.% CaO) is classified into the following types:

- Low cement castable (LCC) containing 1.5–2.5 wt.% CaO
- Ultralow cement castable (ULCC) containing 0.2–1.5 wt.% CaO
- No cement castable (NCC) containing less than 0.2 wt.% CaO

Figure 9.4 represents the influence of CaO content on various properties. High CaO enhances the moisture content (*curve 1*); thus castable demands ultrafine particles, complex binders, and additives to minimize the water requirement for adequate rheological (thixotropic) properties for casting. An increase in content in concretes of very fine powders (*curve 2*) makes it possible to reduce the CaO content, but this simultaneously leads to a reduction in strength at 100°C, σ_{100} (*curve 4*). A significant effect of CaO content (*curves 4, 5*) on concrete mechanical strength is due to hydration and dehydration processes. A sharp drop in concrete strength with a marked CaO content is caused by an increase in porosity formed due to dehydration (*curve 3*). Concretes with a reduced CaO content correspond to a more finely capillary structure (*curve 6*) caused by both the grain size composition and increased density. There are also typically higher temperatures for maximum usage and corrosion resistance (*curves 7, 8*). This effect is achieved with increased concrete refractoriness and their fine capillary structure.

Concrete mixes require intensive mixing followed by optimum vibration treatment to acquire the so-called activated fluidity, making it possible to fill a concrete volume well. The effect of thixotropic dilution in a mixture with vibration molding and subsequent (at rest) their structure formation under conditions of low volume fraction of liquid and certain hydration of HAC makes it possible to prepare a monolithic lining in a comparatively short time.

Some monolithic formulations containing microsilica make CA-cements less desirable as there is a possibility of forming calcium aluminum silicates that melt at low temperatures. In certain metallurgical applications, eliminating both silica

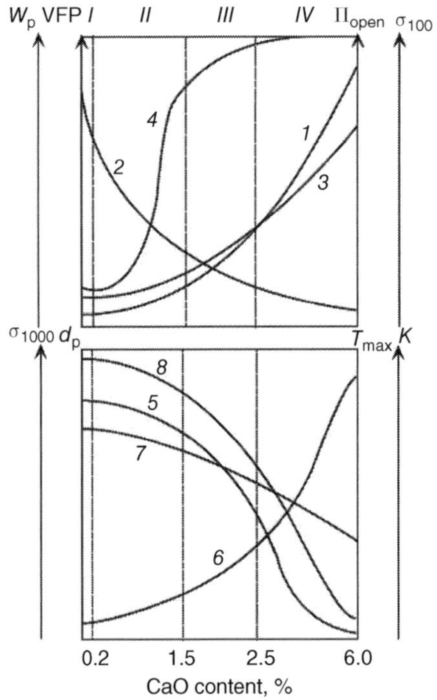

Figure 9.4 Classification of refractory concretes with HAC with respect to CaO content within them (I is cement-free; II is ultralow cement; III is low-cement; IV is medium-cement) and general nature of the dependence of concrete properties on CaO content: 1) concrete mix production moisture content W_p; 2) content of very fine powder VFP in mix; 3) open porosity in dehydrated condition Π_{open}; 4) concrete mutilate strength in compression after drying at 100 °C σ_{100}; 5) concrete ultimate strength in compression heat treated at 1000 °C σ_{1000}; 6) average pore diameter d_p after dehydration; 7) maximum application temperature T_{max}; and 8) corrosion resistance K [4].

and calcia is desired to preclude Ca and Si pick-up by molten metal. For these applications, calcia free, pure alumina binders have demand that usually forms transition alumina of high surface area, through their reaction with H_2O and develops boehmite and binds other particles. Drying dense cast bodies bonded with hydratable alumina binders need special care to preclude the possibility of explosive spalling.

Despite extensive use of both CAC and HA, nanostructured colloidal suspension is used as a promising binder for making refractory castable. Colloidal silica is nothing but a stable water suspension containing 50 wt.% spherical amorphous 8–15 nm diameter silica nanoparticles. The presence of other solid particles can form a network in the branch chain, known as gelation, and water removal can facilitate the formation of solid mass. The surface hydroxyl groups (Si-OH) generate siloxane bonds (Si-O-Si) during drying, resulting in a three-dimensional network. The gelation process is further induced by pH alteration and gelating agents comprised of a water-miscible organic solvent.

Thermomechanical properties, including refractoriness under load (RUL) and creep behavior, are analyzed of three different concretes separately made of equal 3.0 wt.% binder CAC, HAB, and CS, respectively [5]. This research work adjusted the particle size distribution of mixtures, including white fused alumina and calcined alumina grading, to theoretical curves based on Andreasen's packing model with a distribution coefficient (q) equal to 0.21. Additives polyethyleneglycol and citric acid are used to mix hydratable alumina (HAB)-based castable and the other two mixes, respectively. MgO (0.6 wt.% of the colloidal silica, CS) was used as a gelling agent for CS compositions. RUL measurements are performed in pre-fired samples (at 550 °C for 12 h) and the analysis carried out under continuous heating (5 °C/min) up to 1600 °C, under a compressive load of 0.2 MPa. Creep samples are first sintered at 1500 °C for 24 h and then subjected to a compressive load of 0.2 MPa at 1450 °C for 48 h.

Figure 9.5a represents the RUL analysis. As the samples are pre-fired at 550 °C, the deformation level on the RUL test also includes sintering. The colloidal silica composition presented higher sinterability, as a high surface area binder promotes the reactivity of the system and increases the sintering rate. The expansion detected close to 1500 °C for the CS's formulation is likely associated with the mullite formation. Creep measurement confirms the deformation under load. The influence of matrix content on the creep behavior of castables may be explained by Figure 9.5b. The grain sliding deformation mechanism is very active when many fine particles are present. The system porosity is also related to the creep behavior, which explains the smaller creep deformation for the CAC composition.

9.2.3 Aggregates Grading

Castables primarily need to flow, and they also require high density and strength, both at ambient and elevated temperatures. It has a unique requirement compared to that of the shaped refractory and is also contradictory as strength develops from compaction, which restricts the movement due to friction. This flowability is

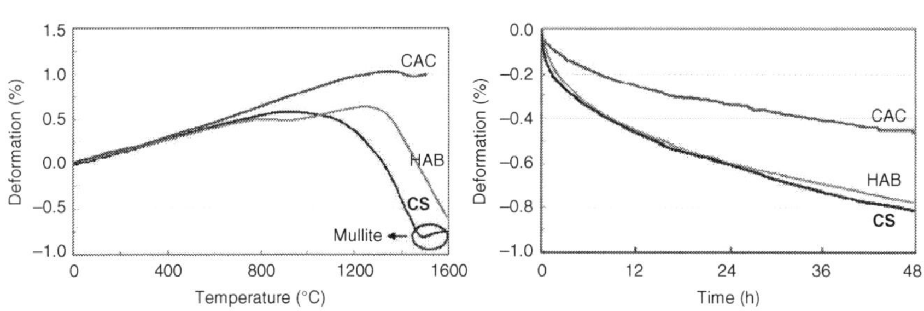

Figure 9.5 (a) Refractoriness under load and (b) creep of pre-sintered at 1500 °C and tested at 1450 °C, three different castables separately made of calcium aluminate cement (CAC), hydratable alumina (HAB), and colloidal silica (CS) based compositions [5].

primarily dependent on the particle size distribution and packing of the castable system. Usually, the packing concept is different compared to shaped refractories. In castable, this concept is known as continuous particle size distribution, and in this distribution, the particle sizes are present in a continuous manner, not in a discrete manner. In continuous particle size distribution, the aggregate fractions are used in many closely sized screened fractions to fill out a continuous distribution curve. Continuous particle size distribution results in excellent rheology/flow characteristics of the system at relatively low water contents with good compaction, low shrinkage, and high strength values. In this aspect, three models, known as Furnas model, Andreasen model, and Dinger and Funk model are usually considered to synchronize the aggregate grading of castables.

Furnas proposed that the best packing occurs when finer particles exactly fill the voids within the larger particles, and the CPFT (cumulative percent finer than) curve for both continuous and discrete are similar. Andreasen proposed that the continuous particle size distribution, particle size variations, and packing arrangements will be exactly similar at any magnification in the distribution. Andreasen assumed such a similarity condition developed in a continuous particle size distribution system and proposed a linear equation relation CPFT plot against particle size. The concept proposed by the Equation (9.2):

$$\mathrm{CPFT}/100 = (\mathrm{D}/\mathrm{D_L})^{q} \tag{9.2}$$

Where D is the average particle size in diameter, D_L is the largest particle size, q is the distribution coefficient. Andreasen proposed the value of the distribution coefficient (q) to be between 0.33 and 0.5 to obtain the optimum packing density. The drawback to Andreasen's approach and equation is that he did not recognize the effect of the smallest particle size. Straight lines on log–log plots continue forever. They can reach huge numbers on the one extreme and can reach especially small values at the other extreme. Dinger and Funk considered the smallest particle size D_s for the calculation of particle size distribution, which is known as the modified Andreasen equation, or Dinger and Funk particle size distribution equation. Dinger and Funk is a modified Andreasen's equation and can be represented as Equation (9.3):

$$\mathrm{CPFT}/100 = \left(\mathrm{D}^{n} - \mathrm{D_S^{n}}\right) / \left(\mathrm{D_L^{n}} - \mathrm{D_S^{n}}\right) \tag{9.3}$$

Figure 9.6 shows such a comparison between these two models with q = 0.37, $D_L = 1000$ μm, and $D_S = 10$ μm. The figure shows that there are no particles present below the 10 micron size for Dinger and Funk's model whereas 18.2% of particles are present in Andreasen's model. By computer simulations, it was also proposed by Dinger and Funk that for a q value of 0.37 or lower, 100% packing efficiency can be obtained for compositions with an infinite number of fractions.

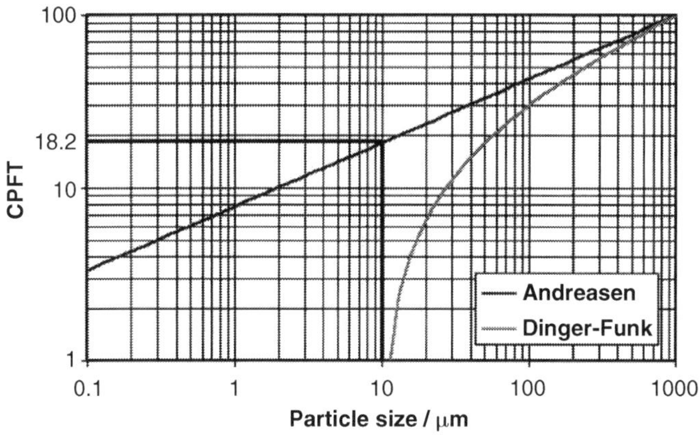

Figure 9.6 Andreasen and Dinger-Funk Particle Size Distributions with q = 0.37, D_L = 1000 μm, and D_S = 10 μm [6].

This is similar to the proposal of Andreasen, where such a packing occurs when the q value is between 1/3 and 1/2.

Also, studies showed that q values close to 0.3 results in vibratable castables (requiring external energy to flow well) and lower q values (<0.25) result in castables with self-flowability that move under their own weight. Higher fineness increases the sphericity of the particles, resulting in better flowability of the compositions and finally producing self-flowability in it. Thus, by changing the q values, one can change the particle size distribution of the castable composition and control the flow behavior and other properties.

9.2.4 On-site Castable Casting

The on-site casting of the backup lining of ladle and tundish is a common practice. The backup lining of the vessel is cast on the target region as a castable mass. First, the alumino-silicate monolithic castable of appropriate composition is mixed with about 5% water and cast through equipment in the desired region fixed with proper anchoring in the shell. Once the casting is done, an appropriate time is allowed for the lining to dry. The amount of moisture must be controlled carefully because higher moisture may lead to porous and high shrinkage, resulting in cracking of the refractory lining.

Anchoring is required and threaded-stud attachments through wall, welded fixes, or bolted assemblies are used to complete the casting process. Anchor design is essential to hold monolithic refractory in its place inside a vessel because the monolithic refractory, in itself, cannot adhere to the walls of the metallic shell (Figure 9.7). Anchors play a vital role in keeping the refractory lining in its proper place during the operation, and an inappropriate anchor design may lead to the failure of the entire refractory lining. The distribution of anchors inside a vessel

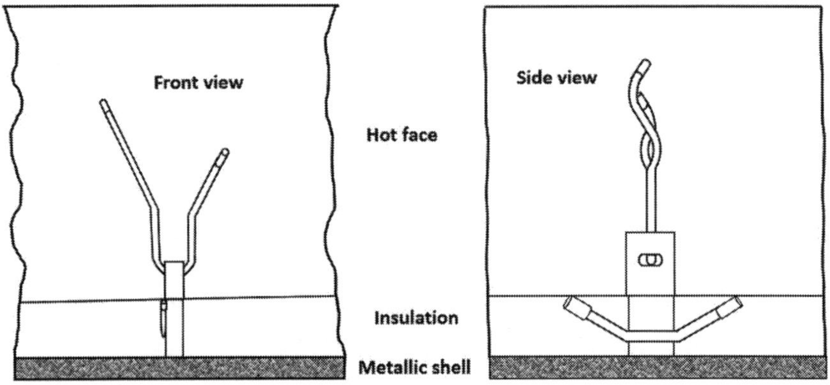

Figure 9.7 A typical shape of anchor.

directly impacts the lifetime of the equipment. A lower number of anchors may lead to poor anchorage, which increases the chances of failure.

On the other hand, adding more anchors may result in poor uniformity of the refractory lining, as the voids among them are too close and may not be completely filled in, resulting in defects, which affect the mechanical properties of the cast refractory. The anchor spacing should be based on the features of each particular need, such as the refractory lining thickness and anchor mechanical properties, especially elastic modulus and strength as a function of temperature. There are mainly two types of anchor, i.e., metallic anchor and ceramic anchor. Metallic anchor is for below 1100 °C and ceramic anchor above 1100 °C.

Prominent well equipped facilities are required to make a successful precast:

- In house mold fabrication with precious dimensional tolerance through CAD generated drawings.
- Large capacity, automated digital control water addition, and high-efficiency mixer machine to mix dry refractory castable and optimum water.
- An effective feeding by conveyors can transport the material and components as required.
- Mixing time controller and adequate lifting and shifting capabilities for large shapes.
- Digital controlled burners capable of heating up to 1000 °C.

Precaution should be taken during casting to obtain the desired performance:

- A blank rotational motion of dry refractory mix prior to water addition facilitates removal of entrapped air and achieves desired flow with optimum water.
- Systematic pouring, employing vibration, removal of the air pocket, and uniform filling essential for another lining contact.

- Project schedule, crew skill, available equipment, job cost, and other demand need to synchronize properly to avoid quality compromise.
- Improper water addition, mixing time variations, under or over vibration, and improper curing drastically change the quality of cast body.
- The drying rate for both initial and final lining is a critical installation parameter.
- Moisture removal throughout the lining thickness may avoid microcracks in shape during field bake-out.

Repair work in service is to be done as if required:

- Permanent and backup lining experience less stress and cracking; however, planning maintenance and repair results in a better life.
- The guniting method can reduce the material consumption, and also reduce the environmental hazards and dusting.
- Anchor attachment points to be taken care of during repairing.

9.3 Precast-Shape Design and Manufacturing

A wide range of precast shapes is an essential and common choice for steel manufacturers. Despite on-site casting or utility of castables, a brief understanding of different precast shapes provides insight into the performance. Before casting and application, the precast shape design is a critical aspect. It demands knowledge of material choice, mold design, mode of fabrication, anchoring (if required), installation, and probable performance during operation. Moreover, the dimension tolerance, manufacturing sequences, component movement, and other issues need to be taken care of for designing precast shapes. For example, a single piece EAF roof delta is a big piece of precast shape. Still, it can be fabricated in three parts and assembled to minimize defect formation during installation. Mold with a specific tolerance limit of $\pm 1/16$th inch is an inherent aspect to impart the desired precast shape, and it is more critical and seeks tight tolerance when the shape becomes complex in nature. Despite mold design, the selection of mold material is necessary for a sophisticated mold pattern, usually made in a machine shop. Proper processing scheduling and sequence have tremendous importance.

9.4 Precast Shapes and Casting

Prime precast shapes are required for the Electric arc furnace (EAF) delta zone, tundish furniture to control the flow, and well block to fix the porous plug. Despite precast shapes, monolithic casting in ladle and tundish covers a major share of the

Figure 9.8 (a) CAD view of EAF Delta, (b) a precast delta, (c) EAF roof assembly, and (d) insertion of electrodes through precast shape hole.

shop floor. In brief, the center roof of the EAF is a delta and made of a precast shape, as shown in Figure 9.8.

Delta zone experiences different stresses, including thermal, chemical, and mechanical. The bottom surface is exposed to temperature by radiation, slag splashes, and corrosive slag and metal oxide fumes. Complicated delta design has to encompass openings for the electrodes and water cooling arrangement that are eventually subjected to considerable mechanical stresses and thermal shocks, causing deterioration of the refractory. Water-cooling panels are lined with high alumina bricks (80% Al_2O_3) or precast blocks fabricated with high alumina castables (70–90% Al_2O_3). Annular space for electrode rings is lined with bauxite base high alumina bricks. Precast made of high alumina castable lining provides higher efficiency than conventional. Despite the water cooling panel, high alumina or burned magnesia-chrome bricks for the delta section and some other outer parts are also used.

The position of different precast shapes is mentioned in Figure 9.1, in which the impact pad, baffle plate, and slag dam precasts are placed to maintain the steel flow.

An impact pad is a precast shape capable of withstanding high impact and wearing resistance when steel is poured from the ladle to the tundish through a shroud that reduces the erosion of the lining and assists in maintaining certain desirable flow patterns into the tundish (Figure 9.9a). The metal stream from the ladle, especially when it is shrouded and has no entrained gas, enters into the tundish at a very high velocity and turbulence. These are made of dense, chemically stable refractory material like alumina castable-based precast. Corrugated impact pads are found to be effective in reducing the turbulence of the incoming stream (see Table 9.5).

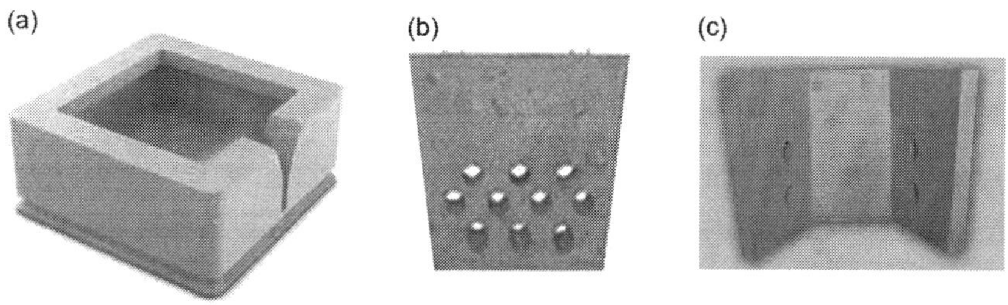

Figure 9.9 (a) Impact pad, (b) MHDB, and (c) slag dam.

Table 9.5 Precast and on-site casting material and performance for tundish. Tundish precast furniture refers as Impact Pad, Baffle plate, Slag Dam.

Type of refractory	Purpose of applications	Materials	Installation/ operation parameters	Performance (No. of heats)	Reason of failures	Reason of less heats
Tundish precast furniture	To maintain flow	LCC-70	Casting and plug with mortar	10 –22 heats	Erosion	Depends on planning, erosion, wrong steel chemistry, different steel grade, floatation, wrong application
Tundish backup lining	To save tundish Shell	LCC-70	Casting	600 heats	Erosion	Impurity content and improper casting facilities low melting phase and porous structure, respectively

Baffle plates control the speed of the liquid metal in tundish and effectively separate out the micro-inclusions that come along with the liquid metal from the ladle. Single precast shapes are usually used as a baffle plate; however, multihole double baffle plates (MHDB) have several advantages, as shown in Figure 9.9b. The multihole array mode diverse the trajectories of the micro-inclusions that control and reduce the molten steel flow and increase the residence time of molten steel, resulting in buoyancy playing a role in the removal of the micro-inclusions (1–50 μm) near the molten steel surface.

Slag dam is used to provide adequate time and maintain a flow that eventually simplifies and increases the chances of larger particles floating out to the surface of the tundish (Figure 9.9c). Therefore, slag dam is a major flow controlling device used in the tundish to control the turbulence of the liquid metal. The most critical aspect related to the slag dam design is the height of the slag dam. Slag dam height can be adjusted depending on the refractory quality of the sidewall. An appreciable slag dam height is beneficial, while sidewall refractory lining quality is relatively poor, as it would bring down the turbulence of the liquid metal to a greater extent compared to a smaller height slag dam. It results in lesser abrasion, corrosion, and erosion of the refractory lining in contact with liquid metal.

Most of the precast is made of high alumina-based castable, in which $MgAl_2O_4$ spinel addition has several advantages, and recent trends follow it. Slag dissolution in precast is often speeded up by spalling; thus deteriorates due to various mechanisms, such as temperature-induced phase transformation and thermomechanical property change throughout the temperature profile. Consequently, cracking occurs parallel to the refractory hot face, leading to lining spalling. The crystal structure of magnesium aluminate spinel ($MgAl_2O_4$) enables it to pick up cation impurities from the liquid. Alumina–magnesia refractories have been widely used in steel ladles and, in particular, as precast castable blocks in considering these aspects. Alumina-rich spinels (90 wt.% of Al_2O_3) dissolve indirectly due to the free alumina available and their ability to form spinel and CA_6 at the grain/slag interfaces, whereas near stoichiometric spinel (78 wt.% Al_2O_3) follows direct dissolution. The local slag liquid (predominantly $CaO–Al_2O_3–SiO_2$) phase contained a higher amount of CaO that could react with the matrix and aggregates, usually leading to CA_6 formation.

As a consequence of the greater amount of Mg^{2+} vacancies for the alumina-rich spinels, more cations from the slag can be accommodated, and then the local liquid becomes silica-rich, increasing its viscosity and reducing the penetration. A greater amount of synthetic spinel or MgO leads to a slight increase in corrosion but a reduction in the penetration by the molten slag. Thus, the spinel amount close to 20 wt.% would optimize both requirements. The corrosion resistance decreases below the optimum amount due to a lack of proper spinel content, whereas above 20 wt.%

reduces the slag infiltration resistance. The spinel grain size also influences the slag penetration and corrosion resistance. The use of fine grains leads to greater effectiveness due to their better distribution throughout the castables matrix. Nevertheless, they could also lead to high sinterability; thus shrinking cracks and lowering the infiltration resistance. Additionally, the corrosion of synthetic spinel containing castables is usually higher than the one attained for in-situ spinel-forming castables.

9.5 Purging Plugs

Secondary steelmaking uses gas stirring in the ladle to improve steel cleanliness, facilitates alloy trimming and homogenization of composition and temperature, and promotes sulfur removal when required through argon stirring from the ladle bottom. Bottom stirring can avoid forming a dead zone compared to lance stirring. Argon gas is purged through a plug at high pressure and a variable flow rate, which varies from 5 Nm^3/h to 50 Nm^3/h. Apart from taking care of high pressurized injected gas, plugs need to be (corrosion and erosion resistant) robust enough to bear the ferro-static pressure in dynamic conditions.

The purpose of purging plug is shown in Figure 9.10a, and their different refractory assembly is represented in Figures 9.10b and 9.10c. High thermomechanical stress is generated when relatively cold argon gas is purged while plug is contacted with hot liquid steel. In consequence, cracking occurs parallel to the material's hot face, leading to lining spalling. Therefore, plug's material needs to be thermal shock resistant. Another significant plug requirement is to ensure no metal infiltration and no leakage to maintain safe operating conditions. Factors affecting plug performance can broadly be categorized into

1) Quality of refractory and plug design
2) Ladle furnace operational practices
3) Maintenance of plug and purging unit

Despite these factors, a detailed understanding of the wear mechanisms of the porous plug is a critical issue for adequate operation. A reliable, cost-effective, and safe purging plug can fulfill the targeted objectives.

9.5.1 Plug Design and Refractory

An adequate balance between quality of refractory (composition, texture, and distribution of grain and matrix phases) and design of porous plug (random porous, directional plug slots type or segment type) are essential to maintain soft bubbling with a bubble diameter of 2 mm. In addition, the installation method altogether has a great impact on plug performance. Figure 9.11 represents the design of

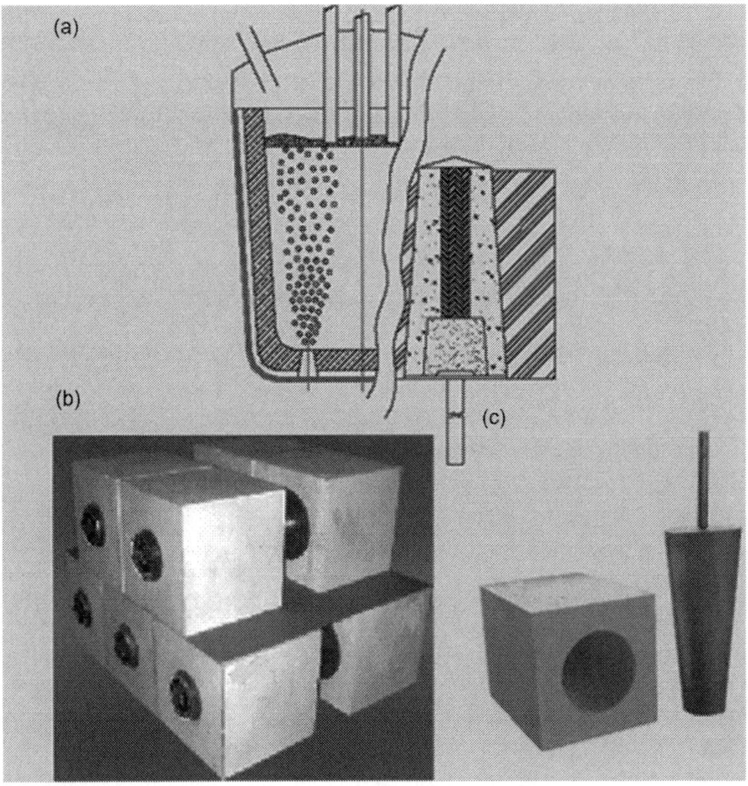

Figure 9.10 (a) A schematic view of the purging plug in LF, cross-section view of the plug and well block assembled together, (b) an assembled product of precast well block and purging plug, and (c) well block and purging plug, separately.

purging plugs, including wear indicator, and their processing and operational features are listed in Table 9.6.

Despite these three popular purging plugs, other purging plugs, including hybrid, labyrinth, and star plugs, are also considered for different purposes. Precast is a common technique to fabricate a high alumina-based porous plug, consisting of nearly 15–18% porosity. The slot purging plug is designed to pass the gas through thin slots placed at a radial angle from the center of the circular plug, see Figure 9.11a. These plugs are cast in one piece and consist of aluminum oxide ceramic material. The slots are made by strips of a polymeric material, placed in the desired location of the slots, later being removed through combustion as the molded piece is being sintered. However, the most advanced segment plug demands multiple components to serve specific purposes. In the segment purging plug, a segment is made of high-grade alumina in addition to Cr_2O_3 and ZrO_2 to

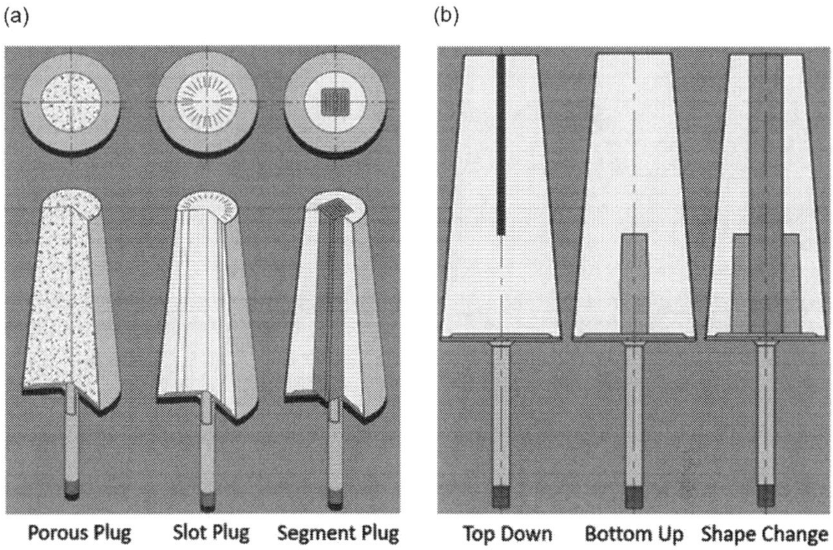

Figure 9.11 (a) Different design of purging plugs; porous plug, slot plug, and segment plug, (b) different mode of wear indicators to predict and replace the purging plug.

Table 9.6 Different operational and performance features of purging plugs.

	Porous plug	Slot plug	Segment plug
Pore structure	Random and high porosity	Directed and separated slots	Random and directed
Processing	Cast	Cast	Pressed and cast
Optimum product for	Stainless steel and foundries	Low-alloyed steel	High-grade steel
Type	Single piece	Single piece	Multi piece
Suitable for application	Continuous and discontinuous	Continuous	Continuous and discontinuous
Initial opening rate	Excellent	Good	Excellent
Flow rate controllability	Excellent	Good	Very good
Flow rate range	Excellent (>2000 l/min)	Excellent (≤2500 l/min)	Very good (≤1200 l/min)
Bubble formation	Pore diameter controls the bubbling character	Bubbling depends on slots	Soft bubbling
Main raw material	Al_2O_3	Al_2O_3	Al_2O_3

Table 9.7 Purging plug life with respect to steel-processing protocols and ladle size.

Plant reference	Primary steel production process	Secondary steel ladle size (tons)	Purging plug (heats)
A	EAF	125	20
B	EAF	85	35
C	EAF	65	50
D	BOF	400	8
E	BOF	170	40

induce thermal stability and corrosion resistance. The surroundings castable made of >90% alumina and $MgAl_2O_4$ spinel enclosing segments and security device is also effective for a better life.

The segment block consists of 4–6 numbered ceramic plates, attached loosely to each other. The groove within the blocks is kept within 0.25–0.35 mm, depending on the gas flow rate and viscosity of liquid steel and to ensure that there is no steel penetration due to back air pressure during cleaning of operation. The difference between the coefficient of thermal expansion of segment blocks and castable should be minimum to avoid the formation gap during operation. A security device is a permeable refractory part that allows gas to flow through it toward the segment. A shape change type security device is also known as "wear indicator" as it is placed at a certain position and indicates plug length in hot conditions, as mentioned in Figure 9.11b. An optical wear indicator is a porous high-alumina refractory with a square or circular cross-section at a residual thickness of around 120–140 mm of purging plug (depending on the plug design). When the purging plug is at residual thickness, the indicator becomes visible during the ladle cleaning after repeated casting. Due to its porous nature, the top section of the indicator in dark color becomes visible when observed from inside the ladle in hot conditions. As soon as the indicator is visible, the operation is to be stopped, and the purge plug has to be replaced by a new one immediately.

A typical performance data of plug concerning different steelmaking processes and ladle capacity is shown in Table 9.7. Extensive data suggests that it is essential to understand the ladle size, holding time, and steel viscosity before designing and selection of the purging plug.

9.5.2 Gas Purging

Argon (Ar) gas is preferred for purging and rinsing the steel as it has extremely low solubility in steel and is inert. In such a process, stirring involves heat and mass transfer phenomena. Many factors could affect this process, including the preferential number of the porous plugs, plug design, plug location, gas flow rate, and slag properties (e.g., slag height).

Purging plugs differ significantly in design and properties and not all of these plugs are optimized for soft bubbling performance. In actuality, soft bubbling at low flow rates with inert gases is the final processing step in steel secondary metallurgy to achieve maximum steel cleanliness. Nonmetallic inclusions (NMIs) are floated up by inert gas bubbles from the steel bath into the slag layer during soft bubbling. As an important precondition, the opening of the slag layer—the so called *open-eye* formation—has to be avoided to prevent slag entrainment into the melt and reactions of the melt with air. Thus, finely distributed gas bubbles are desired.

Regardless of the plug design, an increase in flow rate always increases the open-eye, whereas an increase in the size of the gas injection area continually shifts to higher flow rates. A typical bubble formation with respect to gas pressure is represented in Figure 9.12a, reflects an optimum gas pressure for a particular pore diameter is suitable for maintaining the soft bubbling and uniform bubble formation to obtain effective stirring and minimum open-eye formation. However, the excess pressure may enhance the jet of the bubble and open-eye formation, as shown in Figure 9.12b. In this regard, recent literature represents the numerical modeling and simulated the open-eye formation for a 150-ton ladle with respect to gas flow rate. The slag open-eye predicted by the numerical model agreed well with

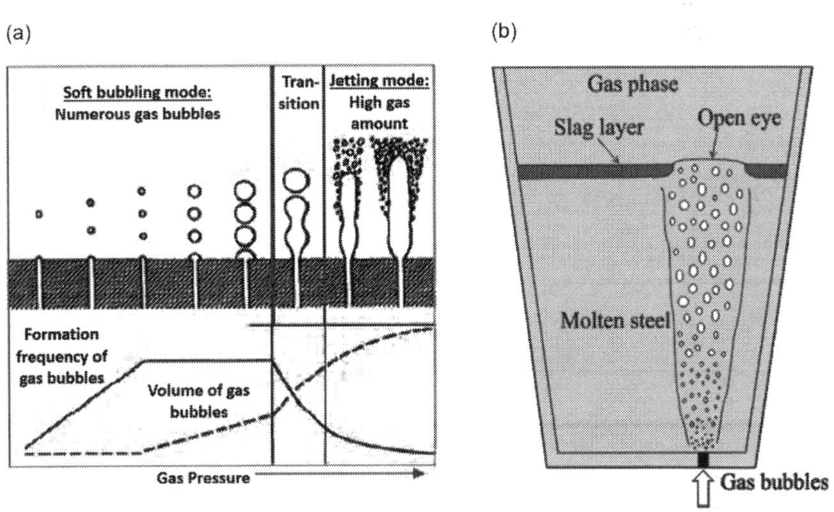

Figure 9.12 (a) Effect of gas pressure on bubble formation, (b) schematic representation of open-eye formation to expose the direct metal–air interaction [7].

experimental results as shown in Figure 9.13a. The slag eye area increases with an increase in the gas flow rate for both experiments and simulations Figure 9.13b.

In purging few essential operational parameters are necessary to control plug performances. These prime factors are:

- Turnaround time (TAT) defines ladle slag traveling time before dumping. TAT is inversely proportional to the opening rate. Lower TAT signifies the probability of a higher opening rate.
- High ladle holding without purging expedites more chance of metal penetration inside the purging plug.
- Low temperature tapping of steel or insufficient tapping may cause chilling and skull formation at the bottom of a hot metal ladle.
- A sudden drop in optimum argon flow rate while purging is on.

9.5.3 Installation and Maintenance

A typical schematic representation of the plug region is shown in Figure 9.14a. Prior to installing the purging plug, a definite thickness near 5–7 mm Al_2O_3-Cr_2O_3 oxide based mortar is coated on the plug as shown in Figure 9.14b, and a graphite coating on the well block inner surface may enhance the precise and quick removal of residual mortar and fast plug exchange. After employing the ceramic coating, the plug is inserted and allowed to set the mortar and correct position (Figure 9.14c), followed by fixing up the requisite number of safety rings and closing the bayonet. Bayonet closure helps to close and open the gas purging system for plug change (Figure 9.14d). Once the fixing is completed, the argon pipe is connected to the plug and the gas flow rate is checked (Figure 9.14e).

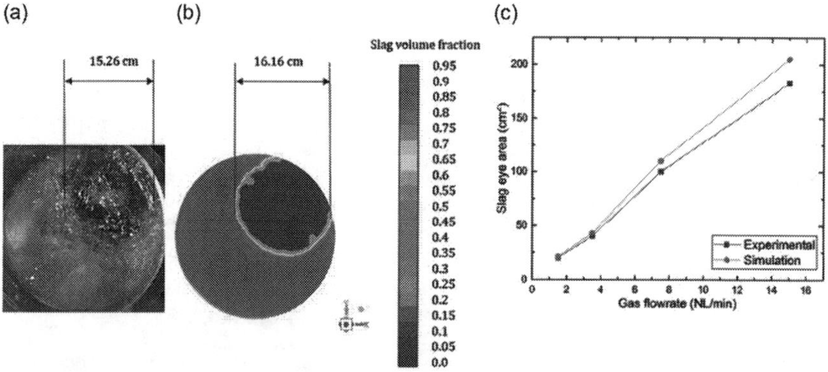

Figure 9.13 (a) The slag eye size for gas flow rate 15 NL/min: (a) experimental, (b) simulation, and (c) variation of slag open-eye with respect to gas flow rate [7], International ASET Inc, 2018.

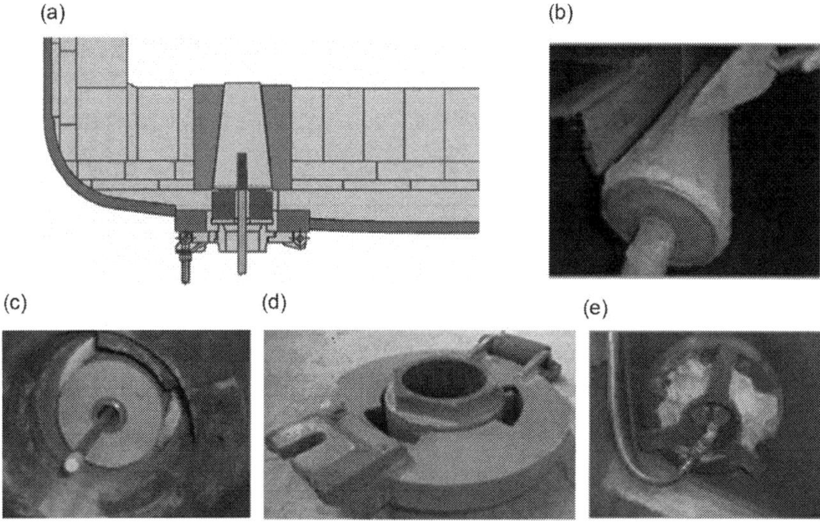

Figure 9.14 (a) Schematic representation of bottom lining including well block, plug, safety ring, and Bayonette closure, (b–e) different stages during plug installation.

The cause for the wearing of the plug is a combination of chemical and mechanical erosion/corrosion, and thermal and mechanical stresses. Effective cleaning is required during idle conditions while the last purging reduces the gas flow. The plug surface is covered by a layer of steel, slag, or both, and the slots are penetrated or infiltrated with steel. Thus, cleaning is an important aspect of operating through oxygen and making sure the slots and plug exposed surfaces are visible. Replacement of the plug should be made when the wear indicator appears. However, the success rate of plug opening depends significantly on the cleaning practice and maintenance of the purging line of transfer car and ladle, auto coupling parts.

Factors during maintenance that affect the plug health can be summarized as follows:

- During cleaning of the plug by oxygen lancing, the back pressure should be sufficient enough to resist the inside metal penetration. Ideally, O_2 pressure should be 4 bars < N_2 backpressure, 6 bars. In most cases, this factor is overlooked, resulting in frequent slicing out of plug material from the cleaning surface.
- During purging, the argon flow should not be stopped suddenly. A sudden gas flow cut may create a vacuum inside the segment, and steel can penetrate segments by suction. The gas flow drop should follow an intermittent path from high to medium to low, then zero.

- Leakages in a gas piping system.
- Improper connection of auto coupler because of misalignment or improper sealing can cause leakages, generally through O-ring.
- Improper ladle sitting on ladle furnace car can again cause the improper connection of auto coupler.
- Malfunction of flow control valve (FCV).

Physical observation of the entire system is essential for the smooth running of the process. In a ladle car, the auto coupler and the gas pipe system must be kept clean and covered. Due to the missing O-ring in an auto coupler, which is responsible for air sealing, purging gets affected and forced to change plug at low life. It must be clean and protected from splashes or other types of dirt to ensure good function and no leakages from the system. Furthermore, the hot repair of well block by castable or self-flow mix to be done with gunning mixes without disturbing plug surface.

9.5.4 Clogging and Corrosion

The porous plugs tend to clog after a while. It is a serious problem that hinders and alters the gas escape path before entering the ladle without contributing to the stirring intensity. Thermal expansion mismatch between the plug and penetrating meal/slag facilitates the cracking of plug. Thus, effective stirring and prediction of flow rate is difficult due to:

- Abnormal lowering of the stirring rate because of porous plug clogging
- Clogging restricts the forward pressure, and backpressure becomes dominant
- Sudden leakage in the gas supply system
- Inconsistent slag viscosity, and thickness mislead to fix the stirring rate
- Lack of data and real time stirring history record on each ladle.

The steel chemistry and refractory lining control the slag composition and change viscosity from shop to shop. In practice, an open-eye is determined by manual visualization that eventually helps to understand the gas flow behavior and probable clogging of the purging plug. Usually, a large open-eye area indicates beyond the optimum limit of gas injection flow rate. Still, too small an open-eye specifies restricted gas flow that eventually causes clogging of the purging plug. However, slag viscosity and thickness may misdirect the gas flow rate. Thus, maintenance or adequate replacement is necessary to ensure no argon gas escapes and uniform stirring.

Gradual interaction with molten material is obvious in corrosion during each cycle. Table 9.6 indicates when steel ladle volume is large, the plug life is predominately low irrespective of steel processing technology. The principal four wear mechanisms of the porous plug are shown in Figure 9.15a. Molten steel peeling

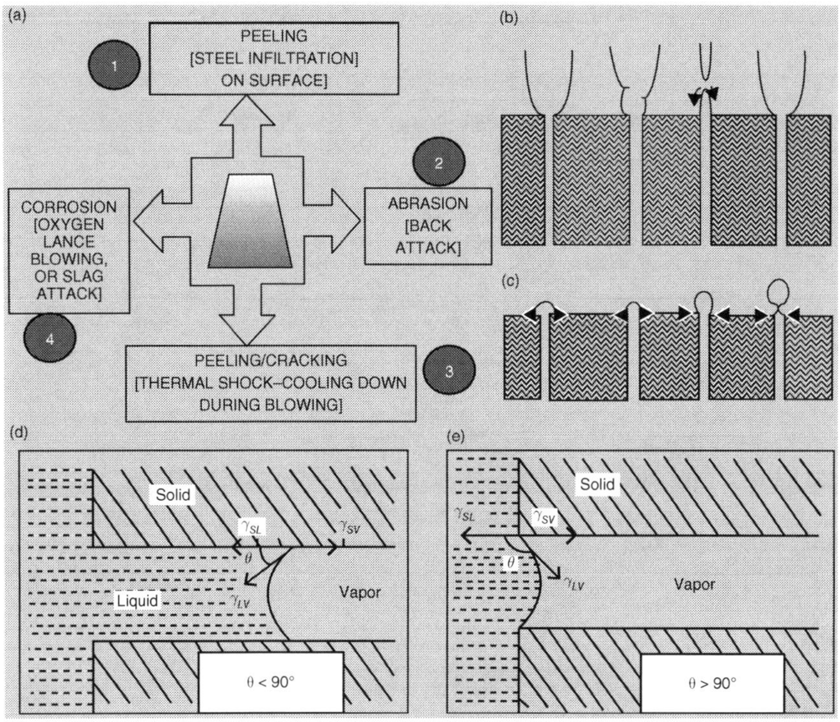

Figure 9.15 (a) Wear mechanisms in purging plug, (b) back attack due to unstable gas flow rate in channel plugs, (c) lower back attack in porous structure, (d) infiltration of molten material in contact angle θ < 90°, and (e) infiltration of molten material in contact angle θ > 90° [8].

off in the temperature range of 1670 ± 50 °C, tapped out from primary steelmaking vessel like converter or electric are furnace may direct contact with the surface of the plug. Molten steel infiltrates the surface in the temperature range of 1670 ± 50 °C during tapping from a primary steelmaking vessel like a converter or electric arc furnace; subsequently, the plug surface may directly come into contact with molten steel.

If the liquid steel infiltrates the refractory due to insufficient gas pressure in the plug, the transformed upper layer develops very different mechanical properties than the unaffected material. This induces stresses and subsequent peeling of the surface.

Abrasion is due to the back attack of the gases during stirring. The intensity of this reaction is dependent on the flow rate and the method of injecting the inert gas. Peeling and cracking due to thermal shock happens when the upper plug surface experience ~1600 °C during steel processing, whereas at the back face and in the gas channel temperature is ~300 °C, such temperature gradient induces very high stresses.

High pressure oxygen from a lance is often used to remove infiltrated steel and slag from a plug after emptying a ladle. While this process removes the blockage, the plug's upper part (1 cm or more) can be removed. There is also a risk of burning a hole in the plug, considerably reducing the life of a device. If the plug design allows, a simple solution is to increase the flow rate to produce a stable flow. For instance, narrow channels or slots with a low flow rate can negatively influence a pulsing, unstable gas flow (see Figure 9.15b). In the case of porous plugs (see Figure 9.15c), bubble movement is preferentially quieter (softer), which results in less wear. The penetration of molten metal into a porous refractory can be described with a model expressed by Laplace–Young's equation (see Section 6.4.3). Liquid wet solid and this wettability are characterized by the angle of contact that the liquid makes on the solid (see Figures 9.15d and 9.15e). The contact angle, θ, is obtained from a balance of interfacial tensions and is defined as:

$$\sigma_{lv}.\cos\theta + \sigma_{ls} = \sigma_{sv} \tag{9.4}$$

where σ_{lv}, σ_{ls}, and σ_{sv} are the interfacial tensions at the boundaries between liquid (l), solid (s), and vapor (v). Here, σ represents the force needed to stretch an interface by a unit distance. The condition $\theta < 90°$ indicates that the solid is wet by the liquid, and $\theta > 90°$ indicates non-wetting. While considering the contact angle, Al_2O_3 and $MgAl_2O_4$ spinel experience contact angles of ~90° and ~110°, respectively. It explains why spinel containing refractory helps prevent steel penetration. The infiltration steel penetration depth depends on the equation, $h = (-2.\sigma.\cos\theta)/(r.\rho.g)$, where ρ is steel density (g/cc), g, the gravitational constant, σ surface tension of molten steel (N/m). Prior to the design of the purging plug it is necessary to remember that

- Below a critical ferrostatic pressure, there is no steel penetration
- Increasing contact angle to reduce the penetration
- Slot size and pore size distribution are required to adjust with steel viscosity.

References

1 Serry, M.A. (2002 August 01). Bauxite based low and ultralow cement castables. *Br. Ceram. Trans.* 101:4, 165–168.

2 Myhre, B. (2011 March 23–24). Microsilica containing refractory castables - the influence of additives on placing properties. Presented at the 47th annual symposium on Additives for monolithics, St. Louis, USA.

3 Madono, M. (1999). Alumina raw materials for the refractory industry. *CN–Spec. Refractories* 6 (3): 54–63.

4 Pivinskii, Y.E. (2020 January). Cement-free refractory concretes. Part 1. General information. HCBS and ceramic concretes. *Refract. Ind. Ceram.* 60 (5). 430–438.

5 Ismael, M.R., Dos Anjos, R.D., Salomão, R., and Pandolfelli, V.C. (2006 July/August). Colloidal silica as a nanostructured binder for refractory castables. *Refract. Appl. News.* 11 (4). 16–20.

6 Dinger, D.R. (2003). Ceramic processing E-Zine. 1 (9).

7 Ramasetti, E.K., Visuri, V.-V., Sulasalmi, P., and Fabritius, T. (2018). A CFD and experimental investigation of slag eye in gas stirred ladle. *J. Fluid Flow, Heat Mass Transf.* 5 (148): 1–10.

8 Tassot, P. (2006). Innovative concepts for steel ladle porous plugs. Calderys. *Millenn. Steel.* 111–115. https://www.academia.edu/34015996/Innovative_concepts_for_steel_ladle_porous_plug

10

Refractories for Flow Control

10.1 Introduction

Longer retention of molten metal in the ladle is required for more sophisticated ladle secondary metallurgy, resulting in better steel quality. The stopper could no longer survive in this severe environment, and a slide gate was introduced to resolve the problem. While secondary steel processing is a concern, two important vessels demand effective flow control: a) Ladle to tundish during steel processing and b) tundish to subentry shroud (SES) during continuous steel casting. In tradition, the two-plate and three-plate systems are used for ladle and tundish, respectively. However, a reverse arrangement can be used on a tailormade basis. Continuous and interrupted flow management is required for high productivity, and it is an effective operation practice for a target of the highest level of performance. It is concerned with converting liquid steel into a solid casted product as efficiently as possible to maximize the achievement of steel quality.

A simple schematic in Figure 10.1 represents the particular location where slide gate systems are essentially introduced to control the steel flow. Other continuous casting refractories are also highlighted in the schematic. However, their details are discussed in Chapter 11. It is obvious several refractory accessories such as well-block, ladle nozzle, fixed plate, sliding plate, and collector nozzle are required to complete the ladle slide gate system. The tundish slide gate is a three-plate system including a well-block and upper nozzle. A schematic representation of both slide gate systems and one of the critical flow control systems is shown in Figure 10.2. The ladle slide gate system controls the liquid steel flow from ladle to tundish during the steelmaking process, as shown in Figure 10.2a. As shown in loop, it requires a slide gate mechanism and other refractory components for its successful operation. The assembly comprises a sliding mechanism, an upper nozzle, a sliding plate, and a lower nozzle. Further, the refractory slide plate is divided into an upper plate and a lower plate. The shape of the plate depends on the required shape of the mechanism.

Fundamental Design of Steelmaking Refractories, First Edition. Debasish Sarkar.

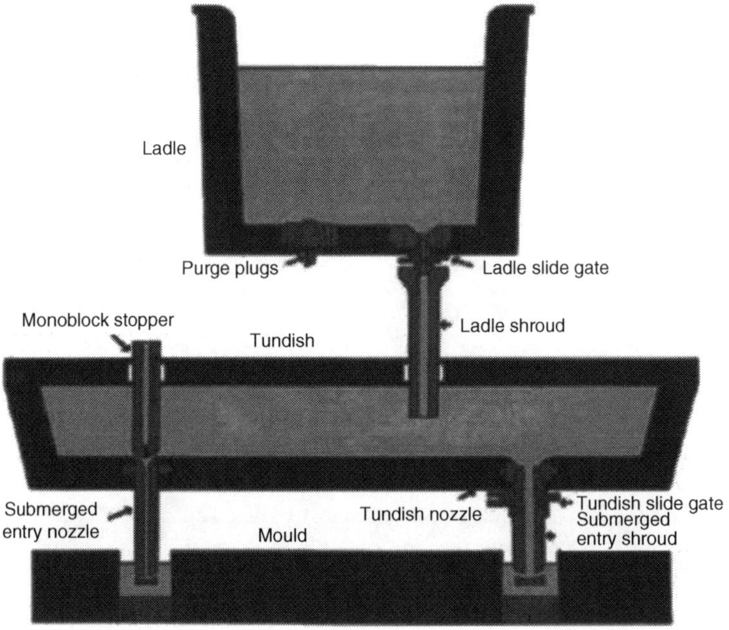

Figure 10.1 Different refractories during flow control and continuous casting. The major attraction is both slide gate systems for *ladle to tundish* and *tundish to SES*.

Mutual exchange of the upper and lower plate is only possible if the external dimensions and bore positions are the same, but usually, they are different in size and have a limitation to be used mutually. In this assembly, the upper slide plate is fixed, and the moveable lower slide plate makes continuous bore alignment to flow steel from ladle to tundish. The bottom part has two types of movements—linear reciprocating and rotary movement. A linear reciprocating mechanism assembly with a sliding nozzle is the most common: a reference plate, a fixed steel frame, a movable frame, and a sliding box.

The lower slide plate is compressed by a spring within the sliding box, and the gap between the lower and upper sliding plate is monitored. A precise mechanical arrangement assembled with hydraulic cylinders and electric push rods synchronizes the reciprocating motion and steel flow control. Despite the slide gate mechanism, the slide gate plate (SGP) refractory quality, steel grades (e.g., Ca killed steel), and plant operating conditions are important for casting and metering the liquid steel.

A three-refractory plate slide plate system is often used to control the flow from tundish to continuous casting mold. In this assembly, these plates are located in the upper tundish nozzle (UTN) and submerged entry shroud (SES), as shown in Figure 10.2b. A hydraulic cylinder is attached with a middle plate to move horizontally and adjust the eye-shaped opening zone to control the flow rate and

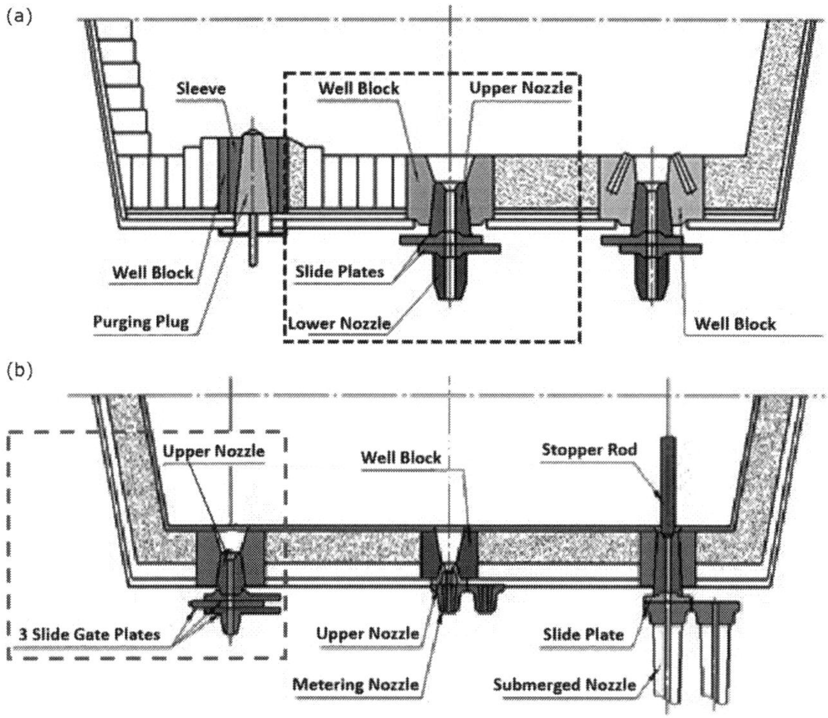

Figure 10.2 Slide gate assembly for (a) ladle to tundish, (b) tundish to SES.

maintain a constant meniscus level in the mold while encountering the measured level by the eddy current sensor suspended above the mold.

Despite a three-plate system, a two-plate slide system is also used. A typical design feature for a three-plate and two-plate tundish slide gate to nozzle system is shown in Figure 10.3. Both slide gate assembly and refractories are necessary to encounter, in which consideration of different slide gate housing mechanisms and refractory compositions are common practice by manufacturers. Table 10.1 represents the refractory system to control the flow, and their details have been discussed in this chapter.

10.2 First–Second–Third Generation Slide Gate

Modern steelmaking process demands continuous development of refractory, and in obvious slide gate assembly, refractory gradually crossed first- to second- to third-generation composition for a definite assembly. The first-generation (1965–1980) slide gate had a massive structure weighing 500–1000 kg, made in auxiliary

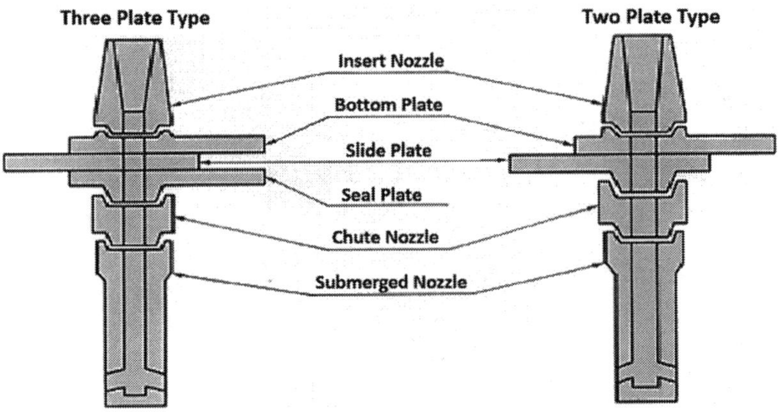

Figure 10.3 Tundish slide system (a) three-plate type, (b) two-plate type.

shops. The gate designs were of the *rigid* type and the plates comprising the gate were pressed against one another by bolts. The plates were installed in the gate with the use of a mortar. The service life of a gate was 70–100 heats. It took more than 2 hours to replace the refractories in the casting unit. Each metallurgical plant developed its drive design (hydraulic, pneumatic, or electrical) without regard for the details of the casting operation itself. The characteristic features of the plates used for the first generation of slide gates were as follows:

- the use of plate materials with a single-component composition based on fused or sintered periclase, mullite, corundum, zircon, or other material;
- the use of unconcentrated raw materials and ceramic binder;
- partial alloying of the plate material with refractory oxides to improve certain properties (Cr_2O_3 was used to reduce the wettability of the surface of the metal, Z_rO_2 was used to improve heat resistance, etc.);
- impregnation of the plates with carbon-bearing liquids (coal-tar pitch, liquid bakelite, etc.);
- the use of a coarse porous structure in which 70–75% of the pores were 20–30 μm in size and 20–25% were smaller than 1 μm; the presence of individual pores with a size in the range 40–70 μm contributed to significant wear of the plates;
- the absence of metal casings and the use of a system of mortars to join the refractories together.

Most of the plates in the first-generation slide gates had the form of an ellipse or a rectangle with beveled corners. The refractory systems of this generation allowed the casting of just a single heat.

The refractories used in slide gates of the second generation could last 2–3 heats. The slide gates were of the *book* design, which made it possible to replace

Table 10.1 An interactive analysis of flow control refractories.

Type of refractory	Purpose of applications	Material design parameters: Composition	Material design parameters: Properties	Installation/ operation parameters	Performance (No. of heats)	Reason of failures	Reason of less heats
Ladle to Tundish							
Ladle upper nozzle	Channel to flow molten steel to slide gate without contacting the machine	High alumina/ spinel	Erosion/ corrosion resistance	Manual with required fixtures	Varies between 4 and 30	Erosion, penetration, crack, void, rat hole	• Refractory failure • Choking • Process low temperature
Ladle slide gate plate	To control flow of steel	Alumina carbon, Mag Car	Compressive strength and erosion resistance	Hydraulically powered, automatic and manual control	Life—more than 9/10 heats	Poor maintenance/ breakout	Wrong operation, hydraulic failure, lancing
Ladle lower nozzle	Guide to connect to slide gate plate to LS	High alumina/ spinel	Erosion/ corrosion resistance	Manual with required fixtures	4–6 heats	Crack, erosion, corrosion, rat hole	Metal overflow, Sticking, can be melting
Tundish to SEN							
Tundish slide gate	To control flow	Alumina graphite	Erosion and corrosion	Thru mechanism	20–30 heats	Erosion and corrosion	Wrong operational practice, wrong steel chemistry, mis-alignment, planning. More slag volume, slag casting
Tundish nozzle	To control flow	Alumina graphite	Erosion	Manual installation	10–22 heats	SEAT area erosion	Wrong operational practice, wrong steel chemistry, mis-alignment, planning. More slag volume, slag casting

the refractories directly on the ladle. Spring-opposed elements made of a heat-resistant metal and lasting 50–70 heats were used to press the plates together. The refractory plates and the nozzles were installed inside a metal casing, which improved the accuracy of the casting unit assembly and shortened the time needed to install new refractories by 10–15 minutes. The above-mentioned structural elements made it possible to regulate the compressive force employed during the casting operation. The service life of slide gates increased to 500 heats. The slide-gate system was equipped with a hydraulic drive adapted for the specific conditions encountered during casting with a slide gate. The characteristic features of the second-generation slide gates are as follows:

- the use of raw materials characterized by a high degree of chemical purity and low content of harmful impurities;
- the use of a secondarily formed ceramic binder (mullite, spinel, silicon carbide, etc.), the binder being synthesized during the firing of the plates or practical use;
- the presence of carbon in the charge composition and carbon formed from the synthetic binder during its coking;
- the use of anti-oxidizing additives to prevent combustion of the graphite;
- the fine-pore structure of the plates (90% of the pores are smaller than 1 μm); average pore size is 0.05–0.20 μm;
- the extensive use of metal casings or shrouds.

The shape of the second-generation plates was chosen with allowance because different levels of thermal stresses may be formed, depending on the design of the specific slide gate.

With third-generation (post 2001) slide-gate systems, it is possible to cast more than 5 heats without changing the refractories. It is also possible to regulate the compressive force on the gate plates. The gate employs reliable dish-shaped elastic elements (with a service life greater than 2000 casts) or gas-filled elements (bellows lasting more than 5000 heats). The gate design allows for automatic adjustments to be made during installation, casting, and refractory replacement. The latest advances are employed in making the refractories:

- high-purity synthetic materials (sialons, nanomaterials, etc.) are in the charge;
- metallic fibers reinforce the structure of the refractory;
- efficient plate designs (bushing, insert, etc.) are used;
- oxygen-free ceramic binders (nitrides, sialons, etc.) are used;
- modern equipment (isostatic pressing, hot extrusion, etc.) is employed;
- the firing operation is performed in a protective or reducing atmosphere (nitrogen, etc.)

The manufacturers of modern slide-gate systems have certain tendencies concerning the design and production of slide gates: the use of openings of the *book* type; the use of dish-shaped elastic elements or gas-filled bellows to compress the plates; hydraulic drives; the fabrication of plates and collecting nozzles inside metal casings; durable gate components; the production of a wide range of different types of slide gates to allow the casting of metal from ladles of different sizes; provision of a complex of services and technological aids for casting metal; provision of spare parts for the original equipment.

10.3 New Generation Ladle Slide Gate System

Slide gate assembly should be user-friendly with a simple, robust, ergonomic design that eventually helps quick fixation during maintenance. Consisting of argon shielding, air cooling capability, and slag detection coil arrangement facilitates the performance of refractory assembly and cumulative life of slide gate system.

Two common designs are common to fix the refractory slide gate plates in which the plate fixation represents (a) the structure of the mechanism is *up-down* to make the gate open or close (Type-1), and (b) the way of opening the mechanism gate is *left and right* (Type-2). Figure 10.4 represents the quick view of such assemblies. Both designs have been illustrated briefly to understand the regions where refractory is being used. In consideration of ladle volume and steel throughput, different sizes of slide plate, stroke length, and slide plate hole diameter are used.

(a) (b)

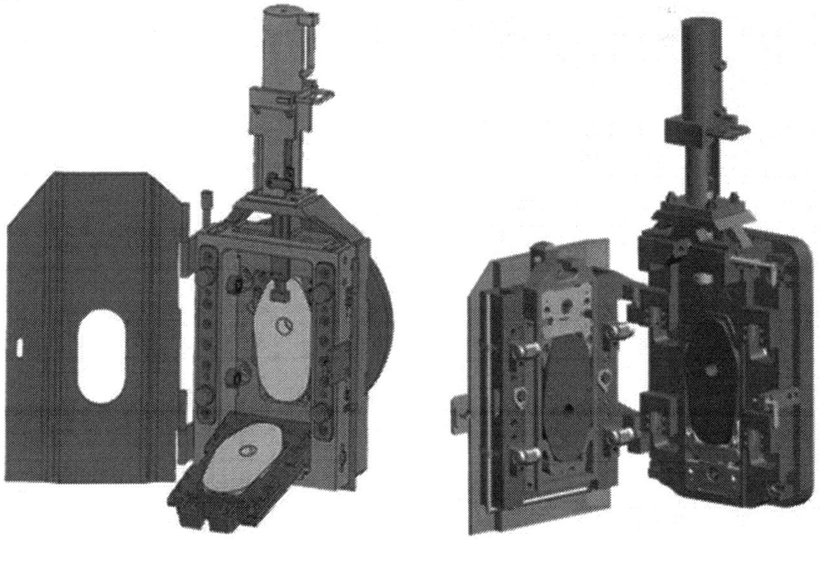

Figure 10.4 Slide gate assembly (a) up-down (Type-1), (b) left to right (Type-2).

The pressure area is constant while considering the *up-down* approach, as shown in Figure 10.4a. The spring-based pressing mechanism provides the alignment that establishes a constant and uniform pressure; thus the entire system is not affected by high temperatures and human factors. Probable changes in thermal expansion or any manual mistake adjustment by the spring results in long service life and can satisfy the dimensional tolerance for many times and continuous operation. Slide interface stability and uniform pressure facilitate the smooth sliding and security of the assembly. The refractory dissembling and installation are very convenient and easy to operate. It has an air cooling device to ensure more reliable operation and service life.

The progressive *left to right* structure (see Figure 10.4b) of the opening gate has a simple mechanism and easy refractory accessibility. Herein, the slider connects with a connecting rod, and the operation initiates by a hydraulic cylinder to establish constant, uniform pressure without any manual operation. The inbuild spring system absorbs alteration of thermal expansion behavior to run smoothly and more securely. The mechanism has an inbuild air cooling device for long spring service life. Both upper and slider plate chamber designs satisfy the desired fixing and high frequency and continuous sliding during operation.

For example, left to right slide gate assembly with a dimension of 1400 mm × 684 mm × 320 mm (L × W × H) usually provides more than 3000 heats. The 410 mm length slide plate (P) consists of a 70 mm plate bore diameter (B) that eventually follows a 150 mm stroke length, as shown in Figure 10.5. Such assembly is attached with either a disk or ring type of spring, and their operating temperature varies from 450–500 °C. The slide gate system assembly has housing, support, springs components, slider, and cylindrical bracket. The ladle nozzle and upper plate are fixed in the housing, and the slider plate and exchangeable nozzle are fixed in the slider. The slide component is the active part in the support component of moving and adjusting the flow control. The Spring component in the plate boundary establishes a uniform and constant interfacial pressure. In working conditions, housing and support are hinged together through a safety pin hinge.

The pressure is transmitted to the plate boundary through the slider slide bar and support slide bar, driven by a hydraulic cylinder to achieve flow control.

10.4 Ladle Slide Gate Plate

Different shape plates are the major part of the ladle slide gate system that controls the metal flow rate during teeming and ensures quality steel. Briefly stated, it should have reliable and robust mechanics, operating regularity, easy dismounting, easy replacement of worn parts, and fast, easy upkeep.

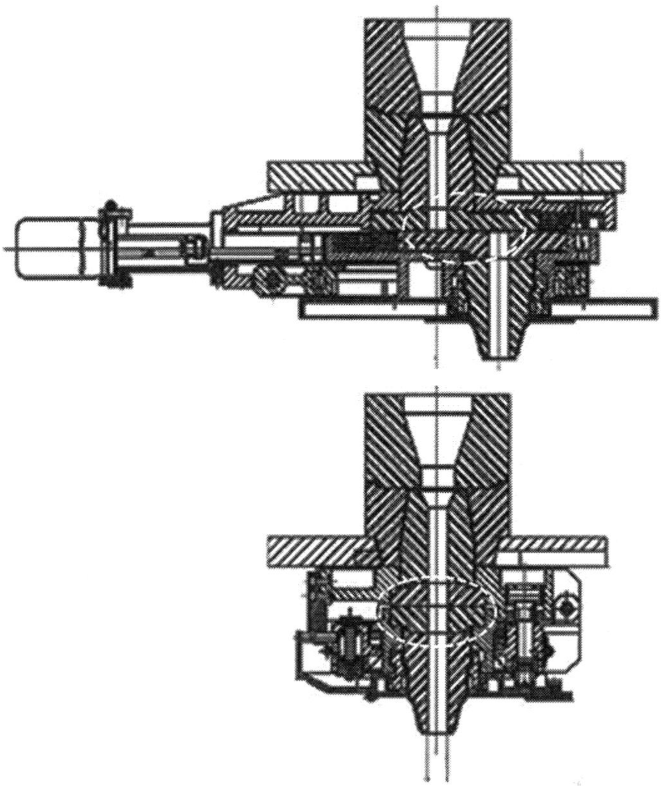

Figure 10.5 Schematic representation of slide gate, top–block, bottom–open condition.

Figure 10.6 represents the refractory assembly in which the bottom fixed slide plate and sliding plate are considered the prime flow control refractories and discussed elaborately. The SGP experiences thermomechanical stresses and severe corrosion in-service conditions, leading to microcracks and extended crack propagation. Such unusual gapping and cracks cause air leakage through the plates with adverse effects on the cleanliness of the steel and the wear of the refractory by corrosion.

Stress concentration varies near the hole, and nonsymmetry geometry expedites radial and lengthways cracks and eventually affects the plate performance. Consequently, the controlled plates wear is encountered due to variable air intake and acceleration of wear between first and last heat because of the deterioration of refractory permeability. This deterioration has repercussions on metal cleanliness (oxygen pick up, inclusions). Thus, in an optimized slide gate design, it would

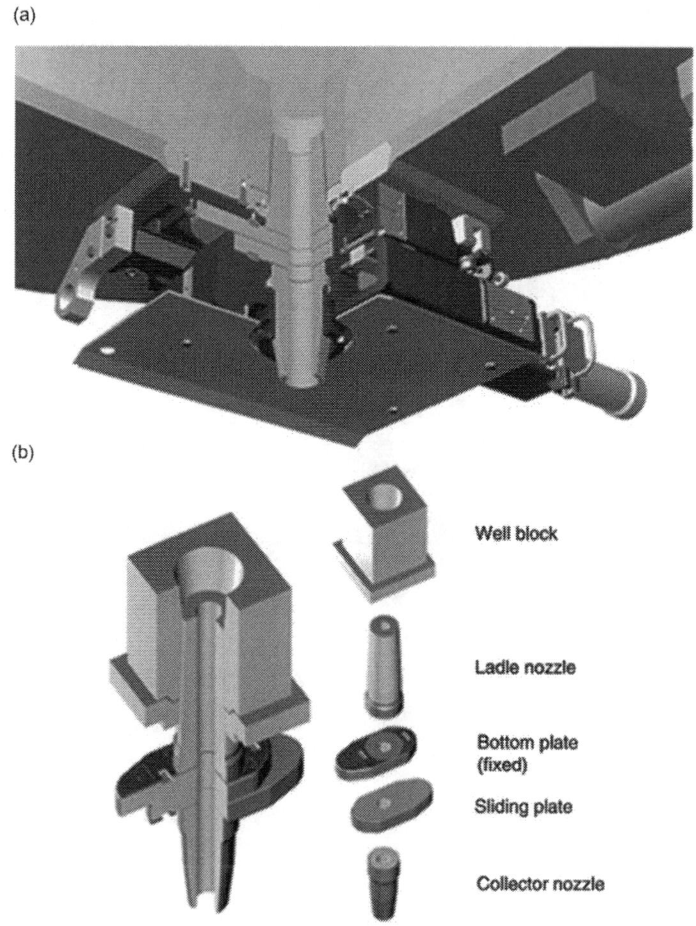

Figure 10.6 (a) Refractory in ladle slide gate assembly, (b) different refractory components [1].

be better to consider the specific plate shape, refractory composition to avoid unwanted corrosion and interaction with steel, and probable mode of failure, including thermomechanical stresses and cracking.

10.4.1 Critical Design Parameters

Design refers to both geometry and composition to achieve the best performance of the device. While geometry is a concern, different shapes and sizes of a plate are shown in Figure 10.7.

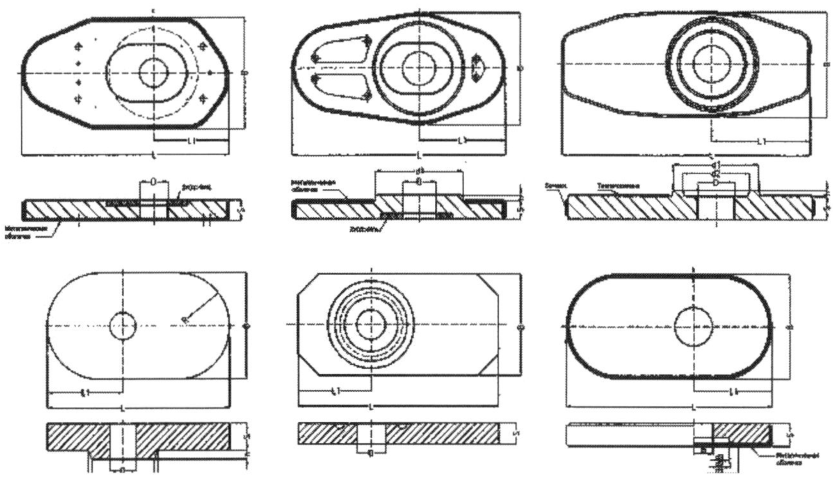

Figure 10.7 Configuration of slide plates by leading manufacturers [2].

Several shapes comprise the traditional elliptical form, symmetrical oval that allows several heats after 180° rotation, a drop shape experiencing uniform thermal stresses from the center of the opening during operation, mushroom shape, 4-point hold, etc. Eventually, FEA analysis and mathematical modeling became an effective tool for designing and developing a new class of slide plates. This analysis facilitates an understanding of the stress distribution and cracks formation in the slide plate and the critical design parameters of the slide gate plate. The mode of finite element analysis of slide plate has been discussed in Section 10.4.5.

It is necessary to ensure that all refractory parts must meet an essential obligation to sustain high temperature, wear, and chemical resistance during casting. Such critical properties eventually assist in absorbing the thermal shock and subsequent wear by molten steel and slag. Although the plate may experience several damaging phenomena like break cuts and peeling, surface roughening, edge rounding, decarbonizing, and cracks, as shown in Figure 10.8a [2]. There are four main types of cracks formed when plates are used in multiple heats, and their schematic representation is given in Figure 10.8b.

The first type of crack is perpendicular to the sliding direction of plates, which originates in the outer sides of it, known as (1) transverse cracks. These cracks often appear in plates used and are difficult to control. However, such cracks may not cause too much impact on the usage of SV plates. The second type of cracks are parallel to the sliding direction of SV plates and are very harmful to them, commonly known as (2 and 5) longitudinal cracks. The third

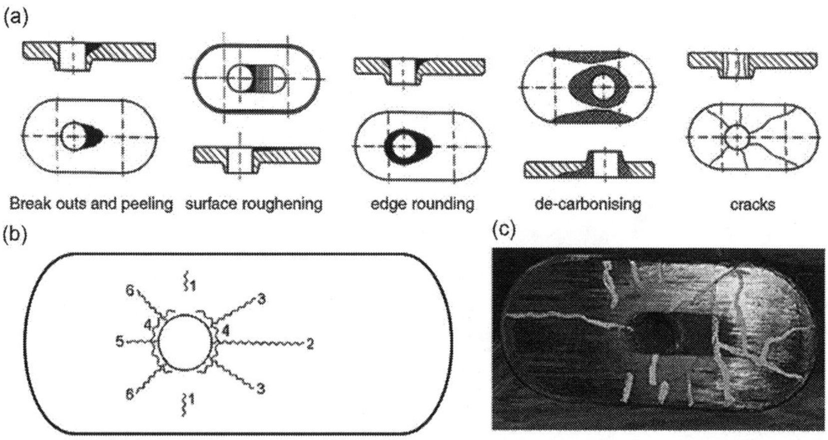

Figure 10.8 (a) Typical wear marks on refractory plates, (b) schematic representation of cracks, (c) real cracks on slide plates [2,3].

type of crack is at an angle to the sliding direction and is harmful; 3 and 6 are typical cracks at various angles. The fourth type of crack is around the cast hole; they are mainly caused by the expansion of the casting hole, known as ring cracks (4). The formation of perpendicular cracks that grow from the edge of the discharge channel to the lateral surface are related to the free expansion of the plate material. The longitudinal crack grows from the center of the opening to the corners of the plate as it moves, and it depends on the compression angle. The compression angle depends on the plate length/width ratio and the discharge opening size. The formation of transverse cracks, which takes place between the longitudinal cracks, depends on the load on the plate (Figure 10.8b) [3].

The growth of these cracks is influenced by the following factors: the critical strength of the refractory material of the plates; the uniformity of the compression of the front and rear surfaces of the plates; the relationship between the *compression angle* and the strength of the material; the method used to compress the plates; and the compressive load on the plates. It is impossible to avoid the formation of cracks during the casting operation, but accounting for the factors mentioned above can alleviate their effects—particularly when refractories are used multiple times.

The most important considerations in choosing the shape of the plate during its design are to have the cracks end up being located and oriented in vertical zones of the plate and to optimize the compression angle and compressive force. For the most part, the dimensions and configuration of the plates are unique to each manufacturer. Several manufacturers continue to use different methods that they have traditionally employed to improve the durability of plates:

- reinforcing the working zone of the plate by using a sealed insert or bushing;
- compaction of the material in the working zone;
- the manufacture of plates and nozzles with a projecting or recessed flange;
- delivery of plates and nozzles in a metal casing or shroud.

Making the shape of the plates more complex significantly increases production costs (due to the use of a mold, complications in charging the plates for the firing operation, the need for grinding, etc.). However, using the plates repeatedly can increase their service life by 15–20%.

A recent article illustrates the influence of plate shape on the strain [4]. Prime consideration was the position with respect to bore diameter. Suppression crack propagation by generating compressive stress around the bore provides better slide plate performance. Different plate geometry has been considered and measured cold experimental data such as strain gauge and plotted with respect to a position from bore diameter (mm). In the conventional shape, tensile strain occurs at a certain part or the whole plate in the throttling direction. Whereas in the 4-point hold (new design), compressive strain is generated over the entire part in the throttling direction, and it is possible to prevent cracks from propagating (see Figure 10.9).

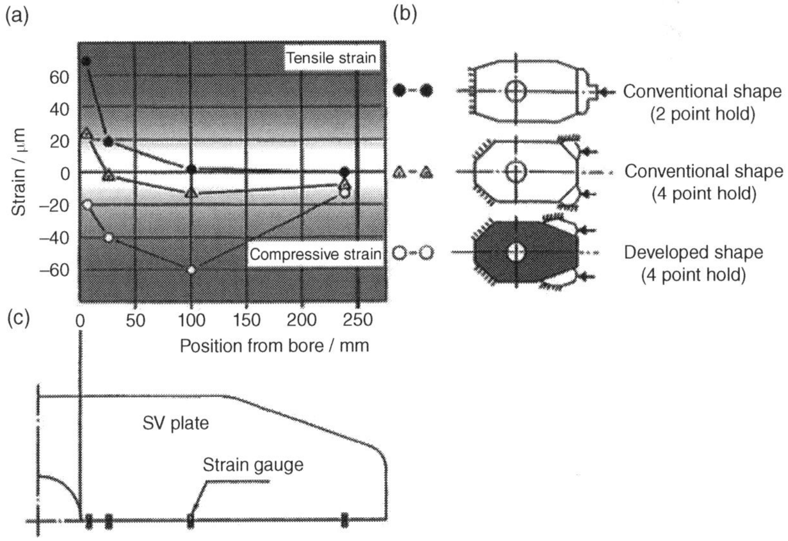

Figure 10.9 Relationship between plate shape and strain in a different position [4].

Conventional 2-point or 4-point geometry experiences tensile stress along the throttling direction that insists on forming longitudinal cracks and affects durability of the plate. Such cracks can be prevented in a developed 4-point shape, and employing this concept, the plate life has been enhanced from 5–7 heats. The plate weight was reduced by 29%, eventually reducing the material consumption by employing the optimized shape.

10.4.2 Selection of Slide Plate and Fixing

The slide gate plate refractory selection is a critical issue during flow control, and it varies from plant to plant, quality of steel, and operational conditions. Some basic guidelines may be encountered for the specific selection and understanding of the reasonable dimension and composition design [5].

a) Steel chemistry

Different metallic and nonmetallic species control the chemical reaction with slide plate refractory at high-temperature casting. Thus, plate refractory must be compatible with steel chemistry. Otherwise, a severe problem may happen during casting leading to disaster. Different steel grades and their relevant refractory are mentioned in Table 10.2; however, a specific selection may demand many more factors for consistent performance.

Table 10.2 Matrix refractory selection based on steel chemistry.

	Mild steel	**Calcia added steel**	**High Mn steel**	**High Si steel**	**High O_2 steel**	**High O_2 and Mn steel**	**Stainless steel**
		Above 30 ppm	Above 0.8%		Above 70 ppm	Free cutting grade	
ALU-ZIR-C	✓	✗	✗	✓	✗	✗	✓
ALU-C	✓	✗	∞	✓	∞	∞	✓
SPL-C	✓	✗	✓	∞	✓	✓	✗
MAG-SPL-C	✓	∞	✓	∞	✓	✓	✗
MAG-ALU-C	✓	∞	✓	∞	✓	✓	✗
MAG-C	✓	✓	✓	✗	✓	✓	✗
ZIR	✓	✓	✗	✓	✗	✗	✗
	✓ = Most suitable			∞ = Suitable		✗ = Not suitable	

ALU-Al_2O_3, ZIR-ZrO_2, SPL-Spinel $MgAl_2O_4$, MAG-MgO, C-Carbon

Table 10.3 Quality selection based on sliding stroke length.

Quality	>200 mm	150–200 mm	<150 mm
ALU-ZIR-C	✓	✓	✓
ALU-C	✓	✓	✓
SPL-C	∞	∞	✓
MAG-SPL-C	∞	∞	✓
MAG-ALU-C	✗	∞	✓
MAG-C	✗	∞	✓
ZIR	✓	✓	✓
	✓ = Most suitable	∞ = Suitable	✗ = Not suitable

b) Sliding stroke length

Slide plate performance also depends on the stroke length; thus matrix selection along with carbon phase is an important criterion. Here, the matrix indicates major oxide phases. For example, MgO is suitable for smaller plates, but Al_2O_3 is most apposite for a bigger plate. A typical list of materials and variations of the stroke length is shown in Table 10.3 to provide some basic guidelines.

c) Caster type

The variety of caster determines the throttling and high abrasion of plates. Thus, the manufacturing of slide plates by an appropriate bonding system leads to the performance concerning caster variation. For example, the slide plates for ingot and billet caster experience a relatively high degree of throttling severity compared to slab and bloom caster. Different bonding systems such as ceramic, carbon, metal, and light burnt metal bond for different plates are suitable for any type of casters. Ceramic bond is the most stable bonding system developed through solid state sintering at high temperature, and carbon bond is the most conventional bonding system with proper firing. Metal and light burnt metal bonds are the advanced bonding systems obtained through suitable heat treatment and specific firing. Table 10.4 is shown as a basic guideline for different bonding systems concerning the casters.

d) Calcium ppm in steel

The selection of slide plate is important when calcium concentration in the steel contains more than >20 ppm. Table 10.5 shows a guideline for selecting the particular refractory considering steel chemistry. Most slide plates are carbon

Table 10.4 Quality selection based on caster type.

Type of caster	Metal bond	Light burnt metal bond	Carbon bond	Ceramic bond
SLAB CASTER	✓	∞	∞	✓
BLOOM CASTER	✓	∞	∞	✓
BILLET CASTER	✗	∞	✓	✓
INGOT CASTING	✗	∞	✓	✓
	✓ = Most suitable	∞ = Suitable	✗ = Not suitable	

Table 10.5 Quality selection based on calcium ppm in steel.

Quality	<10 ppm	10–20 ppm	20–30 ppm	>30 ppm
ALU-ZIR-C	✓	∞	✗	✗
ALU-C	✓	✗	✗	✗
SPL-C	✓	✓	∞	✗
MAG-SPL-C	✓	✓	∞	✗
MAG-ALU-C	✓	✓	∞	✗
MAG-C	✓	✓	✓	✓
ZIR	✓	✓	✓	✓
	✓ = Most suitable	∞ = Suitable	✗ = Not suitable	

bonded alumina or magnesia matrixes capable of withstanding abrasion, thermal shock, and oxidation compared to other bonding systems.

In practice, the same plate quality is used to cast different steel grades; thus proper selection for a particular shop is critical before manufacturing and supply. Most common alumina carbon plates are used but experience high corrosion in the presence of high Ca >15–30 ppm, as shown in Figure 10.10. MgO-based plate is a choice for Ca-treated steel due to excellent corrosion resistance for smaller plates. Zirconia added alumina carbon plate is considered the most common composition for bigger plates to overcome this problem, although it experiences a specific abrasion and corrosion during casting of various steel grades like high Mn, Ca, and oxygen steel [6].

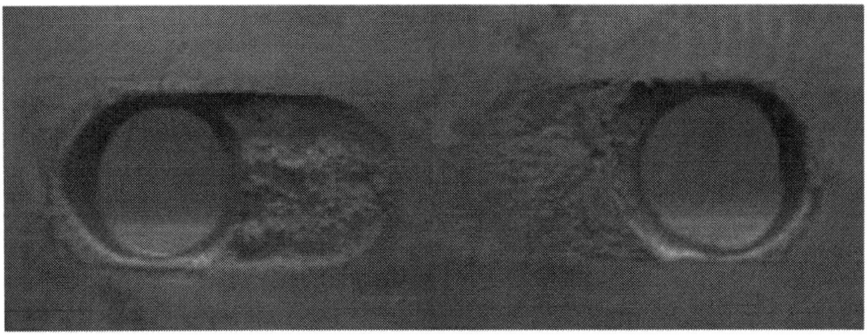

Figure 10.10 Damage of Al_2O_3-C plate in Ca alloy treated steel [6], Ganguli et al. TRLKROSAKI REFRACTORIES LIMITED.

Moreover, the direction of crack growth also depends on the method used to secure the plates in the slide gate. A plate can be secured either by point-to-point contact or by plane compression. Thermal expansion of the plate from the center of the product in the zone occupied by molten metal causes the entire product to expand by several millimeters. The plate may crack if it is rigidly fixed in position and moves inside the body of the gate if it is not strongly secured. The method used to fasten the plate should be chosen with allowance for the coefficient of linear expansion of the refractory comprising the plate. Plates are often secured with special bolts, eccentrics, or *cramps* that are the same shape as the plate. Each fastening technique has advantages and is an important element of the slide gate design.

10.4.3 Materials and Fabrication of SGP

Prior to fabrication of one of the most important flow control refractories, consideration of several parameters may promote balanced properties, including erosion resistance against molten metal and liquid slag, thermal spalling resistance, anti-abrasion resistance with sufficient strength against the metal stream, and resistance of free oxygen either in metal or ambient air that may adsorb through the surface of the plates. The production of slide plate differs in important ways from the mass production of refractory materials, and some are:

- the use of synthetic or highly concentrated raw materials exclusively;
- the use of sintered or fused materials or combinations of different types of materials;
- the simultaneous use of different modifications of the same material in the charge (crystalline or scaly graphite, anthracite, carbon black, or organic binders);

- the use of different alloying, antioxidizing, or other types of additives;
- the use of the latest advances in refractories in manufacturing technology (diamond grinding and drilling, the production of carbon and organic binders, electro smelting, etc.).

The main factor determining the durability of slide-gate plates is the type of binder used. The binder forms the material's structure, determines its properties, and eventually differentiates from one another.

The refractory products are obtained using high-purity raw materials in fused or sintered form by conventional mixing, pressing, and baking technology. Usually, these plates are made based on refractory raw materials, either single or a combination of a few materials: Al_2O_3, MgO, ZrO_2, mullite ($3Al_2O_3.2SiO_2$), or spinel ($MgO.Al_2O_3$). The choice of other materials in the matrix is required to fulfill the target of low thermal expansion, high thermal conductivity, and resistance to high thermal shock. Despite several considerations of carbon sources, the natural scaly graphite is primarily used along with an organic binder. The primary role of the organic binder is to ensure that the ceramic material has adequate strength [3].

Al_2O_3-C refractories are most common; however, some alternate composition or modification is required concerning steel chemistry, as shown in Table 10.2. This material is used to make Al_2O_3–C refractories because it has the following properties: high melting point; good thermal conductivity; low coefficient of linear expansion; durability at high temperatures; low friction coefficient and consequent good compressibility; low wettability by molten slag; and thus superior resistance to corrosion or erosion; high resistance to oxidation compared to other forms of carbon, with oxidation in an oxidizing atmosphere beginning at 500 °C.

The organic binder, such as phenolic resins used as binders for Al_2O_3 refractories, undergoes pyrolysis during coking. The material should have a 55–65% resinous coke residue to obtain the desired properties. Phenolic resins form the strongest bond with other carbon-based materials and exhibit the highest degree of compatibility, but they are less resistant to oxidation than other binders. Carbon-based binders decrease the ability of the working surface of the plate to be wetted by metal and slag.

Corundum crystal is bound quite firmly to graphite particles in plates having a carbon binder. The structure of the plates is distinguished by its high density of fine pores smaller in magnitude in carbon binder-based plates than the pores in plates having a ceramic binder. Thus, the total surface area of the pores is smaller, making the plates more resistant to corrosion.

However, corundum-graphite plates can undergo more rapid wear when used to cast special steel grades. The plate material is corroded by the action of traces of different components and inclusions in the liquid steel: CaO, MnO, and FeO readily react with Al_2O_3 to form low-melting compounds that accelerate the corrosion of

Al_2O_3 refractories. During special steels (bearing steels, high-manganese steels, steels treated with calcium-silicon, etc.) casting, these compounds elevate the corrosion rate of Al_2O_3-C plates. Consequently, corundum-zirconia- graphite plates improve the durability of slide-gate plates that have a carbon-based binder. The manufacturer makes such plates using a specially synthesized zirconia-bearing raw material, and the addition of 2–6 wt.% improves the corrosion resistance of the plate and the heat resistance of the refractory product. Refractories based on this material can be used 2–4 times in modern slide gates. Specific raw materials like synthetic mullite, andalusite, corundum, and silicon carbide are added to enhance the thermal shock stability and corrosion resistance of the non-burning Al_2O_3-C slide plates. Special alloy additions and chemical treatment in steel ladle, in particular, Ca deoxidation process, enhance the severe chemical environment and slide plate refractory attack during steel flow. Adding oxide-bonded zirconia may reduce the erosive and corrosive attack [7]. Optimum mechanical response in a cold and hot environment is desirable to maintain the necessary sliding integrity during service. For example, a slide plate with cold strength (~140 kg/cm^2) and hot strength (~56 kg/cm^2) may perform better casting. Calcia (CaO) stabilized ZrO_2 usually has high cold strength, but a dramatic drop in strength is noticed at high temperatures. It exhibits relatively low hot strength in the range of 12–28 kg/cm^2, not adequate for long-term slide gate service during casting. In actuality, the impurities present in CaO stabilized ZrO_2 migrate toward grain boundaries and form a glassy phase that eventually reduces high–temperature strength. Thus, a suitable refractory composition may enhance the erosion and corrosion resistance of the slide gate valve or insert for such plates. High purity magnesia doped partially stabilized ZrO_2 85 wt.%, Al_2O_3 7 wt.%, Si metal 4 wt.%, graphite 4 wt.%, and 4–7% carbonaceous binder mix may press into the desired shape and fired in a reducing atmosphere above 1000 °C to produce carbon boded zirconia plate. It eventually provides superior hot strength but is relatively expensive because of the high cost of zirconia.

Carbon-bonded Al_2O_3-C and Al_2O_3 (85 wt.%)-ZrO_2 (7 wt.%)-C (8 wt.%) are common slide plate refractory but do not have satisfactory Ca corrosion resistance during Ca-treated steel casting.

MgO-graphite refractory shape has a unique combination of hot strength at the application temperature (1500 °C range), thermal shock, abrasion, oxidation, and hydration resistance. Thus, it can meet the requirement of Ca-treated steel and high O_2 steel to a certain extent. However, the high coefficient of thermal expansion of MgO expedites thermal spalling during casting; thus further improvement is required.

Magnesia-carbon refractory shape with a carbon bonding system including a magnesia-alumina spinel and aluminum carbides precipitate the bonding phase forming in-situ during the high-temperature firing treatment. The resultant properties are especially suitable for slide gate plates, inserts for such plates, and pouring nozzles.

Despite the wide variation, usually, finely divided graphite (6 wt.%), aluminum metal powder (3 wt.%), silicon metal powder (1 wt.%), an effective amount (5.5%) of carbon bond forming resins, and the balance of coarse and fine magnesia grains plus incidental impurities such as boron, calcia, and silica are used for this refractory [8].

Carbon in the surface layer of the plate undergoes oxidation and combustion by the following reactions:

- the carbon and the finely ground part of the periclase in the product react with the pyrolytic carbon formed as a result of coking of the organic binder

$$MgO\,(solid) + C\,(pyr) = Mg^{2+}\,(gas) + CO \tag{10.1}$$

Above reaction expedites the formation of gaseous Mg^{2+} that moves from the pores to the surface layer of the refractory;

- oxidation of metallic magnesium;

$$2Mg^{2+} + O_2 = 2MgO\,(secondary) \tag{10.2}$$

Many studies have shown that the formation of a dense secondary layer of MgO on the hot surface of the refractory has a protective effect. An antioxidant such as Al and Si has a significant role in forming in-situ phases and contributes to resultant properties. During service, a secondary refractory binder (SiC, Al_2O_3, SiO_2, etc.) is formed. The exact binder that is developed depends on the refractory and antioxidant used. The secondary binder fills the surface pores and increases the service life of plates. The metallic additions are oxidized during the casting operation, which increases the refractory volume and decreases its porosity and gas permeability. The following reactions take place when additions of metallic aluminum are used in periclase-spinel plates:

- combustion of the metallic aluminum $2Al + 1.5O_2 = Al_2O_3$, with the formation of active alumina Al_2O_3 being accompanied by a 38.5% increase in volume;
- the formation of spinel $MgO + Al_2O_3 = MgO{\cdot}Al_2O_3$ is accompanied by an 8% increase in the volume of the material;
- the overall reaction resulting in the formation of the spinel $MgO + 1.5O_2 + 2Al = MgO{\cdot}Al_2O_3$.

The amount of metallothermic spinel that is formed in this case is several times greater than when the introduction of pure Al_2O_3 powder forms it. For example, the introduction of 5% metallic aluminum leads to the formation of 13.2% $MgO{\cdot}Al_2O_3$, while the introduction of 5% Al_2O_3 powder leads to the formation of just 6.9% spinel. Spinel formation from metallic aluminum begins at 700 °C. Eventually, the metal powder (3.0–5.0%) alters periclase-graphite refractory density within 700–1600 °C. Adding metallic aluminum increases the strength and heat resistance of MgO-C refractories by a factor of 2–3, making them 3–5 times more resistant to slag. The following additional reactions take place when metallic aluminum and silicon are added to the refractory together:

- the formation of SiC, which begins at 800 °C, $Si + C = SiC$;
- the oxidation of SiC; $2SiC + 3O_2 = 2SiO_2 + 2CO$.

The oxidation of SiC results in the formation of finely dispersed SiO_2, which reacts with finely ground alumina in corundum-graphite refractories to form a dense layer of secondary mullite. This secondary binder prevents the oxidation of graphite from the matrix of the plate and makes it considerably more durable.

MgO-Al_2O_3 spinel-carbon slide plate has less thermal expansion coefficient than pure MgO; and thus provides relatively higher spalling resistance than MgO. However, a certain reaction probability between spinel and Ca in the steel forms a low melting phase and affects service life.

Modern technologies are adapted to develop slide plate through cermets, metallic fibers, SiAlON, etc., eventually useful for repeated heat. A sialon binder can be obtained by introducing sialon (Si–Al–O–N) or synthesizing it during production stages (firing in a protective medium, etc.). The use of composite products containing certain regions in which one of several properties (heat resistance, etc.) is predominant makes it possible to achieve a balance between corrosion-resistance and heat-resistance characteristics over the entire plate volume. Combining finely dispersed chromium oxide and additions of metallic aluminum and molybdenum facilitates both the oxidation and heat resistance of the plate. The chromium in the metallic phase of the composite has a melting point of 1800 °C and, when oxidized, forms a dense oxide film that provides a high degree of resistance to further oxidation. Molybdenum (Mo) has a melting point of 2600 °C and is highly heat-resistant. The resulting composites possess a good combination of heat resistance and corrosion resistance. A nanostructured matrix is formed within their structure by using aggregate-type carbon particles in the form of hybrid graphite soot and a high-quality hybrid binder to increase the service life of the plates. This not only makes the plates more resistant to heat, but it also improves the bond between the particles [3].

ZrO_2 has good high-temperature performance, but high cost restricts the making of the entire plate; thus industry generally prefers to use zirconia ring embedded around the casting hole of Al_2O_3-C slide plate and make a composite slide plate. Thus, sliding gate plate inserts are an excellent solution to increasing the life span of sliding gate plates. A composite plate with an insert made of hot-pressed corundum-graphite and zirconia refractories is widely used for casting aggressive grades of steel. A typical CAD design of such assembly is shown in Figure 10.11. Once a sliding gate plate becomes worn, the worn material can be removed, and a high density zirconia plate insert can be cemented/cast in its place, extending the life of the sliding gate plate.

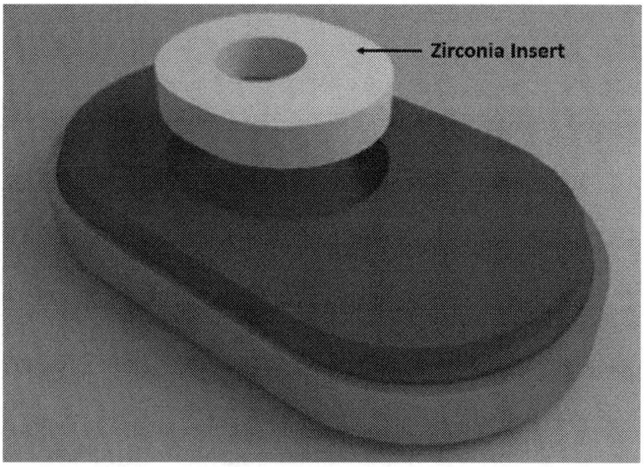

Figure 10.11 The zirconia insert in slide plate.

10.4.4 Mode of Failures

Usually, the slide plate experience four types of damage, and in obvious that there are several reasons for such damage. These damages can be classified as crack formation, edge damage, bore enlargement, and sliding surface damage, as shown in Figure 10.12. Prime factors of such damages are:

- Refractory properties: Mechanical and thermal properties, thermomechanical properties, and chemical properties
- Structural factors: Shape of the plate, fixing method, a pressure between sliding surfaces
- Operating conditions: Steel grade, casting temperature, casting time, intermittent use
- Maintenance: Oxygen cleaning, opening check

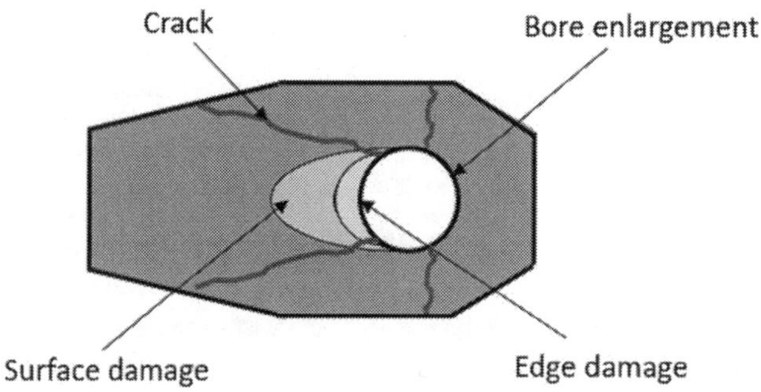

Figure 10.12 Typical damages in slide plate.

Crack formation: External load and thermal stress are the prime resources for cracking formation. The crack pattern predominantly depends on structural factors like plate shape and curvature and the fixing method of the assembly. Usually, a longitudinal crack forms in the throttling direction from the bore and becomes the edge damage origin that results in deterioration of durability. Opened cracks allow oxidation, corrosion, and molten steel leakage. Resistance of longitudinal cracks is an important issue to prevent premature failure and enhance the service life. Despite making the crack sensitive by forming a dense matrix, slide plate geometrical design parameter alteration reduces cracking. A recent article conducted a competitive stress analysis and found an optimal geometrical design, as shown in Figure 10.13 [9]. FEA reveals that a new shape prevents the tensile stress zone and restricts the formation of longitudinal cracks.

While considering the fixing issues, 4 methods can be adopted: i) 2-point fixing by 2 individual bolts, ii) 4-point fixing by 4 individual bolts, iii) 4-face fixing by 1 bolt, and iv) mortar setting in a slide gate housing. Four-face fixing with one bolt secures the plate more effectively to prevent and reduce longitudinal and other cracks at certain degrees from the sliding direction.

Despite the structural design, composition design is an important step as it experiences severe thermomechanical erosion.

Slide plate refractory experiences a distinct temperature gradient as it has a low temperature before steel tapping; however, a sudden rise as high as 1600°C of the

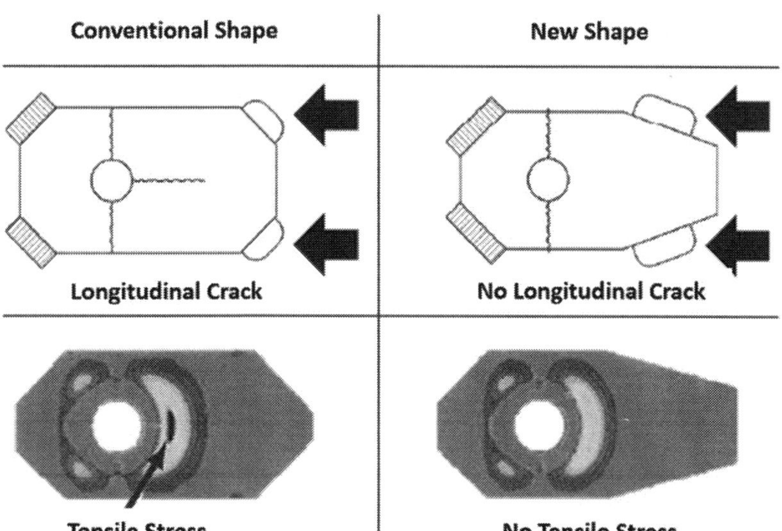

Figure 10.13 Shape variation clearly indicates the formation of tensile stress followed by longitudinal crack [10], Douglas et al. (2016), ECREF - European Centre for Refractories gGmbH.

hole is due to molten steel, a common practice. Rapid temperature change ($\Delta T = 1400\,^{\circ}C$) facilitates thermal shock and a substantial amount of tensile stress and releases as microcracks when exceeding the slide plate strength. It forms radial microcracking centered on the casting hole and is prone to diffusion, accumulation, penetration of foreign impurities, and expediting the chemical reaction. Within due time, the gradual reaction switches to a corrosion reaction that promotes the crack enlargement and expansion, eventually expanding and damaging the casting holes of the plate. Despite stress factors, bore diameter experiences severe friction due to molten steel flow, and cause peeling and block loss.

Edge damage: Edge damage is predominantly initiated by chipping because of spalling cracks, air entrapment through large cracks and corrosion, texture deterioration, and direct corrosion by molten steel interaction. A judicious material design can prevent this incidence: zirconia-based insert or in the matrix is an excellent choice to overcome this problem.

Bore diameter enlargement: Oxygen contamination with carbonaceous refractory at elevated temperature is always detrimental, enhancing corrosion. Thus, bore diameter is significantly enhanced by oxygen cleaning during maintenance despite corrosive grade steel. Bore diameter enlargement expedites the probability of further contact between the sliding surface of the plate and the molten steel as the plate has to move in the closing direction to maintain an equivalent flow rate. Thus, increased contact area leads to a damaged edge and sliding surface.

Sliding surface damage: Generally, surface damage limits plate life. Gradual development and surface damage expansion may insist on difficult flow control as it restricts the sliding difficulty and stoppage failure. It promotes inter-plate leakage of molten steel and enhances the damage width and distribution. In a study, the morphological change in the sliding surface damage width 1 mm around bore diameter is predominantly due to abrasion, corrosion, and peeling. The causes of these include corrosion due to eutectic reaction, texture deterioration due to thermal shock, and/or decarburization. Figure 10.14 shows the surface damaged part of the Al_2O_3-C material.

Decarburization loosens the binding between matrix particles; and thus expedites the metal penetration and alteration of refractory particles in the vicinity of the hot region. In the case of carbon-containing refractory, it is important to protect and suppress the *texture* deterioration to improve the durability. In a much more elaborate way, the following fundamental reasons can be encountered for thermochemical erosion.

Oxidation of graphite and carbon in the presence of oxygen in molten steel, especially high-oxygen steel, facilitates C oxidation to form pores and subsequent iron penetration and adherence between the plate and solidified steel. More vulnerable oxygen from air oxidizes the carbon and forms low-melting material. The low-melting material continues to corrode and penetrate along the pores.

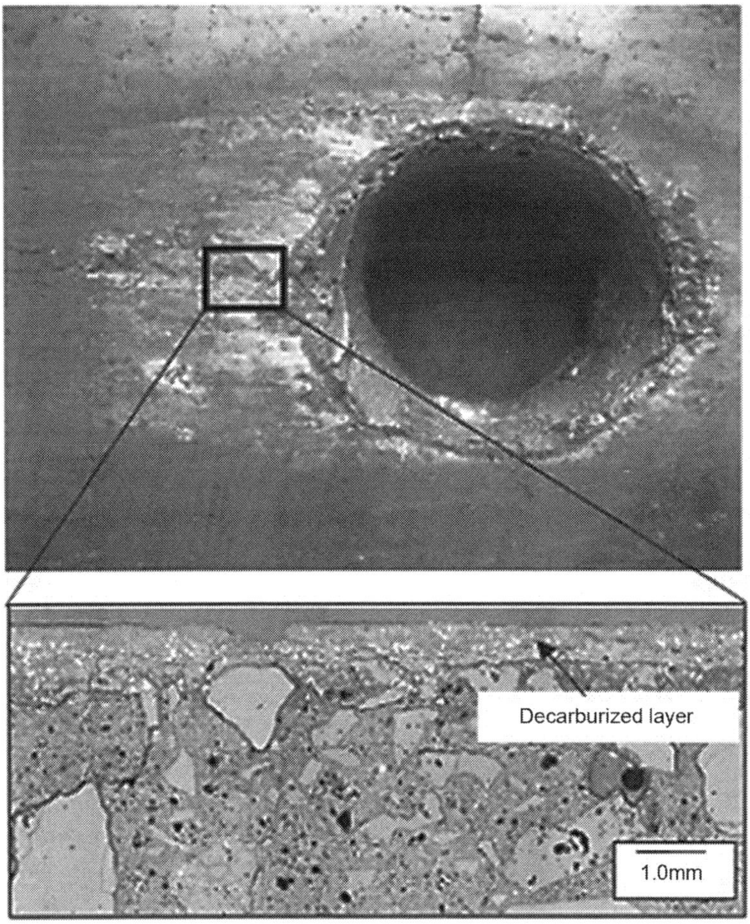

Figure 10.14 Surface damage of Al_2O_3-C plate after use [11], SHINAGAWA REFRACTORIES CO.,LTD., 2019.

Operation Factors: Operating factors, including unreasonable installation of the plate, unreasonable pouring flow control in production, and unreasonable oxygen burning operation, are prime aspects to damage slide plates. Any presence of warping or loose clamping during slide plate installation may also generate external stress that eventually promotes the slide plate damage. Sometimes unavoidable incidence are the cause of unreasonable pouring process including prolonged operation time, frequent movement, and other causes are responsible for premature slide plate refractory. A manual operation may facilitate damage caused by human factors. Uninterrupted flow during poring process demands ladle nozzle burning by oxygen, and unreasonable serious oxygen burning may deteriorate early stage of failure.

The plate is repeatedly exposed to high-temperature molten steel during use, and the use conditions are harsh. Therefore, the plate is required to have the following properties to ensure that there is no leakage of molten steel between the plates during the casting process. Its structure and performance should generally meet:

- The sliding surface is smooth and flat.
- High mechanical strength.
- High corrosion resistance in contact with molten metal and slag.
- Minimization of molten steel adherence to the plate.

It is necessary to pay attention to cleaning up the nozzles, inner region, and door mechanism and avoid using any deformed or cracked plates. High-speed oxygen is mandated to blow from outside to inside to avoid clogging the inner channel.

10.4.5 FEA for Stress and Cracking

Considering slide plate use and different 4 modes of cracks (Type 1—perpendicular to sliding direction, Type 2—parallel to sliding direction, Type 3—an angle with sliding direction, Type 4—around the cast hole) formation is shown in Figure 10.8a. Finite element analysis is employed to develop a new slide plate that eventually experiences higher and more evenly distributed compressive stress and can prevent the formation of the cracks along the sliding direction of the plate effectively. Before establishing FEA, the cracking phenomenon is considered on plate cross-section, as it is much larger than plate thickness, so the analytical model can reduce to a two-dimension plane-stress problem. The simplified finite element model is shown in Figure 10.15a.

In operation, the plates experience severe thermal stress; and thus considered thermal loads and boundary conditions are: (1) before pouring steel, the initial temperature of the plate is 573 K; (2) when pouring, the surface of the casting hole is heated up to 1873 K in 3 seconds; and (3) the air temperature around the plate is 343 K, and the heat transfer coefficient between them is 15 W/mK. Different material properties are considered to perform such analysis and mentioned in Table 10.6.

In the first type of crack, thermal stress distribution changes with time. The thermal stress being more significant and changing more acutely is near the straight line UV, as this region has minimum cross-section area in the plate. While pouring the steel through the cast hole, maximum tensile stress is located at point U, and the highest stress is noticed at the 402nd second, then it begins to decline gradually, as shown in Figure 10.15b. Due to thermal contact, expansion expedites tensile stress and releases the thermal stress beyond the critical crack limit. When the thermal stress is no higher than the tensile strength of the plate, crack

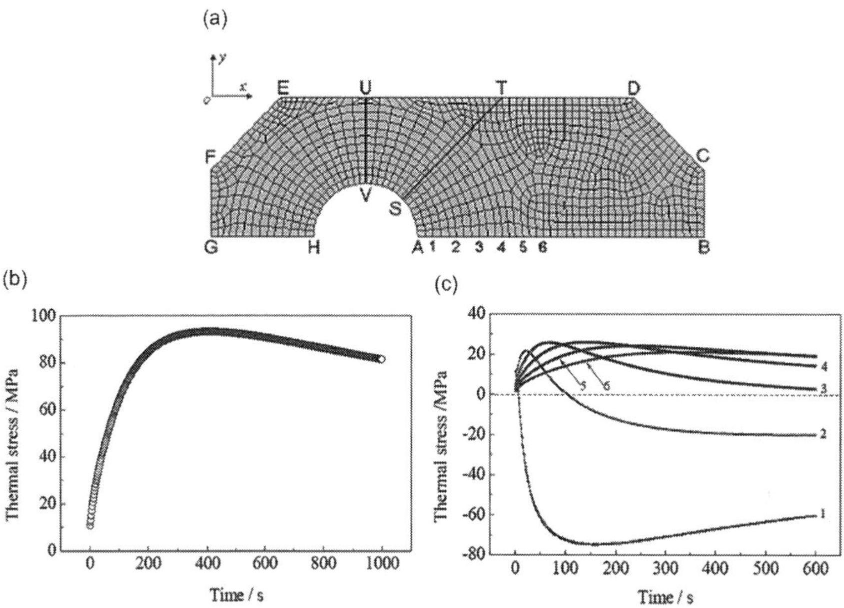

Figure 10.15 (a) 2-dimensional plane-stress FEA model, (b) change in thermal stress with time at point U, and (c) changes of the thermal stress with time at points 1–6 [11].

Table 10.6 The parameters of physical properties of SV plate.

Material properties	λ [W/(m.k)]	Cp [J/(kg.K)]	ρ [Kg/m^3]	E [GPa]	μ	α × 10^6[1/K]
Values	9.5	1200	3330	22.5	0.2	7.0

λ = Coefficient of thermal conductivity; Cp = Specific heat; ρ = Density; E = Elastic modulus; μ = Poisson's Ratio; α = Coefficient of thermal expansion

propagation is restricted. It is expected that the maximum tensile stress position changes during or after the generation of the first type of crack and locates in line with AB in the plate. This stress shows that the maximum stress point gradually moves from the cast hole to the right side along line AB, and the maximum stress increases first and then decreases with time. Figure 10.15c shows that the thermal stress of points closer to the cast hole is tensile stress at the beginning and gradually transforms to compressive stress with time (points 1–6); points far away from the cast hole suffer tensile stress all along. Theoretical analysis shows that the formation of the third type of crack is behind the second type of crack. However,

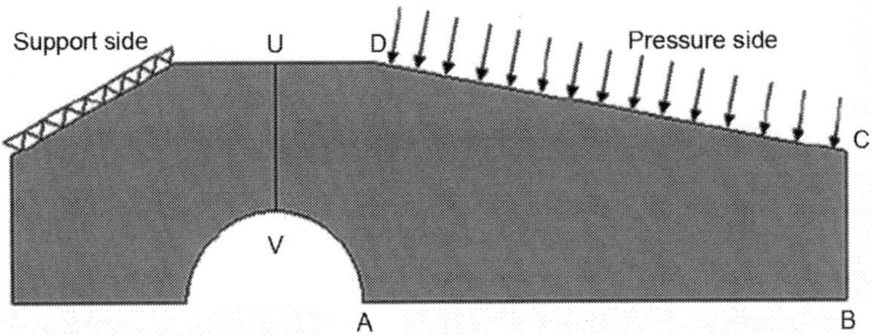

Figure 10.16 Design alteration to enhance the compressive stress and reduce the longtidunial crack in slide plate [11]. Credit: RHI Magnesita NV

in general, the maximum stress value gradually reduces with time, and the point of the maximum stress moves along the vertical direction of arc VA, in accordance with the location where the third type of cracks forms. While considering the fourth type of cracks, very high compressive stress exists around the inside of the cast hole, but tensile stress is a little far from it.

This is mainly caused by the thermal expansion of the material around the cast hole in the thickness direction. The figure also shows that the tensile stress zone along *z*-direction is around the cast hole, which is consistent with the formation of the fourth type of crack. Based on the analysis above, the formation of the plate cracks is because the tensile stress exceeds its tensile strength. An effective method to avoid cracks or reduce their hazards is to preimpose compressive stress in SV plates, which has been applied in actual production through the new model highlighted in Figure 10.16.

10.5 Tundish Slide Gate and Plate

A continuous casting facility includes a long nozzle, tundish refractories, slide gate plates, and submerged entry nozzles, as shown in Figure 10.2b. Because the refractories for continuous casting are used in the last stage of the steelmaking process, the stability and the durability of the refractories affect the productivity of the continuous casting operation and the steel quality.

Slide gate plates are one of the functional refractories. They are used for controlling the flow rate of steel, i.e., from the tundish to the mold.

Figure 10.17 shows a schematic view of a slide gate plate. Slide gate plates consist of two or three plates, with a central hole that is the channel for steel flow. By sliding one plate and shifting the position of the central hole, the slide gate plate can control the flow rate of the steel.

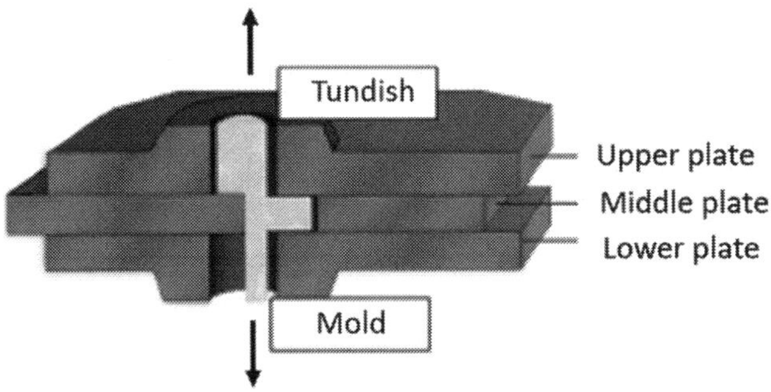

Figure 10.17 Schematic view of tundish slide gate plates.

10.5.1 Modern Slide Gate and Refractory Assembly

The slide gate is assembled with the required support, clamping ring, spring and drive components, and relevant refractories. The throttling plate separates the upper fixed plate and submerged entry nozzle (SEN) plate. The upper fixed plate is installed with an inner nozzle or monoblock tundish nozzle plate (MTNP) for accurate location, and the rest refractory system may have different permutations and combinations to maintain the flow through SEN.

The throttling plate is connected to a hydraulic cylinder and moves in a horizontal direction to regulate and maintain the eye-shaped opening to control flow rate and constant meniscus level in the mold. The automated eddy current sensor suspended above the mold sends feedback and controls the measured level. Under this circumstance, the surface pressure is established accurately and stably before casting the molten steel. Different slide systems commonly, 5, 4, or only 3 refractory components are designed to assure steel quality, as shown in Figure 10.18. The closed 3 system assembly ensures no more joints; thus no air filtration. In addition, inert gas purging through MTNP or inner nozzle restricts alumina clogging during casting.

Vesuvius coined an innovative design including MTNP, a one-piece tundish nozzle that includes argon injection through zirconia insert to avoid clogging, as shown in Figure 10.19.

10.5.2 Materials and Fabrication

Cumulatively, the slide gate refractories should be equipped with suitably designed features, machinery precision durability, and severe operating conditions such as thermal shock and abrasion. In addition, it can control the pouring rate of molten steel to guarantee safe operation, higher productivity, energy saving, better quality, and safety with reduced cost. The competitive difference between the ladle slide gate plate and the tundish slide gate plate is shown in Table 10.7.

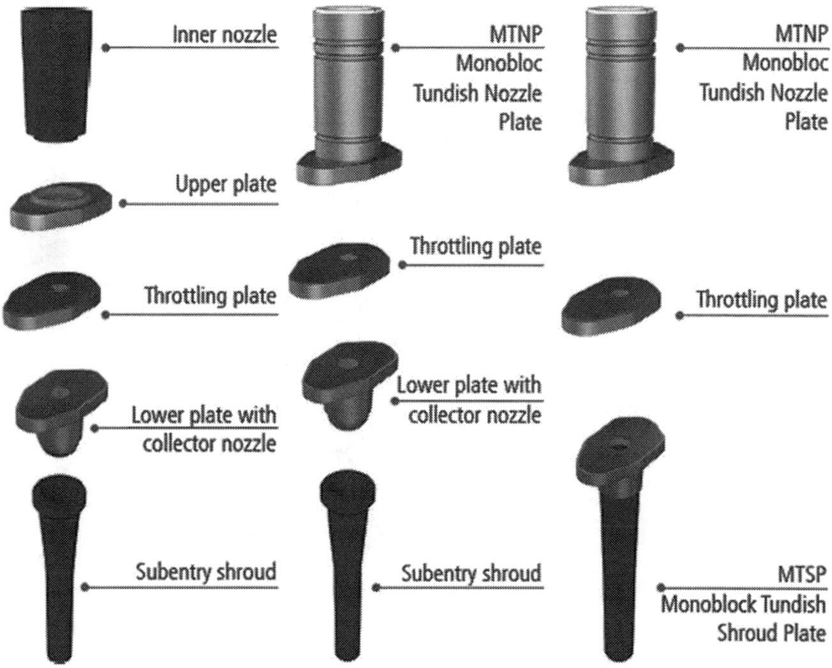

Figure 10.18 Different 5, 4, and 3 slide plate systems for tundish [13], Vesuvius, 2013.

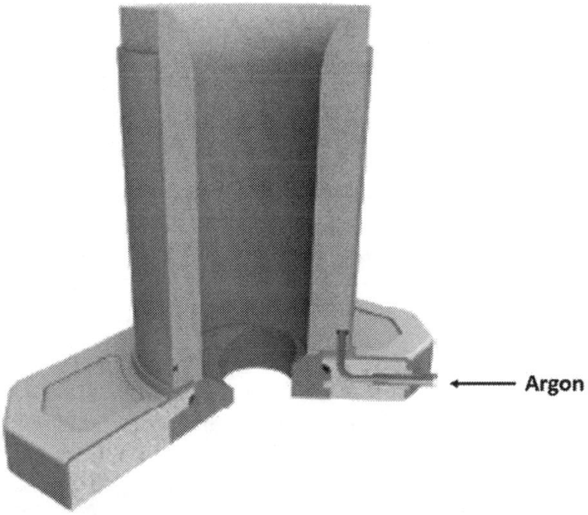

Figure 10.19 Latest technology of Ar injection attachment zirconia inserted tundish nozzle, developed by Vesuvius. Credit: Vesuvius

Table 10.7 Type of sliding plate for different zone and steel quality [14], Chosun Refractories, Co. Ltd.

Technology	General ladle	Closed start	Special steel
Plate geometry	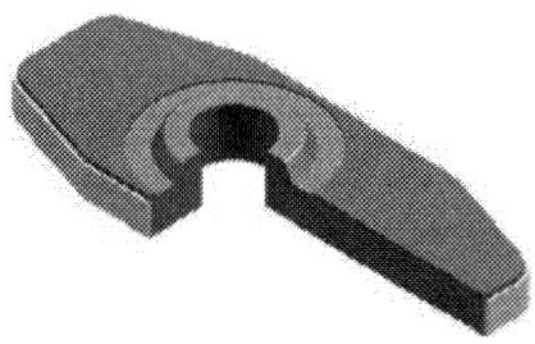		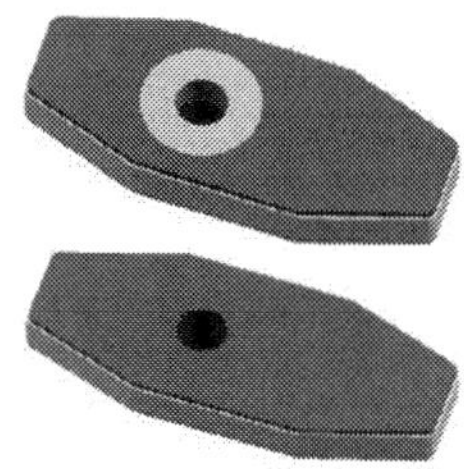
Application zone	General ladle and tundish	Tundish	Special ladle and tundish
Material base	Al_2O_3-ZrO_2-C Al_2O_3-C	Al_2O_3-C Al_2O_3-ZrO_2-C	Al_2O_3-C one body MgO-C insert, ZrO_2 insert
Argon blowing	-	Upper and middle	-
Life	5~10	5~15	4~8

Considering steel quality, the utility of several classes of refractory systems are common practice. The ladle gate slide plate system already discusses different materials, processing, and their pros and cons.

10.5.3 Cracking and Corrosion Phenomena

Cracking the slide-gate refractories is a significant problem because it poses a tremendous safety hazard (steel leakage), steel quality, and subsequent productivity problems. Usually, crack initiation along the thickness may lead to re-oxidation by penetrated air and facilitate insertion of inclusions in the steel. Figure 10.20a represents the transverse direction radial cracks along with the thickness, and either this crack may propagate perpendicular to the longitudinal direction as seen as lower crack, or crack initiation consisting of a certain angle as seen in an upper crack. Most slide gate plates experience such cracks and are referred to as *common cracks*. Figure 10.20b–c show close-ups of a common crack. The white, inverted triangular-shaped area dotted circle on the fracture surface of the through-thickness crack in Figure 10.20c results from the graphite oxidation in the slide plate that eventually expedites the oxidation of steel and formation of

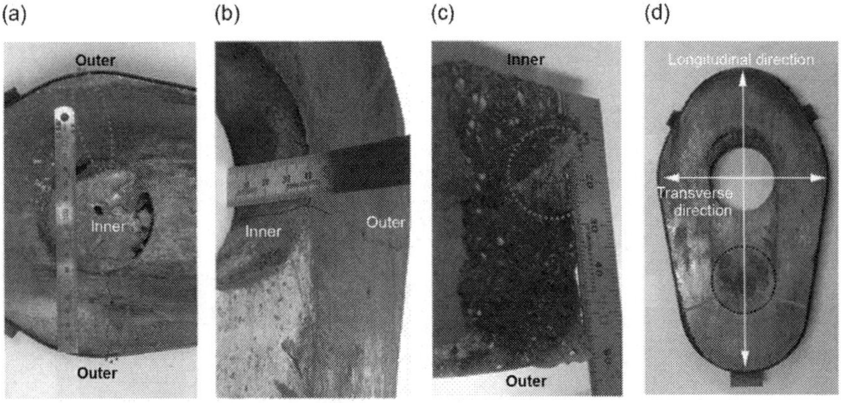

Figure 10.20 Type of cracks: (a) bottom view of the used middle plate and common cracks, (b) photo of common through-thickness crack, (c) fracture surface of common cracks, and (d) schematic view of probable radial/transverse crack regions and the contacting area of molten steel during pouring [15], HYOUNG-JUN LEE et al. (2016), Springer Nature.

inclusions or clogging during casting. Figure 10.20d illustrates the location of less common longitudinal and transverse cracks.

This crack facilitates oxidation of both plate and steel, contaminating inclusions and clogging. The possible mechanisms of formation of such common cracks are:

- Thermal stress induced by temperature distribution of sliding plate with preheating and molten steel temperature.
- Mechanical stress while plates and cassette are tightening by bolts and tight contact with the guide bumps may form non-uniform and high localized surface pressure.
- Friction force caused by horizontal (back and forth) movements of the middle plate to stabilize the mold meniscus level.
- Ferro-static pressure is due to the height difference between the tundish-free surface and sliding gate location.

Thus, considering the above features, a three-dimensional thermal-stress model by FEA has been developed to explore the influence of the thermal and mechanical behavior of slide gate installation and operation like preheating, pouring, and tundish filling, and casting and cooling, eventually assisting in investigating the cracking mechanisms [14]. The highest preheat temperature of the tundish shell allows the control of hydrogen content in steel during the tundish process; thus it is a critical index parameter to synchronize the temperature.

Before the stress model initiation, the analysis encountered all of the parts preheated to 750°C and held for 3.5 hours at 100% fully opened in position. The middle plate is then moved to 0% opening at 25 mm/s in 4.8 seconds and kept for 12.5 minutes during molten steel fill-up of the tundish. Immediately, the middle plate is moved to open 60% for steady continuous casting for 3.5 hours, and the opening is moved back to 0% and allowed to decrease the temperature to room temperature for 4.5 hours.

The predicted temperature and stress behavior at critical locations on the plate surface is shown in Figure 10.21, considering all preheating, tundish filling, casting, and cooling stages. With and without bolt preload are represented as lines and dots, respectively. The predicted temperatures for all three plates are shown in Figure 10.21a, c, and e to guide their influence along with mechanical fittings

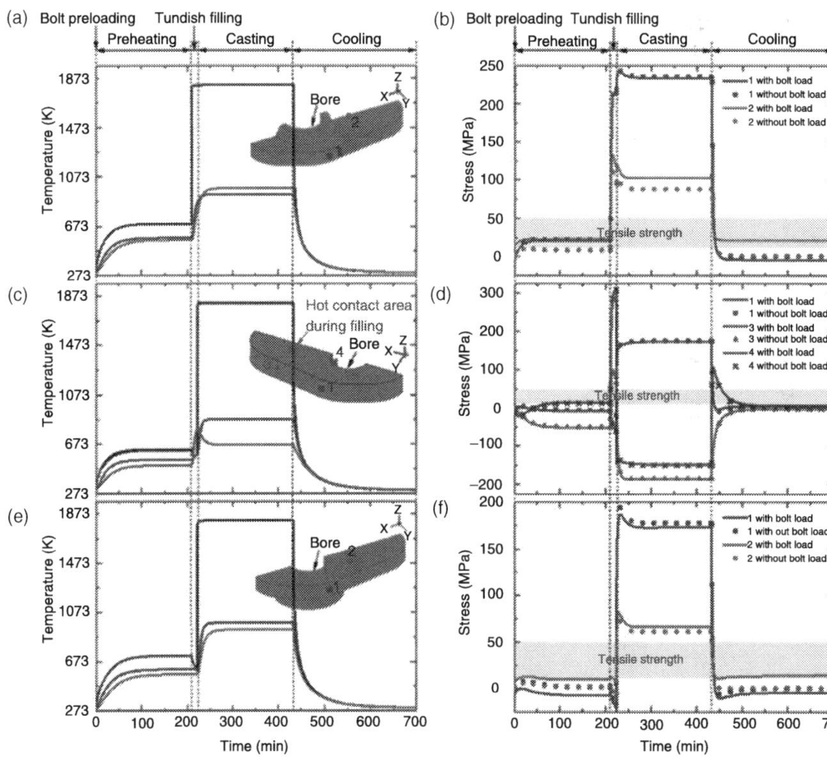

Figure 10.21 Predicted temperature and stresses in tangential (~X) direction (1 = blue, 3 = purple) and Y-direction (2 = red, 4 = brown) at critical surface locations, during preheating, tundish filling, casting, and cooling: (a, b) upper plate, (c, d) middle plate, (e, f) lower plate [14].

on the stress behavior of plates that eventually help to understand the probable failure regions of plates in operation.

Maximum principal hoop stress develops at location 1 for all 3-plates, $\perp^r$ to the growth direction of common through-thickness cracks. Usually, stress-induced radial crack forms along Y-direction initiates in location 2 for the upper and lower plate and location 4 for the middle plate. Transverse crack initiates at the outside edge of the middle plate near location 3, and it is due to active stress along the tangential direction of the X-axis of the plate.

Initially, the initial principal stresses do not exceed the tensile strength of 12–50 MPa in the middle plate while applying bolt preload only (see Figure 10.21d). Hence there are no cracks in the middle plate due to bolt preload, but if a piece of sand or grit is trapped between the plates, it may lead to cracks. However, the cassette frame bends about the X-axis due to initial bolt loading, which causes high tensile stress in Y-axis at location 2 i.e., rare radial crack location along the bisecting line of the upper plate's top and the lower plate's bottom as shown in Figure 10.21b and f. This may create the possibility of inducing radial cracks in the lower and upper plates along the X-Z symmetric plane, which is 90° to the Y stress direction at the time of tundish filling or casting, preheating, when the nozzle bore gets heated up by the hot gases or molten steel.

During the initial 10 minutes of preheating, the principle tensile strength exceeds the tensile strength at marked location 1 in the upper plate, leading to common through-thickness radial cracks appearing at that location. Thus, during preheating, tension is induced due to the thermal expansion of the bore, which may lead to common through-thickness radial cracks. Steel plant engineers used to observe this cracking time consistently [15]. During the first couple of minutes of tundish filling, the molten steel at a higher temperature heats the nozzle bore, and higher tension is generated within the outer surface of the upper plate due to the expansion of the bore (blue line in Figure 10.21b). It continuously exceeds the tensile strength vary at the outside edges to cause common through-thickness radial cracks in each upper plate if those are not already shaped throughout preheating.

Additionally, a spherical space on the prime of the middle plate is contacted by the steel in the molten state to become extremely hot and compressed. Due to this compression, a tension is generated within the inner bore (location 4), as shown in Figure 10.21d. This mechanism could initiate rare radial cracks that propagate outward on the X–Z symmetric plane and perpendicular to the existence tensile stress. The compression within the hot circular contact space conjointly generates a tension hoop stress within the encompassing refractory. This includes the closest surface at location 3, where ~ X-tensile stress throughout tundish filling exceeds the refractory strength (Figure 10.21d). Thus, initiation of transverse cracks at the outer surface (location 3) is rare, and it may propagate inward throughout tundish filling. This mechanism can explain middle plate cracks formation. The molten steel does not influence the lower plate throughout this filling stage (zero per centum opening). In the casting stage, the overall heating of the whole assembly may lead to thermal expansion that can generate a large magnitude of localized compression within the guide bump region of all three plates. Moreover, the overall expansion significantly increases the bending stresses induced from the bolted cassette, which causes the Y-tension at location 2, as shown in Figure 10.21, to extend significantly within the outside surfaces of the lower and upper plates. It causes an opportunity for rare radial cracks within the lower and upper plates.

Rare radial cracks occur in the lower and upper plate due to the tension within the Y-direction caused by bending from the bolt preload along with the best thermal expansion of the plates throughout the casting. Thus, a high bolt preload on the cassette assembly exacerbates rare radial cracks through the lower and upper plates. Additionally, throughout the casting, the highest temperature difference between two sides may lead to hoop-direction tension at location 1, which expedite the throuhout of liquid steel for all 3 plates. It offers another likelihood to initiate rare and customary through-thickness radial cracks in the proximity of location 1, if not prepared already during tundish filling and preheating.The hottest space, that is, the inner bore of the plates, which is in touch with steel in a molten state throughout the casting, cools down to the ambient temperature during the ultimate cooling stage. The thermal contraction of the inner bore and close surfaces of the middle plate at location 4, generates tensile stress within the Y-direction, as shown in Figure 10.21c–d. Consequently, a rare radial crack could initiate and propagate while cooling.

In addition, locations 1, 2, 3, and 4 experience utterly different degrees of stress in numerous incidences are as follows:

- The common through-thickness radial cracks are present in all 3 plates at (location 1): Thermal growth because of radial temperature difference through the plate is the most significant mechanism inflicting common through-thickness radial cracks in nearly every plate in industrial applications. In the first stages of preheating or throughout tundish filling, heating of the nozzle bore from the new gases or melted steel causes internal compression, generating tensile stress at the outer surface of the plate. This cracking mechanism may be synchronized by enhancing the fracture toughness of the refractory, removing surface imperfections, particularly at the outer surface, or tightening the outer steel band to reduce the expansion.
- The rare radial cracks in the lower and upper plate (location 2): The vertical loads applied to the cassette assembly from modification of the bolts exacerbate rare radial cracks through the lower and upper plate because of mechanical bending stresses. These bending stresses increase throughout the casting process because of thermal expansion, generating dynamic bending tension even

when the bolt is not loaded. In general, the geometric style of the cassette and plate perimeter size seems to constrain the plates to apply extra pressure on the slide-gate plates due to the bending of the cassette throughout the operation. Therefore, shape and geometry should be optimized to accommodate the thermal expansion and reduce the risk of excessive bending forces, and bolt tightening should be optimized (lessened) to stop rare radial cracks because of excessive bending.

- The rare transversal and radial cracks within the middle plate (location 3, 4): Compressive stresses are generated throughout tundish filling within the hot circular contact space, wherever the new melted steel impinges on the center of the middle plate, causing tensile stresses within the encompassing refractory. It may be why rare radial and transversal cracks occur within the middle plate. The preheating temperature and total operation time must be increased to avoid this mechanism.

In addition, thermal contraction in the cooling inner bore leads to more tensile stress because the plates settle down when continuous casting is over. It can be another mechanism that may help propagate rare cracks radially outward from the inner bore. A slower and uniform cooling rate is usually recommended to alleviate this mechanism.

In this aspect, it is worth mentioning that the Al_2O_3-ZrO_2-C slide plate is a potential candidate for enhancing fracture toughness and is successively used for ladle and tundish. Adding calcium (Ca) to the steel prevents clogging because Ca reacts with alumina inclusions, forming calcium aluminates, which have low melting points and do not form precipitated layers. Similarly, the Ca treatment for high Si content welding steel may initiate and enhance welding cracks. However, it is observed that the corrosion of Al_2O_3-ZrO_2-C (AZC) slide gate plates increased when used for Ca-added steel casting. Figure 10.22a shows the working surface of the middle plate after use in such conditions. The forms of wear are the change of the inner hole diameter by corrosion, the surface abrasion by the metal stream, and the thermal spalling cracks. The lifetime of the slide gate plate is determined, especially by the diameter change of the inner hole by corrosion and/or thermal spalling.

Figure 10.22b shows the diameter change of AZC slide gate plates used for Ca-added steel casting at Kimitsu steel works of NSSMC [16]. The diameter of the slide gate plate hole increased as the number of heats increased. This phenomenon is much sharper in the case of Ca-added steel casting. Low-melting calcium-aluminate inclusions are formed when the added-Ca reacts with the Al_2O_3 aggregate in the refractories, as Al_2O_3-CaO phase exhibits a eutectic point of 1713 K (1440 °C). So a liquid phase is formed on the working surface of the slide gate plates. Thus, removing the low-melting point layer by the molten steel stream is the primary wear mechanism of the AZC slide gate plates by Ca-added steel.

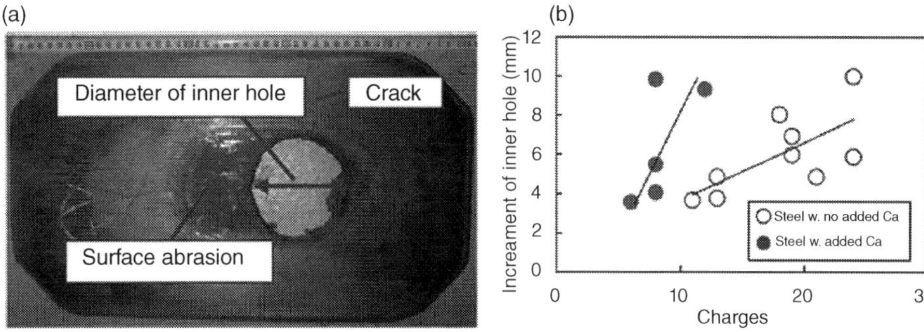

Figure 10.22 (a) Working surface of the middle plate after use, (b) the diameter change of AZC slide gate plates used for casting steel with and without added-Ca at Kimitsu steel works of NSSMC [17], Tadashi et al. MundoDeCongresos.com.

In this study, the microscopic analysis explains the formation of many more pores and accelerates the calcium aluminate penetration into the slide gate plate structure, resulting in the corrosion of the AZC slide gate plates. Thus, the slide gate plate must have balanced properties for corrosion resistance against molten steel and molten oxides and good thermal spalling resistance.

10.6 Short Nozzles for Ladle and Tundish

Refractory nozzles are summarized, including short length, so-called ladle inner nozzles, ladle collector nozzles, and tundish nozzle and tundish metering nozzle. Despite these obvious long nozzles, including ladle shroud and tundish submerged entry nozzles, and monoblock stopper are part of the flow control refractory; however, these have standard shaping and fabrication protocols and are thus extensively discussed in Chapter 11. A typical commercial-grade 4 type of the short nozzle is represented in Figure 10.23.

Figure 10.23 Different shapes, assembly, and projection of (a) ladle inner Credit: ShengHe Refractories, (b) ladle collector nozzle Credit: Tangshan Shichuang High Temperature Material Co., Ltd, and (c) tundish nozzle (open casting) and metering nozzle (close casting), respectively.

10.7 Nozzle Diameter and Gate Opening in Flow

Nozzle clogging is usually due to inclusions aggregating, a casting speed often decreases, and even an entire cast is supposed to cancel in severe cases. Calcium treatment is a common and economical process to avoid clogging of Si-treated steel, and in this aspect, recent literature validated the nozzle clogging phenomena [17].

In this study, a typical dimension of a ladle is considered, where, top diameter = 3800 mm, bottom diameter = 3200 mm, and height = 4006 mm. The bottom and metal zone are lined with Al_2O_3-MgO-C and slag zone lining made of MgO-C, respectively. The slide gate refractory is made of Al_2O_3-C, and the upper nozzle and well-block are high alumina precast shapes. All typical dimensions, including nozzles, and slide gate refractories, a crucial component to flow control from ladle to tundish and shroud to flow from ladle to tundish, are shown in Figure 10.24a. The inclusions features like types, shapes, sizes, and distribution in steel samples are analyzed after casting 10 heats when ladle nozzles are clogged.

Nozzle diameter and opening slide gate significantly influence flow rate and resultant clogging of Al_2O_3 and its composite during continuous casting. In order to understand the effect of diameter and flow rate, some theoretical calculations can be done. In consideration of a fully open slide gate system, a theoretical liquid steel flow rate (Q_m, kg/s) can be calculated by Equation (10.3):

$$Q_m = \rho_l \frac{\pi D^2}{2} \left(\frac{3agh}{6 + \alpha l} \right)^{1/2} \tag{10.3}$$

ρ_l = molten steel density, 7000 kg/m^3; g = acceleration gravity; D = nozzle diameter, 0.075 m for the casting; and α = constant, 1 for the turbulent flow; l = friction loss factor, 0.5; h = remaining steel height from the ladle bottom (meter). A contemporary flow rate through the nozzle (Q_{mc}, kg/s) during slab casting can be expressed as Equation (10.4):

$$Q_{mc} = w.t.\rho_s \frac{S_c}{30} \tag{10.4}$$

In this study, the w = width (1.3 m), t = thickness (0.23 m) of the slab; ρ_s = solid steel density, 7797 kg/m^3; S_c = casting speed for two strands, 1.1 m/min for the target. Thus, Q_{mc} becomes 85.48 kg/s or 5.13 ton/min to fulfill the target casting speed through 2-strands of the continuous casting. As shown in Figure 10.24b based on Equations (10.3) and (10.4), the flow rate decreases with a decrease of the heights of the liquid level, and also depends on the diameter to achieve the targeted casting speed. For example, 50% flow rate (Q_m) reduction is noticed when h decreases from 4 m to 1.5 m. While the nozzle diameter is 90 mm (D), the

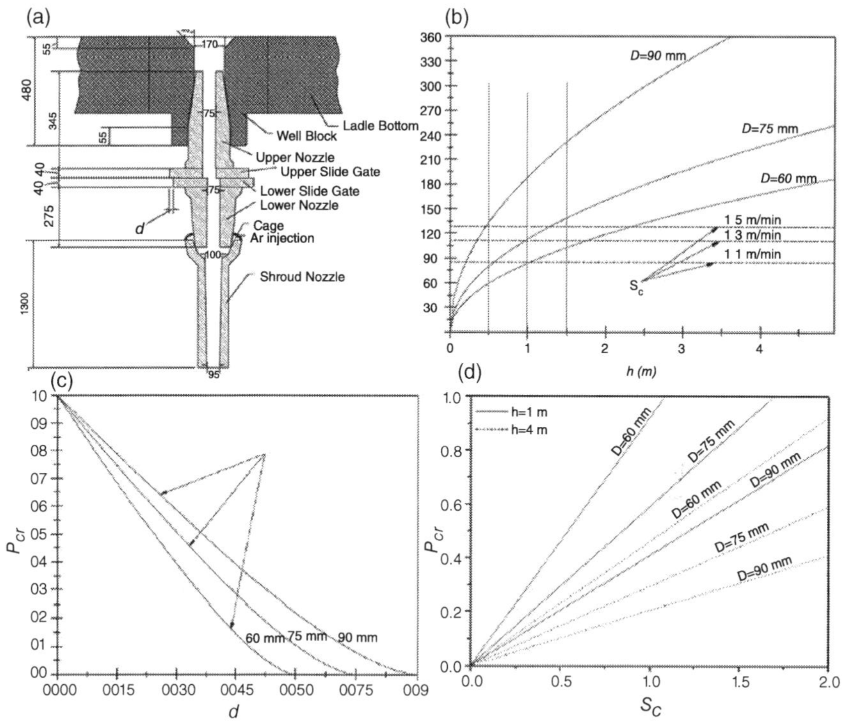

Figure 10.24 (a) The dimension of the slide-nozzle system and the shroud system, (b) the variation of *Qm/Sc* with *h* during casting, (c) the relationship between *pct* and *d* during casting, and (d) the relationship among *pct*, *Sv*, *D*, and *h* [17].

targeted casting speed 1.1 m/min can be achieved at $h = 0.5$ m. However, smaller nozzle diameter 60 mm maximum casting speed can be achieved 0.98 m/min i.e., 84.14 kg/s even $h = 1$ m (double of 0.5 m).

It is indicating that the bigger diameter of a nozzle could effectively generate an enough flow rate for CC. In actuality, the percentage slide gate opening is a determining factor and can draw a relation in Q_{mc} and Q_m, and is represented by Equation (10.5):

$$Q_{mc} = p_{ct}.Q_m \tag{10.5}$$

The p_{ct} = percentage of a slide gate opened and can be estimated by Equation (10.6):

$$p_{ct} = \frac{a}{\pi D^2/4} = \frac{2\left(arccos\frac{d}{D} - \frac{d}{D^2}(D^2 - d^2)^{1/2}\right)}{\pi} \tag{10.6}$$

Specifically, a = slide gate opened area m^2, d = lower slide gate movement with respect to upper one (see Figure 10.24a). It is essential to control the slide opening distance to adjust the flow, especially when the nozzle diameter is less than a bigger one. The variation of a closing slot with the percentage of slide gate opening bore diameter is shown in Figure 10.24c. The higher the nozzle diameter, the more portion is closed of the slide plate. In consideration of the above equations, the modified p_{ct} can be rewritten as:

$$p_{ct} = \frac{w.t}{15\pi D^2} \cdot \frac{\rho_s}{\rho_l} \cdot S_c \left(\frac{3agh}{6+\alpha l} \right)^{-\frac{1}{2}} \tag{10.7}$$

Equation (10.7) can be reduced to Equation (10.8) by considering all relevant values for this particular study:

$$p_{ct} = 3.32\times10^{-3} \cdot \frac{Sch^{-\frac{1}{2}}}{D^2} \tag{10.8}$$

Based on Equation (10.8), the relationship among P_{ct}, S_c, D, and h could be shown in Figure 10.24d.

The interrelation between slide plate percentage opening and casting speed is analyzed in the presence of three nozzle diameters, 90, 75, and 60 mm, when h = 4 m and h = 1 m. At the beginning of casting, 22.54% slide gate opening of 90 mm nozzle diameter can fulfill the casting speed target of 1.1 m/min. However, when the liquid level decreases to 1 m, the slide gate must move to open 45.09% (double the beginning) to reach the same casting speed target. Let's say the nozzle diameter is clogged by certain chemical interaction with steel and plate refractory in the presence of oxygen, and reduced to 60 mm during liquid level h = 1 m. In this circumstance, an almost fully opened slide gate can only meet the casting speed 1.1 m/min.

Further, if the h decreases to 0.5 m, 100% slide gate opening cannot reach the target speed; however, at the same time, the casting speed has to decrease to keep a steady liquid steel level in the tundish. Thus, prior to slide plate design for a shop, it is critical to know the probable variation of casting speed. Let's consider a fixed slide gate bore diameter 75 mm is used for the casting speed variation of 1.1 m/min or 1.5 m/min. The P_{ct} variation is 32.46% or 44.27%, as estimated from Figure 10.24d, to maintain sufficient liquid steel in the tundish. Thus, more slide gate opening compares to theoretical estimation indicates the probable clogging in the ladle nozzle during casting.

References

1 Stopinc Aktiengesellschaft, Bösch 83a, CH-6331 Hunenberg, Switzerland "Linear valve and it's refractory parts".
2 Kononov, V.A., Kononov, N.V., and Vasilenko, V.P. (July 2011). Main trends in the development of slide-gate systems. *Refract. Ind. Ceram.* 52 (2). 118–125
3 RHI AG, Wienerbergstraße 9, A-1100 Wien, Austria, "wear marks on refractory plates".
4 Yamamoto, K., Osada, M., and Takata, A. (2009). Shinagawa technical report. 52: 63–68.
5 Chaudhuri, J., Choudhury, G., Kumar, S., Rajgopalan, V.V., Banerjee, G., Sarda, K., and Bajoria, P. (June 2007). Achieving higher performance & longer service life of slide plate. *Iron Steel Rev.* 86–91. IFGL Refractories Limited.
6 Ganguli, S., Satpathy, S., Samanta, A.K., and Panda, P.B. Design of high performing chemically bonded slide gate plates for casting of alloy steel and mild steel. TRL Krosaki Refractories Limited, Belpahar, Odisha (INDIA). 2016.
7 Rancoule, G.I. (1994). Zirconia graphite slide gate plates, US Patent 5335833.
8 Gilbert Rancoule, Duane L. DeBastiani, (1993). Magnesia-carbon refractory compositions for slide gate plates and method of manufacture, USA Patent 5250479.
9 Pontes, J.C.D., Jr., Galesi, D.F., Souza, L.F.M., Leão, C.F., Morii, H., Yamamoto, K., Takata, A., and Horiuchi, T. (2016). Shinagawa technical report new slide-gate system for long refractories life, Vol. 59.
10 Hamamoto, N., Matsunaga, T., and Iida, M. (2019). Shinagawa technical report, improvement of slide valve plate, Vol. 62.
11 Liu, H.M. (2012). FEM analysis of the cracks formation and control in SV plates. *Adv. Mater. Res.* 557–559: 1298–1303.
12 Tundish slide gate for billet and bloom casters, vesuvius technical report, C-52, 2013.
13 Innovation for the global customers, PR brochure, refractories for steel making, Chosun Refractories, Co. Ltd.
14 Lee, H.-J., Thomas, B.G., and Kim, S.-H. (April 2016). Thermal stress cracking of slide-gate plates in steel continuous casting. *Metall. Mater. Trans. B* 47B: 1453.
15 Kim, S.K. (2010). POSCO Gwangyang Works. Gwangyang, Jeonnam, Korea, Private Communication.
16 Kato1, Y., Ikemoto, T., and Goto, K. Investigation of the corrosion mechanism of Al2O3-ZrO2-C slide gate plates for the casting of Ca-added steel. Nippon Steel & Sumitomo Metal Corporation, Futtsu, Japan.
17 Kong, W. and Chen, Y.-F. (2019). Da-Giang Cang Ladle Nozzle Clogging during casting of silicon-steel. *High Temp. Mater. Proc.* 38: 813–821.

11

Refractories for Continuous Casting

11.1 Introduction

Continuous casting is a casting method for continuous and high volume metal production with a constant cross-section in a water-cooled copper mold. Upcoming steel demand is adopted high and effective output through the continuous casting (CC) process. Tundish is the last reservoir in continuous casting with maintaining desired composition, temperature, and defect-free product at an optimum casting speed.

Figure 11.1 represents the schematic view of refractories for continuous casting (CC). The refractories for CC are mainly classified into the following types: tundish refractories, flow control system refractories, and teeming system refractories. Tundish is an intermediate vessel installed between a ladle and continuous casting molds and has four major functions.

- Molten steel distributing function

Ladle shroud used to transfer the steel to TD. By installing multiple submerged entry nozzles onto a tundish, multistrand continuous casting from a ladle is made possible via a tundish.

- Flow control function

The flow control of the molten steel teemed to each mold is required to solidify the molten steel uniformly in a stable manner. For molten steel flow control in a mold, flow control refractories are used for a sliding nozzle (SN) plate and a stopper.

- Cast steel material quality improving function

Fundamental Design of Steelmaking Refractories, First Edition. Debasish Sarkar.

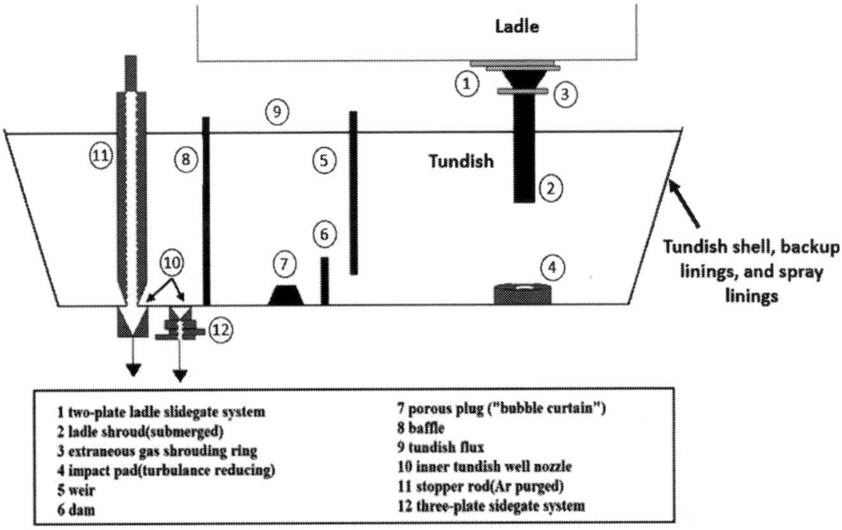

Figure 11.1 Refractories to synchronize the flow during continuous casting of steel.

Nonmetallic inclusions such as oxides and sulfides are formed in the steelmaking process. When coarse inclusions are existent in the molten steel and cast into steel materials, they develop surface defects in the subsequent process of rolling and exert adverse effects on steel product quality. To suppress the penetration of the inclusions into cast steel materials, floating of the inclusions is promoted by holding the molten steel in a tundish. To this end, the floating inclusions can be increased by installing dams in a tundish and extending the flow from the ladle to submerged entry nozzles. Furthermore, the floating of the inclusions is promoted more efficiently by improving the condition of the molten steel flow in the tundish by improving the arrangement of the dams.

- Compensation of molten steel temperature

By applying plasma-heating using a graphite torch or induction heating (IH) and raising the molten steel temperature, solidification of the molten steel in the tundish and/or in the submerged entry nozzle is prevented.

Among several classes of refractories, herein, prime focus has been given on the monolithic tundish working lining, ladle shroud (2, LS), monoblock Stopper Rod (11, MBS, Ar purged), and submerged entry nozzle (SEN). The rest of the refractories are already elaborately discussed in different chapters. For example, the cumulative understanding of slide gate pate for ladle (1), tundish (12), tundish well nozzle (10) are discussed in Chapter 10, porous plug (7) and precast (4,5,6,8) in Chapter 9, etc.

11.2 Importance of Long Nozzles in Steel Transfer

During the transfer from furnace to ladle, ladle to tundish, and tundish to mold, oxygen, nitrogen, and hydrogen absorbed from entrained moist air can be destroyers when modern process is accompanied by low sulfur and oxygen content in steel. An important consideration during steel transfer operations is the formation of *vortices* in the furnace, ladle, tundish, or mold through molten slag from one vessel into the next. As a general principle, although there are exceptions, furnace slag is inappropriate in the ladle, ladle slag is inappropriate in the tundish, tundish slag is unacceptable in the mold, and mold slag is always problematic in the final product. Slag, a refining agent in one reactor, may become a contaminator if transferred inadvertently to the next reactor. In this context, vortex formation, which draws slag from one vessel to another, can be a severe destroyer of quality; and thus needs an apposite device to flow from one vessel to another [1].

11.2.1 Furnace to Ladle Transfer

An uncontrolled and variable amount of air is entrained by the tapping steam, enriched with nitrogen, hydrogen and oxygen, and transferred to ladle during furnace tapping. In obvious air entrapment directly influences fluid flow pattern in the ladle that eventually determines the probability of alloy dissolving efficiency in steel or carried toward the surface as slag through oxidation due to contact with the atmosphere. For this reason, timing, location, and method of deoxidizer and other alloy additions all influence alloy recovery and final steel quality.

Elements like Ca or Mg have a high affinity for oxygen and sulfur; thus ladle slag composition is strongly influenced by the added elements. Furnace slags are oxides and enriched with oxygen that can transfer to the steel during gas stirring. The continuous stirring process promotes fresh contact between metal and slag. While steel is pouring from the ladle, the refractory side wall interacts with the descending slag layer and produces further different oxide products, including glaze and serves as a pool of oxygen and probable sulfur.

These elements can then transfer into the next batch of steel during ladle processing and have an adverse effect on steel quality. Suppose there is a substantial carryover of slag from the furnace containing high iron oxide and manganese oxide content. In that case, a reducing agent may need to be added to minimize the use of an expensive deoxidizing agent or loss of alloying elements that may be added to the steel. Calcium carbide can be used as a reducing agent for the slag phase. Suppose phosphorus oxide is present in the slag carried over from the steelmaking vessel, then during slag reduction or steel deoxidation. In that case, phosphorus may revert to the steel from the slag phase and act as a destroyer of quality.

Highly basic slag is generally used concerning slag compositions for ladle desulfurizing treatments. While such slags have a high capacity for absorbing sulfur, they also have a strong ability to transfer hydrogen faster from a moisture-containing environment to molten steel compared to acidic slag. Steel with a very low inclusion content is less able to tolerate hydrogen. Thus, several processing steps such as inert gas shrouding, vacuum treatment, stream protection, and controlled cooling may be required if problems attributable to hydrogen are to be avoided.

The control over the slag carryover from one vessel to another is always a challenging task. Substantial improvement in furnace tapping operation promotes the formation of a reliable compact tapping stream, including slide-gate valves and bottom tapping systems. With more compact streams, it is easier to protect the steel from contamination with air, by gas shrouding the stream with a protective atmosphere.

11.2.2 Ladle to Tundish Transfer

Tundish is an intermediate metallurgical reactor between ladle and casting molds that facilitates the continuous flow of steel with several monolithic, precast, and shaped refractories. However, tundish slag has a vital role and has an essential requirement:

- To protect the steel from the atmosphere and avoid chemical reaction
- To absorb nonmetallic inclusions from the molten steel
- To behave as thermal insulation and protect heat loss from the upper surface of molten steel

A solid powder layer is an effective option for thermal insulation, whereas a liquid layer prevents reoxidation, N_2 and H_2 absorption from the atmosphere by the molten steel, and nonmetallic inclusions from the steel. An optimum viscosity of tundish flux is expected, as high viscosity limits the flux's ability to absorb inclusions, but very low viscosity of flux may penetrate in the mold during the unstable condition of ladle operation. Thus, tundish slag design is critical and can be synchronized through its optical basicity. It helps to synchronize viscosity and chemical properties that eventually ensure tundish function as a refining vessel. Extended refractory nozzles and gas-shrouding devices are well-established practices to prevent reoxidation during the metal transfer from ladle to tundish and tundish to mold [2]. Despite such precaution, a few seconds of steel–air exposure is still a common incidence during ladle change operation and expedites inferior steel over the next several minutes. A few ppm level nitrogen increments in steel is an excellent indicator of the reoxidation process, as the oxygen dissolution rate is ~200 times faster than that of nitrogen. Picked-up steel sample analysis exhibits several hundred ppm oxygens in the presence of a few ppm nitrogens. Such phenomenon facilitates excessive reoxidation and subsequent generation of oxides as inclusions that separate and collect in the tundish flux. Because of this conversion and entrapment, iron oxide and manganese oxide content influx may be increased remarkably.

A different thermal effect is noticed in liquid steel flow during ladle change operation and steady-state conditions [3]. Residence time distribution analysis predicts that the steel flow during ladle change is radically different from steady-state isothermal conditions. A nonuniform temperature profile is expected during ladle change because the new steel entering temperature is always higher than the already existing steel temperature. The new higher temperature steel follows the reverse phenomenon compared to steady-state conditions during previous steel displacement by new steel. Such incidence continues for a significant period depending on the tundish capacity and casting rate. These effects influence alloy additions, inclusion flotation, and mixing that eventually consider the tundish design and subsequent steel casting and quality.

As the metal level decreases in the ladle, vortex formation can draw the ladle slag into the tundish, resulting in contamination and partial interaction with tundish flux, alteration of viscosity, and a certain portion slag may flow toward the mold. Such impurity adversely affects the mold flux performance and the formation of defects in the final product. Moreover, the ladle change operation provokes a change in the total oxygen content in the tundish. In an approximate estimation, a 2 min ladle change may affect the steel quality as tundish can extend for over 15 min.

11.2.3 Tundish to Mold Transfer

The surface quality of cast product depends on the fluid flow behavior and metal meniscus in the mold, and extensive work established several concepts [4, 5]. When the metal delivery system involves refractory submerged entry nozzles (SEN), the behavior of fluid flow in the mold and at the meniscus is extremely important. The design of the nozzle and the depth of submergence directly influence fluid flow within the casting mold, behavior at the meniscus, vortex formation, and the distribution and location of inclusions in the final product.

Argon gas is frequently injected down the submerged entry nozzle to minimize nozzle blockage and enhance the flotation of nonmetallic inclusions. It can have a prominent effect on fluid flow behavior. If argon injection is excessive, mold slag can be drawn into the molten steel and end up as a defect in the product. Even in the absence of argon injection, if the submerged entry nozzle has not been properly fired and is porous, or if cracks are present in the nozzle wall caused by damage during handling, or if cracks form during use, air can be drawn into the nozzle, the oxygen from which will cause reoxidation. At the same time, the nitrogen will behave essentially like argon and cause considerable disturbance of the mold powder at the meniscus. In addition, if the formation of reoxidation material is excessive, the composition of the mold flux can change, which can adversely affect subsequent performance.

The behavior of synthetic fluxes in the casting mold is a critical parameter for producing quality steel. It is the last slag with which the molten steel has contact before solidification. A substandard casting mold and process may destroy the quality steel after exhaustive effort during the steelmaking process. For this reason, the mold can be considered as the *heart* of quality and the mold slag must be designed to perform a range of functions, which are considerably greater than that required of any of the previous slags in the furnace, ladle, or tundish. Such functions include prevention of reoxidation, provision of thermal insulation, availability as a flux to dissolve impurities, performance as a lubricant, and possession of good heat transfer characteristics to enhance solidification. Since the flux forms a glass during the casting process, this glass can also influence heat transfer in the spray chambers. The interaction of the mold flux with refractory nozzles and the behavior concerning hydrogen absorption from the atmosphere and subsequent hydrogen pickup by the steel are also important considerations when designing the most appropriate flux composition for a particular application.

11.3 Tundish Lining

Tundish has different shapes, and their capacity depends on ladle capacity and casting rate. For tapping steel from a ladle, the tundish capacity for 130–160 tons ladle is 36–40 tons and for 300–320 tons ladle, it is 70–85 tons, at any moment of casting time. The inner lining of a tundish is generally comprised of three layers: a permanent refractory lining, a wear refractory lining, and a coating refractory, all lined in said order from a steel shell. An insulation material may be used between the steel shell and the permanent refractory lining to minimize the temperature drop of the molten steel in tundish.

Promising coating material magnesia (MgO) system has excellent detachability concerning the wear lining refractory used. The coating material is required to simplify the removal work of the residues of scull and slag in the tundish after casting and the protection of the wear lining refractory. Furthermore, the coating is dismantled after each cast, and the tundish is recoated. Cleanliness in the area contacting the molten steel is maintained and therefore, the coating serves to prevent the contamination of the molten steel. Coating work is conducted by trowel work or by spraying.

11.3.1 Lining and Failure

The tundish usage cycle after the construction and drying of the permanent lining refractory and the wear lining refractory generally consists of spraying of coating material, preheating, online steel casting, cooling, and removal of the skull (dismantling of coating) and repair, as shown in Figure 11.2. After dismantling

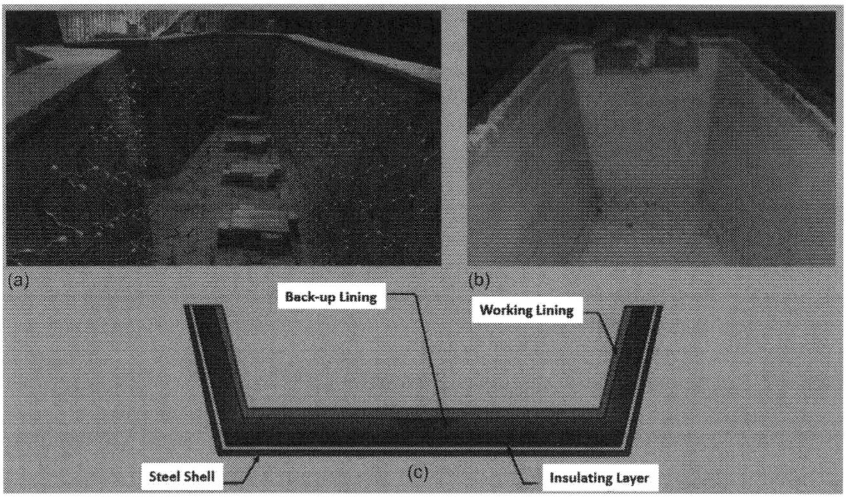

Figure 11.2 (a) A typical anchoring process in tundish, (b) back-up or wear lining refractory by Al_2O_3-based castable, and (c) lining pattern of predominate wear lining (gray color), and working lining coating.

the coating, the wear lining refractory is repaired, and a new coating is constructed. At the same time, the refractories of the teeming system and the flow control system are exchanged.

The general practice is to cool the tundish once after the completion of casting and then to exchange the refractories. However, hot state recycling is usually done by removing residual steel and slag, and quick measure is taken through the exchange of refractories and flow control system. A maximum of about ~500 charges of continuous casting can be achieved by applying the recycling of a hot tundish. The merits of saving the workload in the maintenance between casts and the reduction of refractory costs are also feasible.

The tundish wear lining refractory is damaged due to mainly three factors:

The first is the occurrence of cracks caused by the heating and cooling of the tundish. In the case that a monolithic refractory is used as the wear lining refractory, an extended refractory body is constructed integrally into one single body, and as the displacement caused by thermal expansion and/or thermal contraction is large, cracks tend to occur readily. In the early stage of operation, cracks perpendicular to the lengthwise direction dominantly occur. Enhance casting frequency and heating–cooling cycle repetition expedite the cracks in the horizontal direction and subsequent progress in addition to the cracks perpendicular to the ground level, resulting in the spalling of the surface layer.

The second is the direct adhesion of coating material to the wear lining phenomenon on the interface between the wear lining refractory and the coating material. For example, direct adhesion does not occur in the wear lining refractory state simply being in contact with the coating material. On the other hand, strong direct adhesion occurs when slag penetrates the coating material. Furthermore, the characteristics of direct adhesion between a castable refractory material of the Al_2O_3-SiO_2 system and SiO_2 bearing coating material of the MgO system facilitate the formation of a low melting point phase on the interface between the coating material and the wear lining refractory, and therefore, results in direct adhesion process. In either case, direct adhesion occurs when the elements such as slag and/or SiO_2 increase the liquid phase ratio on the interface between the coating material and the wear lining refractory richly exist. In the case of direct adhesion, the wear lining refractory is also damaged when the coating material is dismantled and simultaneously progresses the damage.

The third is the erosion caused by the tundish slag. The tundish slag mainly consists of ladle slag and the joint sands that flow into the tundish and the heat insulating material charged from above the tundish for retaining the molten steel temperature. Because a coating material of the magnesia system has high corrosion resistance compared to tundish slag is coated on the surface of the wear lining refractory; thus the corrosion of the wear lining refractory is not a serious matter. However, the erosion due to the tundish slag becomes problematic when the number of casting increases in hot tundish operations.

While considering coating as a working lining, two class of mass, say, *spray mass* and *dry vibrating mass*, are popular coatings during casting. The use of magnesia-based *undish spray mass* in continuous casting of steel is common in the Steel Industry. The main constituent of the tundish spray mass is Magnesia (MgO). Dead Burnt Magnesia (DBM) and other MgO-bearing minerals like dunite are major resources to make spray mass. Dunite is a natural forsterite–serpentine or olivine group of minerals. The major chemical composition of dunite is MgO and SiO_2. Dunite is used to partially replace DBM in tundish spray mass composition.

Adding a high large amount of dunite to a good quality MgO source has an initial liquid formation temperature of over 1600ºC. Usually, bulk density at 1550°C decreases with an increasing percentage of dunite in spray mass. Higher bulk density does not improve the performance of spray mass, and lower density material is desirable for easy deskulling and slag removal after liquid metal casting. In order to achieve the proper deskulling, a requisite amount of low dense paper fiber assists in developing porosity in the early stage and high-temperature bonding within ceramic grains through the addition and melting of glass fiber. Powder resin is commonly used in this spray mass to achieve green strength, followed by polymerization at elevated temperatures. Coating shrinkage in the range of 1% is desirable for easy deskulling of the spray mass. However, the high content of the low melting phase facilitates more shrinkage and cracking, resulting in liquid metal penetration and damage to wear lining. Simultaneously, an optimum cold crushing strength is required to sustain its

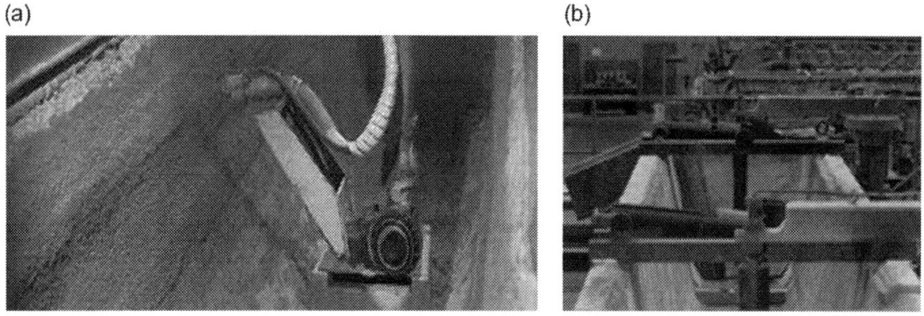

Figure 11.3 On-site tundish coating process (a) spray coating Credit: Dakduklu Mining Company, (b) dry vibrating mass. Credit: WEERULIN GmbH

own shape and tills the liquid metal pouring into the tundish. Preferentially, the dunite in coarser form with DBM in formulation of tundish spray mass shows better properties. A typical coating process of both spray and dry vibro mass is shown in Figure 11.3.

Cumulatively, the spray mass should have the following characteristics to maintain high performance of wear lining during continuous casting:

- Low installed density
- Easy deskulling
- Low thermal conductivity results in reduced temperature loss from the steel
- Reduced hydrogen pickup concerns
- Minimal alkali content to maximize permanent lining service life
- Simplified ramping and dry-out procedure
- Low thermal improvement insulation

The conventional wet spray application is common, but hydrogen pick up by the lining demands an alternative coating mass during the steelmaking process. In this aspect, dry vibrating mass with and without resin, heat, and cold setting has been developed as a coating material for tundish.

11.3.1.1 Dry Vibrating Mass (DVM)

MgO-based tundish vibratable material for both non-preheated and preheated tundish is a start-up practice. Dry tundish working linings permit:

- A fast and simplified installation that can be performed with less manpower than other technologies
- Dust free ultra-fast installation without harmful organic fumes is developed through a rapid bond. The latent heat of the backup lining facilitates this bond formation
- Eliminates moisture as a source of hydrogen, and steel cleanliness by a strong reduction of the hydrogen pickup

- Low organic binder content to reduce possible hydrogen and carbon pickup
- Improvements in availability and costs:
  - Very long casting sequences (e.g., 10–20)
  - Very easy deskulling–installation
  - Equipment is nearly service-free, low-cost
  - Minimization of tundish circulation
- Additional operations like drying or preheating can be avoided or reduced dramatically
- Increased slag resistance due to low porosity
- Consistent internal tundish dimensions
- Reduced energy requirements for curing dry versus spray linings
- Relatively high wear resistance
- It provides excellent stabilization of the steel temperature and enables temperature decrease upto 10°C by tapping
- Uniform lining to reduce steel pollution and maintain a tundish flow pattern

An additional advantage is that the life of backing lining is prolonged due to smaller thermal shocks. The sequential development of DVM can be classified and discussed in two modes;

11.3.1.2 DVM Resin Bonded

Most DVM is made of magnesia and resin-bonded and installed with a simple steel mandrel or a more sophisticated vibrating mold. It has several advantages, including:

- Usually, less than one hour is required to coat the DVM before drying.
- Drying time is less than an hour at close to 350°C, allowing complete resin hardening up to 3–5 cm thick lining at 180°C.
- Resin setting assisted soft drying enables to reduce the energy consumption and noise by 10 dB in operation.
- It is an alternative choice to reduce hydrogen pickup at the start of the sequence casting.

However, resin-bonded DVM has some drawbacks, including environmental aspects concerning health and safety or technical aspects.

11.3.1.3 Resin Free DVM

Resin free mineral bonded dry vibrating mix (MB-DVM) is an excellent substitute for resin-bonded DVM. MBDVM is a promising material made of mineral bonds compatible with the magnesite matrix and dry coating method. Interestingly, this new generation tundish coating provides sufficient strength without preheating for removal or stabilization.

11.3.1.3.1 Hot Setting Resin Free DVM

Emissions of polycyclic aromatic hydrocarbons and other toxic compounds may be a hazardous environmental issue during their thermal treatments. In this aspect, a metallic salt can be utilized to react with basic particles like calcium hydroxide to form gypsum-type minerals below 200°C. A typical reaction between magnesium sulfate and $Ca(OH)_2$ in the temperature range of 80–200°C because of different activator efficiency degrees:

$$MgSO_4{\cdot}7H_2O + Ca(OH)_2 \rightarrow 2\ CaSO_4{\cdot}2\ H_2O + Mg(OH)_2 \tag{11.1}$$

Despite the temperature, certain minerals called activators and inhibitors have also influenced adjusting to activate and postpone the reaction. The adjustable bonding behavior of MBDVM is illustrated in Figure 11.4. It shows how activators and inhibitors allow imparting mechanical strength both in cold conditions (using certain heating equipment: case 1) and in case of hot installation (using the only residual heat of residual lining: case 2).

Usually, a dwell time of 1–2 h at an optimum temperature of 120°C is required to attain working strength that eventually develops relatively high cold crushing strength at elevated temperatures through the removal of water from the mix without forming any obnoxious and irritating vapors.

11.3.1.3.2 Cold Setting Resin Free DVM

Resin-free cold setting DVM uses a mineral bond that hardens without heating to develop sufficient strength for handling before preheating. It is an energy and cost-saving material. The inorganic chemical reagent becomes hardened in the presence of a specific catalyst. In practice, the binder and accelerator are

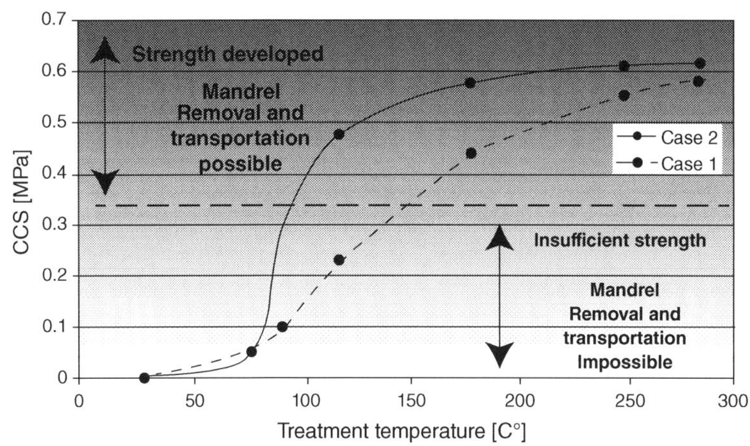

Figure 11.4 Evolution of the strength in MB-DVM [6].

simultaneously mixed in liquid solutions with the matrix. The working and hardening time of the final mix is essentially controlled by the particle size distribution of the aggregate, accelerator/mineral bond ratio, and fineness of the entire mass. An automatic mixing arrangement allows the distribution of the final mix in tundish and final adjustment manually as required.

Herein, no heat is required during the hardening period and the formwork can be usually removed in less than 5 min after the end of the installation in hot tundish and 20 min in case of cold tundish. Interestingly, the initial stage *green* mechanical strength is relatively high compared to resin-bonded materials. It allows an easier demolding and transportation with the crane up to the caster. No dedicated specifications for final preheating before cast are needed. The main properties of a different class of coating materials are summarized in Table 11.1. It is worth mentioning that preheating management breaks the hydrocarbon binder deeper in the lining, slowly releases hydrogen, and picks up in the steel. Moisture release from working lining may also facilitate tundish bubbling during operation. Hydrogen control in steel is essential to produce rail-grade steels or pipes.

Table 11.1 Competitive properties of spray mass and different grades of DVM.

Material properties	Temperature	Units	Spray	DVM hot setting resin-bonded	DVM hot setting resin-free	DVM cold setting resin-free
MgO		%	87	90	88	-
SiO_2		%	5.2	3	8.5	7
CaO		%	3.1	1.2	2.5	1.8
Fe_2O_3		%	1.6	1.6	0.4	1.8
Water requirement		%	21	-	-	-
Material requirement		g/cm^3	1.7	1.95	1.9	1.77
Bulk density	150°C	g/cm^3	1.51	1.92	1.8	1.60
Weight loss	150°C	%	19	0.2	1	0.8
Weight loss	1000°C	%	3.1	3	3.4	2.2
CCS	1000°C	MPa	0.8	0.5	0.2	8
CCS	1500°C	MPa	5	3	2	14
PLC	1000°C	%	−0.5	0	0	−0.15
PLC	1500°C	%	−2.2	−2.4	−2.2	−2.3

11.3.2 Lining Improvement and Maintenance

Wear lining life can be improved by steel fiber addition. Sato et al. evaluated, with respect to the wear lining refractories for a tundish, the effects of the amount of the metal fiber addition and its distributive situation on the resistance to fracture value after the occurrence of cracks, and pursued the optimum addition amount of metal fiber (see Figure 11.5) [7].

The resistance to fracture value increases proportionally with the increase of the metal fiber addition in the range of 0–4 mass%. Beyond the optimum addition, there is no further improvement. Furthermore, the distributive situation of the metal fiber does not vary greatly in the range of 2–6 mass%. The metal addition expedites 15% life, and the consumption of repairing material could be reduced by 10%.

A major cause of damage to tundish wear lining refractory is spalling, which is caused by the mechanical shock at the time of dismantling the coating and the extension of cracks due to heating and cooling. An apposite dismantling of the coating can prevent the direct adhesion of coating material and wear lining refractory, eventually reducing the mechanical shock to the wear lining refractory. Usually, the direct adhesion is caused by a reaction between a coating material and a wear lining refractory and reduces the impurities to suppress the growth of a liquid phase in the coating material (Table 11.2). Such effort can reduce the wear rate of the lining, and refractory could be improved by 7%.

Furthermore, to suppress the extension of cracks of the wear lining refractory, due to heating and cooling, expansion characteristic optimization may enhance the tundish wear lining refractory life. At 1500°C, the permanent linear change ratio reduction from 1.45% to 0.41% can reduce the expansion in the vicinity of the working face. By increasing the permanent linear change ratio at 1000°C

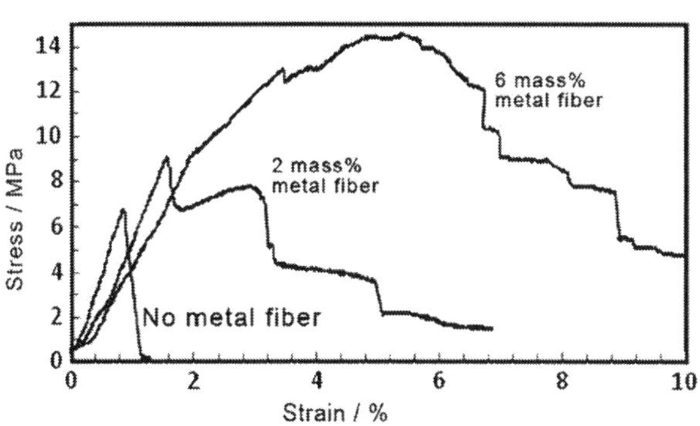

Figure 11.5 Stress–strain curve by a difference among the amounts of metal fiber addition.

Table 11.2 Chemical composition of the coating.

		Conventional	Advance
Chemical composition (mass %)	Al_2O_3	2	1
	SiO_2	12	5
	MgO	74	86
	CaO	5	2
	Fe_2O_3	4	1

from −0.21% to 0.06%, the expansion difference reduces between the working surface and the wear lining refractory, and the cracks are suppressed. As a result, the damage progress rate of wear lining refractory preferentially reduces by 28%.

Careful tackling of dry coating applications can reduce the direct adhesion of coating material. Usually, the spray coating technique is popular to prepare the tundish working lining, but trace compositions in the binder may cause direct adhesion with the base material. Dry coating can overcome this incidence and a competitive application protocol with tie variation is shown in Figure 11.6a. While maintenance is a concern, the dry coating time is always relatively more time consuming than spraying. In a typical industrial application, the maintenance time is extended to 15.5 h due to the employment of the dry coating method from that of 12 h by the conventional spraying method; thus an apposite promotion may reduce the maintenance time. Herein, the conventional feeding from a fixed tank to tundish demands more time; thus employment of a mobile type of tank may enhance the feeding rate from 40 kg/min to 100 kg/min and cumulative construction time reduces by 110 min, even less than spray coating (Figure 11.6b).

The offline evaluation confirms that the hardening of the dry coating material is completed at over about 100°C. Based on this knowledge, an optimum preheating condition is required. The hardening temperature is reached at any point between the heated side and its opposite side within a shorter heating period, eventually reducing

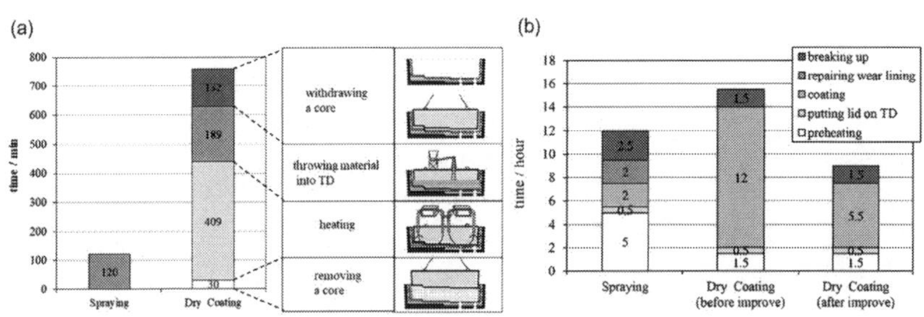

Figure 11.6 (a) Change in maintenance time by applying dry coating technique, (b) improvement of maintenance time [7].

the construction time by 15 min. The maintenance time reduces from 15 h to 9 h due to other improvements. Due to these improvements, reduction of the direct adhesion and the improvement of the construction efficiency becomes competitive.

Furthermore, in the dry coating, as water is not used as in the case of spray coating, the hydrogen pickup reaching as high as 1.5 ppm further minimizes to below 0.3 ppm. Thus, the effect of lowering the hydrogen level is also expected. Tundish boiling, frequent bubbling in molten steel in tundish is common when atmospheric humidity is higher and coating materials are not dried properly before steel pouring.

11.4 Ladle Shroud (LS)

A refractory device facilitates the transfer of molten steel from ladle to tundish through a confined space, shielding the teeming stream from atmospheric contamination, and plays a vital role during continuous casting and tundish metallurgy. A ladle shroud is also called a long nozzle, shrouding pipe, or pouring tube. A series of steel quality issues like reoxidation, air-pick up, and formation of macro inclusions during flowing may be avoided through ladle shroud. Furthermore, it serves to maintain the laminar flow of the molten steel in a tundish and thereby prevent the trapping of slag therein. Improved composition of Alumina-graphite composites, modern structural design, and coating technology serves better life. Moreover, the continuous argon injection and adequate sealing/holding technologies optimize the shrouding effect and minimize the air intake toward the steel flow. Cumulative design and steel flow, processing and materials development, installation protocols, and probable failures are discussed.

11.4.1 Design and Geometry

Effective design and maintaining ladle shroud geometry during fabrication and operation is a challenging task. It is connected to the collector nozzle of a ladle slide gate employing a simple counterweight or a fully automated mechanism. In a ladle shroud, two important sections can be distinguished with respect to their functions; these are the *bell* and *barrel*.

11.4.1.1 Function and Design of Bell

This part of the ladle shroud has the following functions (Figure 11.7):

- To assure a good connection to the ladle slide gate collector nozzle.
- Having a suitable shape for positioning against the collector nozzle and adequate handling and proper connection to the collector nozzle is important when considering the shroud design. There are two main families of connection to the collector nozzle, conical connection and butt connection, each having advantages and disadvantages.

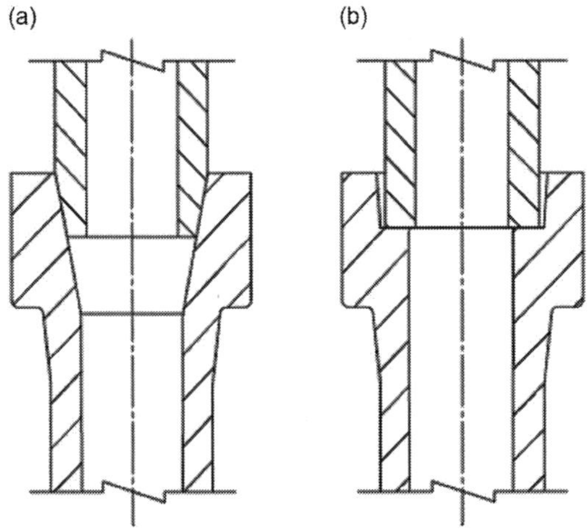

Figure 11.7 Design of bell portion of shroud, consists of (a) conical, (b) butt connection.

11.4.1.1.1 Conical Connection

This type of connection between two conical-shaped refractories (Figure 11.8) offers several advantages. There is a large area of surface coupling between the two refractory components, which helps the transportation of the LS during the slide gate movement and ensures good sealing.

It is self-centering. This property comes from the conical design itself. The coupling operation is simple because the smaller dimension of the collector nozzle matches the bigger bore of the ladle shroud. In such a connection, forces used for the coupling are released against the internal surface of the ladle shroud. In this condition, the refractory material works under tension. This type of load is emphasized during use by dilating the collector nozzle when manufactured from a material having high thermal expansion (MgO and similar). Most nozzles in India are; however, made of alumina. This fact is a limitation to using the conical connection because the refractories exhibit a low strength in tension. For this reason, the conical coupling is typically used when the force for the coupling does not exceed 250 kg.

11.4.1.1.2 Butt Connection

This connection provides the coupling between the collector nozzle and the ladle shroud through a flat surface. The coupling between the two refractory components is not as easy to achieve as the conical one. The two surfaces in contact are not oriented to facilitate the LS transportation during the slide gate movement. High force can be applied to assure the coupling between the ladle shroud and the collector nozzle because the two refractory components work in compression, one against the other, as shown in Figure 11.9.

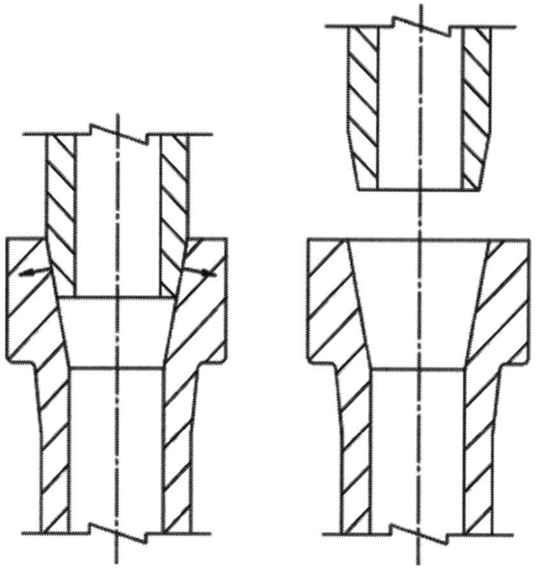

Figure 11.8 Schematic representation of conical-shaped refractories for ladle shroud and best sealing of collector nozzle in the shroud.

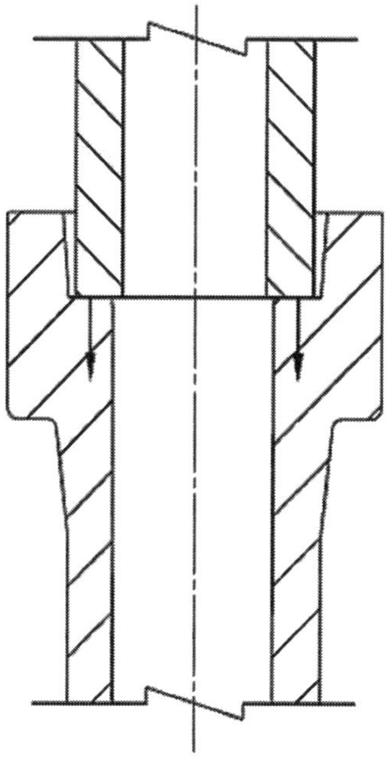

Figure 11.9 Schematic representation of butt connection in a shroud.

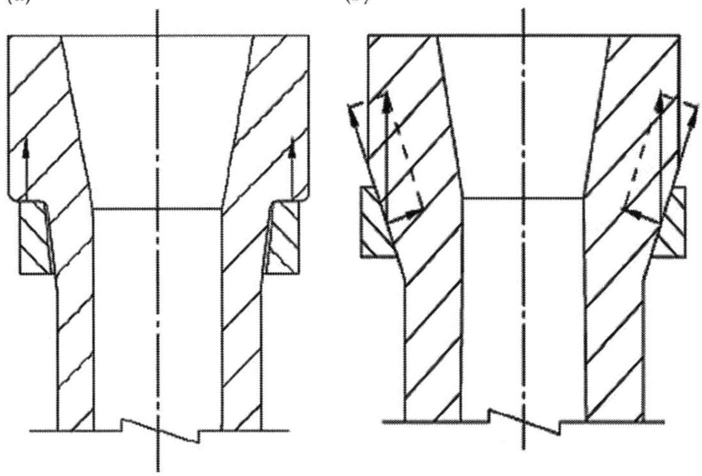

Figure 11.10 Support ring assisted shroud (a) first shape, (b) second shape (Y-shape).

Connection to the supporting ring is another important aspect as the bell shape needs to be designed to keep the ladle shroud coupled with the collector nozzle and handling operations. These functions are performed using a metallic ring that supports the LS when in position and facilitates handling. This support ring is connected to the lever arm of the mechanism designed for the LS handling (for detail, see later). The first shape looks like Figure 11.10a.

This shape, the first designed when the LS was developed and introduced into the market, is still the most used in steelmaking. The force applied to the support ring is transmitted to the refractory flange of the LS that experiences shear. Because the strength of the refractories to this type of load is generally low, this shape should not be used when the LS is supported with a counterweight exceeding 250 kg or with a hydraulically operated mechanism. The second shape (called Y-shape) shows a different solution to distribute the force applied as shown in Figure 11.10b. Due to the conical connection, it is possible to partially release the load through a component that works in compression on the LS. Such design and component, working to shear is reduced and can be minimized using a correct angle for the taper section. Shear component minimization expedites the total applied force to LS without any strength problem. For this reason, this flange shape is used when the force applied exceeds 250 kg or when a hydraulically operated mechanism is used. It is recommended to use this type of shape for every new design.

11.4.1.2 Canned Flange

It is expected that the flange in LS experiences mechanically stressed by different types of loads, and one of those is critical for the refractories (shear). The design calculation for the flange determines the dimensions and the wall

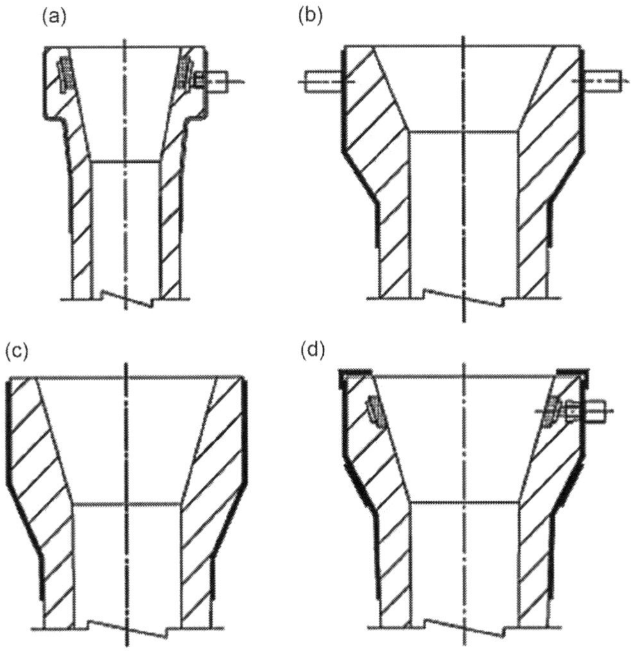

Figure 11.11 Canned flange (a) type 1, (b) type 2, (c) type 3, (d) type 4.

thickness of the LS able to resist the operating conditions. Among the several designs, the correct design may lead to dimensions that dramatically increase the weight of the LS, as shown in Figure 11.11. For example, the design in Figure 11.11d may experience high thermal shock. When this happens, a solution to the problem is offered by using a can at the flange section. The mechanical strength of the can itself partially absorbs the load applied to the flange so that, even in heavy operating conditions, we can have a ladle shroud with reasonable dimensions. Below we can see different patterns showing the canned flange application.

11.4.1.3 Function and Design of Barrel

There are two main functions:

- to be a shield against the reoxidation of the steel
- to distribute the steel into the tundish

The barrel has the shape of a ceramic tube having an internal bore bigger than the bore of the plates of the slide gate to which the LS is connected. This practice avoids a possible skull formation that can limit the steel flow. Because of this over-dimensioning, the LS system works with an internal negative pressure once the barrel is submerged. The value of this negative pressure depends on:

- the ratio between plates-bore and barrel internal diameter
- type of conicity of the barrel
- length of the ladle shroud

11.4.1.4 Tapering of the Barrel

A prerequisite tapering of the barrel is an essential design aspect as it has an important role in achieving ladle shroud performance. In the drawing below the definition of taper is shown. When the value a–b is positive (a > b), it is recognized as direct taper; when negative (a < b), it is designated as reverse taper.

LS with direct taper: It is the most common shape and easy to manufacture. The direct taper (Figure 11.12a) does not increase the negative pressure inside the ladle shroud from the predetermined value given by the ratio between the plate bore and barrel internal diameter.

LS with reverse taper: Most customers want to open the ladle gate with the ladle shroud already connected and with the bottom part submerged into the steel. This

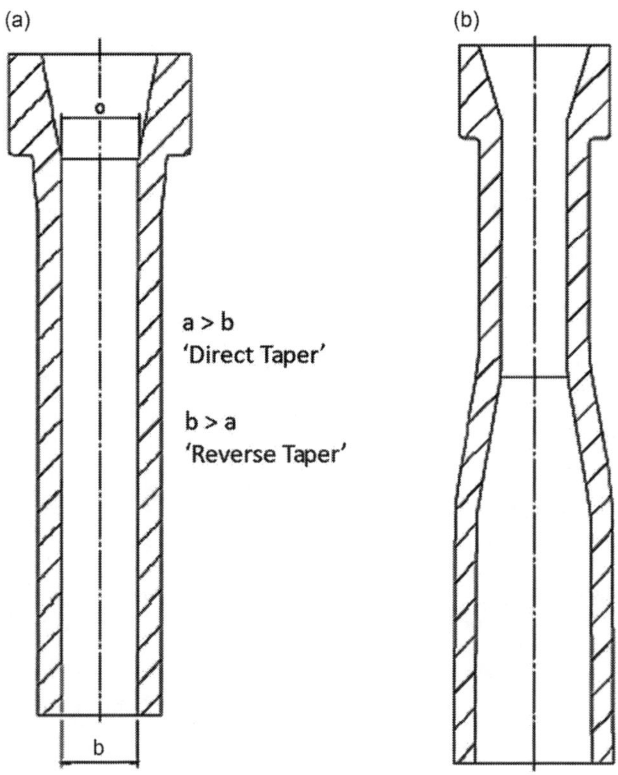

Figure 11.12 Tapering of ladle shroud (a) direct, (b) reverse.

practice requires an increased internal volume of the submerged part to avoid plugging caused by the powders used into the slide gate and skull formation, which cause blowback of steel from the connection.

The way to increase the internal volume of the submerged part of the ladle shroud is to increase the internal bore, leading to the use of the reverse taper (Figure 11.12b). The bell ladle shroud is a special example of this application. Since an LS with a reverse taper works as a diffuser, there is increased negative pressure inside the ladle shroud. Thus, a negative pressure inside the ladle shroud causes air suction through the joint between the collector nozzle and ladle shroud. Air suction is also a result of the venturi effect caused by the flow of steel.

11.4.1.5 Ladle Shroud with Argon

When the connection between the collector nozzle and the LS is not properly fix or tight, the possibility of air infiltration is made worse by:

- the venturi effect caused by the steel flow;
- the fact that the LS has a negative pressure inside the barrel.

Air suction means reoxidation, Al_2O_3 formation, and increasing N_2 pickup. These are not desirable, especially if Al-killed steel is cast.

To minimize air suction, different designs for argon injection through the LS are used to:

- compensate the negative pressure inside the LS Barrel
- create a sealing with an argon barrier

11.4.1.6 Ladle Shroud for Argon Injection (LS-A)

The configuration of this LS when connected to the collector nozzle is shown in (Figure 11.13a). Argon is injected through the annular cavity in the top flange of the LS-A and then passes through a series of six slots into the steel stream below the collector nozzle.

Such operation expedites positive argon pressure inside the ladle shroud and provides two benefits as;

- Firstly, argon flows down the LS-A as bubbles are entrained in the steel stream. These bubbles rise quickly to the surface in the tundish and assist with the floatation of inclusions.
- Secondly, the argon pressure ensures that argon will flow out of any possible gap between the LS-A and the collector nozzle and prevents any possible air ingress at the joint.

The argon supply is connected to the LS-A employing a simple coupling fitted into the top flange. In the above drawing, the coupling embedded into the LS-A

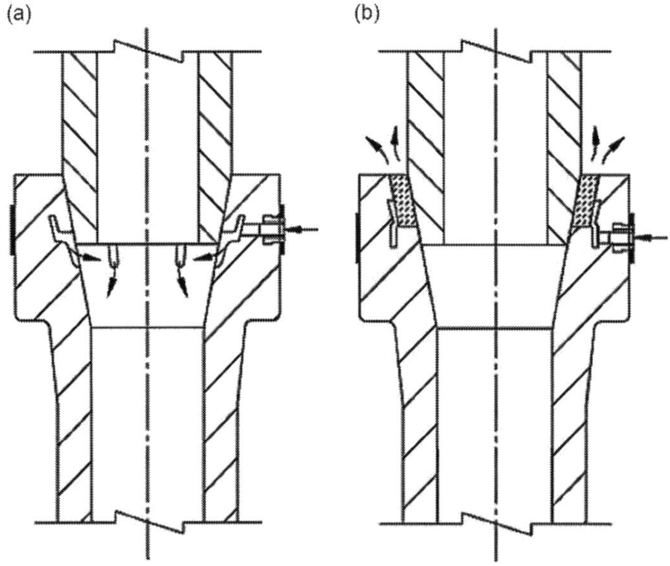

Figure 11.13 Argon injected ladle shroud (a) through the cavity of top flange, (b) through the refractory permeable ring.

flange is supported by a metal band around the flange circumference and assists in keeping the coupling in place.

11.4.1.7 Ladle Shroud with Argon Shielding (LS-A Co-pressed Porous Ring)

In the drawing below, the configuration of this LS when connected to the collector Nozzle (Figure 11.13b). Argon is injected into a slot and then passes through an alumina graphite permeable ring, ensuring uniform gas flow. The permeable ring is located in the coupling area of the sliding gate collector nozzle. The function of the permeable ring is to ensure that at the coupling between the refractories, air cannot be drawn in, but only argon. The top of the permeable ring creates an argon atmosphere around the junction, which avoids air infiltration. The permeable ring is made from Al_2O_3–graphite material, and isostatically co-pressed with the LS body. The manufacturing technique permits the location of the permeable ring in various positions of the ladle shroud top flange (Figure 11.14).

In the drawing below (Figure 11.15) you can see how the argon sealing can be created when the steel can is used. The gap created on the top surface of the ladle shroud will allow the diffusion of argon all around the collector nozzle creating an atmosphere from where only argon can enter the ladle shroud.

This type of argon injection is becoming more and more popular to overcome shop difficulties. There could be some fiber between the can and the flange to distribute the argon around the circumference evenly. This design is known as

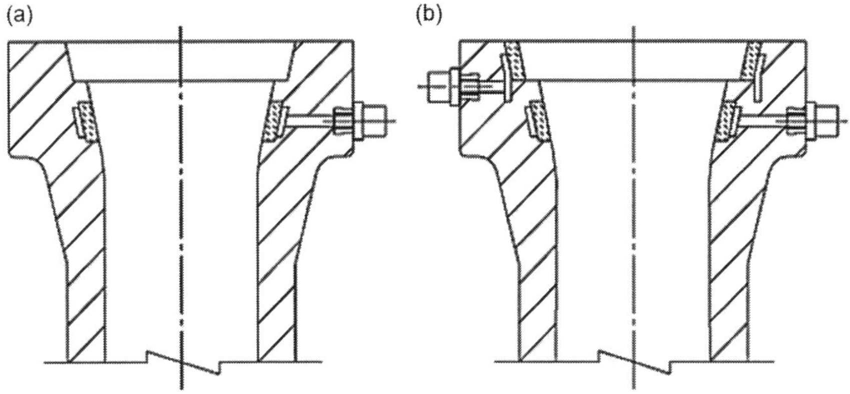

Figure 11.14 Position of permeable ring, (a) type 1, (b) type 2.

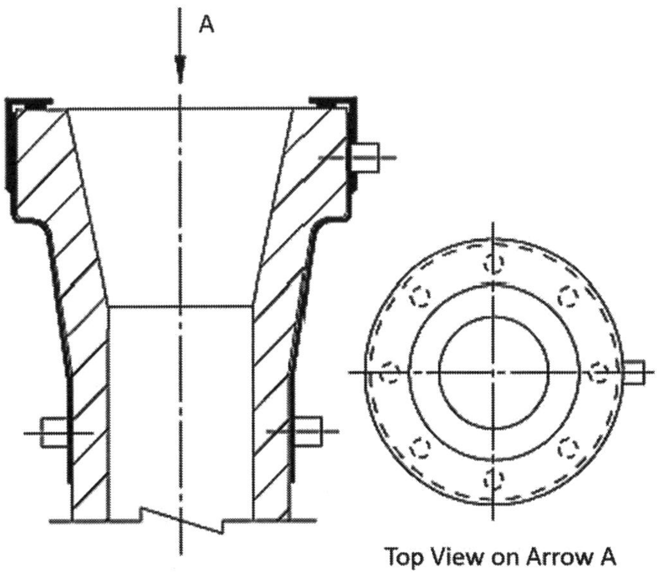

Figure 11.15 Argon sealing management, side, and top view.

top-can argon system and is widely recognized for giving the best performance. In reality, the flange shows many shapes, connections, and steel-can configurations. Each solution depends on the operational condition and customer satisfaction.

11.4.1.8 Gaskets

The pressure on steel producers to continually improve quality has developed a large market for effective sealing products to reduce air aspiration in LS and SES applications. Most new LS designs are now intended to be used with a certain gasket to improve the seal at the mating surface with the ladle collector nozzle. Many

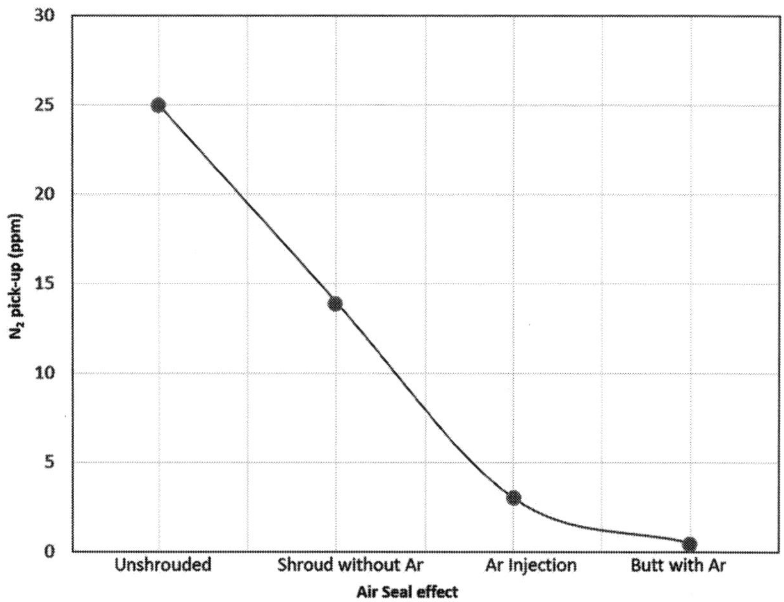

Figure 11.16 Effect of air sealing effect on the nitrogen pickup from air.

steelmakers now use the ability to reduce air aspiration (by measuring the N_2-pickup between the ladle and the tundish), an important criterion of LS performance. To minimize N_2-pickup, most customers are now placing a new gasket into multiple use LS's before each heat. At some plants, a structurally sound LS may be replaced if the N_2-pickup levels are exceeded. The proper gasket can be a key to achieving improved overall LS performance.

It is common practice that worldwide quality steel producers prefer a gasket for each ladle cast. A competitive N_2 pickup concerning different shroud systems is shown in Figure 11.16, in which a typical butt shape with Ar injection may facilitate the lowest N_2 pickup into steel from the environment during operation.

A typical ladle shroud and function of their respective portions is shown in Figure 11.17, and one can understand why the design of both geometry and material are essential to achieve maximum heat.

11.4.2 Failures, Materials and Processing

The choice of the material and the process used to manufacture ladle shroud is usually dictated to fulfill the need of:

- minimize thermal shock
- introduce the least stresses possible during manufacturing
- have erosion resistance of the material to an extensive range of steel-grades

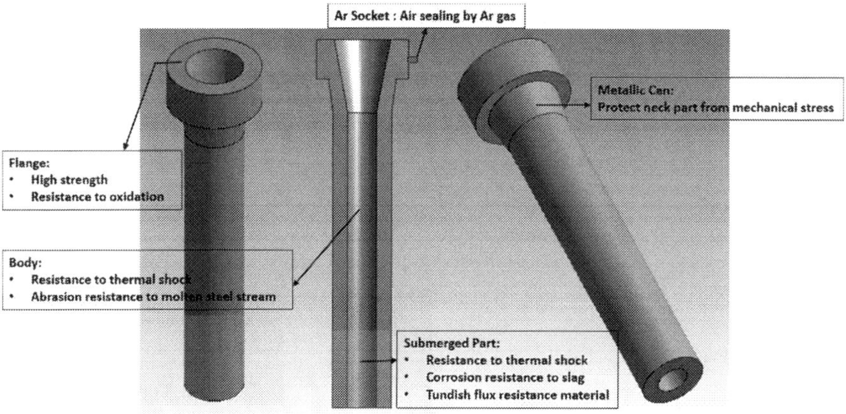

Figure 11.17 Details overview of typical design and geometry of a complete Ladle shroud. Without care of submerged part may follow premature corrotion.

The ladle shroud family, comprising different designs, has a body made from isostatically pressed Al_2O_3–graphite. Unique surface and body glaze treatment ensure resistance to oxidation during use. A specially designed firing process can assure the possibility of using the ladle shroud without preheating. A newer but more expensive design uses a liner mix to overcome the thermal shock. This liner is also very erosion-resistant; thus reuse can be envisaged (although with certain limitations—see further for detail).

The steel flow through the LS causes:

- erosion by the mechanical effect
- corrosion by chemical attack

The first is mainly crucial in the impact area of the stream at the bell and in the submerged area. The second is important for the body of the ladle shroud and is emphasized at the steel level in the tundish, where the tundish fluxes cause the chemical attack. The design of ladle shrouds is becoming more complex as different materials are now used in the various parts of the shroud (zoning) so that the more wear resistant mixes and generally the more expensive ones are only used in those areas subject to excessive wear. In some cases, a ladle shroud can have as many as five materials used in its design. Hereunder is a schematic drawing of a typical ladle shroud with some of the most occurring erosion/gouging/thermal shock and chunking problems (Figure 11.18a). Mostly the solution will be a combination of several.

Figure 11.18b shows the symptoms of different zones after damage through; A. Cracking through Flange, B. Horizontal Break at Collar, C. Thermal Shock Crack,

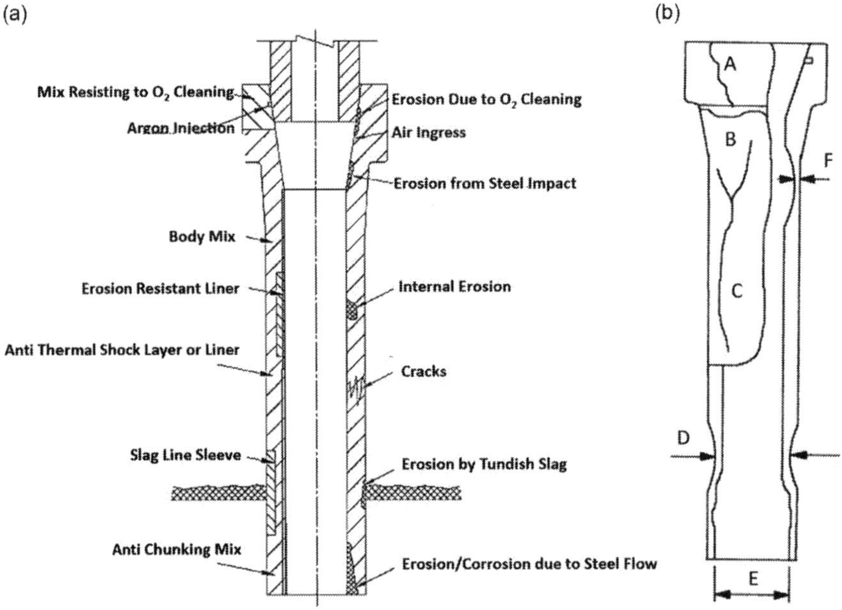

Figure 11.18 (a) Critical area and their probable failure process, (b) schematic representation of distorted or damaged shroud because of several reasons.

D. External Corrosion, E. Internal Corrosion, F. Internal Erosion. Such failure may have one or several causes. These are:

A1 Excessive Counter-Weight Force
A2 Ladle Nozzle Mismatch
B1 General Condition of Holding Ring
B2 Insufficient Clearance between Nozzle and Ring
C1 Insufficient Oxidized Layer
C2 Excessive Moisture Pickup
D1 Aggressive Tundish Cover Powder
D2 Slag Carryover
E1 Slag/Tundish covering Powder Entrapment
F1 Shroud Not Vertically Positioned

While considering the different failures of a shroud, the following causes can be encountered and developed further, either through design alternation or material selection, or a combination of both. In prominent installation and operation parameters are important causes to continue the best performance of Shroud.

11.4.2.1 Choice of Materials for the Barrel

There are several choices of mixes for the barrel, depending on the application. For longer casting times and for casting more erosive steel grades like low carbon, Ca treated grades or high oxygen steels, the mix **T-1** is suitable. For applications that are prone to thermal shock, the body mix should be **T-2**, which has a better thermal shock resistance than the other body mixes. In some extreme cases where internal gouging is a problem, then a liner of mix **T-3** is used in the barrel. Liner mix **T-4** is used as an anti-spalling layer. This liner as well provides a cold start-up as a cold reuse, but only when enough liner remains after the first casting(s). This solution (**T-4** liner) also provides an extended service life for a ladle shroud. The normal solution for thermal shock (all body mixes) is to decarburize a substantial layer of the internal bore of a ladle shroud. This is typically 20% of the wall thickness, where the layer is quickly washed away during the cast, so basically, it loses a significant part of the ladle shroud service life. The **T-4** liner has a good erosion resistance, so it adds to the life span of the ladle shroud. Note, it does NOT add to the overall wall thickness mechanical strength. The **T-4** liner technology requires a double pressing formation and has more expensive raw materials. Thus, usually, it is considered a premium grade.

11.4.2.2 Choice of Materials for the Seating Area

Materials for the seating area have to be resistant to abrasion and the effects of oxygen cleaning of steel skull, which builds up just below the collector nozzle in the seating area. The mix normally used for this area is mix **T-5**, which can be pressed with the main barrel, as summarized in Table 11.3. In the case of a flat seat connection, a high alumina ring can be used, which has to be cemented in

Table 11.3 Different zones for type of applications and relevant properties on the product.

Mix use	T-1 Body	T-2 Body	T-3 Liner	T-4 Liner	T-5 Flange/ Seat	T-6 Exit	T-7 Sleeve	T-8 Sleeve
Application	Aggressive steel	Standard	Anti-gouging	Long life/ cold reuse	Antioxidation	Anti-chunking	Anti-Slagline	Anti-Slagline
Properties	Higher thermal shock loses MOR in time	Improved thermal shock	-	Good thermal shock, good erosion	Good resistance to O_2 with improved strength	-	Good resistance against slag attack	Excellent performance; manufacturing difficulties

position, and in some cases, this ring can be fixed using a gasket material and even replaced during the cast. A typical material **T-5** also helps to reinforce the flange as it has exceptional good mechanical resistance properties.

11.4.2.3 Choice of Materials for the Outlet Area

As alumina graphite composition changes to different phases during extended use, parts of the outlet (immersed) may crack (horse shoe cracking or chunking). In this circumstance, the **T-6** can be a solution for that.

11.4.2.4 Choice of Materials for the Slag Line Area

The mix available for the slag line is mix **T-7**, which is suitable for basic slags and is based on zirconia. In theory, **T-8** work as well and provide a purer (read: more resistant) zirconia sleeve, but protection of this material during manufacturing is (more) difficult due to the high content of zirconia and very pure graphite.

11.4.2.5 Choice of Materials for Gasket

Glued ceramic fiber gaskets were initially used in the LS bell to assure a good sealing between the shroud and the collector nozzle. Ceramic fiber paper was the first gasket material used. The sheet product (usually 3 mm thick) is cut to shape and then glued into the product using sodium silicate. This material produces the least effective gasket, and it has the further disadvantage that it cannot be replaced between ladles in a sequence cast. For these reasons, ceramic fiber is being phased out in most applications. The ability to form this product into irregular shapes keeps this material active in a few products.

Thermoseal: Thermoseal gasket is an upgraded material that is rigid enough to make good conical shapes for LS applications. Grafoil (a rolled graphite) is applied to the outside surface of the gasket to minimize sticking at the LS-Carbon interface. A new gasket should be used for each heat. This product must be sandwiched between two surfaces under continuous pressure to make it effective sealing. Its poor erosion resistance and hardness exclude this gasket for SES applications. Overall, this is a fair gasket material for LS.

Tempseal: This high-temperature plastic sealing compound, having the consistency of putty, is extruded in sheet form and die-cut into a flat gasket shape. Tempseal works well for sealing SES as long as it is kept under compression.

11.4.2.6 Processing

Cold isostatic pressing (CIP) is the prime and foremost machinery to obtain the long tube consisting of uniform body strength. A schematic representation of the processing and planning of continuous casting refractories preferentially made by

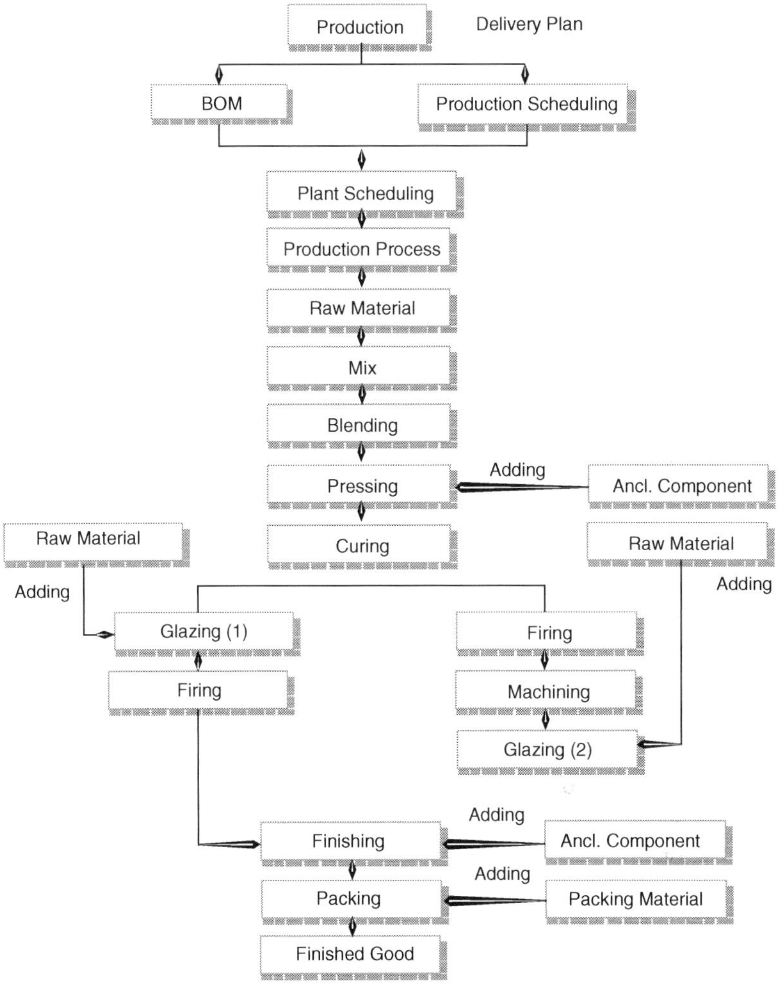

Figure 11.19 Process and planning flowsheet of continuous casting refractories. BOM = Bill of materials; Ancl. = Ancillary.

CIP is shown in Figure 11.19. The left side of the schematic represents the process of ladle shroud (LS) and monoblock stopper (MBS), where pressing, glazing, and firing are done. Right side process protocol highlights the precious machining preferentially CNC is required for SEN followed by glazing and firing. The glaze slurry composition and viscosity are carefully controlled to perfect sticking and avoid premature failure due to a mismatch between the coefficient of thermal expansion (CTE) of pressed body and glaze composition.

11.4.3 Operational Practice

Ladle shroud has several advantages to obtaining high grade steel; these are:

- Controlled steel flow from ladle to tundish.
- Stream protection to prevent reoxidation of the steel.
- Cold start-up without preheating.
- Prevention of steel splashing during casting.
- Prevent N_2 pick up during casting.

Despite material and design, installation practice monitoring is needed to achieve maximum performance. The precious wall thickness and dimension make it susceptible to fracture; thus it is critical to pack in a polyethylene bag, handle and store carefully in a dry place to avoid moisture contamination. Shroud fixing in the manipulator is a tricky task, and one needs to ensure that it should not be scrapped. Shroud should not drop into the holding ring as this may cause damage. Prior to fitting in the manipulator, the following checkpoint should encounter on-site:

- Adequate clearance of about 5 mm in a shroud and holding ring need to maintain and avoid any mishappen.
- The holding ring should be free from any steel splashes to avoid point loading and subsequent stress on the shroud and restrict movement.
- Pivot points of the shroud ring should be free moving with good lubrication.

A perfect vertical shroud fitting onto the exchangeable nozzle is important to ensure good sealing between the shroud and exchangeable nozzle and prevent premature erosion of the shroud due to air ingress and immature failure because of the generation of point loading stress in operation, as shown in Figure 11.20a. Erosion preferentially happens if it deviates vertical position and steel flow, impacting the same region of the bore of the shroud.

The shroud removal is essential for a non-free opener ladle while the ladle is lanced. After establishing the steel stream and tundish filling up to the desired

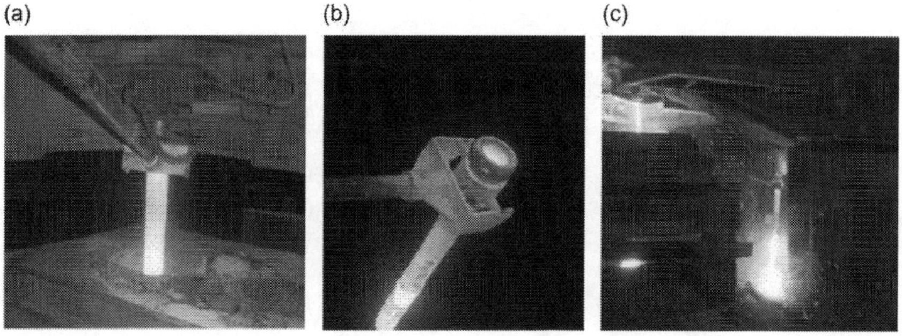

Figure 11.20 (a) Vertical position of shroud during operation, (b) checking of the connecting area of exchangeable shroud, (c) O_2 lancing in vertical position of the shroud.

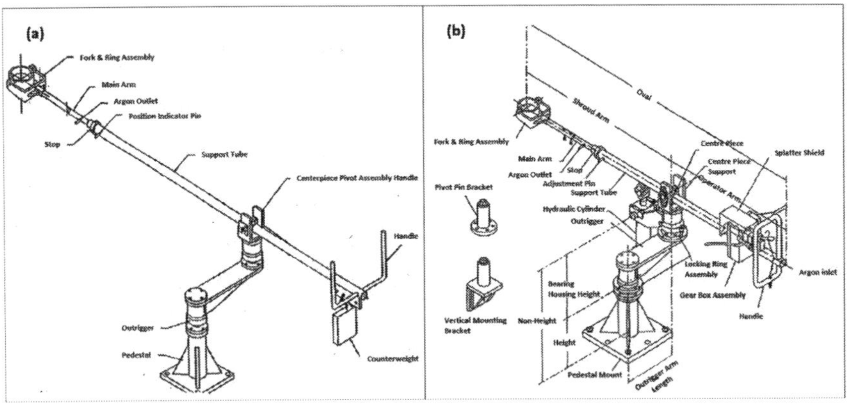

Figure 11.21 (a) Manual and (b) hydraulic ladle shroud manipulator.

level, the shroud again refits to the exchangeable nozzle (Figure 11.20b). However, a couple of precautions are beneficial before refitting:

- Throat and bore areas must be adequately cleaned for good sealing between shroud and nozzle that eventually prevents blowback by the shroud.
- The replacement of a damaged fiber gasket is necessary.
- Ensure the slide gate is fully closed during refitting to prevent steel dribbling from entering the throat area of the shroud, any additional steel between shroud and nozzle content may lead to point loading stress and prevent perfect sealing.

In the cast, a shroud is preferentially immersed in the steel bath a minimum of 200 mm to prevent excess turbulence in the tundish and sufficient tundish powder cover that avoid tundish powder entrapment in the shroud bore. Slide gate throttling expedites shroud movement, and it should not be limited by the tundish powder crusting over; instead, enough is required to ensure free movement of the shroud. The shroud is cleaned during ladle change by employing an oxygen lance with a vertical shroud (Figure 11.20c). It reduces the risk of the molten metal blowing back toward the operator and prevents the running back of molten metal onto the shroud sealing face. Proper sealing, gasket replacement, and bore cleaning remove steel skull that facilitates steel flow during slide gate opening and avoids blowback. Figure 11.21 shows some typical manual and hydraulic ladle shroud manipulators that are essentially required during the installation and exchange of shroud.

11.4.4 Flow Pattern

Slide gate opening and closing adjustment eventually control the steel flow rate through the shroud. The lower nozzle inlet area from the slide gate depends on the available crossing area between two eccentric circular faces, as shown in Figure 11.22a. Usually, the vortices form near the mainstream and path deviation

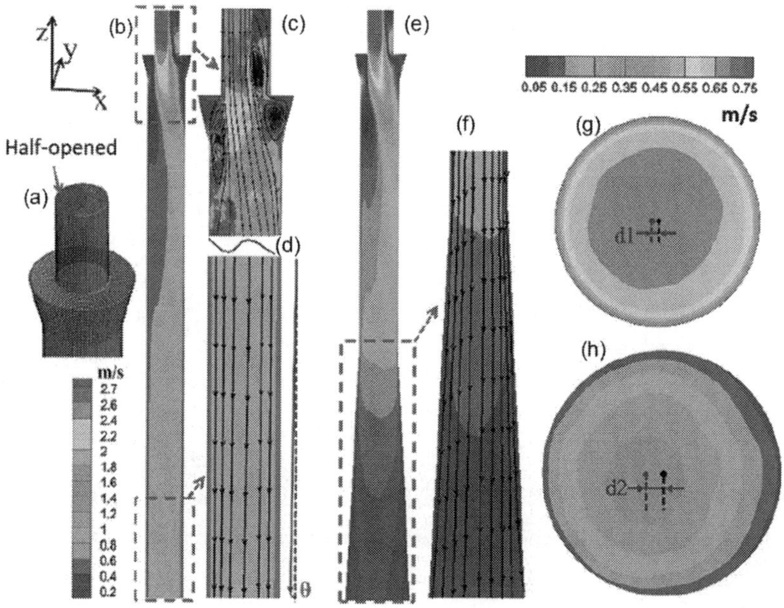

Figure 11.22 (a) The meshing of the ladle shroud with a half-opened inlet. (b) The time-averaged velocity contour of the whole conventional ladle shroud (CLS) for 30 s. (c) An instantaneous flow profile at the inlet part of the CLS. (d) The mean velocity contour and trajectories at the outlet part of the CLS. (e) The time-averaged velocity contour of the whole trumpet-shaped ladle shroud (TLS) for 30 s. (f) The mean velocity contour and trajectories at the outlet part of the CLS. Mean velocity contours at the outlet of (g) CLS and (h) TLS [10].

when the slide gate is half-open, as shown in Figure 11.22b–c. An extensive flow pattern between the lower nozzle and ladle shroud is available in recent literature [8, 9]. The flow behavior and inner lining wear pattern strictly depend on the opening; a biased flow results only when a one-side opening is available and impacts that particular side. A biased inflow is not completely dispersed, and the outlet jet tends to flow with a biased degree of θ as illustrated in Figure 11.22d. Thus, such opening provides a non-concentric outlet velocity profile when the slide gate is open partially; different simulated flow pattern for conventional and trumpeted shape shroud is shown in Figure 11.22b–f.

A distinct shift between the flow center and geometric center is noticed in conventional ladle shroud (CLS) and trumpet-shaped ladle shroud (TLS), as d1 and d2, respectively [10]. Such biased flow is analogous to the formation of eccentric tundish open eyes (TOE) because of the argon bubble emission rate is more than the optimum flow rate [11]. In this aspect, a theoretical study with ¼ and ¾th opening area and different casting speeds can be considered to understand the casting parameter effect on the flow pattern.

11.5 Mono Block Stopper

Monoblock stoppers (MBS) are used to control steel flow through the subentry nozzle (SEN) or the tundish nozzle. It is made of magnesia carbon or alumina carbon quality with special compositions at the tip and slag zone to suit customer requirements.

Usually, SEN is used in combination with MBS–SEN. However, subentry shroud (SES) is preferentially used for MBS–tundish nozzle–SES combination. SES further demands a manipulator during installation. Some critical features are:

- MBS clean shut off at the end of casting
- It should have thermal shock resistance
- Erosion of steel and resistance to tundish powder and slag
- Tip design should facilitate argon flow through MBS to prevent the Al-clogging
- Stopper tip material customized to steel grade such as MgO-C nose for high Ca ppm steel
- Tip liner with ultralow carbon material and consist of slag resistant material band

A typical cross-section of MBS is shown in Figure 11.23, in which critical parts include a metallic nut hanging system, slag line flux resistant composite sleeve, body structure, argon metering plug, and stopper tip coexists.

11.5.1 Preheating Schedule

The existence of an inner layer in the ladle shroud is capable of preventing thermal shock on initial start-up and can be used in cold start without preheating. The monoblock stopper (MBS) is heated up on the casting platform by the burners that are usually used to heat up the tundish working or wear lining. It is important to maintain a time gap up to maximum 5 min in between *end of preheat* and *start of casting* to prevent thermal shock cracking of the stopper and steel stream freezing due to the stopper sticking in the submerged entry nozzle (SEN) throat. Prior to preheating the thin slab SEN, it should be wrapped by cerawool to prevent

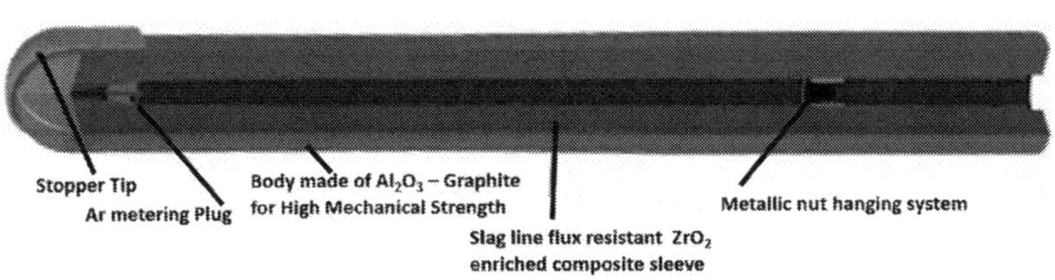

Figure 11.23 A typical design and structure of MBS.

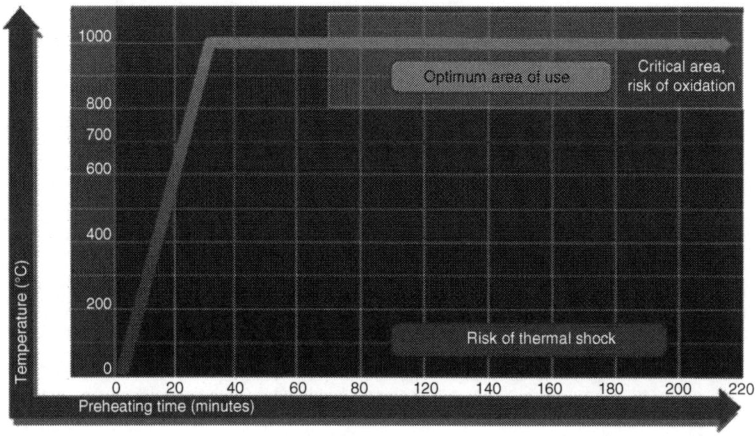

Figure 11.24 Preheating schedule of SEN.

excess heat loss after switching off the burner prior to casting. Burners have to be placed on both sides of the port to maximize stopper nose and SEN body preheating. A typical preheating schedule for SEN is illustrated in Figure 11.24. Similar *off gas to start cast* protocol is needed to maintain, where the time gap between preheating completion and starting casting is 5 min maximum. Submerged Entry Shrouds (SES) are pre-heated either off-line or on-line. On-line pre-heating can be done using external burners or by venturi pre-heat. Offline pre-heating can be done by using external burners or by placing the SES in an oven. The SES should be pre-heated for at least 60 min to achieve proper heat soaking. The SES body temperature is preferentially kept from 800–1150°C.

11.5.2 Installation

The brief installation process can be summarized theoretically; however, expertise and a sense of work environment conditions may facilitate the best performance of components. A few essential installation steps are:

- Before fixing MBS inspect the monoblock visually. Ensure the workstation is clean and insert the steel rod only when the MBS is in a vertical position (Figure 11.25a)
- Before fixing the stopper pin in a stopper, blow the air inside the monoblock (to remove any dust / foreign particles).
- Fix the stopper pin inside the monoblock and fully tighten.
- After fixing the pin in the stopper, put the MBS pin in the arm of the tundish mechanism. Place the MBS into the MBS control arm and adjust the bottom nut to allow downward movement (Figure 11.25b).

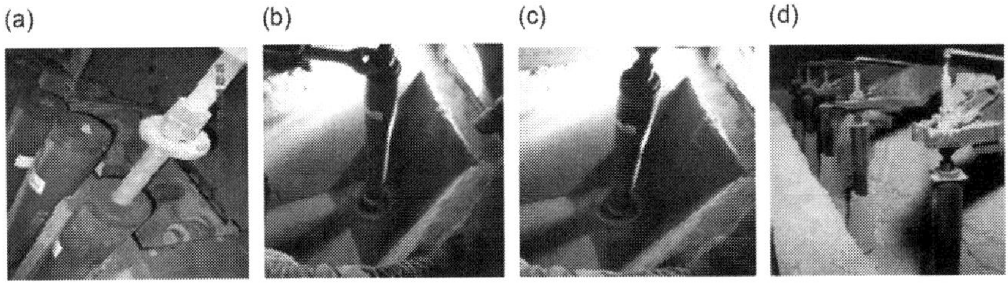

Figure 11.25 Installation practice of MBS.

- Set the working gap between SEN seat area and MBS (open and closing). Then tighten the arm of the mechanism.
- Tighten all the bolts and arm bolts of tundish mechanism and check the opening and closing movement (need to check the MBS nose should be properly set in the seat area of SEN). Tighten the top nut and test the position of the MBS to the Nozzle. Off-set the MBS by a few millimeters to allow for distortion (Figure 11.25c),
- Do final alignments, check, and tighten the nuts above and below the cross arm. Put the tundish covers and send them to the casting floor for pre-heating (during pre-heating MBS should be in the opening position) (Figure 11.25d).

11.5.3 Failures

Refractory life depends on both product quality and operation features. A schematic representation of some typical cracking in MBS is shown in Figure 11.26. During operation, breakage and wear influenced by erosion and corrosion are common failure symptoms found in different MBS zones.

These failures may be due to several causes:

- Breakage initiation by crack and formation of crack has several reasons: fixing end problem due to misalignment, loose fixing and defective fixing method; main body misalignment and bend product; nose defect due to misalignment, sticking and incorrect nose shape.
- Erosion and corrosion are due to steel chemistry, MBS material quality, misalignment, and steel flow behavior. For example, the uncontrolled turbulence of steel in tundish may enhance the corrosion rate of MBS. Misalignment provokes unwanted MBS-steel interaction in misaligned zones that eventually facilitate corrosion.

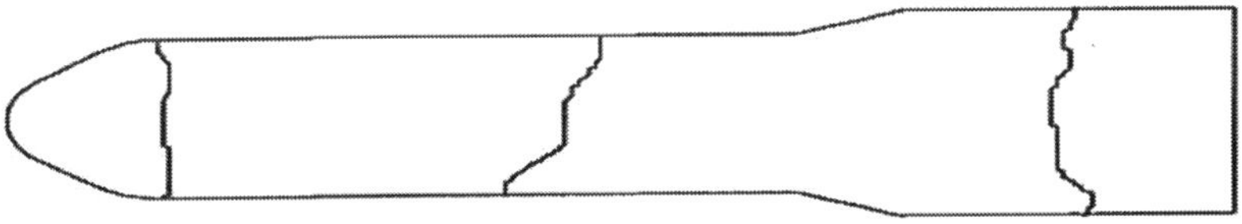

Figure 11.26 Schematic representation of typical cracking of MBS.

11.5.4 Glazing

In general, we can split the glaze applications into two major categories: glaze we will vitrify (bring into the glass phase) during the manufacturing process (open firing), and glaze we send out to the customer in non-vitrified condition. There are two methodologies involved in this system:

- For open firing vitrification done during CCR process—the flooding technique
- Vitrification is done at the customer end—dip glaze technique

The vitrification process is usually completed during CCR manufacturing process, and quality is checked after firing. Very thin glazing may exhibit oxidation beneath the glaze layer, whereas too thick expedite blistering or balling up. Furthermore, the lousy flood glaze application method facilitates pinholes and other defects. The entire process is related to keeping the flooding process under control (wash–clean–flooding practice), and keeping the glaze properties within the ranges typical for the specific plant–flood machine (pickup and density). Also, kiln malfunction can interfere with the final quality of the vitrified layer.

11.5.4.1 Process Chart for Flood Glaze of MBS:

Material received from curing → Loaded at flood glaze station → Washing thoroughly with jet spray → Glaze flooding → Air drying → Force drying @ 150–180°C → Loading in a kiln car with base plate and pin → Open firing → Firing Temp 940–960°C → Soaking time 40 min Max → Total cycle time 4 h max.

11.5.4.2 Important Properties of Flood Glaze during MBS Glazing:

GPU: usual target 9–11 gm; Density: 1.80–1.83 gm/cc; Moisture: 26.5–28 Max; Thickness: using comb gauge, 0.45 mm (minimum) to 0.55 mm.

After glazing, it is essential to place the comb gauge after precisely 3 min (180 s) ± 10 s on the middle of the body to measure the glazing thickness. It is a tentative indicator of surface tension or flowability and stickiness. Herein, handheld timer with an alarm (chronometer with beeper of the cooking clock with alarm signal) is required to control the process. MBS with an alumina nose does not need any extra protection coating for oxidation, but MBS tip with magnesia nose should have an extra low melting glaze coating to prevent any skin or complete oxidation during curing or firing.

11.6 Submerged-Entry Nozzle

A submerged entry nozzle (SEN) is preciously used to transport the molten steel from the tundish to the mold, a critical refractory for the steelmaking process. The prime objective to use of SEN is to maintain desired flow for a particular shape of a cased product and prevent oxygen and nitrogen pick up by molten metal from the atmosphere. Thus, it prevents the reoxidation of steel and air aspiration, and eventually, the smooth steel flow toward the caster. Frequently, fiber insulation is used to prevent thermal shock and cracks and reduce the aluminum clogging.

11.6.1 Installation and Failures

It is mandated to clean the tundish bottom with compressed air before fixing the SEN, and follow the steps (Figure 11.27):

- Check the well block area (inside and all around).
- Put the SEN inside the well block without cement (to ensure it is going smoothly).
- Put the cement all around the neck area of the SEN (cement thickness should be 10–15 mm). Insert SEN in the well block and push the SEN inside the well block as per the length required for casting.
- Check the straightness of SEN, and it must be at 90° (to check the straightness of SEN may check to use a plumb / right angular scale).
- After fixing the SEN seal, the seating area and bottom area of SEN need to be checked.

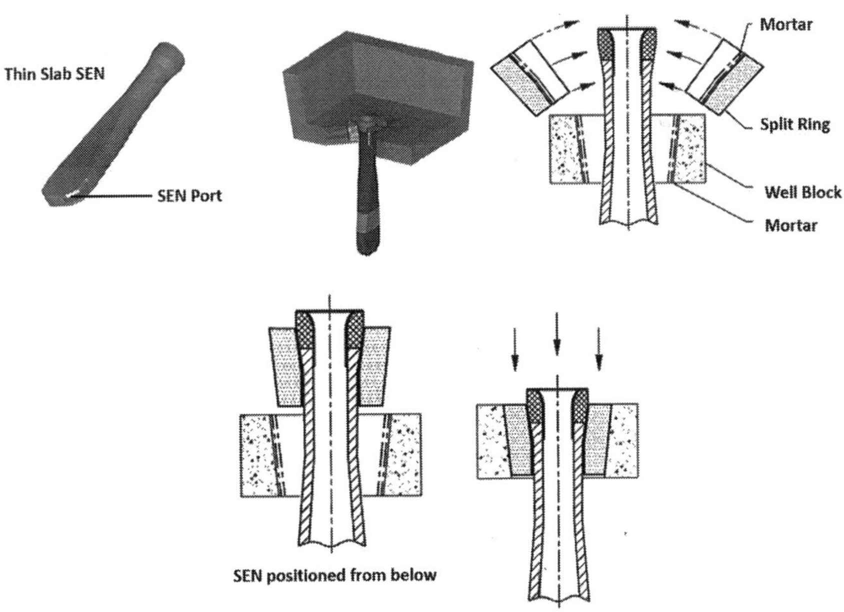

Figure 11.27 Schematic representation of SEN and their installation protocol.

11.6.2 SEN Fixing for Thin Slab Caster

Certain parameters are essential to consider while designing the component and fixing it. Split rings coated with mortar are preferentially used to fix the SEN in the well-block during thin slab casting. Figure 11.28 shows the annular space of well-block that can accommodate split rings, fix the SEN perfectly from the bottom and ensure the tight-fitting. Before fixing the SEN, it is essential to ensure the well-block is perfectly clean and free from any tundish debris to avoid misalignment of split rings and SEN. During fixing of SEN by ceramic mortar, there should not be any cavity or channels allowed that may draw air and produce non equilibrium pressure difference. While fixing the component, SEN should not be hammered or struck by an external object. It is necessary to confirm that there is no movement of split rings and perfectly flush with one another on the top or exposed surface in tundish. Critical dimension and alignment check is mandatory to avoid mishap during operation. The applied mortar cures and hardened during preheating of SEN.

11.6.3 SES Installation and Failures

Sub Entry Shroud (SES) can be connected in the tundish gate system and a bottom part of the tundish nozzle, whereas SEN is directly used in tundish well-block only. Subentry shrouds like the ladle shroud are also susceptible to fracture; thus careful handling and storing in a dry place is a critical issue.

For on-line heated SES (Figure 11.29):

- Position and mount the new SES to the tundish slide gate bottom plate or to the tundish nozzle.
- Ensure that the SES top face is in complete contact with the bottom plate or tundish nozzle upon locking the holder ring mechanism.
- For venturi heated SES—open the venturi exhaust pipe only after the tundish has been heated sufficiently.
- For burner heated SES—heating by burners can start immediately.

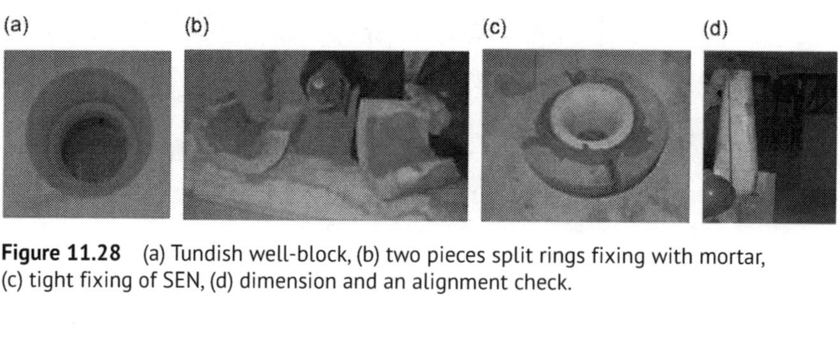

Figure 11.28 (a) Tundish well-block, (b) two pieces split rings fixing with mortar, (c) tight fixing of SEN, (d) dimension and an alignment check.

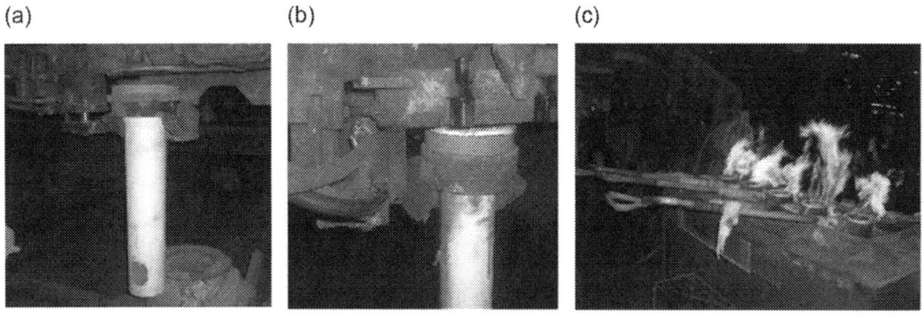

Figure 11.29 Online heating of SES.

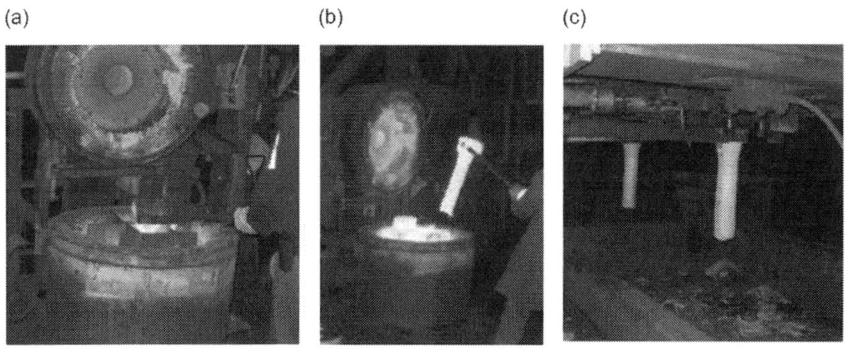

Figure 11.30 Off-line heating of SES.

For off-line heated SES (Figure 11.30):

- Position and mount the new SES to the burner stand or burner oven.
- Heat-up the SES according to the pre-heating guideline.
- Around 5 min from target start cast, remove the SES from the stand or oven.
- Install the gasket.
- Position and mount new SES to the tundish slide gate bottom plate or to the tundish nozzle.
- Ensure that the SES top face is in full contact with the bottom plate or tundish nozzle upon locking the holder ring mechanism, and check the SES alignment.

SES failure symptoms as shown in Figure 11.31.

A Cracking at Tapered Area
B Internal Erosion
C Thermal Shock Crack
D Excessive External Corrosion
E Internal Erosion
F Cracking across Port Area

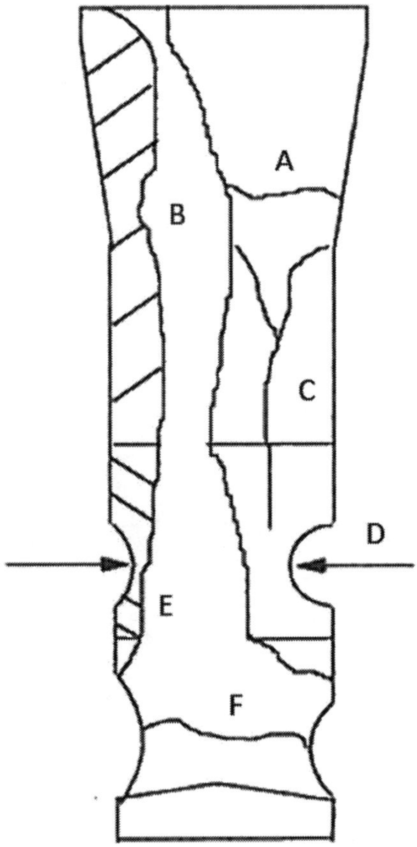

Figure 11.31 Type of failure symptoms of SES.

Causes

A1 Fitting problem with the SES holding ring
B1 Steel Quality
B2 Irregular Steel Flow
C1 Insufficient Preheat Time/Temperature
C2 Long Delays between Preheat Off to Cast Start
D1 Oxidation
D2 Casting Position/Ramping Procedure
E1 Steel Quality
E2 Oxidation
E3 Casting Procedure/Powder Entrapment
F1 Internal/External Erosion
F2 Vibration

11.6.4 Corrosion and Clogging

The occurrence of submerged entry nozzle (SEN) corrosion is often the phenomenon determining the duration of the continuous casting sequence. The flow of liquid steel through the inner wall causes erosion and finally leads to corrosion of SEN. This problem leads to lower productivity and higher production costs. To avoid corrosion problems or extend the lifetime of SEN during casting, the SEN must be appropriately prepared before it is installed into the casting machine. Suitably preheating temperature and preheating time are considerations for the preparation of SEN. Too low preheating temperature leads to clogging problems, and fracture of SEN may occur. Too high preheating temperature leads to oxidation of SEN itself. Termination of sequence continuous casting due to clogging and shape design of SEN to prevent any clogging problem are common issues and widely reported. Different field studies may provide better insight into corrosion and clogging phenomenon that eventually assist in designing the SEN.

11.6.4.1 Slag Zone Lining, Preheating Temperature, and Time

In a recent article, a competitive study was carried in between two different qualities of SEN, in which the entire slag zone is made of zirconia (Type A), and the outer surface is made of zirconia (Type B). With preheating at 750–900°C, SEN Type A exhibits corrosion occurring at the outer surface of SEN around slag zone. The inner surface of SEN Type A showed only shallow local corrosion as shown in Figure 11.32 [12]. SEN Type B shows the intensive corrosion at the inner surface by the same preheating temperature, and finally, leakage of SEN appears and eventually stops the casting sequence.

The competitive performance of both Type A and B are shown in Table 11.4. The analysis indicates the importance and effect of zirconia on SEN slag zone, in particular high corrosion resistance against slag. This incidence was further confirmed while leakage data collected for 13 casting sequences for 65 numbers of SEN. In this exercise, 7 of 48 pieces or 14.6% of SEN Type A, and 7 of 17 of 41.2% of SEN Type B appear to fail through leakage, confirming that Type A has more resistance and life than Type B.

Despite material selection for a particular design purpose, the mode of preheating temperature and schedule significantly influences the corrosion of SEN. SEN Type A has high corrosion when preheated lower than 750°C and higher than 900°C, and corrosion increases with increasing temperature; some field data has been given in Table 11.5.

During casting, the outer surface of SEN is in contact directly with liquid steel and slag that expedites the interaction between slag and SEN body at high temperature in long sequence casting leading to corrosion of the SEN materials. At the tip of SEN, the boundary of SEN materials and slag zone are often found at the higher corrosion level. It is found that the critical area always exists from the

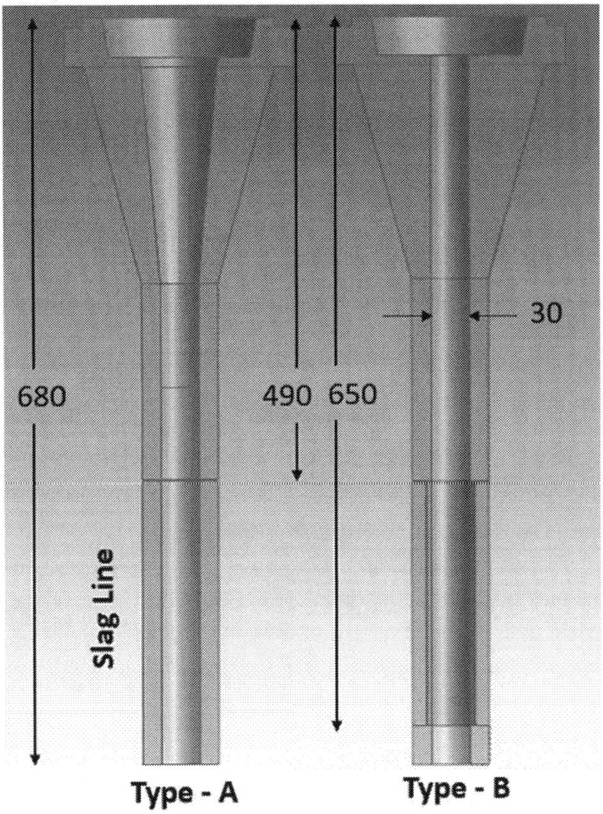

Figure 11.32 Shape and size of SEN type A and B. Adapted from reference [12].

Table 11.4 Competitive operational parameters and performance of Type A and B.

	Strand number and SEN type				
Data of casting	**1 Type A**	**2 Type A**	**3 Type A**	**4 Type B**	**5 Type B**
SEN preheating temperature (°C)	760	820	790	850	780
Casting time (h)	17.59	17.57	18.04	15.18	14.22
No. of heat in sequence	24	24	24	21*	21*
Corrosion appearance				SEN tip lost	SEN tip lost
Height of SEN after stop sequence (mm)	680	680	680	470	460
Inner diameter after stop sequence (mm)	36	37	35	72	70

*Leakage before the completion of casting

Table 11.5 Effect of preheating temperature on corrosion behavior of SEN Type A [12].

Casting details	Strand No.–SEN Type 1 Type A	 2 Type A	 3 Type A	 4 Type A	 5 Type A
Preheating temperature (°C)	640	1000	900	810	730
Preheating period (h)	11.1	5.1	11.1	11.1	8.6
No. of heat in sequence (heat)	13	7*	13	13	11*
Appearance of corrosion	SEN tip is lost 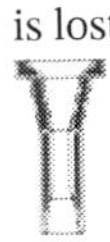	SEN tip is lost 	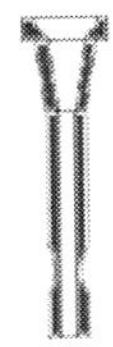	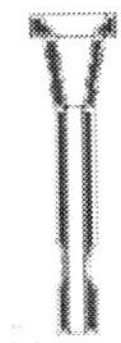	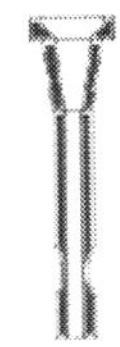
Height of SEN (mm)	570	665	680	680	680
Inner diameter (mm)	60.0	51.0	32.8	32.6	34.0
Outer diameter (mm)	-	-	63.0	66.0	62.0

*Leak before the completion of casting.

boundary of different SEN material. This area is composed of different chemical compositions and easily reacts with liquid steel or slag at high temperatures, and the corrosion of SEN originated in this area predominately. Another critical corrosion area of SEN is around the slag zone. The corrosion mechanism of this area is controlled by the dissolution of the refractory oxide phase dissolved in the slag. This led to break of SEN and finally stopped sequence casting. A typical relation of preheating temperature with time is shown in Figure 11.33. It indicates a rapid

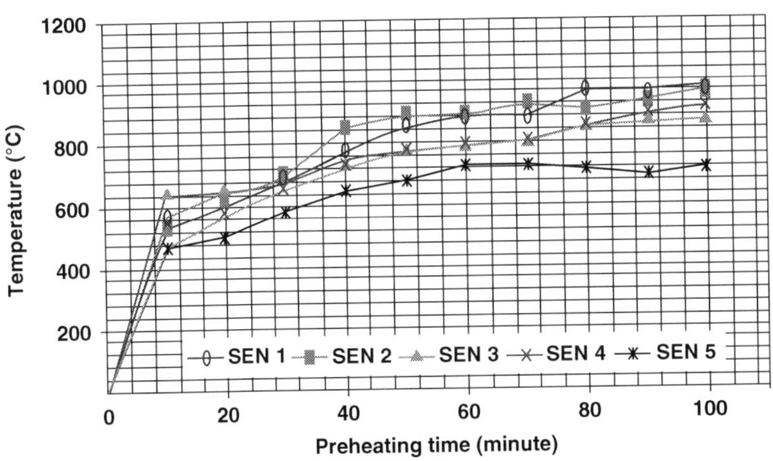

Figure 11.33 Preheating curve of SEN.

temperature increment of SEN within 10 min, followed by a relatively slow increment. However, after 40 min, the temperature rises further very slowly and flattens. Therefore, to ensure the homogenization of the SEN temperature, it is expected that the duration of the preheating process is at least 40 min, but it does not exceed 90 min. However, too long preheating period may cause oxidation of the SEN itself.

11.6.4.2 Clogging and Decarburization

In general, the interaction between the SEN and molten steel may be categorized into three different mechanisms:

- Chemical reactions between refractory materials and impurities in the nozzle and molten metal.
- The attachment of immersed nonmetallic inclusions in the molten steel to the surface of the nozzle.
- The erosion of the nozzle refractory materials.

The first two mentioned mechanisms may cause clogging during the casting process, limiting the productivity by interrupting the casting process, restricting the number of charges per tundish, affecting the quality of the produced steels, and consequently increasing the customer rejections.

The stainless steelmakers have started to produce more complex steels such as ASTM S35315 to satisfy the requirement of customers. The austenitic stainless steel grade resists oxidation at temperatures up to 1423 K [13]. Furthermore, it can provide moderately good service in carbon, nitrogen, and sulfur containing atmospheres due to REM (Rare Earth Metal) alloying elements. However, adding the REM alloying elements such as cerium may lead to severe nozzle clogging that can interrupt the steel processing. SEN accretion post-mortem analysis may ensure the formation of delta ferrite during solidification due to the REM alloying inclusion in molten stainless steel. In addition, REM-alloying metal may react with alumina in SEN refractory and expedite clogging. The clogging may continue by agglomeration of REM oxides on the SEN internal surface in both cases. Despite such incidences, a probable reaction in between nozzle silicon oxide and REM-alloying metals in the molten steel leads to silica and transformation of REM materials. Furthermore, the presence of REM oxides, silicates, aluminates within SEN internal surface eventually attracts and reacts with the existing SEN pores. The attraction forces between REM oxides may vary drastically due to their sizes. Thus, a new generation glass and silicon powder coated SEN is commercially available. It is intended to protect the SEN inside surface from graphite oxidation during the SEN pre-heating.

Graphite in the SEN base refractory materials without glass/silicon powder coating may be oxidized directly in an oxidizing atmosphere in oxygen or a carbon dioxide-rich environment. Graphite oxidation starts at 873–973 K in air to form carbon monoxide or carbon dioxide by an indirect oxidation process as shown in Equation (11.1).

$$C(s) + SiO_2(s) = CO(g) + SiO(g) \tag{11.2}$$

Thus, an antioxidant addition is an excellent choice to inhibit the decarburization of alumina–carbon refractories, similar to MgO-C refractories. An antioxidant is an element or a compound with a high oxidation affinity. Aluminum is the most conventional antioxidant, but the formation of Al_4C_3 after sintering and their hydration expedite the formation of $Al(OH)_3$ as illustrated in Equations (11.2) and (11.3), attribute to volume expansion and cracking of the refractory. Despite hydroxide formation, the volume expansion during oxidation of aluminum is also essential to consider during the designing of refractory composition.

$$4Al(l) + 3C(s) = Al_4C_3(s) \tag{11.3}$$

$$Al_4C_3(s) + 12\ H_2O(g) = 3CH_4(g) + 4\ Al(OH)_3(s) \tag{11.4}$$

Several literatures highlighted the utility of different materials like $Al_8B_4C_7$, Zn, silicon-containing boron glass coating, making the decarburization resistance. Coexisting liquid phase and additives may form oxidative products that eventually inhibit further oxygen penetration. However, extensive volatilization of glaze components during sintering and preheating of the SEN above 1273 K may have an adverse effect on the decarburization resistance.

11.6.4.3 Wear Profile

Wearing SEN impacts the continuous casting process because it can limit the number of heats per sequence and eventually may induce quality problems in the cast product. Excessive erosion may promote nozzle cracks or breakages that can promote steel reoxidation or mold slag entrainment. In general, one of the most affected zones of the submerged entry nozzles is in contact with the mold powder. Therefore, it is customary to use a more resistant material in this area, like zirconia-graphite. Nevertheless, in the case of casting small sections, the thickness of this insert (and of the nozzle itself) could be limited by the available space. Because of these reasons, it is of great importance to define strategies that maximize the life of the nozzles without affecting their structural integrity [14]. Thus, wear profile evaluation may provide better insight to monitor and develop the next generation SEN. A recent article developed a methodology to estimate the remaining wall thickness and the shape of the wear profile, and the amount of refractory loss. In this research work, Table 11.6 represents the main features of continuous casting machines where SEN used.

In order to evaluate the wearing rate for a wide range of casting process variables, nozzles were used in sequences with different duration and operative

Table 11.6 Main features of two continuous casting machines.

	CC2	CC3
Bloom size (diameter)	148–170–215 mm	215–254–290–310 mm
Number of strands	4 (curved types)	4 (curved types)
Electromagnetic stirring	In-mold	In-mold
Mold-level control	Radioactive Co 60	Radioactive Co 60
Tundish capacity	15 tons	18 tons

Table 11.7 Main process conditions in those heats where nozzle erosion was evaluated.

Number of analyzed nozzles	70 (45 sequences) from both CC machines
Sequence duration	2–19 heats (100–800 min)
Electromagnetic stirring	Heats with and without EMS
Cast steels	Steels with low and medium carbon content
Casting powder	$CaO/SiO_2 = 0.8$ viscosity (1300°C) = 5.7 poise

conditions (see Table 11.7). In all the cases, the tested nozzles were composed of an alumina-graphite (Al_2O_3-C) body and a zirconia-graphite (ZrO_2-C) insert in the contact region of the casting powder.

Nozzle erosion is mainly concentrated in the contact region in contact with the mold slag. Figure 11.34a presents a longitudinal section of an actual nozzle after the operation, showing the typical erosion pattern and the location of the ZrO2-C insert. As an example of the results typically obtained from the measurements, Figure 11.34b shows the wear profile determined in one of the evaluated nozzles.

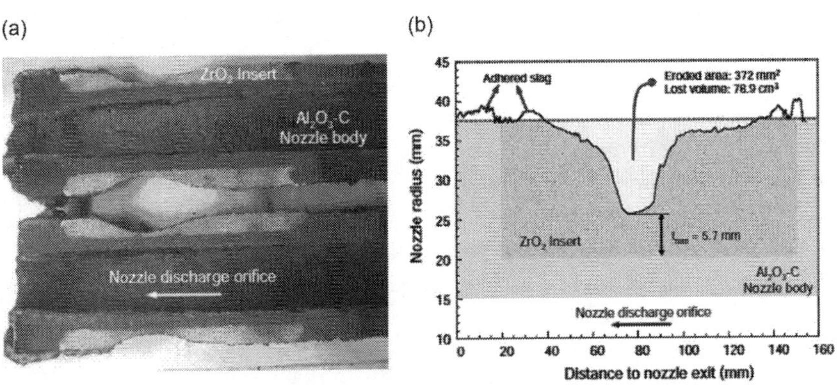

Figure 11.34 Characterization of nozzle wearing (a) typical wear profile of actual nozzles, (b) example of the measurement results.

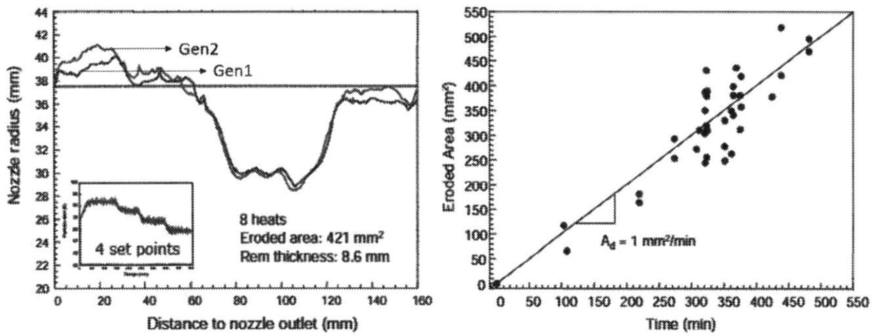

Figure 11.35 (a) Wear profile measured in two nozzles with different operating conditions, (b) relationship between total eroded area and casting time for the different studied cases.

Additional results, like the total eroded area and the remaining wall thickness are also obtained from these measurements. Figure 11.35 shows the wear profile measured in two nozzles used in different sequences of the same continuous casting machine. A good agreement between the wear profiles measured for the different generatrices (Gen 1 and 2) is observed in both cases.

It is worth mentioning that the nozzle used for a more significant number of heats (Figure 11.35a) exhibits a larger eroded area and a thicker remaining thickness. This is mainly attributed to the strategy used to change the steel level in the mold along the sequence (four steps instead of three). While the total eroded area is related to the time that the nozzle spent in operation, the remaining thickness depends on the relative position between the steel level and the nozzle during the sequence. Therefore, to compare the erosion produced by different strategies, it is more convenient to take the total eroded area rather than the remaining thickness. Accordingly, Figure 11.35b shows the relationship between casting time and the total eroded area measured for all the studied cases. Although some scatter is observed in the results, it is clear that the eroded area increases with the casting time. An estimated average wearing rate of 1 mm^2/min is calculated.

A method was developed to quantify the continuous casting submerged entry nozzle wear profile. With this procedure, it is possible to determine the remaining refractory thickness, the total eroded area, and the location of the eroded zone. These results help greatly to evaluate different operation conditions and perform comparisons between different refractory materials. Cumulative performance analysis of continuous casting refractories is shown Tables 11.8 and 11.9. This collective information may provide insight and a brief idea about the chemical composition, physical properties, and performance of different continuous casting refractories. Sometimes, low heat or premature failure may be the reason for refractory imbalance quality or faulty operation; several reasons are discussed next in Chapter 12.

Table 11.8 An interactive analysis of continuous casting refractories.

Type of refractory		Purpose of applications	Material design parameters		Installation/ operation parameters	Performance (No. of heats)	Reason of Failures	Reason of Less Heats
			Composition	Properties				
Ladle shroud		To guide molten metal from ladle to tundish	Alumina graphite	Erosion and thermal shock resistance	Manually or manipulator	10/12 heats	Crack/ erosion/ corrosion	Off cast time more, slag casting, slag volume more, oxygen penetration
Tundish working lining	Spray mass	To provide better life	70% MgO	Erosion	Spraying	10–22 heats	Erosion	Depends on planning, erosion, wrong steel chemistry,steel grade difference.
	Dry vibrating mass	To provide better life	70% MgO	Erosion	Vibration	10–22 heats	Erosion	Depends on planning, erosion, wrong steel chemistry,steel grade difference.
Mono block stopper		To control flow	Alumina graphite	Erosion	Thru mechanism	10–22 heats	TIP erosion	Wrong operational practice, wrong steel chemistry. misalignment, planning. More slag volume, slag casting.
Submerged entry nozzle		To control flow in mold	Alumina graphite	Erosion and corrosion	Manual installation	10–22 heats	Z band erosion .	Wrong operational practice, wrong steel chemistry. misalignment, planning. More slag volume, slag casting.

Table 11.9 Chemical composition and properties of flow control and continuous casting refractories.

Type of refractory	Composition as it is (wt.%)							Properties						
	SiO_2	Al_2O_3	CaO	MgO	ZrO_2	LOI	Fe_2O_3	MOR (kg/cm^2)	CCS (kg/cm^2)	AP (%)	BD (gm/cc)	ASG	PLC (%)	LBD (gm/cc)
Ladle upper nozzle	0.05	96.3	3						590		2.91			
Ladle slide gate plate insert	1.8	0.9	0.1	2.5	94		0.1			14	4.7		+0.1	
Ladle lower nozzle	0.1	95.7	4.1	0.05			0.1		615		2.94			
Ladle shroud	7	55				32		76		18	2.45	2.9		
Tundish Backup lining	25	70	2				1		800		2.75		−0.2	
Spray mass	28			60		2			1.9					1.15
Dry vibro mass	12			85		4								1.62
Precast furniture	21		2	73			0.8		1000		2.8		± 0.1	
Porous plug	0.2	91	2.6	6			0.1		856	18.2	3.03		−0.09	
Mono block stopper	7	54				32		81		18	2.4	2.85		
	1	84				14		87		17	2.84	3.3		
	1.4			87		5		90		14	2.56	3.05		
Tundish slide gate insert	1.8	0.9	0.1	2.5	94		0.1			14	4.7		+0.1	
Tundish nozzle	7	54				32		81		18	2.4	2.85		
Submerged entry nozzle	4	65			5	24		75		17	2.4	2.9		
			3.59		92	14.6		60		14	3.7	4.4		

References

1 Daya, Z., Yang, Y., Sommerville, I.D., and Mclean, A. (2013). Liquid steel transfer — the delivery of quality. *Can. Metall. Q.* Volume 38, Issue 5, December 1999, Pages 285–293.

2 Yang, Y.D., Daya, Z.A., Sommerville, I.D., and McLean, A. (1998). *J. Iron Steel Res. Int.* 5 (2):7

3 Lowry, M.I. and Sahai, Y. (1992). *Iron. Steelmaker.* 19 (3): 81. Thermal effects of the flow of liquid steel in continuous casting tundishes.

4 Brimacombe, J.K. (1999). *Metall. Mater. Trans. B* 30B: 553. The challenge of quality in continuous casting processes.

5 Samarasekera, I.V., Brimacombe, J.K., Kumar, S., and McLean, A. (1995). Proc. Vol. International Symposium on Near-Net-Shape Casting in the Mini Mills, Met. Soc. C.I.M., p 135.

6 Tassot, P. and Turrel, C. (2011). Actual trends for the tundish refractory lining. *Refract. Worldforum.* 3 (3): 93–98.

7 Sato, M., Sukenari, S., and Shimpo, A. (2008). Submitted to the 86th refractory subcommittee. The Iron and Steel Institute of Japan.

8 Guthrie, R. and Isac, M. (2018). *Extraction 2018*, 729. New York: Springer International Publishing.

9 Zhang, J., Yang, S., Li, M., Chen, Z., Jiang, Z., and Li, J. (2017). *Ironmak. Steelmak.* 44: 732.

10 Zhang, J., Liu, Q., Yang, S., Chen, Z., Li, J., and Jiang, Z. Advances in ladle shroud as a functional device in tundish metallurgy: a review. *ISIJ Int.* Advance Publication by J-STAGE. 2019.

11 Zhang, J., Yang, S., Chen, Y., Chen, Z., Zhao, J., Li, J., and Jiang, Z. (2018). *JOM* 70: 2886.

12 Yingnakorna, T. and Khumkoaa, S. (2014). Corrosion behaviour of submerged entry nozzle (SEN) during continuous casting of steel. *Full Paper Proc. MISG-2014* 1: 62–69.

13 Memarpour, A. (2010). A study on the submerged Entry Nozzles (SEN) respecting clogging and decarburization, Licentiate Thesis Stockholm 2010. Department of Materials Science and Engineering, Division of Applied Process Metallurgy Royal Institute of Technology, SE-100 44 Stockholm, Sweden, ISBN 978-91-7415-664-5.

14 Capurro, C., Cicutti, C., Galliano, P., Fuhr, F., Torga, G., and Scoccia, J. (2008). Optimization of submerged entry nozzles wearing. June 2008 Conference: 6th European Conference on Continuous Casting, Riccione, Italy.

12

Premature Refractory Life by Other Parameters

12.1 Introduction

A conversation between a steel melting shop manager (Mr. X) and an application engineer (Mr. Y) of the refractory manufacturer is the starting point of this chapter.

MR. X: Why does your refractory not have consistent performance?
MR. Y: Sir, what happens? Please inform.
MR. X: Ladle-1 had 125 lives, more than expected, but Ladle-8 is only 60 with the same brick. Please supply quality products; otherwise, your company will be blacklisted.

Mr. Y was afraid and immediately enquired to his host organization and found that both ladle bricks were from the same consignment. Both the personalities were right from their viewpoint, productivity, and business perspective. The question is *why such difference* and *what are the reasons for such premature refractory life?* We have to look at several reasons from both parties to find the solutions, sometimes not encountered and invisible. Herein, the prime motivation is to pick up the probable factors that eventually affect the performance either by uncontrolled processing defects, packaging and transport from the refractory manufacturer side or delay in procurement and unusual installation practice, and operational mismanagement from the steel sector side.

From the understanding of the heat transfer, refractory processing and properties, and their mode of application for vessels and continuous casting, it is clear that all steel vessels and continuous casting refractories constantly operate under extremely harsh conditions characterized by intense levels of heat, chemical attack by molten steel/slag, and thermomechanical stress. Due to the high strengths and thermochemical resistance, refractory linings in different vessels have a crucial role in proper functioning in the steelmaking process. Thus, critical refractory

Fundamental Design of Steelmaking Refractories, First Edition. Debasish Sarkar.

manufacturing parameters are essential issues to control over the desired properties for achieving consistent performance, including:

a) High purity and minimum impurities in natural or synthetic, or recycled raw materials.
b) Large crystals and type of phase and minimum porosity of starting raw materials.
c) Maintain proper particle size distribution to achieve the desired compressibility index and Hausner ratio to predict the compressibility efficiency of shaped refractory.
d) To control the adequate particle size distribution using a statistical approach to achieve flow and compressive strength for castable.
e) Control over the properties like elastic modulus, coefficient of thermal expansion, thermal conductivity, strength to synchronize the thermal shock resistance, and crack arrest.

Experience has, however, demonstrated that several operational and installation issues might result in costly and premature refractory failures. In such a failure, a series of independent investigations must establish the incident's root cause and develop ways to avoid future recurrences. In a nutshell, the investigations are often designed to eliminate or validate each of the following critical refractory aspects that, among many others, are prevalent known causes of premature refractory failure:

a) Excess furnace internal temperatures: reaching refractory material melting point results in their softening and deformation.
b) Spalling/Thermal shock: cracking of refractories due to stress.
c) Slagging: chemical reactions on the refractory surface or within the refractory.
d) Hydration: resulting in massive volume changes and cracking.
e) Shell deformation.
f) Faulty refractories or faulty installation.
g) The unexpected delay in any step of the operation.
h) Over or under preheating.
i) Throttling zone and slag interaction.

In cumulative, several issues were raised and discussed to understand and avoid early life that eventually relies on both manufacturers and consumers.

12.2 Refractory Manufacturing Defects

Refractory performance and competence depend on the design, composition, and manufacturing strategy. Extensive research and expertise synchronize all together to complete the loop, as illustrated in Figure 12.1. Selection of raw material resource and processing protocols, refractory manufacturing, failure analysis, and feedback to R&D

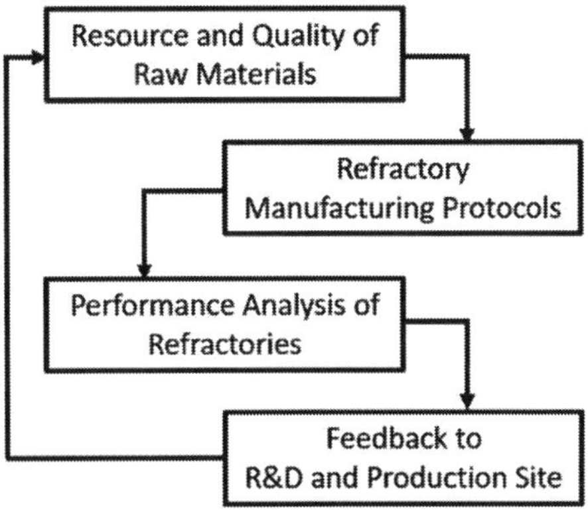

Figure 12.1 Schematic representation synchronizes quality refractory production and consistent performance.

and production site eventually develop a successful follow-up and product. While refractory manufacturing defects are a concern, it is mandated to emphasize these responsible critical issues for the immature life of refractories. Consistent refractory performance depends on installation and steelmaking operational conditions despite adequate properties and quality and is discussed in the following sections.

12.2.1 Consistence Raw Material

High crystallinity, low impurity, proper particle size, and packing provide a quality refractory for steady performance. The ore demands a different level of processing to remove impurities, enhance purity content, and high-temperature processing to make it a worthwhile refractory for a particular application. Figure 12.2 shows the significant natural and synthetic raw materials that are essential for steelmaking refractory.

Geographical region to region of natural resource, raw material processing machinery and processing protocols, and high-temperature processing (calcination, sintering, fusion) are critical considerations to achieve the desired specifications of raw materials. Any deviation may affect refractory performance, and thus correct feedback from the application site is compulsory to synchronize the raw material processing, which is a first measure to produce quality refractory. Gradual raw material scarcity and cost restrict natural resources; thus a paradigm shift toward waste utilization of steelmaking refractory is worthy of achieving quality raw materials, which is a new generation of research horizon.

Increased urbanization and industrialization have resulted in considerable residential and commercial structures, especially in rising economies such as India and China. According to the Global Construction Perspective and Oxford Economics, the global construction market will expand by 85% to USD 15.55 trillion by 2030

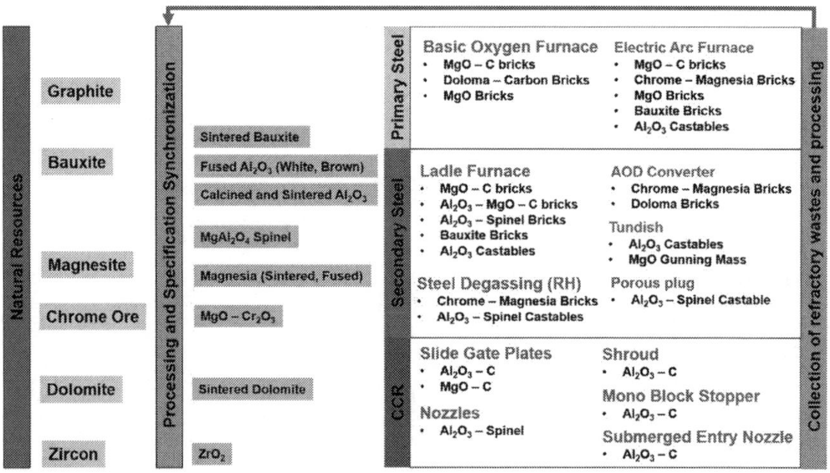

Figure 12.2 Brief about natural and synthetic raw materials, different steelmaking refractories, and the potential utility of waste utilization.

worldwide, and three nations, China, India, and the US, account for 57% of the world's growth. China is one of the fastest developing economies; increasing demand for travel calls for a steady improvement of the country's trains and roads. The growing factor has also bolstered the car sector in China. Intensified construction of residential and commercial structures has led the rising countries to increase their infrastructure development, which is anticipated to lead to demand for refractories in the iron and steel, and cement industries. The glass sector, which would lead to a refractory market, is also likely to be derived by increasing construction activities in emerging economies.

The refractory materials utilized are known as refractory grogs and are further recycled to make secondary raw materials. There are two sorts of refractories: refractories of high grade and low grade. The utilization of recycled refractories is based on the quality of the actual refractory, while only 0–30% of the recycled refractories created by low-quality refractories can be utilized in other processes. As raw material costs, the cost of goods, and environmental pressure increased, the requirement for recycling refractory products increased, forcing businesses to search for more sustainable options for refractory uses.

Refractory recycling is an essential step and depends on several factors. Between 2008 and 2010, was a volatile time in the refractory industries worldwide, during which certain commodities were virtually impossible to obtain due to high demand and low supply. Pressure has been stemming companies from their respective carbon footprints to become ecologically friendly. There is enforcement of higher fees and emissions on landfills and tax incentives to encourage enterprises to enhance their efforts in trash recycling. Recycling refractories contributes to reducing manufacturing costs and hence provides a good potential for the growth of this market.

12.2.2 Processing Parameters

Grading is imparted to develop effective particle size distribution of raw materials by several unit operations. The prime motive is to achieve the desired packing density (PD), referred to as the solid volume ratio (V_S) to the total volume (V_T) of arrangement. The total volume is defined as the cumulative volume of solids and voids. The packing efficiency depends on random particle packing and its morphologies. Multiple particle fraction sizes facilitate decreased void density, and high-dense refractory demands a few millimeters to submicron particles. Recently, nanoscale research in refractory is at the forefront to enhance the properties further. Simple calculation emphasizes the significance of different particles and their relation with packing density. Let us consider that any mono-size particle helps 65% random packing, which means the system has a 35% void. In an attempt to reduce the void, relatively smaller particle results in 87.75% (65% + 65% of 35%) packing density; thus another smaller particle results in 95.7% (87.75% + 65% of 12.25%) density. Thus, another smaller particle may enhance the density, but different and inhomogeneous particle grading reduces the packing density. Sometimes, the following are probable reasons to escape grading during production:

i) Adequate raw materials crunch
ii) Sudden failure and maintenance of raw material processing units
iii) Disturbances in the supply chain within the factory
iv) Improper handling, weighing, and transport of raw materials

Consequences on the refractory properties:

i) Segregation and nonuniform microstructure
ii) Density and strength variation batch to batch
iii) Ball mill fine (BMF) carry relatively more impurities in ppm level compared to coarse grains; thus the difference in chemical analysis from batch to batch
iv) A relatively high fraction of smaller particles results in low density and pinhole if any impurity is present

Despite the achievement of particle size selection by tap-density, binders facilitate coating on the particles and subsequent flow during pressing that eventually enhances the green strength of refractories [1]. Due to thick coating, the prerequisite binder content and their homogenous mixing are compulsory to avoid reverse impact in packing density. Thus, mixing particles and binders has a vital role in synchronizing the density. The selection of binder content with having a viscosity and a proper mixer machine is critical to increasing the physical and chemical homogeneity of the refractory mix. The mixer machine efficiency is monitored by considering effective particle movement from one region to another, active shear interaction, and microscopic interchange of particles during the final mixing stage. Homogenous mixed mass without composition variation from batch to batch and silo position is essential to avoid drying, segregation, and nonuniform green strength. Aging or curing at desired humidity, temperature, and retention time provide adequate polymerization and green strength of resin-bonded mixed mass. Green density is the first stage achievement after pressing or casting, but the strength variation is more significant in unshaped refractories like castable. A deviation in cement resources and content, aggregates, and humidity eventually reduce the castable properties.

Few incidents can be summarized from the research outcome. The compressive strength of fire clay grog castable is analyzed concerning the type of cement, hydraulic cement content 10–30%, cement: water ratio, curing time, and humidity [2]. The chemical composition of different types of cement is shown in Table 12.1, which has a significant influence on the hydration and resultant strength. Any alteration may affect the performance on site. Figure 12.3a shows increased strength with an increase in curing time between 6 and 24 hours.

Herein, the specimens made from domestic or imported cement exhibit strength increment even after 6 hours of hardening. The force gain is rapid up from 10 to 24 hours for both the cement; however, a distinct difference may be observed concerning change in cement quality. Figure 12.3b displays the compression strength of specimens when the cement in mixtures is kept to 25%, and its water content has changed in the range of 1.4–2.1. A significant influence on humidity and temperature on strength is also observed when cement: water ratio varies in this range. With the increased water volume, the shattered specimens have more voids or pores visible, which may be responsible for the above phenomenon.

Table 12.1 Chemical analyses of calcium aluminate cements (%).

Source	Insoluble	Al_2O_3	CaO	FeO	SiO_2	Fe_2O_3	TiO_2	MgO	MnO	S
Imported	0.3	39.8	37.2	10.5	2.8	6.6	1.5	0.1	0.0	0.0
Domestic	0.4	39.2	36.9	5.8	9.3	5.4	1.1	1.0	0.1	0.1

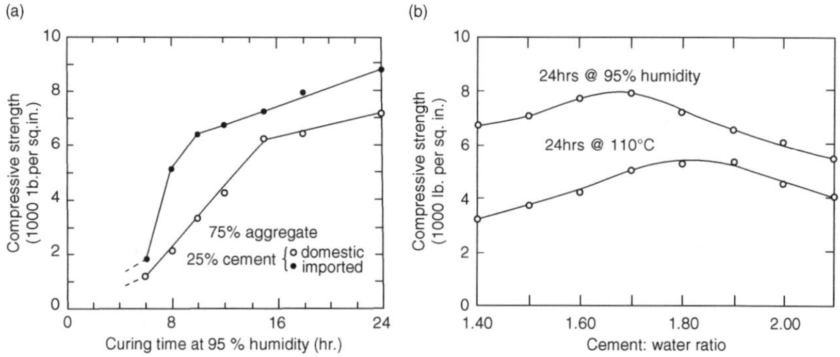

Figure 12.3 Effect of (a) curing time on the strength of castable and (b) humidity and temperature [2].

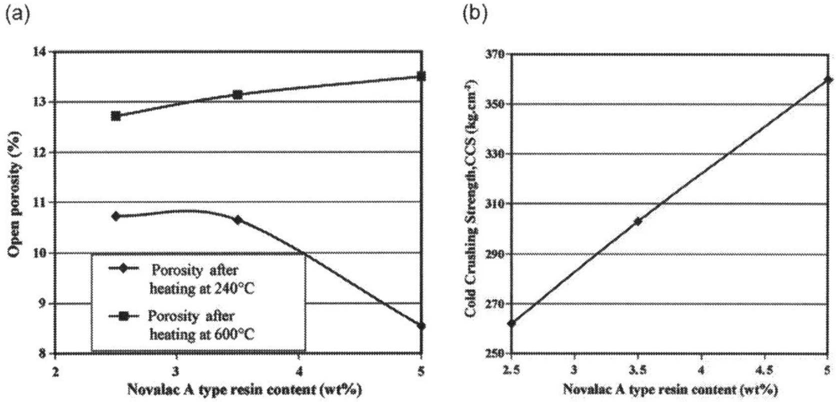

Figure 12.4 (a) Variation of open porosity after heating at 240°C and 600°C of different binder content (b) Strength of same MgO-C brick concerning binder content [3].

Binder imparity is another critical aspect to control the strength for both shaped and unshaped refractories. A typical result for shaped refractory is shown in Figure 12.4a. The open porosity modifications of samples are compared with their resin content (Novalac A-type). The compaction can be improved while the pores are pressed and occupied as resin content increases.

Porosity reduction is evident with an increment of resin in tempered bricks, but the porosity during preheating increases at 600°C due to the burning out of the complete organic fraction of the resin. Higher resin content results in more significant weight loss or burn-out value and hence increased porosity. On the other hand, as illustrated in Figure 12.4b, the tempered samples are raised in cold crushing strength when their resin content increases due to reduced porosity and improved resin bonding. Thus, preferable binder and their content are critical selections for a particular zone of interest.

12.2.3 Pressing and Firing

Compaction depends on grading, mixing, and binding per millimeter and the load applied. Despite well-known theoretical ideas, essential components include die design, materials, manufacture, maintenance, and press availability, and delaying processing before pressing are critical components. Inappropriate or manual stroke can decrease pressure and produce weak zones while being applied, which appear hard to identify as a specific product in a batch, but such bricks might reduce performance and develop pockets in the vessel. While the material is punched or broken, energy is released, and reverse loading conditions arise, the volume and the laminated cracks of the pressed product are expanded beyond a significant level, and the density reduction in the spring back as Figure 12.5a.

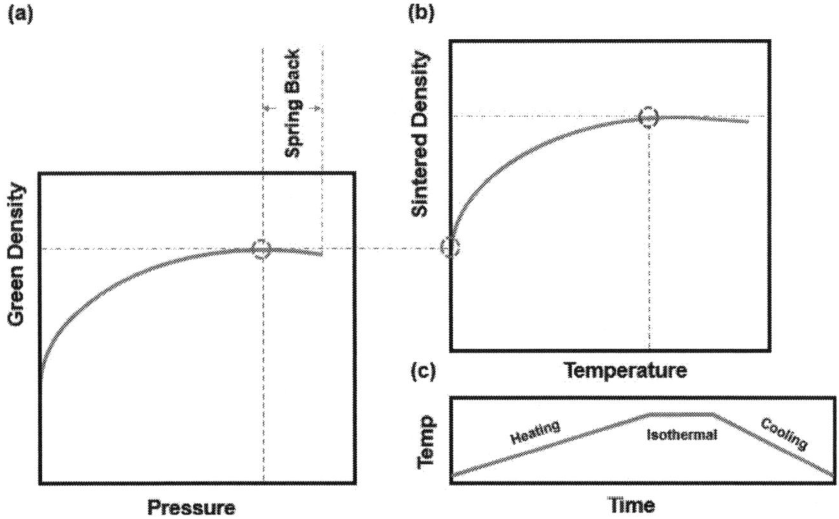

Figure 12.5 Variation of green density and sintered density concerning (a) load and (b) temperature, respectively. An indication to pick the optimum pressure and temperature.

When dietary wear begins due to harder grain, a typical problem emerges and gradually worsens with time, deviating from a specified limited tolerance of green bricks. The worn particles may also be imprisoned in the green product. Therefore, this problem can be foreseen through adequate die material selection and load distribution analysis, with differing shapes for the same shipment without manufacturer and quality control knowledge. Green products achieve sintered density at high firing temperature, as Figure 12.5b shows, and cannot further increase density over critical temperature, rather pore coalescence reduces a certain degree of density. The firing process is a vital step in manufacturing and controlling ceramic products.

Structural alterations in the materials attain the proper ceramic structure responsible for the final attributes of the product with permanent structural modifications. Figure 12.5c shows a typical profile for peaking temperature followed by isothermic treatment according to the materials and the intended product that eventually depends on firing circumstances, firing temperature, firing rates, and duration in the kiln. Details of the relevance of temperature and the ceramic bonding affect the properties, as stated earlier. Its primary aim is to identify the likely causes and faults that may lead to immature impacts and decrease the performance. Although few reasons have been considered, a) uniform heat transfer over the kiln and failure of thermocouples may mislead the display temperature, b) changes in fuel and its heat value may not be enough to attain the critical temperature in a predefined timetable, c) improper refractory lining and leakage facilitate heat loss, d) improper loading pattern of refractory, and e) faulty car feeding accessories, etc. Ultimately, unexpected firing schedules may cause refractory defects, including over burnt, bubbled, cracked, scaled, blown, laminated, size, deformed, etc.

12.3 Packing and Transport

The unreliable production process may facilitate defects, including invisible, micro-cracks, and corner breakage in the refractories, and reduce the performance in customer sites. Adequate packing, storage, and transportation of refractories, services, and information to meet the customer's requirements are additional issues to moderate the rejection. An extensive supply chain management is maintained logistically by which critical planning, coordination, and expert management assistance in the flow of products is achieved from origin to consumption. Cargo is usually received in the packaging for any transit. Transportation, load, and reloading activities ensure the safety of cargo. In addition to providing the security of the shipment, the packaging is a prerequisite for forming load transport units (based on the volumes and weight), checking cargo details during the loading and reloading process, delivery and dispatch processing, optimization of the efficient transportation volume, and providing conditions for the cargo loading and reloading, packaging and labeling.

12.3.1 Packaging and Packing Material

An adequate packaging may facilitate transporting the commodity through any freight like the sea, air road, or railway. Thus, several factors must be maintained during packaging:

- Protection requirements of the goods to be packaged
- Means of transport to be used to make vibration-free consignment
- Prediction and estimation of the transport mechanism and transshipment
- Center of gravity and weight issues

Classification of the package is based on the form. The most common containers used for the refractory industry are pallets, cartons and cardboard boxes, wooden crates, bags, and drums. Usually, monolithics are packed in bags, and solvent/binder are in the drum. However, several critical parameters must be encountered when shaped refractory transporting from manufacturer to consumer sites. Improper packing and material selection may affect the customer requirements and product-own specifications in specific sizes and shapes. Thus, it is essential to find out the most suitable conditions for all refractories:

a) Goods characteristics: nature, number of units, size, loose or mobile parts, and additional protection if there is a possibility of rust or contact of external agents.

b) Packaging characteristics: dimension, base characteristics, type of jacking and latch, packing material to be used, physical/chemical and mechanical protection, etc.
c) Essential to know the mode of transport, details of the route, details of warehousing, marking instructions, etc.

While material and packaging are concerned, refractory manufacturers are following specific protocols. However, any deviation damages the packed refractories and sometimes there may be an accident. Thus, particular concerns may provide safe and effective refractory packing:

a) Quality wood with relevant treatment and properties to carry the effective load.
b) If pallets are packaged with shaped or precast refractors, these shall be solid double-deck pallets that provide enough support for the load during transportation and storage. The pallet should be able to carry the loaded mass on the pallet dynamically enough.
c) Pallet structure should be handled by any mechanical device or loading and stored on any surface. The entire packing should be wrapped with a good synthetic or steel strip.
d) Proper fixing of beams to avoid any internal movement of the components enable them to withstand stacking. This practice eventually helps to protect any fragile component.
e) Shock-absorbing material should be used to avoid damage by vibration and shock.
f) Airtight bags containing desiccant material (e.g., silica gel) within crates are required to protect the material from moisture. Waterproof plastic wrapping is needed to avoid moisture from outside contact.
g) All openings on the products must be closed with wooden or plastic covers to prevent damage to the openings and interiors.
h) High-quality styrofil or pelaspanpack is the recommended filling material, and avoid straw, hay, wood wool, etc.
i) Additional components or spare parts should pack separately with proper manual instruction.

During packing, an inspection of the packing and marking before shipment is also essential to settle the proper condition of the product. Important symbols are shown in Figure 12.6.

12.3.2 Vibration-free Packaging

The packages experience a certain vibration at different frequencies in every transport mode. It is crucial to evaluate the natural frequency of any component prone to harm from accessible free space and hence vibration to preserve vibration-related items and then compare them with the vibration properties of the package

Figure 12.6 Different important symbols are necessary to remember during packaging, transport, and unloading refractory products.

design. Refractory manufacturers should deal with such efforts when packaging shaped refractory either individually or by third parties. The most severe risk of damage happens when vibrations creating resonant frequency vibration stimulation coincide with the product's natural frequency (and somewhat the crisp component in the product). For most vibration-sensitive products, it is sufficient to prevent vibration damage during transport to ensure that the packing is not designed to amplify vibrations at the product's natural frequency. However, reducing the frequencies of concern in very vibration-sensitive products may be required. As with shocks, the vibration features of a particular design should preferably be evaluated, but the knowledge of the vibrational characteristics of the compressor material can be estimated. Companies may make optimum use of the space in containers with the correct packaging technology for bulk items while minimizing packing materials, significantly cutting transportation costs. Whether containers, crates, big bags, octabins, or classic boxes—the vibrating tables, are remedies to restrict vibration and are always individually tailored to the container being used and adapted to the client's needs or the production environment. Space management has to be taken care of to reduce the cost of transport and adequate packing.

12.3.3 Loading, Transporting, and Unloading

Considering the push–pull system, the pull system always prefers well-organized loading, transportation, and unloading up to stock maintenance. It is apparent that these practices become beneficial to maintain continuous production. However, sometimes production schedule situations may force toward *push* system that eventually disturbs the entire supply chain and generates additional costs to logistics. In a hurry, product quality may affect, and delay transport may discontinue production. Overall transport process may affect components because of:

- average distance cargo transport,
- localization of distribution points and cargo transportation conditions,

Table 12.2 Different critical parameters during loading and unloading of products.

Loading	Unloading
• Checking the packing according to the specification • Suitable vibration-free packing with all proper marking • Adequate vehicle and accessories preparation for loading • Checking the quality and quantity of cargo • Proper loading and stacking to avoid jerks	• Vehicle and accessories preparation for unloading in consumer site • Adequate time and space for unloading and storage • Note both the quantity and quality of the delivered goods • Proper vehicle/forklift to unload the box or component to the application area • Inspection of product before installation

- the capacity of the vehicle,
- technical speed of vehicle movement,
- technical vulnerabilities of cargo transport,
- the level of mechanization degree of loading mechanisms construction,
- construction of highways surface and other.

Simultaneously, loading and unloading procedures and time management play a crucial role in handling the materials.

Reliable, fast, and safe transport depend on freight loading unloading with minimum points of intermediate halt. To continue effective logistics, the human factor (e.g., the operator's behavior) is an essential and unavoidable issue that may affect the process chain and depend on management. In general, a few factors are encountered during loading and unloading, as shown in Table 12.2.

12.4 Procurement and Lining Failures

Procurement and maintaining refractory inventory are essential to sustain and operate vessels in time with high productivity. One booming steel sector should have a strong refractory purchase department knowing both refractories and procurement protocols. Strong knowledge of the following may help refractory application engineers and steel production people to continue the quality steel; any purchase and order release delay may result in bad quality refractory in a hurry:

a) List of refractory suppliers and cost of individual items
b) List of inventory and consumption schedule of products
c) Performance statistics and readiness in the perspective of supply and service
d) The time required to supply refractory for a particular supplier

e) Procurement of refractory based on the steel chemistry (e.g., Si or Al or Ca killed)

Once refractory stored in user's site, application engineering may be either appointed by the supplier or from steel sector and has to know the following;

a) Is the stored refractory fulfilled and the specifications maintained for the proper storing to avoid self-degradation of refractory?
b) Is workmanship proper to install both shaped and unshaped refractories?
c) Are preheating and refractory maintenance proper and time prior to/during steel processing?
d) Is there an adequate supply of materials as when required the refractory maintenance?
e) Collect the information about installed refractory from operational department.

Although recent refractory purchase depends on the number of heat guarantee and total refractory management basis; however, steel processing and operation people should be aware of:

a) Specification of refractory and chemistry of both slag and steel
b) Operational conditions to synchronize the molten temperature and their chemistry
c) Total processing time, including actual operating and idle time
d) Pass over the information to refractory application engineer to monitor the life in time
e) Make ready the vessel before the next operation

Excellent coordination with the application–operation–supplier module is required to design the purchase process that may eventually boost the work environment, refractory life, and smooth steel production without refractory failure and disaster.

12.4.1 Total Cost of Ownership Concept

The so-called TCO (Total Cost of Ownership) approach is a standard methodology used to evaluate the actual value of goods and services acquired. The model is defined as indicated in Figure 12.7 as adapted to procurement operations. The idea explores the iceberg hypothesis, where only a tip of the iceberg appears in the deep water, and the large piece of the iceberg swims underneath the surface. The buying price for the goods or services is the dominant factor over the water surface. Below the surface must consider other elements synthesized as *internal business cost* and *value in use* costs.

A similar concept can also be used for refractory procurement for the steel sector. The model used for calculating the TCO was updated concerning logistics and

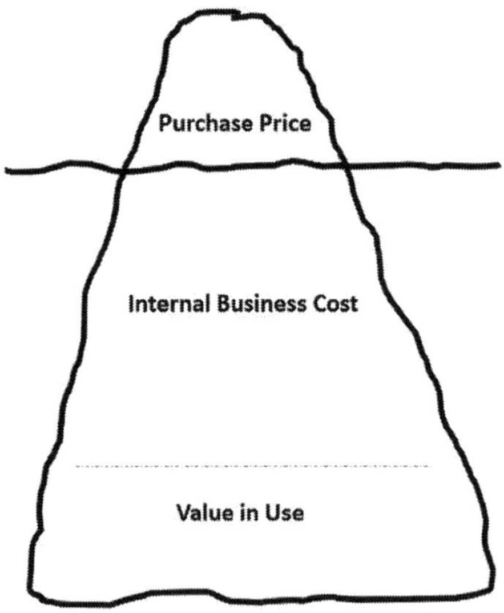

Figure 12.7 Fundamental aspects to procure refractories through TCO (total cost to ownership) concept [13].

operations in a steel plant to be appropriate for the functioning of the steel plant. The purchase price is usually included in the so-called life cycle cost, and all values used in the usage parameters are summarized in the so-called *operational* cost. The TCO model utilized does not reveal direct savings. It should be noted instead, it defines merely a refractory-influenced cost volume, but it displays the impact on the total operational cost of the refractory material performance. Total cost to own for steelwork process can be synchronized in two modes; a) life cycle cost, and b) operational cost. Usually, 25–30% of life cycle costs include refractory purchase and payment, freight and external storage, internal logistics and storage, mechanical maintenance, service (demolition, refining, processing), claims, waste, and disposal. However, 70–75% of the cost is associated with operational cost, including fluxing agents, slag splashing management, the inefficiency of bottom stirring, refining losses, maintenance losses, yield losses, and quality losses.

12.4.2 Preliminary Features of Lining

Before procuring refractory, a basic understanding of the necessity, different properties, and probable refractory failures may help select the suitable refractory during effective and standard steel operational conditions. Refractory materials, lining reliability, and operation timing and sequence are essential to increase the performance of vessels and casting output. Moreover, it can optimize production yield, minimize energy costs, and enable consistently high performance over the life cycle of the refractory lining. Preventing corrosion by using a lining is desirable for several reasons:

- To extend the life of the equipment
- To prevent shutdowns
- To prevent accidents resulting from corrosion failures
- To avoid product contamination

The possible life of a protection system is unlikely to be fully realized unless the right choice of the system is made:

- a decisive assignment is required to supply refractory as required
- there is no irreparable damage to the integrity of materials or coatings by the handling, transportation, and storage of all materials and lining components
- no lining damages which the erection process cannot completely restore, and such restitution of the damaged areas leads to protection at least as good as the entire areas

Common failures include cracking, erosion, corrosion, pocket formation, brick fall out, and molten metal/slag penetration. A mutual understanding between end-users and refractory material suppliers can minimize the risk of failure and improve reliability. Thus, brick properties, lining installation, and operational practice can synchronize together. Before summarizing the reason for failures, a concise and common idea on the lining design for both shaped and unshaped refractory may provide insight into the subject.

For the lining contractor to provide recommendations, the buyer should give accurate details of the fitting or service requirements. The buyer should supply the following information in the latter case:

- Design and fabrication details of the equipment to be lined, including drawings if possible
- Details of gasket materials as when required
- An exhaustive analysis of materials including steel and slag
- The design temperature
- The design pressure
- Any mechanical properties required of the lining, including suitability for machining
- The methods of heating, cooling, or both
- Cleaning methods, such as water washing, solvent washing, boiling, or steaming
- The location where the lining work will be done
- Site conditions that may affect the work, such as availability of services
- Preparation of surface

At the same time, the contractor should provide the following relevant information:

- Any special requirements for the materials and system used in surface preparation
- The type and specification of material to be used for the lining
- Different degree of lining thickness as required during operation
- The provision of test plates
- The types of solvents, with their flashpoints and threshold limit values, when hazardous may be at risk
- The specified sheet or rubber lining
- The minimum and maximum allowable temperatures and allowable humidity, which are required for the correct application of the lining
- The minimum temperature required during the vulcanization of the lining (for rubber linings) and the vulcanization schedule, where applicable
- Inspection techniques and procedures to be employed, acceptance levels, and stages of inspection
- Handling, transport, storage, and installation procedure
- Any particular information needed or recognized concerning inorganic linings

From the above information, it is understood that the refractory manufacturer is responsible for collecting the data from both the purchaser and contractor. It is mandatory to provide data by the purchaser to refractory manufacturers, but contractor selection has some criteria discussed in the next section.

Most of the time, the refractory failure cause is not immediately known. An investigation is required, and often, even then, a *conclusive* reason may never be found. When a refractory failure occurs, the steel processing operators usually assume the cause is one of two things:

- It may be lousy installation practice
- Refractory materials are terrible without consistent performance

The human and manual effort may provoke improper installation and affect the performance of faulty lining. Usually, incorrect installation is relatively easy to determine; however, installation is not an issue if the same vessel is managed by the same contractor or installation operator early without any deviation. As far as harmful materials are concerned, while this is possible, it is not probable. Reliable and reputable refractory manufacturers pay close attention to material selection, process technology and quality control (QC), packing, etc. Thus, some other factors should be considered when both materials and installation are in good practice. Major factors are:

- Design of vessels and process equipment
- Detail engineering
- Refractory material selection and specification concerning steel composition
- Refractory anchor type, spacing, and welding

- Lining installation
- Joint minimization as required
- Dry-out (curing), preheating of the refractory
- Start-up the operation
- Operational practices
- Regular maintenance practices

Thus, it is obvious that the *chain-of-events* are responsible for leading the failure; in other words, multiple variables impact the refractory material in service.

The refractory materials are only one component in the overall design of process equipment and industrial furnaces that is eventually sensitive and influenced by time, temperature, and operation factors. For example, a typical temperature profile with a variation of idle time and operational conditions is shown in Figure 12.8. A minute variation in the temperature range of 1555 ± 5 °C is common during continuous casting; however, several processing steps are involved enough to provide thermal and mechanical stress during operation. Unexpected hanging and idle time expedite a drop in temperature, resulting in refractory life degradation. We need to minimize and synchronize it. Cumulatively, the culprit may be one or a combination of the following, and the first three are most suitable for kiln and furnaces:

- Improper combustion (fuel, ratios, atmosphere, flow, etc.)
- Flame impingement (flame geometry, improper setup, etc.)
- Improper furnace controls (PLC's controls, circuity, thermocouples, dampers, etc.)
- Operational change (cycling, throughput, load, etc.)

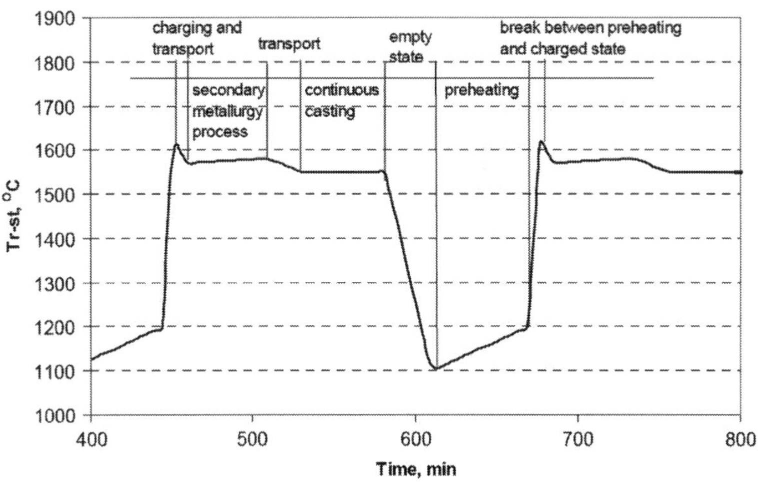

Figure 12.8 Temperature variation during the technological cycle [4].

- Operational procedures
- Maintenance practices and relevant automation
- The different thermal expansion rates of shell and lining movement
- Mechanical stresses
- Environmental factors
- Variation of molten metal chemistry in the same vessel

While considering the refractory, all these factors may have impact and causes are:

- Excessive temperatures (shrinkage, etc.)
- Hot spots
- Excessive cracking due to
  - Spalling
  - Thermal shock (a result of temperature fluctuations)
  - Mechanical shock
  - Impact, bending, linear change
- Design factors and the entire structure
- Refractory damage preferentially by erosion and abrasion
- Failure of the anchoring system
- Shell damage by high-temperature deformation or molten material
- Mechanical damage in transport

All of these might lead to the failure of refractory materials sooner than planned. The selection of the refractory utilized may have to be rethought when the circumstances change. If a product has new variables to deal with, upgrading or altering the material may be warranted.

12.4.3 Workmanship

Several degrees of expertise and automation are required to install the refractory; thus frequently, refractory manufacturers depend on professional third-party contractors as working personalities for the installation of refractory. However, multiple issues need to be considered when a job has to be completed through a contractor, and sometimes it isn't easy to select or secure available workmanship.

Refractories are not an example of a *one-size-fits-all* product. The lowest priority for maintenance managers and production and plant engineers is in selecting and installing refractories. However, steel sector management must remember that "Refractories are like face masks (in Covid-19 pandemic), financially insignificant compared to steel cost management, but the most effective material in preventing a disaster." Typical refractory installations create extended plant downtime, cost millions in lost production, and even worse, safety problems are sometimes substantial hurdles. Before selecting the proper workmanship through a third party contractor, one should remember the following points to balance between quality and job costs:

- *The track record* of any contractor is essential to provide consistent service, any *taillight guarantee* eventually reduces the refractory life in operation.
- *Uninterrupted service and alteration* as required with technical knowledge, research and development, and other reasons behind the scenes on the spot.
- Should provide *Guarantee* of their best job. One expert can pick up the suitable refractory concerning the mode of applications, the zone concept of vessels and processing unit, operational temperature, probable abrasion in the process, and other variables as maximum as possible.
- On-time completion of lining as well as repairing helps plant engineers to make a plan in a systematic manner that helps to boost the faith of the supplier.
- The safety record is one of the biggest concerns to sustain in steel shops for a prolonged time. It becomes more vulnerable when sacrificing safety for profits that eventually threaten contractors, employees, and consumers. Just some contractors cut bottom-line costs with low standard consumables.
- Despite quality refractory manufacturing, technical expertise is also mandatory to serve quality installation, reduce downtime, and stop production.
- *Total lifetime job cost* should also be considered in the production cost. Regardless of initial material and service price savings, recurring problems will invariably cost more than a premium job done right the first time.

However, a longstanding relationship with the manufacturer–contractor–customer and intricate knowledge of the material, designing concept, and operational features allow completing the complex work properly for a lower cost of the lining.

12.5 Preventive Maintenance in Operation

Advertisement has a significant role in motivating and changing our thought processes. If we look in our vehicle, it demands maintaining regular intervals for consistent performance for an extended period. Sometimes we follow the exact schedule, and occasionally deviating may reduce the engine life. Preventive maintenance during operation protects the investment and is also a safety precaution. In the same way, different types of refractory fitting require varied schedules for maintenance. Unfortunately, it is not simply to maintain and recommend vehicles that the required maintenance activities and time intervals are advised. The refractory engineer should thus monitor and examine the heat-processing equipment by physical inspection and develop a specific maintenance plan for the repository and installation method. This maintenance may be the partial replacement by bricks or monolithics to boost the vessel's performance; otherwise, a new lining consumes more time. While certain boats are constructed and built to remain online throughout extended campaigns, specific general criteria may contribute to maintaining good refractory linings health:

- Predefined schedule downtime for refractory-lined equipment for regular lining inspection
- Regular maintenance or repairs during the inspection
- Unexpected performance need monitoring to restrict the catastrophic failure and find out the probable reasons for failures to avoid the exact incidence in the future

12.5.1 Professional Service

Practical engineering and design of vessels and refractory lining can reduce preventive maintenance. Refractory lining purpose and predicted process conditions may provide better insight to design. It is essential to create adequate refractory anchoring systems, perform structural analysis of metallic supports and heat transfer calculations to ensure operating at peak performance. Thermal, mechanical, and chemical properties and interaction and their relation with microstructure are critical criteria before designing refractory, as discussed in Chapters 4, 5, and 6, respectively.

The success of the refractory fitting when putting vertically or overhead depends on the performance of the system. A refractory anchor must be suitable for use like the refractory material selection. The selection criteria of a particular anchoring system might be as complex as that for the material itself. In engineering and design, the requirements are often disregarded. All crucial elements that need to be evaluated meticulously include length, diameter, distance, and orientation of the anchor. Numerous anchors have to be employed depending on the operative conditions, the material utilized, and the liner thickness. The choice of anchor material, shape, quantity, and dimensions considerably affects the refractory casting lifespan. The anchoring system installation is also essential for lining life. The anchor system frequently entails welding a thermal anchor onto a carbon-steel shell; thus, creating a precise welding method in the design phase required for definite application.

Heat transfer analysis, including a database of refractory materials and properties, can ensure the optimal and quality lining that should be done through design and procurement of refractory, eventually reducing preventive maintenance. This exercise helps customers reduce breakdowns and costs through the more efficient lining. Without understanding the entire drawing for both vessels and refractory, long time lining performance is difficult; thus, extensive engineering and design capability can overcome it. CAD is essential for communication to customers and provides suggestions for upgrading an existing design and creating an entirely new lining system to reduce preventive maintenance. Professional attitude always likes to analyze why refractory is experiencing premature failure, and for this, extensive

ANSYS software-assisted analysis can sort it out to some extent. However, continuous inspection, understanding the operational parameters, slag chemistry, installation problem, startup/dry out an issue, and the material problem can conclude in the precise matter to continue the maintenance.

A typical error occurs when using refractory datasheet values for necessary lining stability calculations with three key wear variables (thermal, mechanical, and chemical) when engineering a refractory lining. The use of erroneous or average material data could lead to significant faults in the construction of the lining and a lower lifespan of the lining. When constructing a refractory fastening, extreme minimum and maximum property values are recommended to account for bad eventualities.

12.5.2 Slag Composition, Temperature, and Viscosity

The growing industry has cleared the way for massive growth in the production of iron and steel. Thus, vast quantities of slag, including BOF slags, electrical arc furnace (EAF) slags, secondary metallurgical slags (e.g., LF, VOD, AOD), tundish slags are generated. Slag—the byproduct of the iron and steel industry—has many functions, including (1) desulfurization, (2) dephosphorization, (3) nonmetallic inclusions, (4) foreign elements removal, and (5) thermal loss minimization. Metallurgical slags vary in their chemical composition concerning steel classification, input materials, and treatment purposes. All metallurgy operations between the slag and the liquid metal rely on thermophysical qualities, viscosity, density, and surface tension. Therefore, it is essential to investigate these features in advance to efficiently control the process and prevent the premature failure of refractories. A concise idea of the influence of temperature and slag composition on the viscosity and their corrosive effect on refractory may explain how to synchronize the slag composition and operation temperature for a particular steel grade.

Viscosity directly impacts the kinetic conditions of processes and is one of the essential physical properties of slag. Mass and heat transport, slag formers and additions, and metal and slag separation are improved with low viscosity. Although several classic sources of literature on viscosities are available [5], a brief on basics of viscosity for the molten slag system provides insight into the subject and assists in understanding the influence of oxides present in the slag and temperature on the viscosity and their resultant effect on refractories. Several popular models are well established to discuss the impact of temperature and slag composition on viscosity. Important models known as Urbain model (CaO-Al_2O_3-SiO_2 system), Modified Urbain Model (CaO-Al_2O_3-SiO_2-FeO system), Riboud Model (SiO_2-CaO-Al_2O_3-CaF_2-Na_2O system), Optical basicity model (NPL), Iida model, modified Iida model, KTH Model is a key consideration for iron and steelmaking production. NPL, Iida, and modified Iida models considered Arrhenius equations, whereas the KTH model is based on the Eyring equation similar to the Wayman-Frankel relationship.

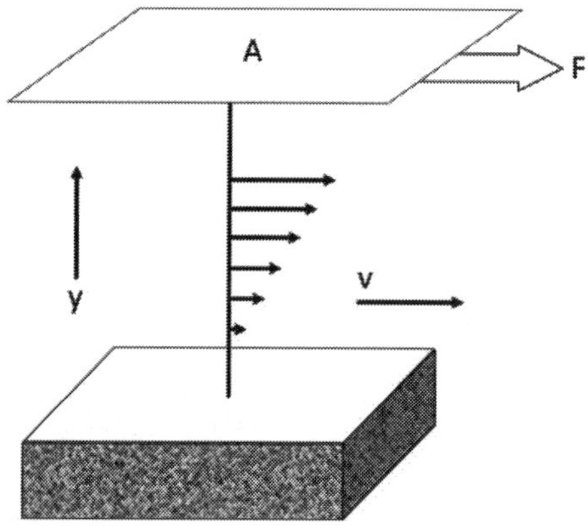

Figure 12.9 Representation of the viscous flow of a liquid with an applied shear force.

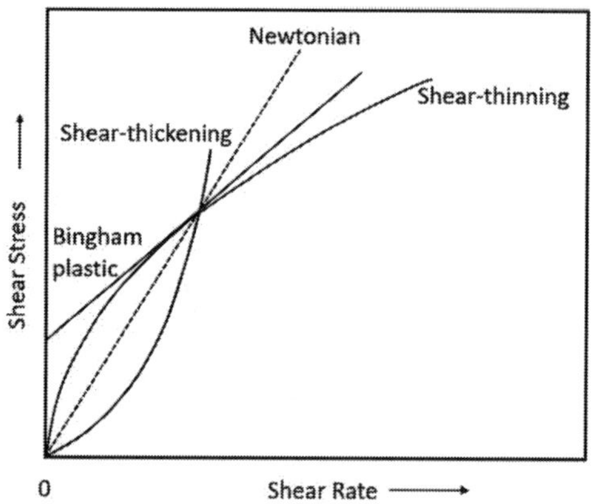

Figure 12.10 Shear stress–shear rate behavior of Newtonian and non-Newtonian fluids.

Viscosity (η) is an actual incidence of fluid's internal friction, which becomes apparent when a fluid layer moves with another layer, as shown in Figure 12.9.

The greater the friction, the more significant amount of force is required to cause this movement, as illustrated in Equation (12.1).

$$\eta = \frac{F / A}{dv / dy} = \frac{\tau}{dv / dy} = \frac{\text{shear stress}}{\text{shear rate}} \tag{12.1}$$

Most metallic melts and molten slags are Newtonian fluids where the viscosities are independent of the shear rate; thus the viscosity is defined according to Newton's law as the proportionality constant between the shear stress and velocity gradient normal to the shear stress (Figure 12.10). Non-Newtonian fluids exhibit a relationship between shear stress and velocity gradient characterized by the viscosity change with a shear rate compared to Newtonian flow's simple linearity. In short, the pseudoplastic shows a decreasing viscosity with a higher shear rate. Shear-thinning is sometimes called this kind of flow behavior. This dilatant increases viscosity and characterizes the dilatant fluid with an increase in shear rate. Shear thickening behavior is also known as dilatancy. Plastic behaves as a solid under static conditions. An optimum amount of applied force is required to induce the fluid flow, known as yield value. Once the yield value is exceeded and flow begins, plastic fluids may display Newtonian, pseudoplastic, or dilatant flow characteristics.

The slag system follows both Newtonian and non-Newtonian behavior. Non-Newtonian behavior is complex and can entail fluid pseudoplastic. Slag system consists of 10–40 vol% solid exhibit Newtonian behavior. The viscosity behavior of slags with a high solids content is more appropriately described as plastics from Bingham. Molten slack viscosity is particularly sensitive to variations in temperature and slag composition due to the internal structure of the oxides melt. The dependency on viscosity-temperature (DI) is commonly illustrated in the form of the relationship with Arrhenius:

$$\eta = A_w \cdot e^{\frac{Ew}{RT}} \tag{12.2}$$

where A_A is the pre-exponential factor, R is the gas constant, E_A *is* the activation energy, and T temperature (K). The molten slags are composed of discrete ionic structural units. Ions and ionic complexes and the interionic forces they generate are directly linked to their activation energy. The activation energy fluctuates at temperature as the types and sizes of ions change. The Wayman-Frankel relationship is occasionally applied to account for this change;

$$\eta = A_w T \cdot e^{\frac{Ew}{RT}} \tag{12.3}$$

where A_W and E_W are similar to the A_A and E_A parameters in Equation (12.2) with different values. In many circumstances, this link is more consistent with experimental evidence that can be reached with the Arrhenius expression (Equation 12.2). High temperature loosens the structure; thus viscosity decreases with increasing temperature. In general, viscosity is proportionate to the reverse temperature exponentially. However, sudden temperature change can lead to substantial variations at some temperatures, called a critical viscosity temperature (T_{CV}, Figure 12.11). The nature of this phenomenon is linked to the solid melting phase crystallization, which impacts the behavior of slag viscosity.

As the slags are ionic in their nature and polymerization changes according to the metal oxide concentration in the slag, it can be assumed that the viscosity of slags is very sensitive to the size of the ions and the electrostatic interactions and therefore to the slag structure. The slag systems can have multiple components in the

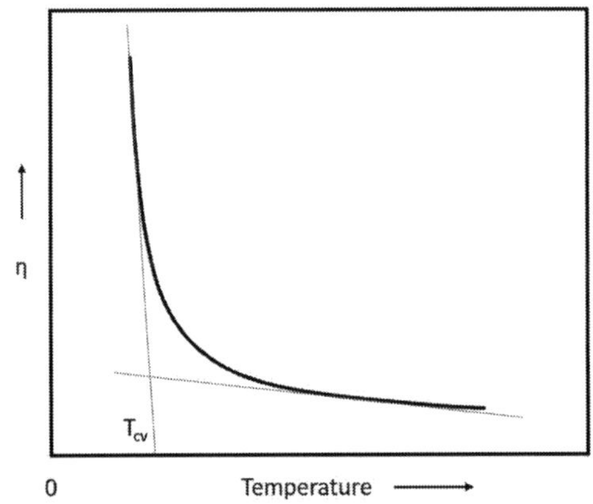

Figure 12.11 Schematic representation of the definitions of critical viscosity temperature (T_{CV}).

metallurgical process. The oxides present commonly are Al_2O_3, CaO, Cr_2O_3, MgO, MnO, SiO_2, and Fe_nO in steel slags. Most of the oxides are also essential components in nonferrous metal slags. Network formers like SiO_2, P_2O_5, and B_2O_3 have strong, covalent connections of metal-oxygen, causing high viscosities of liquids. The network breaks down and, accordingly, viscosity lowers progressively by adding alkali and alkali earth oxides, Li_2O, Na_2O, K_2O, MgO, CaO, together with other divalent oxides like MnO and FeO. The elemental size effect depends on the components present in the slag and their quantity. After the slag composition, amphoteric oxides such as Al_2O_3, Fe_2O_3 could act or break down the network. Oxide systems are technically highly essential in a wide variety of applications, including ironmaking, steelmaking, and ceramics processing, including the ternary CaO-Al_2O_3-SiO_2 slag system. The compositional changes of these systems have significant implications for the thermochemical and thermophysical characteristics, including thermal capacity, density, and viscosity. A recent analysis shows the CaO/SiO_2 ratio and temperature effect on CaO-Al_2O_3-SiO_2 slag system study [6].

The dynamic viscosity comprehensively depends upon the chemical composition of oxide systems. The CaO/SiO_2 mutual ratio, i.e., the fundamentality, significantly affects viscosity. Viscosity first decreases with the increasing basicity while achieving its minimum value at the CaO/SiO_2 ratio = 1.19 (1723 K); 1,14 (1843 K), as shown in Figure 12.12. At the higher values of the CaO/SiO_2 ratio, the values of viscosity again increase. The effect of basicity on the viscosity value of slags is much more pronounced at lower temperatures. The internal structure determines viscosity in the slag systems. At low CaO content, modification effects are prominent, in which the silicate bonds are generally enlarged or destroyed with a drop in viscosity. Defragmentation already reached the degree to which other increased levels of CaO are not decisive in this regard at specific critical concentrations, and a partial viscosity of CaO is becoming apparent, leading to a higher viscosity in the system.

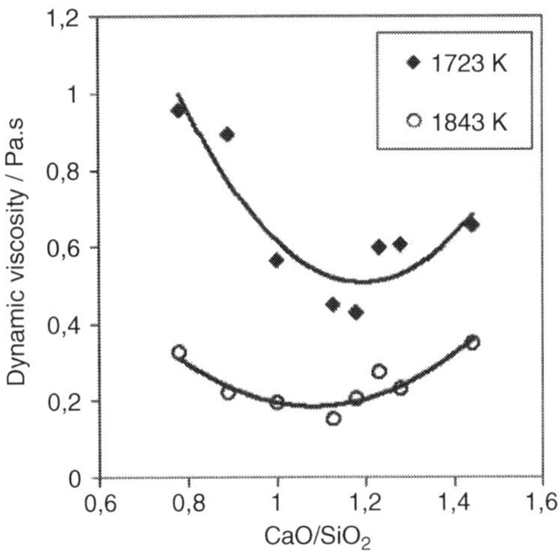

Figure 12.12 Viscosities of $CaO-Al_2O_3-SiO_2$ slag as a function of basicity and temp [6].

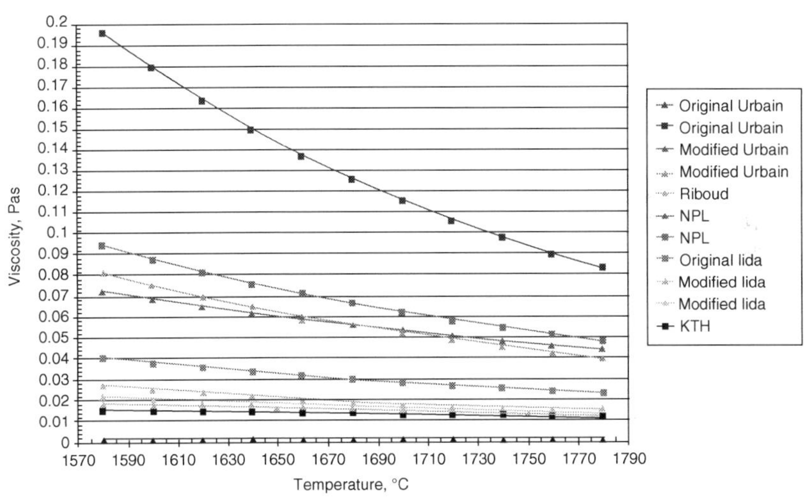

Figure 12.13 Calculated viscosity values of BOF-Slag [7].

The transition metals (e.g., Fe, Cr) have variable oxidation states (Fe^{2+} or Fe^{3+}, Cr^{2+} or Cr^{3+}), and they have a distinct effect on viscosity. In addition, oxidation states are crucially influenced by the oxygen partial pressure, temperature, and bulk composition, and any such changes can cause significant phase-balance alterations and alterations in slag viscosity. In this aspect, two crucial primary and secondary steelmaking slags are analyzed and estimated from proposed models and represented in Figures 12.13 and 12.14 [7].

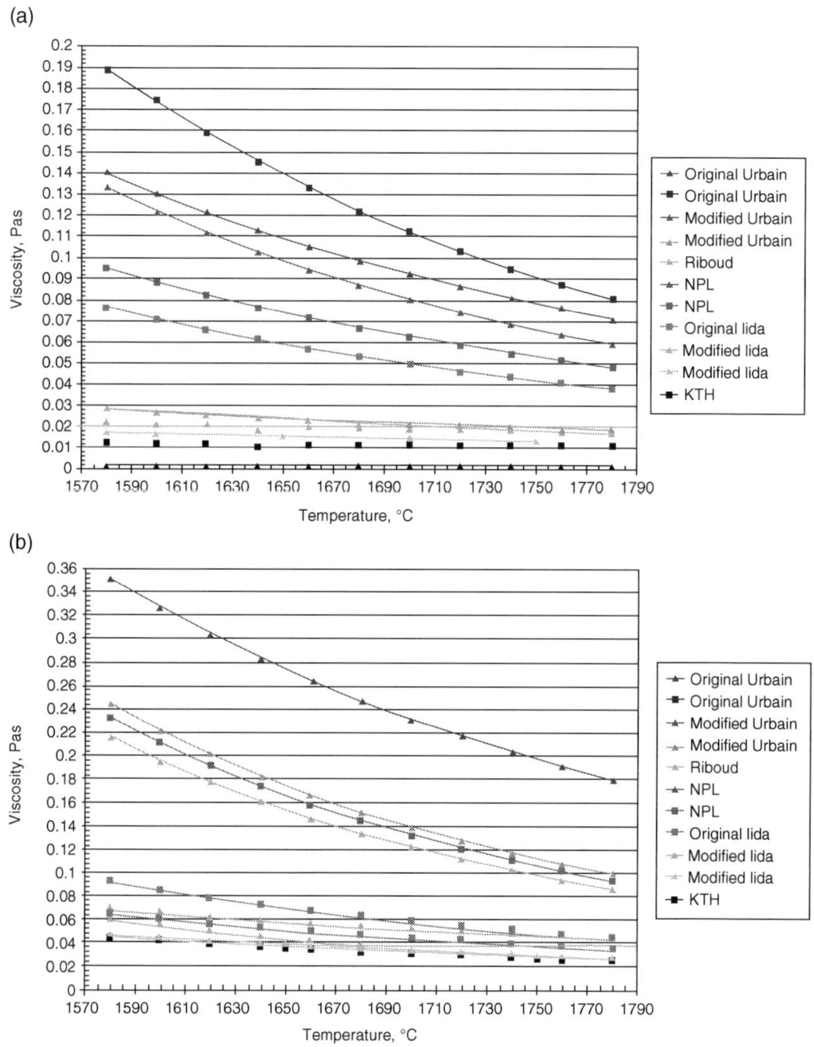

Figure 12.14 Calculated viscosities of AOD-slag after (a) oxidation and (b) reduction period [7].

12.5.2.1 BOF Slag

The Basic Oxygen Steelmaking (BOS) process aims at refining hot metals made in the blast furnace into raw liquid steel, which is further refined at the secondary steelmaking facility subsequently. Basic Oxygen Furnace (BOF) has the primary function of decarburizing and removing the phosphor and other impurities in the hot metal and optimizing the temperatures of the steel to make any additional treatment with a minimum of reheating or refrigeration of the steel before casting. Exothermic oxidation reactions generate high heat energy during BOS—it is more

Table 12.3 Typical composition of BOF slag (wt.%).

Compounds	CaO	SiO_2	MgO	Al_2O_3	MnO	FeO	TiO_2	V_2O_5	P_2O_5
%	54.6	11.4	1.6	1.2	3.5	21.7	1.8	3.1	0.8

than necessary to achieve the desired steel temperature. This additional heat is utilized for melting scrap and adding iron ore. Usually, 60–100 kg BOF slag is produced per ton of steel. Oxides SiO_2, MnO, FeO, V_2O_5, P_2O_5, and TiO_2 are present in the BOF Slag, with additional fluxes CaO and MgO. The typical composition of the BOF slag is shown in Table 12.3.

BOF-slag viscosity variation concerning temperature has been calculated from different proposed models and plotted in Figure 12.14. Viscosity values are virtually comparable throughout all temperature ranges computed by Iida and amended Urbain models. The ~0.021 Pa · s at 1580 °C changes to ~0.013 Pa · s at 1780 °C. Compared to the other Urbain models, the adoption of the Urbain original resulted in greater viscosity values. The main benefit of the modified Urbain models is that they make it possible to consider the chemical differences of individual components; hence they must be more accurate than the original Urbain model when forecasting slag viscosities.

In addition, the original Urbain model and Riboud model fail to distinguish between the different cations. Both models so obtained comparable viscosity values. In this high basic slag, the values computed by the KTH model appear to be too high, while the NPL model reveals excessive viscosity values. It is rare to acquire accurate viscosity forecasts from slag composition, which have adjusted optical fundamentality above 0.73.

12.5.2.2 AOD Slag

The decarburization of stainless steel is done using the mixture of oxygen—inert gas in the AOD (Argon oxygen decarburization) converter. The AOD technique exploits the chromium oxidation dilution concept. In the beginning, the molten metal has high oxygen to inert gas content. The oxygen-to-inert gas ratio falls with the carbon concentration in the bath. Inert gas dilution reduces the bathing partial carbon monoxide pressure, hence facilitating the removal of carbon. Such a process increases the elimination of carbon, but the oxidation of metallic chromium is diminished. Oxidized chromium is recovered simultaneously with desulfurization from the slag during the reduction cycle. After oxidation and reduction periods, typical AOD-slag compositions are presented respectively in Tables 12.4 and 12.5.

Table 12.4 Typical composition of AOD-slag after oxidation period (wt.%).

Compounds	CaO	SiO_2	MgO	Al_2O_3	MnO	FeO	Cr_2O_3
%	42.1	5.3	4.2	1.1	10.5	10.5	26.3

Table 12.5 Typical composition of AOD-slag after reduction period (wt.%).

Compounds	CaO	SiO_2	MgO	Al_2O_3	MnO	FeO	Cr_2O_3
%	57.0	31.1	9.3	1.1	0.5	0.5	0.5

(a) (b)

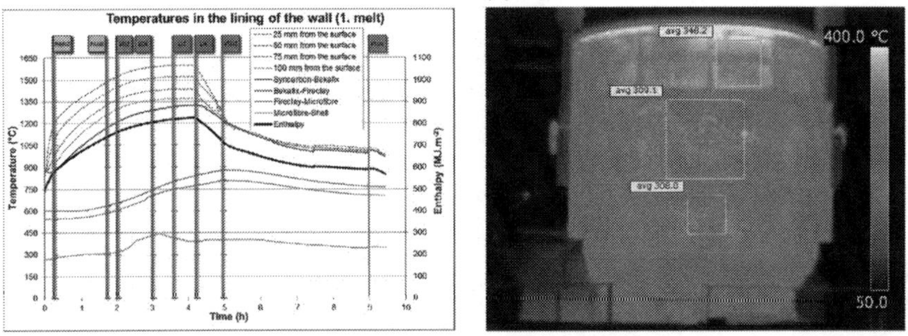

Figure 12.15 (a) Detailed temperature profile on 1st melt and (b) thermovision view of ladle [8], Vlček et al. (2016), Polish Academy of Sciences.

Comparisons are provided in Figures 12.15a–b, respectively, of estimated viscosity values obtained using various AOD-slag models based on the temperature following oxidation and reduction periods. In all, temperature ranges are pretty similar patterns to each other. Different models derive the viscosity values. In addition to existing Al_2O_3 and CaO, FeO, and SiO_2, the influence of both chrome oxides and magnesia are also analyzed. They have high fusion points, and the chromium is found as Cr^{2+} and Cr^{3+}, with the corresponding oxides CrO and Cr_2O_3 at two oxidation stages in metallurgical slacken ($CrO_{1.5}$). These two oxides have a different viscosity impact, and consequently, the amount of CrO and Cr_2O_3 in the slag must be known to understand the slag penetration behavior. The reduced AOD slag condition exhibits relatively high viscosity compared to oxidized slag. Thus, process conditions and environment impact the slag–refractory interaction phenomenon and performance.

12.5.3 Monitor and Maintenance of Lining

Temperature, time, corrosive gases, slag, and operational practices may influence the overall life of the heat-processing equipment. Additional culprits leading to the degradation of a refractory lining can be incorrect combustion controls, improper flame setup, anchor failure, or thermal shock resulting from severe temperature fluctuations. A refractory expert should collect samples of the existing refractory material during the process to help identify the causes of failure. For example, glazing and excessive shrinking indicate the possibility of experiencing extreme temperatures, sudden cracking or shearing away of loaded refractory evidence of thermal spalling, and many more. Continuous monitoring before regular maintenance is a good idea that eventually reduces the maintenance cost and downtime. Key benefits of constant monitoring are:

- Improved safety
- Prevention of liquid steel breakouts
- Early identification of areas with increased wear
- Cost savings through maximization of the refractory lining service life without safety risks
- Optimized refractory maintenance schedule and strategy
- Comparative evaluation of different refractories

A ladle, one of the critical vessels, may be considered for such discussion, where continuous monitoring is essential to achieve consistent output. Ladle pins support the safety of steel production and maintenance. Ladle check and refractory monitoring system minimize the possibility of liquid steel outbursts while the durability of the refractory lining can be maximized. Liquid metal breakouts from a faulty ladle lining represent a common hazard in the manufacture of steel. The known repercussions of the surrounding equipment are harm, high cost of repair, and loss of revenue caused by delays in production. In addition to the economic damage, the potential of injury or loss of life is always significant. The ladle check system is essential to design to prevent heat breakdowns efficiently. By ongoing monitoring of the ladles during production, areas with faults or depletions must immediately be discovered in the refractory lining long before they become crucial.

Monitoring also contributes by increasing the life of the ladle refractories to minimize the manufacturing expenses. The state of the fireproof lining deteriorates each time it comes in touch with liquid metal until the ladle has finally been relegated. In general, the refractory maintenance schedule is established on experience. The time intervals are often highly conservative for safety reasons.

Ladle control enables a reliable evaluation of the remaining life span of the refractory body. Therefore, the number of heats can be maximized, and the relining can be planned more precisely without any safety problems. Because of the high cost of refractory maintenance, the system can significantly reduce steel production costs.

—PANZ (PANK)–start (end) of steel processing in ladle furnace
—VDZ (VDK)–start (end) of steel processing in cofferdam
—LIZ (LIK)–start (end) of casting
—VTOZ (VTOK)–start (end) of ladle high-temperature heating

We can consider an example of a 70-ton ladle, where lining damage and the following condition to steel breakage in the ladle shell is not uncommon. The ladle's thermal profile consisted of magnesia-carbon bricks as working lining, followed by the successive wall coating layers including magnesia dust dilatation powder, castable layer, fireclay layer, microfiber 1000 material isolation layer, and a steel shell. Thermocouple sensors of K and B types were inserted about half a ladle height during the bricking of the ladle. A data logger GRANT Squirrel was used to record the data measured and placed on the shell's outside in the cooled thermo-insulator box. In this setup, the lining temperature was continuously recorded during all technological operations, including ladle carriage, performed on the ladle. Temperature measurement sites were both at the interface of each material and in the working layer of the fitting.

A total of 11 thermocouples have been installed for measurement. During an operational experiment, a thermal lining field was observed during the new lining's preheating process and over five circulation cycles, including reheating between cycles in the lining. Each cycle of circulation includes tapping, secondary refining in the furnace of a ladle, vacuum station, casting, inspection, and maintenance. A detailed perspective of the 1st melt is provided in Figure 12.15a. The thermovision camera FLIR ThermaCAM P25 with adjustable emissivity at the value 0.95 measured the ladle surface temperature concerning the time of ladle. The ladle immediately after taking out the vacuum station, the thermovision view is presented in Figure 12.15b. The findings of thermovision measurements are compared with operating circumstances in the case of the ladle fitting and the enthalpy-calculation, facilitates ladle lining:

- high temperature inside lining
- a wide interval of working layer temperature change caused either by high tapping temperature or by insufficient heating of ladle, or combination of factors
- frequent total cooling of lining
- insulating layer damage
- shell deformation

With attention to the above analysis and incidence, some important parameters can be considered to boost the ladle life through refractory management and operational alterations:

- To reduce the number of total cooling down of ladle lining cycles,
- To alter the working layer material (a significant disadvantage of materials with high MgO content is their low resistance to cracking and flaking) composition,
- To minimize the time of no covered ladle with lid, the significant reduction of ladle enthalpy between the end of heating and tapping occurs.

Thus, such monitoring exercise has a significant impact on achieving better performance.

Given refractories' importance to operations, it is essential to remember that they are consumables and will wear out. Such a phenomenon is significant because without proper maintenance, the processing equipment may be inactive at the most crucial time; thus downtime for a high-temperature vessel may result in hundreds or thousands of dollars. Therefore, take care of refractory—the unsung hero of the manufacturing process—and it eventually takes care of safety and efficient work environment, reduces energy loss, minimizes downtime, and avoids critical, catastrophic failure of heat processing equipment.

12.6 Consistent Supply and Time Management

Demand from refractory raw materials depends on the steel market, and delayed requests interrupt the supply line. Natural disasters or erratic markets may reduce the steel required for all sectors. However, refractory manufacturers must maintain and offer refractory quality for ongoing business under all such constraints. To achieve the final customer needs, a supply chain as a sequence of (decision making and implementation) activities and (material, information, and money) flow in and between various phases of the supply chain.

The supply chain involves manufacturers and their suppliers and carriers, warehouses, merchants, and consumers (depending on their logistical operations). It encompasses new product development, marketing, operations, distribution, financing, and customer support but is not limited to these. The generic supply chain for the entire network supply chain is presented in Figure 12.16.

At least one supply chain belongs to any company: it generally consists of many suppliers and clients. There are several dependencies to produce both shaped and unshaped refractory and supply to the customer site. The supply chain maintains an uninterrupted sequence and flow to fulfill the customer demand. Usually, two modes are popular in supply chain management: cycle and pull/push the concept.

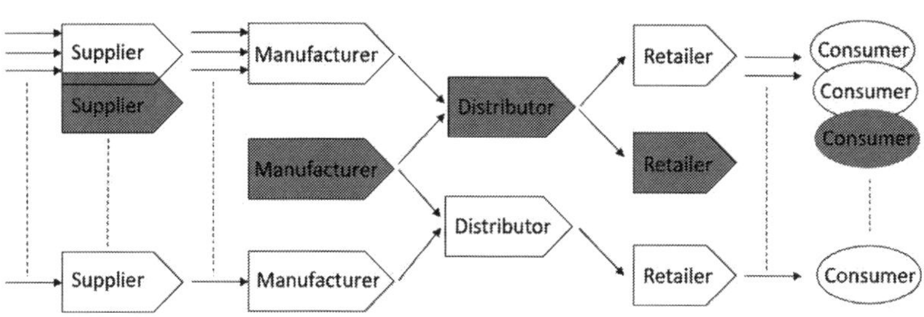

Figure 12.16 Schematic diagram of a supply chain (shaded) within the total supply chain network.

12.6.1 Cycle Concept

- Point out the relevant processes and corresponding responsibilities
- All functions in the supply chain divide into a series of cycles, a practical approach for operational decisions
- A cycle is supposed to perform within the interface of two successive stages of a supply chain
- A systematic supply chain may be simplified into four steps for smooth operation—customer order, replenishment, manufacturing, and procurement
- Such efforts specify the roles and responsibilities of all stack holders of the supply chain
- The predefined process helps to design the infrastructure to execute an uninterrupted process

12.6.2 Pull/Push Concept

- The supply chain depends on the pull concept (response to customer order) or the push concept (anticipation of a customer order)
- Fixing process execution time to fulfill the customer's mandate
- The degree of certainty of customer demand should be apparent during the execution of the pull process
- During push process execution, the process demand is not known and must be forecast
- Pull concept is a reactive process, but the push is a projected concept
- Push/pull boundary in a supply chain separates push process from pull process
- Very useful when considering strategic decisions relating to supply chain
- Forces more global consideration of supply chain processes as they relate to a customer order
- Better supply chain when pull process is more significant

An apparent breakdown or partial shutdown time may not impact the refractory performance, but, in reality, any disaster and discontinuing production can hamper the refractory lining. Despite this, operational people should consider and maintain probable guidelines to avoid disaster or premature failure. In a recent article, important guidelines are tabulated to control the hazard and accident-free operation [9]. A known processing time in each step helps to predict the entire operation process. Such time management for the steel industry is given below.

1) Primary Steel Processing TIME in BOF for all relevant steps, including idle time
   - Time in BOF: Scrap and HM charge: 4–8 minutes
   - Blowing time: 16–18 minutes

- Temperature and chemistry check: 5 minutes
- Final blowing: 4 minutes
- Tapping: 5–7 minutes

2) Steel Tapping time from BOF to LF: BOF operation, 30 heat/day
3) Transport time from BOF area to LF Processing area: 10 minutes–1 hour, depends on crane, car, if temperature drop, then need additional OLP (open lance purging for 8–14 minutes) in LF to maintain the desired temperature of steel
4) Secondary Steel Processing TIME in LF for all relevant steps, including idle time
   - LF operation
   - Lime additions: 4 minutes
   - Arcing: 8–10 minutes
   - Purging: 12–16 minutes
   - Chemistry check: 5 minutes
   - Correction if required: 5 minutes
   - Tapping: 4–6 minutes
5) LF to Tundish transfer time: 45 minutes, after the transfer, subsequent use may need 3–4 hours (i.e., downtime)
6) Casting time through Tundish including idle time: 45 minutes

The value chain of steel manufacturing is starting from the steel melting shop from which customization started despite product differentiation. In actuality, steel production is associated with several safety hazards such as exposure to dust, excessive heat, fire and exploration, noise, a sudden drop of unwanted material from height, confined space in the operational zone, electrical shock, gas leakages, etc. Thus, the entire process should be foolproof to identify the hazards associated with the steel melting shop process and ensure accident-free working.

References

1 Sarkar, D. (2019). *Ceramic Processing: Industrial Practices*. New York USA: CRC Press. ISBN 9781138504080.

2 Heindl, R.A. and Post, Z.A. (1950). Refractory castables: preparation and some properties. 33 (7): 230–238.

3 Hashemi, B., Nemati, Z.A., and Faghihi-Sani, M.A. (2006). Effects of resin and graphite content on density and oxidation behavior of MgO-C refractory bricks. *Ceram. Int.* 32: 313–319.

4 Petkov, V., Hadjiski, M., Boshnakov, K., and Mihailov, E. (2013). Diagnosis of metallurgical ladle refractory lining based on non-stationary on-line data processing. *Cybern. Inf. Technol.* 13 (2): 122–130.

5 Braithwaite, J. (2017). Chapter – 4, *Viscosity*. Morgan & Claypool Publishers.

6 Rehackova, L., Rosypalova, S., Dudek, R., Kukutschova, J., and Dobrovska, J. (2015). Effect of CaO/SiO_2 ratio on viscosity and structure of slag. *Metabk* 54 (3): 455–458.

7 Kekkonen, M., Oghbasilasie, H., and Louhenkilpi, S. (2012). Viscosity models for molten slags. Aalto University publication series, SCIENCE + TECHNOLOGY 12/, Finland.

8 Vlček, J., Jančar, D., Burda, J., Klárová, M., Velička, M., and Machovčák, P. (2016). Measurement the thermal profile of steelmaking ladle with subsequent evaluation the reasons of lining damage. *Arch. Metall. Mater.* 61 (1): 279–282.

9 Recommended guidelines for iron and steel sector by Ministry of Steel, Govt. of India, steel melting shop. 1–32. Doc No: SG/39.

Index

Fundamental Design of Steelmaking Refractories, First Edition. Debasish Sarkar.

b

c

e

f

g

q

r

S

Made in the USA
Columbia, SC
17 September 2026

06662cbf-3913-427e-b0dc-a84b34b6fe9aR02